THE NEW NATURAL

A SURVEY OF BRITISH N/

ALIEN PLANTS

THE NEW NATURALIST LIBRARY

ALIEN PLANTS

CLIVE A. STACE

and

MICHAEL J. CRAWLEY

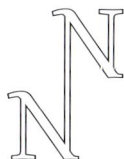

N

WILLIAM
COLLINS

This edition published in 2015 by William Collins,
an imprint of HarperCollins Publishers

HarperCollins Publishers
1 London Bridge Street
London
SE1 9GF

WilliamCollinsBooks.com

First published 2015

A CIP catalogue record for this book is available
from the British Library.

Set in FF Nexus

Edited and designed by
D & N Publishing
Baydon, Wiltshire

Printed in Hong Kong by Printing Express

Hardback
ISBN 978-0-00-750215-8

Paperback
ISBN 978-0-00-750214-1

Contents

Editors' Preface

The title of this new addition to the New Naturalist Library might lead to the expectation that we are revisiting and updating a subject that was treated in the early days of the series. However, in truth, Sir Edward Salisbury's *Weeds & Aliens*, published in 1961, was essentially a natural history of weeds. The 'Aliens' half of the title largely came about as half a century ago most 'weeds' were assumed to be 'aliens'; in Sir Edward's own words, 'the subject of weeds cannot be naturally separated from that of alien species and hence the dual character of the subject matter of this work'. Fifty years on, the subject of alien species is high on the natural history and international conservation agenda. Increasing numbers of alien species are establishing themselves as members of our flora. Three years prior to the writing of *Weeds and Aliens*, J. E. Dandy's *List of British Vascular Plants* estimated 626 species to be alien, representing 29 per cent of the flora. The current comparable figures revealed by the authors in the present volume have risen to 2,068 alien species, constituting 59 per cent of the British flora. There are now more alien than native members of our flora.

It is thus more than timely for the New Naturalist Library to address this important aspect of our natural history. *Alien Plants* is the result of a collaboration between two of our most distinguished botanists, a taxonomist and an ecologist. Both have been naturalists since youth. Clive Stace, Emeritus Professor of Plant Taxonomy at the University of Leicester, is author of the standard work on the British flora. Indeed, he is a member of that elite galaxy of authors who achieve distinction by producing a work that comes to be better known by the author's name than by its title, hence *New Flora of the British Isles* is universally referred to simply as 'Stace'. Mick Crawley, FRS, is Emeritus Professor of Plant Ecology at Imperial College London. Editor of *Plant Ecology*, our most influential standard textbook on the subject, he is particularly interested in the interaction between grazing herbivores and vegetation. This partly explains the annual phenomenon of his famous disappearances during March and August, when he retreats to the most remote outpost of the British Isles, St Kilda, to pursue his long-term research on the relation between the island vegetation and its Soay sheep.

The British flora has been, and is, greatly enriched by interesting and attractive alien species, which significantly increase the nation's biodiversity and form an important part of its ecology. On the other hand, a few have come to pose serious economic, ecological and conservation problems, such as Japanese Knotweed and New Zealand Pigmyweed. Not only does *Alien Plants* provide the most comprehensive account of this most important element of our flora, but it also contains much material that is new and appearing in print for the first time. This includes the comprehensive list of alien species set out in the appendices. Also a first for the New Naturalist series is the coordination with *The New Naturalist Online* website, where the appendices are set out in an interactive format in Excel, enabling readers to edit the lists as they wish. We are particularly pleased to be able to welcome and present this latest innovation to the series.

Authors' Foreword and Acknowledgements

The study of alien plants has, in the past half-century or so, progressed from an eccentric hobby, enabling amateur botanists to increase the total of wild plants that they could record, to the full-blown sciences of invasion ecology and alien genetics, in which the factors controlling plant immigration and evolution are researched, particularly with reference to possible deleterious and beneficial effects on native and cultivated species and their habitats. Alien species no longer present an optional extra, but must be accepted as an integral part of mainstream botanical investigation. The amount and breadth of data that have been accumulated on alien plants in the British Isles is exceptional. The number and proportion of alien plants in our flora is extremely high – arguably the highest in Europe – and to some extent compensates for a relatively poor native flora. The subject has also become one familiar both to naturalists and the general public, due to exposure via the media to such diverse topics as damage to the environment by Japanese Knotweed and New Zealand Pigmyweed, the attraction of bees and butterflies to cities by such plants as Butterfly-bush, the court cases involving Leylandii hedges, the threats to the purity of our native Bluebell by the mass planting of its Spanish relative, and the cultivation of new sorts of Christmas tree. We therefore felt that this was a good time to write a book on alien plants, and that it would be appropriate to base it on the flora of the British Isles.

We must mention *Weeds & Aliens* (1961) by Sir Edward Salisbury, number 43 in the Collins New Naturalist series, as this covers a subject area quite close to our own. The emphasis in the two books, however, is quite distinct; Salisbury's book is foremost about weeds, while ours is about aliens. Our book is in no way intended to compete with, update or replace the 1961 work.

Writing a book of this nature is a new venture for both of us, but one that we were keen to undertake. We have not previously collaborated as authors, but were convinced that the subject demanded a breadth of expertise that neither one of us possessed. Both of us have spent more than 40 years teaching and researching plant biology in English universities, CAS in taxonomy and evolution and MJC in ecology. Inevitably each chapter or section was written primarily by one author, but each of us commented freely on the other's efforts and all chapters are to varying degrees joint efforts, in some cases equally shared. Because of this we have not designated separate authorship to each chapter. We have also both maintained a keen interest in wildlife and the natural world from an early age, going back to the immediate aftermath of the Second World War, and trust that our enthusiastic love of natural history and dedication to plant science are equally evident in this book, as is the relevance of alien plants to everyday life. Inevitably, two authors will favour different styles; we have not sought to disguise these, but hope that they complement rather than clash. The use of the first person (I, my, etc.) always refers to CAS, except in the section on seaweeds. We found that in some instances we hold different opinions, mainly in the degree of emphasis that should be placed on various aspects, but such differences were, perhaps surprisingly, very few.

We have endeavoured to cover the most important aspects of plant alien studies, including the threats they impose from both ecological and genetic angles, and their value to humans and wildlife. Much of the modern research on aliens would have been more difficult or impossible if it were not for the activities of the early alien-hunters, of whom we provide a brief account.

Natural history now embraces all the latest technological advancements, with the result that some of these fields of investigation, such as high-power microscopy and molecular biology, are not available to the average amateur. There is, however, still much of immense value that enthusiasts can discover, and in some places we have indicated suitable areas for such contributions. The successive volumes of the New Naturalist series have faithfully reflected these changes. Nevertheless, it is quite difficult to pitch the subject at a level that utilises the modern approaches yet remains comprehensible to a wide range of readers. We hope that we have achieved a reasonable compromise, and in striving to do so have been constantly encouraged by the similar attempts by E. B. Ford in *Butterflies* (1945), the first New Naturalist volume of 70 years ago. But, to cite just one example, it remains a fact that it is impossible to understand the evolutionary impact of aliens without discussing chromosomes. There is a danger, otherwise, that natural history pursuits will become confined to taking photographs and adding dots to distribution maps. We have followed the New Naturalist tradition of using English names, where they exist, as well as the scientific equivalents. The scientific names used in this book are those found in the most recent standard reference works of the various groups; they will be familiar to keen naturalists and we have not attempted to substitute them with very recently proposed but as yet little-used updates.

We have put much effort into providing definitive lists of the alien plants of the British Isles (i.e. the United Kingdom, Republic of Ireland, Isle of Man and the Channel Islands). The lists of angiosperms are presented in Appendix 1 (neophytes) and Appendix 2 (archaeophytes) near the end of this book. They are also available in Excel on the New Naturalist website (www.newnaturalists. com/AdditionalResources), enabling users to manipulate, analyse and add to the lists as they wish. These appendices, and the lists of the various groups of non-flowering plants provided within the text, are intended to represent a standard up-to-date (post-1986) list of aliens of the British Isles, based on objective criteria. The lists have not been previously published. They include some other data: the categories of neophytes (naturalised, casual, persistent) and archaeophytes (denizen, colonist, cultivated); the number of hectads (10 × 10 km squares) in which each species has been recorded since 1986; the life form and mode of reproduction; for the neophytes additionally the year first recorded, area of origin, and the vector(s) involved; and for the archaeophytes additionally the main uses to which the plant has been put.

GENERAL ACKNOWLEDGEMENTS

Most of the distributional data presented in this book were collected by the dedication to field studies by members of the BSBI, mainly over the past half-century. In 2013, the Botanical Society of the British Isles changed its name for legal reasons related to its charitable status to the less accurate Botanical Society of Britain and Ireland, but for simplicity we have used the acronym BSBI throughout.

We owe a big debt of gratitude to Ian Tittley, former marine algologist at the Natural History Museum, who has a special interest in alien seaweeds, for contributing the section on marine algae (seaweeds) in Chapter 10, a subject beyond our capabilities.

Many people very kindly agreed to lend us photographs for use in this book, in all cases without any recompense. They are, of course, acknowledged in the relevant captions, but we name them all here: Ian Atherton, John Bailey, Ken Butler, Bryony Chapman (Kent Wildlife Trust), Tony Child (Thanet District Council), Ian Denholm, Norah Eastwood *per* Geoffrey Wilmore, Trevor Evans, William Gloyer, Alistair Godfrey, Peter Greenwood, Diana Grenfell, Peter Hall, Stephen Hendry (Forest Research), Michael Hollings, David Holyoak, Celia

James, Stephen Jury, Geoffrey Kitchener, Peter Llewellyn (UK Wildflowers), Ted Lousley *per* Chris Boon, Philip Oswald, Ron Porley, Kathleen Pryce, Tim Rich (National Museum of Wales), Mary Sheridan *per* Chris Boon, Jeanette Sitton, John Somerville, Margaret Stace, Richard Stace, Ann Steele (National Trust for Scotland), Ian Tittley, David Tomlinson, Mario Vallejo-Marín, Ruud van der Meijden *per* Nelleke van der Meijden and Goronwy Wynne. Their generosity has contributed greatly to the appearance of the text. In the event, some photographs could not be used due to the need to reduce the length of the manuscript. In addition, permission to reproduce images was kindly granted by: Botanical Society of the British Isles *per* Lynne Farrell, California Department of Food and Agriculture *per* Fred Hrusa, Defra *per* David Pearman, Emorsgate Seeds *per* Mark Schofield, Gallery Oldham *per* Patricia Francis, Hancock Museum Newcastle *per* June Holmes, Royal Botanic Garden Edinburgh *per* Leonie Paterson, Royal Botanic Gardens Kew *per* Christopher Mills, John Wiley & Sons and *New Phytologist*, and Yorkshire Naturalists' Union *per* John Bowers. The Collins 'team' and David Streeter have been very helpful and supportive throughout, for which we sincerely thank them.

CAS ACKNOWLEDGEMENTS

All naturalists owe much to the older people who helped and advised them in their early days. As a young teenager I learnt most of the common plants from members of the Tunbridge Wells Natural History Society, particularly Aline Grasemann and Lionel Langmead. In 1951 (when I was 13), the latter bought me Ford's *Butterflies*, the first New Naturalist, as a gift for learning my first 100 scientific names of wildflowers in that year. The former bought me Clapham *et al.*'s *Flora of the British Isles* ('CTW') in 1953. After the Kent Field Club was formed in 1955, I had a wider group of tutors, in particular Francis Rose. However, none of the above had a special interest in alien plants, although they did identify for me species like Japanese Knotweed, Slender Speedwell and Small Balsam, which had defeated me. In my late teens I joined the Botanical Society of the British Isles, after which I had the cream of British field botanists at my disposal. Among them I was fortunate to encounter Ted Lousley, Duggie Kent and David McClintock, at that time our most distinguished alien experts, who became my good friends. If it were not for this background of assistance, I would probably have not become a plant taxonomist, and certainly could not have written this book.

I am very grateful to the following people who very kindly read drafts of some of my chapters, and pointed out errors, omissions and other shortcomings:

Richard Abbott, John Bailey, Arthur Chater, Eric Clement, David Pearman, Petr
Pyšek and John Richards. Any mistakes that remain are, of course, my own. There
are also about 140 other individuals who helped me personally in relation to this
book, mainly by answering innumerable emails, letters and telephone calls, to all
of whom I am extremely grateful. David Pearman in particular answered many
questions, and also generously allowed me to use his personally researched first
date of record for all the neophytes (Appendix 1). Brian Wurzell and Mike Wilcox
proudly escorted me around their local patches (Tottenham and Bradford,
respectively) to show me many of their local aliens. Chris Preston and Sam
Bosanquet helped ensure that the data on bryophytes were up to date. Quentin
Groom and Tom Humphrey painstakingly researched the databases of the BSBI
to provide me with the most up-to-date and relevant data. I am very grateful to all
of them.

MJC ACKNOWLEDGEMENTS

My interest in alien plants began with childhood holidays on the gravel banks
of the River Breamish in Northumberland, discovering the spectacular flowers
and fantastic colour patterns exhibited by the monkeyflowers and their hybrids.
My professional interest in alien plants began with ecological studies on tropical
islands like Hawai'i and Mauritius, where entire native ecosystems had been
replaced by alien plant species. The fascination was that these plants were
growing in communities where they were surrounded by species with which
they shared no recent evolutionary history. The really important lesson that was
brought home by these island travels was the importance of alien animals like
pigs and goats. They created the disturbance regimes and competitive conditions
under which the alien plants were able to thrive.

Three scientists were particularly inspiring. I never met Charles Elton, but
his 1958 book *The Ecology of Invasions by Animals and Plants* was the cornerstone of
the discipline of invasion ecology. Two close friends, Bob May and the late John
Harper, are outstanding theoretical ecologists who demonstrated that mastering
plant demography was the key to understanding biological invasions.

CHAPTER 1

What Are Aliens, and How Many Are There?

I have been an alien in a strange land.
(Exodus 18: 3)

The word 'aliens' is used today in many contexts, conjuring up a range of sentiments including fear, dislike, fascination and sympathy, but rarely indifference or disinterest. It is used to describe strange beings from Mars, human migrants, and non-native plants and animals. The third category alone concerns us here; the term alien is employed universally by biologists to indicate organisms that have been introduced by people to new territories. A more precise definition will be provided shortly.

Alien plants are today the topic of interest of many professional research teams and amateur enthusiasts around the world. As a group, these plants are common, conspicuous, pestiferous, beautiful, edible, and otherwise useful or harmful. In the British Isles there are very roughly as many species of naturalised or frequently recurrent alien flowering plants as natives and, besides that, there are numerous other less common ones. Many species have been an integral part of our wild flora for such a long time that we can no longer be sure whether they are in fact native or alien. Even some recently discovered species are similarly problematic. Aliens are proving to have such wide interest simply because they cannot be ignored, and because they add diversity to our otherwise rather limited flora. Many of them have profound effects on the environment by competing with native vegetation or by populating empty ground (Fig. 1). Others have altered the course of evolution by their genetic interactions with natives. Alien plants in the British Isles, whether

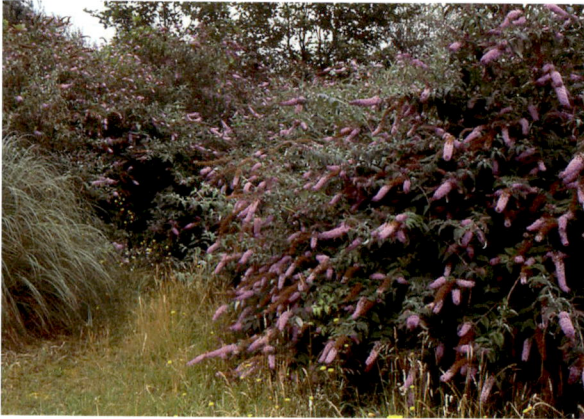

FIG 1. Butterfly-bush (*Buddleja davidii*), one of the most attractive and familiar neophytes in the British Isles, was not recorded in the wild here until 1922. Seen here are Butterfly-bush shrubs on waste ground, with a tussock of Pampas-grass (*Cortaderia selloana*) on the left. (CAS)

they be food-plants or pests, are a major and largely measurable factor in our economy.

It is an indisputable fact that alien plants exhibit all the same structural, behavioural and utilitarian characteristics as native species, which begs the question of why so much fuss is made of the distinction between the two categories, especially since it is sometimes impossible to decide between them. Why don't we treat them as equals and simply ignore their different origins in our flora? There are actually very good reasons.

An important feature that makes alien plants so interesting is that they evolved somewhere else, and that they left behind many (if not all) of their co-evolved species when they were moved to the British Isles. Their genotypes were forged during interactions with a different set of plants, fungi, micro-organisms and animals from those with which they now cohabit. In contrast, the genotypes of native plant species evolved while interacting with roughly the same set of plants and animals they interact with today.

Furthermore, when investigating the history of our vegetation, its development and evolution since glacial times (the Ice Ages), whether this be in ecological, taxonomic or genetic terms, it is absolutely essential that we consider only our native flora. Inclusion of an alien element in the subject of our study would skew our results and lead to false conclusions. For example, what would be the point of investigating possible migratory routes of the small number of mainly North American species occurring in the extreme west of the British Isles if, in fact, they were put there by humans? The need to differentiate carefully between alien and native genetic stock is not born of chauvinistic elitism, but of scientific necessity.

DEFINITIONS

One could write long essays on the precise meanings of each of the two words of our book title. While this would represent an unreasonable overindulgence, some explanations are necessary. We shall explain what we mean by 'aliens' here; 'plants' will be defined in Chapter 10.

Native and alien

Over the years there have been many definitions of native, alien and so on proposed by botanists, often involving unnecessarily difficult terms (e.g. autochthonous, which effectively means native) and complex hierarchical classifications. Much argument and many column inches are still expended on this topic, which we shall not replicate. Fortunately, the increase in studies of alien plants and invasion ecology by research groups across Europe and beyond has meant that generally agreed terms became necessary, and these have now become entrenched in the literature and more or less universally adopted. Fortunately for us, most of them are based on the English language, rather than Latin, Greek or German. These agreed terms have the advantages of simplicity and unambiguity, and we shall use them throughout our text. They have been clearly defined along very similar lines by Macpherson *et al.* (1996), Richardson *et al.* (2000), Pyšek *et al.* (2004) and others, working as informal groups or within organisations such as the BSBI and Delivering Alien Invasive Species Inventories for Europe (DAISIE), but the current definitions have evolved over many years of discussion. In all cases they must be seen as relating to a predefined geographical area, be it a continent, a country, or a local area of any size.

The two most fundamental terms are *native* and *alien*, defined as follows.

Native

A native taxon is one that has originated in the area without human involvement, or has migrated to the area without human intervention from an area in which the taxon is/was native. Despite claims to the contrary, providing they meet the above criteria, native taxa do not need to have been present in the area for a long time; theoretically, a new native plant could arrive next year, e.g. if it were brought here from a native area by a bird. Synonym: indigenous.

Alien

An alien taxon is one that has migrated to the area with human involvement (whether intentional or unintentional), or has migrated there without human involvement but from an area in which the taxon is/was also alien. Alien plants can have arrived any time since the beginning of the Neolithic period (6,000 years ago), when man's activities first made such migrations possible. Synonyms: introduced, non-native, non-indigenous.

Categories of alien plants

We recognise two main categories of alien plants, *archaeophyte* and *neophyte*, each with three subcategories.

Archaeophyte

An archaeophyte is an alien plant that has been present in the area in a wild state since before 1500 CE (Fig. 2). These are the subject of Chapter 2, where choice of the year 1500 is justified. Archaeophytes are sometimes known as 'honorary natives', because they have been here for so long.

We recognise three subgroups of archaeophytes:

Colonist – 'A weed of cultivated land, by road sides or about houses, and seldom found except in places where the ground has been adapted for its production and continuance by the operations of man' (Watson, 1847).

Denizen – 'At present maintaining its habitats as if a native species, without the direct aid of man, but liable to some suspicion of having been originally introduced by human agency, whether by design or by accident' (Watson, 1847).

Cultivated plants – plants introduced here by people as crops, or developed here as such from species that arrived here by chance with them or their crops, and then escaping or persisting.

FIG 2. The archaeophytes Common Poppy (*Papaver rhoeas*) and Corn Marigold (*Glebionis segetum*) in a cornfield. (CAS)

Neophyte
A neophyte is an alien plant that has arrived in the area in a wild state since 1500 CE (Fig. 1). In the past, when archaeophytes were more often considered as natives or 'doubtfully native', the terms neophyte and alien were often equated.

There are also three subgroups of neophytes:

Naturalised plants – those plants that are self-reproducing or increasing from year to year by sexual (seeds) or vegetative (rhizomes, tubers, bulbs, fragments, etc.) means. Often a minimum number of years is stipulated before a plant is considered naturalised; Macpherson *et al.* (1996) suggested five and Pyšek *et al.* (2004) stated ten, but in practice many botanists make their own judgements on local field experience. In order to qualify for inclusion in *Flora Europaea* (Tutin *et al.*, 1964–80), a species had to be naturalised in one locality for at least 25 years or 'in a number of widely separated localities', but that work was dealing with a very large geographical range for which a different approach is to be expected. Plants that have arisen as offspring of garden or cultivated specimens (often from wind-blown or bird-sown seeds) but which are not themselves reproducing, we consider as either survivors or casuals, according to how long they survive (*see below*). Synonym: established plants.

Casual plants (casuals) – those plants that are not self-reproducing from year to year and whose continued existence in the area relies on repeated introductions. These introductions may be of imported seed, or as offspring from planted individuals. Synonyms: adventive plants, waifs.

Surviving plants (survivors) – individual plants that are persistent in the area from year to year simply due to longevity, but do not reproduce sexually or increase vegetatively. Most of these are woody species, especially trees, without means of vegetative propagation, and were originally deliberately planted. Sometimes termed 'planted', 'relict' or 'cultivated'.

The relationships between these categories are summarised in Figure 3. The definitions used in *New Atlas of the British & Irish Flora* (Preston *et al.*, 2002) differ in one small but very significant way: both archaeophytes and neophytes are there defined as plants that '*became naturalised* [our emphasis] between the start of the Neolithic period [4000 BCE] and 1500 CE'. In practical terms this means that casuals are not a sub-group of neophytes, but that the primary division of aliens is between naturalised plants and casuals. The need for naturalisation does not appear in any of the major Continental definitions of neophyte or archaeophyte

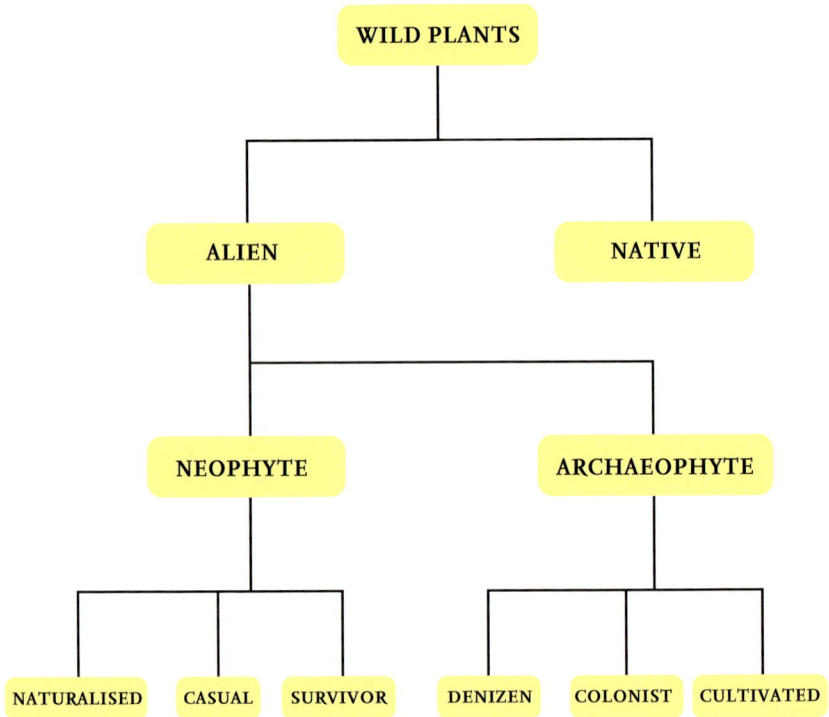

FIG 3. Classification of plants adopted in this book according to their native/alien status.

(from 1903 to the present), and it was specifically rejected by Pyšek *et al.* (2004), who expressed the opinion that residence status (archaeophyte, neophyte, etc.) and invasion status (naturalised, casual, etc.) should be kept as separate concepts. In our view also, archaeophyte and neophyte should be the primary division of aliens and both of these categories can be naturalised or casual. Casual archaeophytes are indicated in Appendix 2 by a dash in the S/V column. Nevertheless, we consider denizen/colonist/cultivated to be a more useful subdivision of archaeophytes than naturalised/casual.

Other terms

There is a further category that has also led to some debate and disagreement, without any compromise or conclusion having been reached. This refers to those plant taxa (hybrids or species) that arose anew in our area, but were derived from hybridisation between a native and an alien species (e.g. Common Cord-

WHAT ARE ALIENS, AND HOW MANY ARE THERE? · 7

grass *Spartina anglica*, Hybrid Bluebell *Hyacinthoides hispanica* × *H. non-scripta*), or between two alien species (e.g. *Mimulus peregrinus*, Bohemian Knotweed *Fallopia japonica* × *F. sachalinensis*). As Pyšek *et al.* (2004) said, 'There is currently no clear agreement on how to treat the products of hybridisation involving alien species.' One can easily see both sides of the argument – in one sense they are natives, having arisen here and in some cases not being found anywhere else; in another, they are alien because they could not have arisen if the alien parent had not been introduced. They clearly belong to a distinct category, for which we propose the new term *neonative*:

Neonative – a plant that has originated in the area without direct human involvement, but has arisen as the result of hybridisation either between a native and an alien taxon or between two alien taxa, or as the result of evolution from an alien or neonative taxon.

The importance of the concept of 'in a given area' cannot be overemphasised. Normally in this book we are referring to the British Isles (Britain, Ireland, the Isle of Man and the Channel Islands) when we state whether a taxon is native or alien. Very many species are native in some part(s) of the British Isles but alien elsewhere, or are archaeophytes or neophytes in different places, or are naturalised in some areas but only casual in others. A number of taxa are, in fact, represented in one area by both native and alien, or both naturalised and casual, plants. Generally, when describing the status of a species in an area, its 'highest' rating is stated. For example, we say that Sea-buckthorn (*Hippophae rhamnoides*) and Jacob's-ladder (*Polemonium caeruleum*) are native, but that only applies to restricted parts of England (Fig. 4); elsewhere in the British Isles they are neophytes (garden escapes or planted).

Four other terms need to be defined here, for reasons that will become apparent; again the definitions relate to any specified area:

Invasive
These are naturalised plants of semi-natural habitats that attain a substantial portion of the biomass of the invaded plant community, e.g. Indian Balsam (*Impatiens glandulifera*) on a riverbank, or New Zealand Pigmyweed (*Crassula helmsii*) smothering pondside vegetation. We would *not* use the word invasive to describe naturalised species that reach high abundance only in man-made or highly disturbed habitats, e.g. Guernsey Fleabane (*Conyza sumatrensis*) on urban waste ground. There are more species in this second category than there are true invasives.

No. of 10 km² occurrences

Native	GB	IR
1987–99 ●	50	0
1970–86 ●	7	0
pre 1970 ●	8	0

Alien		
1987–99 ●	352	43
1970–86 ●	24	5
pre 1970 ●	76	11

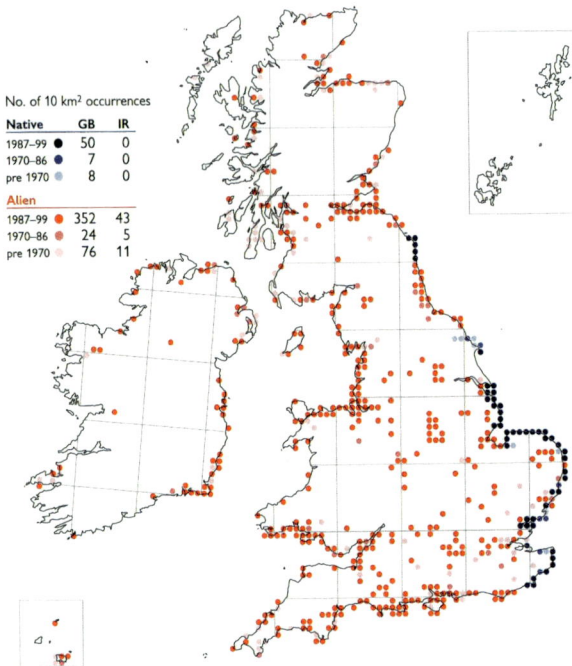

FIG 4. Distribution of Sea-buckthorn (*Hippophae rhamnoides*) in the British Isles, showing contrasting native (blue) and alien (red) distributions. (Reproduced by permission of Defra *per* David Pearman, from Preston *et al.*, 2002)

Endemic

Endemic plants are those that occur as a native in *only* the area concerned, e.g. Scottish Primrose (*Primula scotica*) in northern Scotland. The term is used in quite a different sense in the field of parasitology, when it means the same as 'native' in the botanical sense. Pedunculate Oak (*Quercus robur*), for example, is a native of the British Isles, but it is not endemic to them, although if it were a viral disease it would be described as such.

Extinct

An extinct taxon is one that was once present but is no longer so in the area concerned. The disappearance of the taxon could be due to its eradication by people or to its natural demise because of some biological or environmental cause. One can speak of 'extinction in England' when it still exists in, say, Scotland, or 'extinction as a native' when it still occurs in the area as an alien. In recent years there has been argument about the precise meaning of the term, but we believe that the above is that most widely accepted.

Weed

A weed is a native or alien plant that is in the wrong place. This is an often-used definition, but of course is highly subjective and is seen entirely through the human perspective. Richard Mabey's *Weeds* (2010) offers 16 pages of amusing and informed discussion on the topic, but does not propose a more satisfactory definition. The term is usually applied to species that infest cultivated ground, but is sometimes used much more widely, e.g. unusual plants in natural habitats. The problem of definition is often summarised as 'the carnation in the cabbage-patch conundrum'.

ENGLISH NAMES OF ALIENS

English names of wild plants are of two sorts: those that have arisen over the centuries, invented incidentally by country folk as they were exposed to the plants in their day-to-day dealings; and much more recent names especially coined in order to provide an English equivalent (and often translation) of the scientific name. The former provide a fascinating glimpse into our association with, and utilisation of, plants, a subject masterfully covered by Geoffrey Grigson (1955) in *The Englishman's Flora*. Many of these 'folk names' are very local, and many common plants have different names in different regions; conversely, the same name might have been applied to different species in different areas. Such names cannot challenge the unambiguity and precision of scientific names unless they are fixed by some agreed standard list, such as that produced by the BSBI in 1974 (with later editions). A high proportion of our archaeophytes have folk names that were ascribed long ago, a number of which are discussed in Chapter 2. This, however, is much less true of neophytes, many of which have arrived here only in the past century or two and have often been provided with newly coined English names. In recent years many non-flowering plants (bryophytes and algae) have been (controversially) treated similarly. Indeed, in order to be referred to in legislation, a plant (or animal) has to have an English name, leading to some frankly ridiculous inventions.

Whether to use English or scientific names often leads to argument, with little prospect of resolution. The pendulum of opinion swings in different directions depending on the particular group of organisms. In the British Isles, birds are almost always referred to by their English names except by professional scientists. For most invertebrate groups and lower plants, however, only the scientific name is used by all, but the Lepidoptera and, in recent years, the Odonata are exceptions. We very much favour the use of scientific names for vascular plants, but have given these at first mention only and thereafter use the English name, as is traditional for books in the New Naturalist series. However, we do not accept the argument that English names are more easily remembered; it is simply a matter of familiarity and custom. In many cases the common name of a plant is in fact its scientific genus name, e.g. *Chrysanthemum, Rhododendron, Anemone, Laburnum* and *Cotoneaster*; names such as these seem to provide no problems, nor indeed do the long scientific names of dinosaurs even to pre-teen children. Indeed, some plants with well-known English names are at least as frequently referred to by their scientific genus name – e.g. *Brassica, Antirrhinum, Narcissus* and *Primula* – by the general public. In a few instances the specific epithet is used as a common name, e.g. Japonica (*Chaenomeles japonica*), Auricula (*Primula auricula*) and Shallon (*Gaultheria shallon*).

Some English names are particularly misleading:

1. Plants with similar names might, in fact, be only very distantly related, e.g. Lesser Celandine (*Ficaria verna*) and Greater Celandine (*Chelidonium majus*) are in different families.
2. An English name might be the same as the generic name of a different species, e.g. the English name Aster is used for *Callistephus chinesis*, but *Aster* is the generic name of Michaelmas-daisy; Nasturtium is used for *Tropaeolum majus*, but *Nasturtium* is the generic name of Watercress; Geranium is used for *Pelargonium*, but *Geranium* is the generic name of Crane's-bill; Syringa is used for *Philadelphus* (Fig. 5b), but *Syringa* is the generic name of Lilac (Fig. 5a); Bacopa is used for *Sutera*, but *Bacopa* is the correct name of another genus; Cineraria is used for *Pericallis*, but *Cineraria* is the correct name of another genus; and Mimosa is used for *Acacia*, but *Mimosa* is the correct name of another genus. Also, Rape is *Brassica napus*, not *B. rapa*, and Fuller's Teasel is *Dipsacus sativus*, not *D. fullonum*.
3. English names can be misspellings of the scientific name for the same plant, e.g. Chicory for *Cichorium*, Aubretia for *Aubrieta*, Alison for *Alyssum*, Brachycome for *Brachyscome*, and Weigelia for *Weigela*.

Other English names result from the corruption of scientific names, e.g. Cherryanthus for *Cheiranthus*, Sparrow-grass for *Asparagus*, Aunt-Eliza for *Antholyza*, and Gallant-soldier for *Galinsoga parviflora*. The last gave rise to the name Shaggy-soldier for the related, more hairy *G. quadriradiata* (Fig. 6), rather akin to the derivation of skyjack from hijack.

FIG 5. Two unrelated shrubs with entwined names: (a) Lilac (*Syringa vulgaris*), whose generic name is sometimes used as the English name for Mock-orange; (b) the European Mock-orange (*Philadelphus coronarius*), the original garden plant sometimes called Syringa, now largely superseded by horticultural hybrids. Both shrubs have a wonderful sweet scent. (CAS)

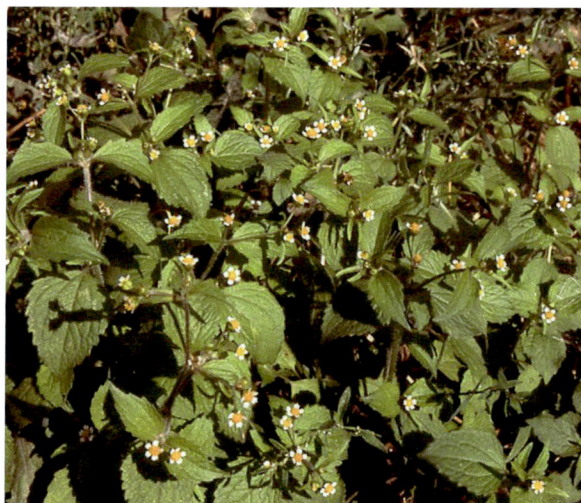

FIG 6. Shaggy-soldier (*Galinsoga quadriradiata*) is one of a pair of species now common on waste ground. Gallant-soldier (*G. parviflora*) was first recorded in the British Isles in 1861 and for more than a century was much the commoner species, but today it has been overtaken in abundance by the former. (CAS)

NATIVE OR ALIEN?

Although the difference between native and alien species can be easily and precisely defined, in practice it is often difficult to decide which of the two terms applies to a particular species. For many species there have been long-standing differences of opinion, some of which are unresolved to this day. The basic reason for these controversial cases is a lack of the requisite information, due to the combination of an imperfect fossil record between the last glaciation (c. 12,000 years ago) and the Neolithic period (c. 6,000 years ago), and the paucity of written or specimen records before the seventeenth century. We shall probably never overcome this lack of early historical knowledge of our flora, and therefore in many cases have to rely on our powers of deduction. This process of deduction has usually been exercised by crudely subjective means, notably the claim that a plant 'looks perfectly wild', or is far from habitation or occurs in abundance. Anyone viewing the vast expanses of Sea-buckthorn on the South Lancashire coast (Fig. 7) would hardly imagine that it is other than a long-standing native, forming an integral part of the ecology of the region, but we know that it was planted there just prior to 1900 and has spread rapidly over the past century (Fig. 4). Chauvinism all too frequently has played its part; botanists have often been too easily persuaded that their local speciality is thoroughly native.

Of the 30 or so native species that have been discovered in the British Isles since about 1900, most are rare and are often found only in rather remote parts of these islands. Others were originally confused either with other species or with introduced

FIG 7. Sea-buckthorn (*Hippophae rhamnoides*) in fruit on the dunes at Ainsdale, Lancashire, where it is known to have been planted more than 100 years ago yet now behaves like a thriving native. It provides abundant food for berry-eating migrant birds. (CAS)

examples of the same species. However, the following might be examples of species that have actually arrived by natural means in the past 120 years:

Welsh Mudwort (*Limosella australis*) – first recorded in 1897.
Esthwaite Waterweed (*Hydrilla verticillata*) – first recorded in 1914.
Ribbon-leaved Water-plantain (*Alisma gramineum*) – first recorded in 1920.
Pigmyweed (*Crassula aquatica*) – first recorded in 1921.
Lesser Tongue-orchid (*Serapias parviflora*; Fig. 8) – first recorded in 1989.
New Forest Bladderwort (*Utricularia bremii*) – first recorded in the mid-1990s.
Narrow-fruited Water-starwort (*Callitriche palustris*) – first recorded in 2000.

FIG 8. Lesser Tongue-orchid (*Serapias parviflora*) is a good example of a plant whose origin – native or alien – at its one (Cornish) site is unknown. (CAS)

Although these seven species are usually accepted as natives, as they are in this book, cases could be made for them being human-aided introductions. A further species, Somerset Rush (*Juncus subulatus*), two patches of which were found in a saltmarsh at Berrow, Somerset, in 1957, can be used to illustrate the difficulty in assigning native versus alien status. Subsequently the patches were recognised in aerial photographs of the site taken in 1954, but it was calculated that they were perhaps only about ten years old at the time and the area of saltmarsh in question had been reclaimed from the sea and vegetated only about 40 years previously (Willis, 1960). The method of arrival of the Somerset Rush is unknown, but marine traffic or birds are the most reasonable suggestions. These two vectors imply alien or native status, respectively; we shall never know for sure, but we have opted for alien.

In an attempt to instil objectivity into these debates, David Webb (1985), formerly Professor of Botany at Trinity College Dublin and the leading Irish plant taxonomist of his generation, proposed eight criteria, the majority of which should be satisfied before native status was credible:

1. The presence of a *fossil record* between the last glaciation and the start of the Neolithic period.
2. The absence of any *historical evidence* of introduction, or of a sudden surge in abundance such as many aliens have shown.

3. Existence of the species in so-called *natural habitats*, as opposed to overtly man-made ones.
4. An overall *geographical distribution* that fits in logically with existence in our area.
5. A lack of *frequent naturalisation* in the area where native representatives are suspected.
6. The presence of detectable *genetic diversity*, particularly divergences between suspected native and known introduced populations.
7. The ability to display *sexual reproduction* as opposed to vegetative spread.
8. The lack of any well-defined *means of introduction*.

None of these criteria in isolation can be used as compelling evidence and, as Webb himself pointed out, there are many well-documented natives that represent exceptions to each of the points. Use of these criteria is particularly inappropriate in the case of numerous species that are natives in a few restricted areas but alien in the great majority of places, often close to native sites. In practice, there remains a huge degree of subjectivity in the interpretation of the criteria, and personal experience inevitably plays a large part. Webb listed 41 species that were then generally or frequently considered native in the British Isles but which he considered were probably or almost certainly alien. The list has less relevance today, because no fewer than 30 of his species are now treated as archaeophytes, the recognition of which as a category removes the need for a lot of the discussion. Seven of the remaining eleven in Webb's list are still accepted as natives by most authorities.

There are still, however, a good number of disputed cases and, because we have attempted to produce a definitive list of neophytes (Appendix 1), these need to be confronted. In *New Atlas of the British & Irish Flora*, Preston *et al.* (2002) distinguished between native and alien occurrences by using differently coloured dots on their maps, and so were similarly forced into a decision for every species. Like Webb, they stated their criteria, by chance eight again, for accepting native status:

1. *Distribution in Europe* is relatively stable.
2. *Distribution in our area* is more or less continuous, or if not then reflects the similarly discontinuous pattern of suitable habitats.
3. Distribution in our area is a *natural extension* of the world range.
4. Distribution in our area has *reached equilibrium* and changes only in response to environmental factors.
5. Usually occurs *in natural or semi-natural habitats*.
6. Persistent in specific areas *for long periods*.

7. Recorded, or likely to have been present, *before 1700* CE.
8. *Continuous fossil record* since glaciation.

The two lists of criteria show some significant differences of approach that probably reflect the authors' different purposes, although, as one would expect, there are no stark contradictions. Nevertheless, Preston *et al.* (2002) found, even after the removal of many problematical species by their recognition of 157 archaeophytes, that there were 42 cases of which they remained uncertain. These were marked as native on the maps (at least for some of the 10 × 10 km grid squares), but designated 'native or alien' in the captions. This is almost the same number as in Webb's list of uncertainties, but in fact only three species (Hairy Buttercup *Ranunculus sardous*; Grape-hyacinth *Muscari neglectum*; and Bermuda-grass *Cynodon dactylon*, Fig. 9) are common to both lists. This lack of correspondence is largely explained by the majority of Webb's species being considered archaeophytes by Preston *et al.*, but is also a reflection of the uncertainty and subjectivity that still bedevils attempts to unravel the history of some of our plants; we believe that this situation will inevitably persist.

FIG 9. Bermuda-grass (*Cynodon dactylon*) has been known in the British Isles since 1688 and is often claimed as a native; we treat it as of uncertain status. (CAS)

Other criteria have been suggested from time to time, e.g. choice of food-plant by insect herbivores. Recently Pearman (2007), in a post-*New Atlas* reappraisal of some of the 42 'native or alien' species, used ten sets of data to draw conclusions: first date in cultivation; first date in the wild; presence in semi-natural habitats; spatial coherence; trends in frequency; persistence; use; European range; archaeological evidence; and genetics. Using these criteria, he reclassified four of the 42 uncertain cases as aliens.

We also recognise a number of 'uncertain cases'; a comparison of Preston *et al.*'s 42 with our 55 is given in Appendix 1, and our 55 are listed there (distinguished by an asterisk).

In this book we are considering in detail only those plants that are not native anywhere in the British Isles, even though many of our rarer native species are much commoner as aliens. Two examples of the latter, Sea-buckthorn and Jacob's-ladder, have already been mentioned, and we shall allude to others from time to time. If we were to cover fully all alien occurrences of plants in our area, we would be dealing with the vast majority of our flora, an impossibly large task.

HOW MANY ALIENS? THE EVER-CHANGING INVENTORY

Not only have the species of alien plant most commonly encountered here altered over the years as the types of imports they have contaminated have differed, but changes in the sources of those materials, and in industrial and agricultural practices associated with them in the British Isles, also determine the frequency and manner of appearance of their contaminants. Changing fashions in horticulture have equally led to the import of different sorts of decorative plants.

One consequence of the changing inventory of aliens is that many species recorded in the past are no longer found here. It is difficult to decide how long one has to wait before considering an alien to be 'extinct', and perhaps that adjective is not appropriate, at least to casual plants that were never naturalised; Pyšek *et al.* (2012) prefer the term 'vanished'. Clement & Foster (1994) and Ryves *et al.* (1996) chose 1930 as their cut-off point – more than 60 years previously. Of the 3,586 taxa in the former book, I counted 640 (*c.* 18 per cent) as unrecorded since 1930. With regard to the grasses, Ryves *et al.* estimated that about 70 (*c.* 12 per cent) out of their total of about 580 have not been recorded in the British Isles since 1930. This gives a total of 710 (*c.* 17 per cent) out of 4,166.

In this book we are aiming to provide a more contemporary account of our aliens, so the list presented in Appendix 1 includes only those species recorded since 1986, 1986/7 being one of the cut-off points in the recording schemes of the BSBI. Choice of a more recent date runs a considerable risk of excluding the many plants that remain unrecorded for several years but are then refound, often in the same location. For example, an alien from southwestern Europe, Lax Viper's-bugloss (*Echium rosulatum*; Fig. 10), was naturalised at Barry Docks, Glamorgan, from 1927 to 1985, but then was thought to be extinct until it was found again in 2007 in the same place by Stephanie Tyler. Sometimes the 'rediscovery' is in a completely different locality, and might be due to a different vector. Fern-leaved Beggarticks (*Bidens ferulifolia*) was found as a casual of unknown origin in some brick-pits in Oxfordshire in 1915, but was not seen again in this country for many decades, until it recently became a very popular subject in hanging baskets; it now self-sows on the ground underneath the baskets in

FIG 10. Lax Viper's-bugloss (*Echium rosulatum*) was thought to have become extinct in Barry Docks after 1985, but it was rediscovered there in 2007. (Trevor Evans)

several urban localities. Disturbance of old ballast tips or wool dumps might re-expose alien seed deposited there decades ago, and there will, of course, always be a chance that a species will be rediscovered, but we think that a quarter of a century is a reasonable time span to cover and that it provides a reliable estimate of the current situation.

The alien species included in the first edition of *New Flora of the British Isles* (Stace, 1991) represented at that time the most accurate list available; it showed many differences, both additions and deletions, from the lists provided in earlier Floras. This was updated to some extent in the second and third editions (Stace, 1997, 2010). More than 20 years have elapsed since the first edition, however, and our alien flora has noticeably changed in composition in that time, not least because of the near disappearance of wool aliens and changing horticultural preferences. We have therefore reconsidered the available records of alien angiosperms, using criteria that are explained in the preamble to Appendix 1.

The revised list in Appendix 1 contains 1,809 angiosperm neophytes and neonatives, of which 1,138 (63 per cent) are considered naturalised, 342 (19 per cent) casual and 329 (18 per cent) survivors. We have listed 57 (5 per cent) of the naturalised species as invasive. Of those 1,809 aliens, we reckon 141 (8 per cent) to be neonatives. These figures exclude the 55 taxa of uncertain status (alien or native) also listed there. To obtain a full list of our current (post-1986) plant alien flora, the totals of the algae, bryophytes, pteridophytes, gymnosperms, and angiosperm archaeophytes, all separately presented in their appropriate chapters, should be added (Table 1).

TABLE 1. Total number of alien taxa in each of the main groups of plants.

Group	Number of aliens	Listed in
Marine algae	38	Table 13
Bryophytes	23	Chapter 10
Pteridophytes	13	Table 14
Gymnosperms	49	Table 15
Angiosperm archaeophytes	197	Appendix 2
Angiosperm neophytes	1,809	Appendix 1
Total	**2,129**	

It is interesting to compare the figures in Table 1 with those of the only two other countries for which equally detailed analyses have been made. The Czech Republic covers about one-quarter of the area of the British Isles, with less than one-sixth of the population, and an alien inventory has been presented by Pyšek *et al.* (2012, a second edition of a 2002 work). The Czech Republic has a much richer native flora than the British Isles (2,945 taxa, compared with our *c.* 1,560), despite lacking a seaboard and any frost-free zone. Pyšek *et al.* (2012) listed 1,454 alien vascular plants, compared with our 2,068, but the numbers are not really comparable since the Czech total includes very rare and extinct aliens, although it apparently does not include survivors. Our total comparable with the Czech figure is probably nearly 4,000. The Czech aliens were estimated to represent about 33 per cent of their total flora, whereas the comparable figure for our aliens is about 59 per cent. However one manipulates the figures, it is abundantly clear that the Czech Republic has many more natives and many fewer aliens. An inventory of the alien plants of Slovakia has also been published (Medvecká *et al.*, 2012). The contrast here with the situation in the British Isles is even more pronounced. Slovakia is much smaller than the Czech Republic, with only about half of its population, but it has even more native and even fewer alien plants, amounting to only 21.5 per cent of the total flora.

CHAPTER 2

Archaeophytes –
Our Ancient Cohabitants

People will not look forward to posterity, who never look backward to their ancestors.
Edmund Burke, *Reflections on the Revolution in France* (1790)

AFTER GLACIATION

Our story begins with the retreat of the glaciers, about 18,000 years ago. The scene is open and bleak. Cold winds whip up whirls of spume from countless shallow pools, gravel spits dissect the landscape, and there are extensive flats of windblown loam. Mature soils have all been scraped away or buried beneath the sands and gravels of glacial outwash, and there are very few vascular plants. There is little to attract hunter-gathering people to this landscape.

There had been people here a very long time before this, but they had been driven back repeatedly by readvancing glaciers, colonising again once the ice had melted away. The first evidence of people is from Boxgrove in West Sussex, nearly 500,000 years ago, where remains of *Homo heidelbergensis* were found in 1994; this is the only post-cranial hominid bone to have been found in Northern Europe. Teeth from another individual were found there two years later. The next record comes from Swanscombe in Kent about 400,000 years ago, of a late *Homo erectus* or an early archaic *Homo sapiens* female (Stringer, 2006). But by 18,000 years ago the slate had been wiped clean of human influence. The entire set of species that we currently regard as our native flora had to immigrate from refugia in southern and southeastern Europe and re-establish self-replacing populations. The challenge is to use the pollen and fossil records to figure out which of the species would not have arrived here without the intentional or unintentional involvement of people.

As the glaciers melted away, primary succession began the slow process of soil formation. Great waves of plant communities crept slowly northwards: first there was tundra, then low scrub and, eventually, woodland. At least some trees were certainly (birches, Juniper (*Juniperus communis*), Aspen (*Populus tremula*)) or probably (Scots Pine (*Pinus sylvestris*)) already present during the late glacial, or even persisted in southern ice-free areas throughout the glacial stage (Huntley *et al.*, 2013).

PEOPLE RETURN

Like the native flora of the British Isles, the local inhabitants spent the last glacial period in refugia to the south and southeast. Modern genetic analysis indicates that many present-day people in the British Isles are descended from those who returned from glacial refugia in the Basque country of northern Spain via an Atlantic, western coastal route. Others came from the east, migrating over the land bridge of Doggerland via Germany and central Europe, probably from refugia in the Balkans. Later invasions by groups that formed ruling elites like the Romans or the Normans appear to have had much smaller effects on the genetic structure of the current populace.

For the first few thousand years after the retreat of the glaciers, the human inhabitants of the British Isles were nomadic hunter-gatherers, and probably had little direct impact on long-distance plant migrations. Of course, we can only speculate as to the plants that were moved about by these early people. They would certainly have carried seeds and fruits as food, and equally obviously they would move plants unintentionally attached to their clothes and footwear, or stuck in their hair. They would have moved about the seeds of plants that shared their living quarters; the camp followers of the botanical world. What we don't know is when people first began planting useful species that were new to the British Isles, or over what kinds of distances these early people traded living plant material.

All of the major native tree species had returned to the British Isles by the time of the climatic maximum during the Atlantic period 7,000 years ago. There is debate about the extent to which the vegetation at this time consisted of a dense, closed woodland (as envisaged by Godwin, 1975), and the degree to which it more closely resembled 'wood pasture', with free-grown mature trees standing in extensive areas of open grassland maintained by herds of large ungulates (as argued by Franciscus Vera, 2000). There is general agreement, however, about the identities of the plant species comprising the woodland. European Beech (*Fagus sylvatica*) was the last of the big trees to arrive back in the British Isles. People valued the plant as a source of Beech nuts, or mast, for their pigs, and humans would eat the nuts (just as they ate acorns) when times were hard. The wood of Beech was prized as firewood, flooring and for furniture making, so it is at least plausible that some pre-Roman range extensions of Beech within the British Isles were attributable to people (Huntley & Birks, 1983).

Although the native status of the Beech has long been denied or doubted by some, Brian Huntley writes:

> *I am aware of no credible evidence of people having played a role in introducing any of the major trees. The nearest one comes, as far as I am aware, is evidence of hazel nuts being carried around and used as food by Mesolithic hunter-gatherers, but taking the step to suggesting that they acted as an agent of dispersal or even, as some have argued, deliberately planted hazel, is a step too far for me in terms of the evidence. It seems much more likely that birds (especially the Jay) and larger mammals (including Brown Bear) played key roles in dispersing larger tree fruits, with birds the potential dispersers for those trees entering the British Isles after sea level had risen sufficiently to isolate them from the remainder of Europe (relevant especially to late-comers such as Fagus and Carpinus). (Brian Huntley, pers. comm., 2013)*

The Neolithic period
Far away, in Asia Minor, Palestine and the eastern Mediterranean, the Neolithic had begun about 14,000 years ago. The pre-agricultural Natufian people had settled around the eastern coast of the Mediterranean to exploit wild cereals such as Emmer (*Triticum dicoccon*) and Two-rowed Barley (*Hordeum distichon*), and here they began to domesticate the first crop plants. Rousseau's vivid evocation of the Neolithic – 'Vast forests were changed into smiling fields which had to be watered with the sweat of men' (Rousseau, 1755) – draws attention to the replacement of a nomadic lifestyle with a sedentary one dominated by hard labour. As people migrated westwards through the Mediterranean, then northwards into central Europe, bringing a new settled way of life based on cereal cultivation and domesticated animals, so they began to import the weeds of their new-found crops. In every consignment, countless seeds of the cornfield weeds from the country of origin came mixed up with the cereal seeds. By 6,000 years ago, a farming economy based on cereal cultivation (mostly barley), limited horticulture using hoes, and animal husbandry was well in place, involving Sheep (*Ovis aries*), Goats (*Capra aegagrus hircus*), Cattle (*Bos taurus*; for transport and fertiliser, rather than for eating at this stage) and pigs (*Sus*), but not the Horse (*Equus ferus caballus*).

As to the spread of the Neolithic culture, there is no clear evidence about the relative importance of direct migration by people, or the movement of ideas and materials across a reasonably settled human landscape. Certainly, human migration was the principal mechanism at first, as people moved northwards up the Atlantic coast from their refugia in the Basque country of northern Spain, through France and Cornwall, and into Ireland:

The Neolithic period is one of remarkable changes in landscapes, societies and technologies, which changed a wild, forested world, to one of orderly agricultural production and settled communities on the brink of socially complex 'civilisation'. It was a period that saw the arrival of new ideas and domesticated plants and animals, perhaps new communities, and the transformation of the native peoples of the British Isles. The Neolithic opened an entirely new episode in human history. It took place in the British Isles over a relatively short space of time, lasting in total only about 2000 years – in human terms little more than 80–100 generations. (Harris, 1996)

FIG 11. Garden Pea (*Pisum sativum*), now perhaps our most frequently eaten green vegetable, was already part of the diet in Neolithic times. (CAS)

It is clear that the principal cereals were Six-rowed Barley (*Hordeum vulgare*), Emmer, Spelt (*Triticum spelta*) and Common Millet (*Panicum miliaceum*). In addition, Foxtail Bristle-grass (*Setaria italica*) played some role. Legumes such as Broad Bean (*Vicia faba*) and Garden Pea (*Pisum sativum*; (Fig. 11) also occurred regularly and supplemented the diet of the prehistoric settlers. It is clear, too, that these settled farming communities had not given up on collecting wild fruits and seeds in season.

At first, the procurement of wild plant food (foraging) was replaced by the production of wild plant food using minimal tillage, then cultivation with systematic tillage gave way gradually to full-blown agriculture. The impacts of people on the landscape changed over this period from burning vegetation, through protective tending, to replacement with intentional selective planting and sowing, along with weeding, draining, irrigating and fertilising, and from land clearance to systematic soil tillage. At the same time, there were innovations in harvesting and storing the products of agriculture and in selective breeding of both animals and plants, processes we now call domestication (Harris, 1996). Throughout, there was increasing assistance of plant dispersal to new habitats, but also intentional introduction of useful alien plants and unintentional introduction of alien weeds: these reflect subtle combinations of local and intrusive developments. As for the intentional introductions, people would have been interested in certain useful woods for building purposes or for tool-making, and, of course, they would have been thoroughly familiar with anything that was edible (fruits, nuts, grains) or medicinally useful. Sadly, we have no records of any of this. The first archaeological remains of identifiable plants date from much later.

Most of the archaeophytes that survive today will have arrived in the British Isles by this route at roughly this period. These early weeds fall into one of two broad categories: those that mature their seeds at the same time as the cereal crop, and that have seeds that are roughly the same size as the crop, pass through the same coarse-grained sieves and so are trapped within the same fine-grained sieves; and those whose leaves, when young, resemble those of the crop so they are difficult to remove by hoeing or hand-weeding. Writing in the nineteenth century about the origin of the annual weeds of arable agriculture, Baker (1863) argued that 'it is probable that a large proportion of them are plants which have their original home in those lands where the Cereal Grasses were first cultivated, and that their seeds have been carried about with corn seed from country to country'.

The Bronze Age
During the Early Bronze Age (4,300–3,800 years ago), plough-based cultivation first evolved. Deeper cultivation required by newer crops like Rye (*Secale cereale*) required more traction, which meant bigger Cattle or more of them, and more traction required more fodder. This initiated the changes that led to the evolution of a dependent peasantry and a protective elite who exploited their labours. By now, the climate was deteriorating; where once the weather was warm and dry, it became much wetter as the Bronze Age continued, forcing the population away from easily defended sites in the hills and down into the fertile valleys. Large livestock farms developed in the lowlands and appear to have contributed to economic growth and to have inspired increasing forest clearances. The pattern revealed by the tree pollen and non-tree pollen diagrams jointly considered is one of slight and temporary forest clearances in the Neolithic and Early and

Middle Bronze Ages, moderate deforestation in the Late Bronze Age, and rapidly extended deforestation in the pre-Roman and Roman Iron Age. Forests were cut to provide fuel for iron smelting and for lime kilns, and there were increasing grazing impacts from domestic livestock and increased clearance for arable cultivation.

Plants in the Roman period

It was almost 100 years after Julius Caesar's first Roman campaigns in Britain in 55 BCE that the successful conquest took place (43 CE), during the reign of the Emperor Claudius. There had been trade links with Continental Europe before this date, but the Roman invasion signalled a massive change in the scale of the movement of goods and people.

Roman roads facilitated the movement of plants, both intentionally and unintentionally, all over the country. The Romans fed barley and oats to the Horses of their mounted regiments, and clearly transported considerable quantities of seeds from place to place. The Romans had substantial environmental impacts, too, and would fell 15 ha of woodland to construct a single fort. The Romans brought with them many plant species new to the British Isles, including at least 50 new plant foods, mostly fruits, herbs and vegetables. These introductions represented a major diversification of the plant component of the British diet at this time, adding important nutrients, variety of flavours, and new ways of expressing cultural identity as well as social status (Veen *et al.*, 2008). The list of vegetables introduced to the British Isles at this time includes Garlic (*Allium sativum*), Onion and Shallot (*Allium cepa*), Leek (*Allium porrum*), Cabbage (*Brassica oleracea*), Garden Pea, Celery (*Apium graveolens*), Turnip (*Brassica rapa* subsp. *rapa*), Garden Radish (*Raphanus sativus*) and Asparagus (*Asparagus officinalis*), along with herbs like Rosemary (*Rosmarinus officinalis*), Thyme (*Thymus vulgaris*), Bay (*Laurus nobilis*), Basil (*Ocimum basilicum*) and Winter Savory (*Satureja montana*). The Romans also introduced herbs that were used in brewing and for medicinal purposes, and instigated new farming practices and more productive grains, so that bread became a much more important part of the British diet. Our knowledge of the plant species that figured in human diets comes from painstaking dissection of human faeces preserved in undisturbed (and hence dateable) ground where latrines had been dug, and from microscopic analysis of charred remains of cooked or burnt plants preserved in dateable fire sites.

Among the trees, Walnut (*Juglans regia*; Fig. 12) and Sweet Chestnut *Castanea sativa*; Fig. 19) were introduced by the Romans (it is likely that Sycamore *Acer pseudoplatanus* was a much later introduction). They also introduced a wider variety of cultivated fruit trees, including Apple (*Malus pumila*; as opposed to Crab

FIG 12. Walnut (*Juglans regia*) is valued for its edible nuts and its high-quality timber, which is used by cabinet-makers: (a) flowers; (b) fruit. (CAS)

Apple *Malus sylvestris*), Grape-vine (*Vitis vinifera*), Black Mulberry (*Morus nigra*), Damson (*Prunus domestica* subsp. *institia*) and Cherry (*Prunus avium*). There was a period when the Romans prohibited the establishment of vineyards outside Italy, in order to safeguard the Italian wine trade, but in the third century CE Emperor Probus granted permission to England, Spain and Gaul to re-establish Grape-vine.

There are several instances where we know that a species was present in Roman Britain, but we do not know whether the Romans imported the fruits, whether they grew the plants as crops, or whether they allowed the plants to escape or even naturalise, in which case the species would be archaeophytes. Several tree species fall into this category, including Sweet Chestnut and Walnut. It is obvious that many of the foods preserved in faeces in Roman latrines were imported and were most unlikely ever to have been grown in the British Isles. Dates are the most obvious example, and even though olive stones were common in archaeological digs of sites from the Roman period, they were never found again once the Romans had left until the post-Columbian era. The truth is that we shall never know the full list of plant species that were deliberately or unintentionally introduced by the Romans rather than with subsequent waves of immigration.

The Dark Ages

The Romans left the British Isles in 410 CE and took the resources and know-how for maintaining their infrastructures with them. Their towns and villas soon fell into disrepair, and their farms reverted to more primitive forms of agriculture. We know next to nothing about plants for several hundred years after this.

Land management was under the control of the monasteries. The monks were considerable travellers and are likely to have brought back colourful plants from their journeys. We know that Benedict of Aniane in Languedoc (*c.* 747–821) exchanged medicinal plants with Alcuin of York (*c.* 735–804) when the latter was visiting France. Seeds and bulbs, the most portable parts of plants, were the most often transported.

It is not too fanciful to imagine travelling monks returning to their English homes with attractive species like Monk's-hood (*Aconitum napellus*), Mezereon (*Daphne mezereum*), Fritillary (*Fritillaria meleagris*), Snowdrop (*Galanthus nivalis*) and Summer Snowflake (*Leucojum aestivum*). But why didn't they write about them? Perhaps garden ornamentals were considered too frivolous compared with crops and medicines? Or perhaps they never did bring them back?

The Middle Ages

By the thirteenth century there was a commercial gardening trade in London and Oxford, selling seeds and plants in large numbers. Seed brought from outside often did better than home-grown seed; this was a discovery made by Sir Walter of Henley before 1280 and promulgated by John of Pontoise, Bishop of Winchester in 1282–1304, so there was a powerful incentive to the trade in plants. The fourteenth-century Dominican friar and horticulturist Henry Daniel had a garden at Stepney in east London boasting 252 species of herbs, of which perhaps 100 were aliens. By 1350, England had already acquired a great many introduced species, including Saffron (*Crocus sativus*), Monterey Cypress (*Cupressus macrocarpa*), Wall Germander (*Teucrium chamaedrys*), Sweet Marjoram (*Origanum majorana*), Pomegranate (*Punica granatum*) and gourds from central and Mediterranean Europe, North Africa and Asia Minor. Chives (*Allium schoenoprasum*), Hollyhock (*Alcea rosea*), Asarabacca (*Asarum europaeum*), Leafy Spurge (*Euphorbia esula*), Sweet Cicely (*Myrrhis odorata*), Garden Peony (*Paeonia officinalis*), Orpine (*Hylotelephium telephium*) and Fenugreek (*Trigonella foenum-graecum*) were all well known as aliens to early botanical writers.

THE ARCHAEOPHYTES

The term archaeophyte, while representing an extremely useful concept, is somewhat problematical. It is very familiar to students of the European flora, having been long used on the Continent (apparently since 1903), but it was first applied consistently to the British flora much more recently by Preston *et al.* (2002, 2004). For convenience, we shall repeat the definition given in Chapter 1: an alien plant that has been present in the area in a wild state since before 1500 CE. This is the definition long accepted on the Continent, the date 1500 more or less coinciding with the end of the medieval period and with the discovery by Europeans of America, and the start of the influx of introduced plants from that continent. As indicated in Chapter 1, Preston *et al.* (2002) modified the definition by insisting that in order to be considered an archaeophyte a plant must be naturalised. Although the majority of archaeophytes are indeed naturalised, or at least have been in the past, we doubt that they must be so, and there is scant evidence in several cases (e.g. Gold-of-pleasure *Camelina sativa* and Small Bur-parsley *Caucalis platycarpos*) that they ever were.

It seems probable that, after their initial, often unwitting, sometimes deliberate, introduction by man, most archaeophytes were at first casuals in the wild, later becoming naturalised. The naturalisation process would have taken very varying lengths of time, some taxa not achieving it until after the medieval period and some not at all. Others, once naturalised, became less persistent, reverting to casual status or even becoming extinct. Four of the 157 archaeophytes listed by Preston *et al.* (2002) are now extinct and there are no modern records of them even as very rare casual reintroductions. These are Violet Horned-poppy (*Roemeria hybrida*), last seen here in 1969; Lamb's Succory (*Arnoseris minima*) and Small Bur-parsley (both last seen in 1971); and Downy Hemp-nettle (*Galeopsis segetum*), last seen in 1975. A few species, such as Thorow-wax (*Bupleurum rotundifolium*), are scarcely less rare now, occurring only as occasional casuals. Some other species, such as Stinking Hawk's-beard (*Crepis foetida*) and Corn Cleavers (*Galium tricornutum*), are encountered only as deliberate reintroductions by conservationists. There are also some former crops, such as several species of primitive wheat (notably *Triticum spelta* and *T. dicoccon*), which are rarely grown now and occur in the wild in the British Isles only as extremely rare casuals, but which were recorded by Godwin (1975) from the Post-glacial period and should therefore be considered as extinct archaeophytes.

The fact that some archaeophytes, once common as cornfield weeds and considered thoroughly naturalised, have now very greatly diminished due to cleaner agricultural methods, suggests that those species were in fact only casuals that owed their persistence to constant reintroduction; Thorow-wax, Shepherd's-needle and Corncockle (*Agrostemma githago*) are good examples of species that became very much rarer in this way in the twentieth century. Thorow-wax (Fig. 13) was last recorded as a cornfield weed in the 1960s, but at about that time it was thought that it had begun to reappear as a casual from birdseed. It was soon realised, however, that the birdseed species was in fact a very similar lookalike from the Mediterranean, *Bupleurum subovatum*, which became known as False Thorow-wax. The two differ mainly in the ornamentation of their seeds. Careful taxonomic analysis is essential if we are to understand our aliens properly.

The same lesson can be learnt from the pimpernels, *Anagallis arvensis*. The well-known scarlet-flowered pimpernel is thought to be a native, but in

FIG 13. Thorow-wax (*Bupleurum rotundifolium*) is strange in both name and appearance. It was once a regular cornfield weed, but is now extinct in the British Isles apart from very rare casual appearances and some deliberate plantings. (Emorsgate Seeds, reproduced with kind permission of Mark Schofield)

addition there is a much less common blue-flowered pimpernel, typically a cornfield weed on light soils (Fig. 14). But two taxa masquerade under the latter: a blue-flowered variety of the native Scarlet Pimpernel, which together form the subspecies *Anagallis arvensis* subsp. *arvensis*; and a separate subspecies, subsp. *foemina*, the genuine Blue Pimpernel, which is an archaeophyte. The two differ almost solely in the type and density of minute glandular hairs on the fringes of their petals. Despite this tiny difference, and the two subspecies having the same chromosome number, they usually cannot interbreed successfully, forming rare sterile hybrids. The blue and scarlet varieties of subsp. *arvensis*, on the other hand, interbreed to form fertile hybrids. There are in fact other colour variants of subsp. *arvensis* (white, purple and a beautiful pink), the genetic relationships of which were worked out via hybridisation experiments by E. M. Marsden-Jones and F. E. Weiss in the 1930s.

The 197 archaeophytes that we recognise are listed in Appendix 2. In the left-hand column (scientific name) we use colour-coding to categorise each

FIG 14. Blue Pimpernel (*Anagallis arvensis* subsp. *foemina*), an archaeophyte of cultivated light soils, is distinguished only with difficulty from a blue-flowered variant of the common Scarlet Pimpernel (subsp. *arvensis*). (CAS)

taxon as cultivated (albeit escaped into the wild to some degree), colonist or denizen, as defined in Chapter 1. Six of the denizens are marked as invasive. There are 45, 65 and 87 taxa, respectively, in these three categories. In addition to the archaeophytes in Appendix 2, a number of those asterisked in Appendix 1 (indicating uncertainty regarding native/alien status) would be archaeophytes rather than neophytes if they are not native.

RECOGNISING ARCHAEOPHYTES

Preston *et al.* (2004) used six criteria for the recognition of a naturalised taxon as an archaeophyte:

1. No fossil record before the Neolithic period (4000 BCE).
2. Restricted to, or much more frequent in, man-made habitats.
3. Recorded in the wild before 1500 CE.
4. Range relatively stable since 1500 CE, except for changes (usually reductions) explicable by environmental change or suitable habitat availability.
5. Native range in Europe likely to be uncertain.
6. Present (as a neophyte) in North America, Australia and New Zealand (the so-called 'neo-Europe').

While these are very useful guides, none either singly or in combination is totally reliable, and there will always be uncertainty in a proportion of cases between archaeophytes and natives on the one hand, and between archaeophytes and neophytes on the other. To that extent, 'archaeophyte' is often treated as a category of uncertainty. In the past, species now considered to be archaeophytes have been variously treated as native or alien, but in the twentieth century in Britain they were most often considered as mainly native, e.g. by Clapham *et al.* (1952), Dandy (1958), Stace (1991), and Sell & Murrell (1996–2014), or as doubtfully native, notably by Druce (1928), Hyde & Wade (1934), and Lousley (1953). Today, they are classed as alien. Of those in our list, Wild Leek (*Allium ampeloprasum*), Greater Burdock (*Arctium lappa*) and Wild Turnip (*Brassica rapa* subsp. *campestris*) are among the strongest contenders for native status.

Preston *et al.* (2004) carefully assessed the status of all candidate taxa, and presented a list of 157 species that they considered to be archaeophytes. They also paid particular attention to the status accorded to candidate species in other European countries. There are other species that might be better considered as archaeophytes, all of them now usually being treated as natives; Barberry (*Berberis vulgaris*), Service-tree (*Sorbus domestica*) and Tansy (*Tanacetum vulgare*) are prime examples. Due to the equivocal nature of the evidence, they are not included in Appendix 2, but they appear (asterisked) in Appendix 1. One taxon that we have added is *Aethusa cynapium* subsp. *agrestis*, a dwarf subspecies of Fool's Parsley adapted to growing in cornfields. Preston *et al.* (2004) did not differentiate the two subspecies of *A. cynapium*; the well-known weed Fool's Parsley (subsp. *cynapium*) was considered by them to be a native, which is supported by evidence from fossil remains.

Preston *et al.* (2004) also compiled a list of 30 additional crop or garden species that are known to have been present here before 1500, but which now 'usually occur only as crop relics or casuals'. Examples from this list are Onion, Bread Wheat (*Triticum aestivum*), Garden Pea, Spinach (*Spinacia oleracea*), Garden Lettuce (*Lactuca sativa*), Coriander (*Coriandrum sativum*) and Flax (*Linum usitatissimum*). All are annuals or biennials. Others, however, should be added; for example, Turnip is an obvious omission, and Grape-vine and Walnut, albeit perennials, are surely in a similar category. Grape-vine is often naturalised in southern England, e.g. along the banks of the Thames near Kew, and Walnut self-sows very readily. We see no reason, other than an insistence that archaeophytes must be naturalised to deserve that title, not to include this supplementary list under the archaeophytes; both Grape-vine and Walnut are included in Appendix 2, producing our total of 197. This treatment is in accord with that of Pyšek *et al.* (2012).

It is among the archaeophytes in our flora that there is the greatest uncertainty of native versus alien status. If Shepherd's-purse (*Capsella bursa-pastoris*; Fig. 156), Red Dead-nettle (*Lamium purpureum*; Fig. 15) and Hemlock (*Conium maculatum*; Fig. 34) are archaeophytes, one might wonder why Common Chickweed (*Stellaria media*), Annual Meadow-grass (*Poa annua*) and Fool's Parsley are considered natives. The doubts concerning Groundsel (*Senecio vulgaris*) are briefly discussed in Chapter 9, and Corn Spurrey (*Spergula arvensis*) falls into the same category. These were the questions assessed by Preston *et al.* (2004) using their stated criteria, and in the great majority of cases they have arrived at decisions that are most likely the correct ones as far as the evidence goes. We have largely accepted their judgements rather than trying to reassess the often equivocal evidence, but further information coming to light might well change matters.

Wild Leek was treated as an archaeophyte by Preston *et al.* (2004). In the wild this is a scarce plant of the southwest on rocky or sandy well-drained soils. It exists as three varieties: *Allium ampeloprasum* var. *ampeloprasum*, with a compact head of flowers; var. *bulbiferum*, with a compact head of flowers and bulbils;

FIG 15. Red Dead-nettle (*Lamium purpureum*) is the most widespread of our five archaeophytes in this genus; it is common in cultivated ground almost throughout the British Isles. (CAS)

and var. *babingtonii*, Babington's Leek, with a loose head of flowers and bulbils. Babington's Leek is not known outside the British Isles, and var. *bulbiferum* is known only in the Channel Islands and a very small part of adjacent France. If the latter two varieties are truly archaeophytes they must have arisen as distinct entities since their arrival here, in which case they should be rated as neonatives. The species was first recorded in 1683. The conundrum of a suspected alien being endemic to this country is discussed further in Chapter 12.

There are also a number of other species mentioned by Godwin (1975) as weeds or crop plants, of which remains have been found dating from Roman times, and that are still found here as aliens. These are potential archaeophytes, but either the evidence of identification is not sufficiently strong or the remains have been in the form of the edible products, so they could have been imported in that form rather than having been grown here. They include Coriander, Buckwheat (*Fagopyrum esculentum*), Fig (*Ficus carica*), Stone Pine (*Pinus pinea*) and Black Mulberry. This subject is discussed in more detail by Philippa Tomlinson and Allan Hall (1996), who used the Archaeobotanical Computer Database at the University of York to summarise the available information. Some additional species they covered include Date Palm (*Phoenix dactylifera*), Olive (*Olea europaea*), Almond (*Prunus dulcis*), and the herb Summer Savory (*Satureja hortensis*). We have not included any of these in our list of archaeophytes.

Our 197 archaeophytes represent 9.8 per cent of the total of 2,006 (1,809 + 197) alien angiosperms, whereas the lower number of 157 recognised by Preston *et al.* (2004) represents 7.8 per cent. These results contrast markedly with the comparable figures published for the Czech Republic by Pyšek *et al.* (2012), who concluded that 350 (24.1 per cent) out of 1,454 aliens found there are archaeophytes.

USES OF ARCHAEOPHYTES

Appendix 2 gives the uses (culinary, medicinal or ornamental) to which the 197 taxa we treat as archaeophytes have been put; if no use is apparent, 'weedy' is entered. With many weeds having a close physical association with human habitation, people are more than likely to have found positive attributes for them. In many cases it is impossible to tell whether a species was introduced intentionally for some use, or was brought here accidentally as a contaminant and subsequently found to be useful. Examples are Common Fumitory (*Fumaria officinalis*), a weed used in folk medicine and even today in some proprietary concoctions, and Field Pepperwort (*Lepidium campestre*), a weed once used in flavouring salads, soups and sauces. Almost certainly some useful plants subsequently became weeds (as indeed some more recent valued introductions still do). There can be no better example than Ground-elder (*Aegopodium podagraria*), which has perniciously invasive white rhizomes that were, incidentally, used in the 1950s by N. G. Ball and T. A. Bennet-Clark in their classic studies of the diageotropic responses (growing at right angles to the force of gravity) of plant organs, much to the frustration of their university gardeners. Other species more obviously introduced for medicinal purposes became noticed for their attractiveness and were developed as garden ornamentals, e.g. Feverfew (*Tanacetum parthenium*), which is available in at least six cultivars, including those with yellow foliage or 'double' (*flore pleno*) flower heads. Feverfew is still valued as a migraine cure, the fresh leaves being eaten in salads or in sandwiches. Thirty-six taxa in Appendix 2 are given dual entries as weeds and non-weeds, and we envisage the usefulness of many species to have been viewed ambivalently by people in the past, just as they are today.

Appendix 2 shows that 71 plants are categorised as culinary plants, 66 as medicinal plants, 20 as ornamentals and 116 as weeds; 9 other minority uses are noted as well. Hence about 38 per cent of the species fall into more than one category. People continue to find new uses of plants, including archaeophytes, especially as fashions and social patterns change, and the list of plants being grown commercially in this country is therefore constantly evolving. Two examples spring to mind.

Borage (*Borago officinalis*) is an attractive blue-flowered garden annual (Fig. 16), which has been used for a variety of very minor culinary and medicinal purposes, such as colouring vinegar blue and as a floating addition to fruit punches. According to Edmund Launert, in his fascinating book *Edible and Medicinal Plants of Britain and Northern Europe* (1981), 'a concoction of flowers soaked in wine was drunk for melancholy and depression'. One wonders how essential the borage flowers were. But in recent years borage oil has been championed as a skin conditioner (usually under the more euphonious name of star-flower oil), and

FIG 16. Borage (*Borago officinalis*), a herb with many uses in the past, is now being increasingly grown for its cosmetic oil and as a nectar source for bees. (CAS)

fields of the plant have appeared in some areas since the 1980s for that purpose and incidentally as a nectar source for bees. In 2011, about 1,000 ha of Borage were harvested in Britain.

Garden Rocket (*Eruca vesicaria* subsp. *sativa*) is an important flavouring for some salads, but it has a very short season, going to seed as soon as it is full sized. In the last few years, 'perennial rocket' or 'wild rocket' has been offered by seed merchants. This is, in reality, Perennial Wall-rocket (*Diplotaxis tenuifolia*), an archaeophyte that apparently had no previous use. It has the advantage of providing fresh green leaves throughout the year, except when frosted; its flavour is similar to that of the annual Garden Rocket but it will prove too strong for some, as will its rather foetid smell. Pot Marigold (*Calendula officinalis*) and Purple Viper's-bugloss (*Echium plantagineum*) are now being grown as sources of specialist oils.

Some plants come in and out of fashion, then return again. Flax (Fig. 17), Hemp (*Cannabis sativa*) and Opium Poppy (*Papaver somniferum*) have all seen

FIG 17. Flax or Linseed (*Linum usitatissimum*) is today grown much less than previously for both oil and fibre. (CAS)

particularly wide variations in popularity. Flax was cultivated by ancient man and was probably introduced to the British Isles by the Romans. Fields of blue-flowered Flax were a common sight after the Second World War, but subsequently they virtually disappeared and even the traditional cultivation of Flax for fibre (linen) in Northern Ireland has died out. In that period Flax was still much grown for fibre just across the English Channel, where the blue fields give way in summer to the distinctive rich brown colour of the cut stems lying in the fields for retting (partial rotting). Today, northern France, Belgium and the Netherlands are still major Flax-growing countries, from where the raw fibres are imported to serve the diminished Irish linen industry. In recent decades, Flax has become commoner again in Britain; the usual blue- as well as a white-flowered variety are grown for seed (linseed oil, and as a health food). In the early 1990s, I asked a local Midlands farmer whether he was growing the crop for oil or fibre; he replied, 'Neither; I push it into a pile in the middle of the field and burn it. It does not have to be utilised to gain the government subsidy!' Flax fibre is also used to make high-quality paper and its seeds are a source of oil for paints and varnishes. After oil extraction, the seeds are used in animal feed mixes.

Hemp, likewise, is becoming commoner again. It was much used from Roman to medieval times as an oil crop, and for many centuries its stem-fibre has been used to make rope. There is evidence of its use as a drug as well as an oil for several thousand years. The hallucinogenic resin is secreted by epidermal glands on the stems, leaves and flowers, the greatest concentration being in the bracts surrounding the fruits on female plants. Today, Hemp is also made into paper and cloth, and it is being investigated as a biomass source of energy. Fortunately, the varieties of Hemp now being grown on a field scale contain extremely low concentrations of the hallucinogen, and the scientific name *Cannabis* is carefully avoided by growers. Potent varieties are, however, much grown illegally in this country, and openly on windowsills and canal-boat decks in the Netherlands. Hemp seed is also a constituent of birdseed, and the plant is quite common as a casual on rubbish tips and where birds have been fed. The genus *Cannabis* is nowadays usually regarded as consisting of a single variable species, *C. sativa*, but there are some who prefer to split it into three or four. This caused problems in the courts of the United States in the 1970s, when some charged with possessing *C. sativa* successfully pleaded that their supposedly illicit material came from one of the other species.

Opium Poppy has for centuries been grown for the remarkable assemblage of alkaloid drugs it contains, many of which (such as morphine) occur only in that species of poppy in the whole living world. The crude latex extract from the poppy fruits is known as opium, from which the separate alkaloids (or opiates) can be extracted. The best-known opiate, the first to be isolated (in 1804) and still the strongest medical analgesic (painkiller) is morphine. Heroin is a related chemical that is artificially synthesised from morphine. Other important examples of poppy alkaloids are codeine, papaverine, thebaine, noscapine and laudanosine. Opium Poppy has always been grown on a very small scale in this country. The drug is used for both recreational and medical purposes, as it has been for several millennia, but the great majority of material used in these ways is imported from hotter countries, notably Afghanistan. Today, fields of mauve-flowered Opium Poppies (Fig. 18a) are quite a common sight in some parts of England, but they mostly contain very low amounts of alkaloids. They are euphemistically called Blue Poppy, from which poppy seed (maw seed) is collected for the baking industry to be used in and on cakes and bread. The seed also produces an oil used in some paints and edible products like salad dressings. In addition, some drug-producing varieties are now being grown in England under licence. The species is a popular annual garden ornamental, modern cultivars with red and/or white flowers (Fig. 18b), and with lacerated petals or double flowers (*flore pleno*), being preferred to the rather dull mauve of the original. Some wild occurrences are of these sorts. After flowering, the handsome poppy fruits are still attractive, and very popular as a food source for Blue Tits (*Cyanistes caeruleus*) and in cut-flower arrangements.

Sweet Chestnut *Castanea sativa* (Fig. 19) has an interesting history of varied usage. Its exact native range is uncertain (most likely the Balkan Peninsula and perhaps from Italy eastwards), but it has been widely cultivated for many centuries for its nuts and timber across much of Europe, probably reaching

FIG 18. Opium Poppy (*Papaver somniferum*), perhaps the most important and remarkable drug plant of all: (a) grown on a field scale for poppy seeds; (b) an attractive horticultural cultivar. (CAS)

FIG 19. Sweet Chestnut (*Castanea sativa*): (a) unusual among our trees in not flowering until July; (b) the distinctive spiny fruiting clusters contain large edible nuts. (CAS)

Britain with the Romans, who used the nuts to feed their army. However, the existence of charcoal, timber or nuts in archaeological deposits does not prove that it was grown at that time; perhaps the products themselves were imported. There is apparently no pre-medieval pollen record of the species. We do not know when it was first mass-cultivated here, let alone when it began to be self-generating in the wild. As late as 1870, H. C. Watson said that Sweet Chestnut was 'always planted'; he was not right, but obviously it was not commonly naturalised at that time.

The nuts are usually stored until winter and then boiled or roasted, and are often used to flavour various soups, stuffings, stews, sauces and puddings. Roasted chestnuts are still sold direct from the brazier on the streets of many English towns, but these are imported nuts. Chestnuts can be eaten fresh, especially when young enough to allow the rather bitter inner skin to be easily peeled from the kernel. In the 1950s, when I was a boy in the High Weald of Kent, 'chestnutting' was an annual ritual in late autumn, and too many were eaten for comfortable digestion. Sweet Chestnut flowers very late in the season (July) for a tree, pervading the air for miles around with its sickly sweet smell, and the British climate is marginal for the ripening of its nuts, which occurs regularly only in the south of England north to parts of East Anglia and the south Midlands. Hence the species can become naturalised only in this very restricted area. In southern Europe the nuts are ground into flour and used to make bread, cakes and pasta, and in some regions – particularly of Italy – they are a major gluten-free source of carbohydrate. In Britain, however, chestnut flour is at best a marginal and imported commodity.

As a timber, Sweet Chestnut is strong and durable; it is one of only four common British ring-porous timbers (the others are oak, ash and elm). It is used for construction in southern Europe, but has been adopted for this purpose to only a very limited extent in the past in Britain. Latterly it has been used extensively only for fencing and garden furniture, being mainly grown as coppice. It was also used in the construction of the royal barge *Gloriana*, built to help celebrate Queen Elizabeth II's Diamond Jubilee in 2012. Chestnut coppice is a characteristic vegetation type in Kent, Sussex and Surrey, the individuals being planted quite densely so that no ground flora can persist when the coppice approaches maturity. Instead, the ground is covered by dry leaves and the spiny fruit cupules from previous years. Traditionally there were mature oak trees (standards) at intervals (coppice with standards). For most purposes the Sweet Chestnut growth is cut off low down about every 12 years, though it is sometimes left longer to obtain thicker poles. When an area of coppice is cut the ground is opened up to the light, and in the following spring one is often rewarded with a magnificent display of spring-flowering plants, notably Bluebells. Over the following years, the stools grow their crop of new poles, each starting with perhaps a dozen to a hundred or so shoots in the first year, these being reduced by competition for light to a few (rarely more than ten, each perhaps about 15 cm diameter at the base) by the time that the cycle is completed.

The poles are used for posts and spiles (the latter with a pointed base), but many are split with the grain into six to ten pales, which are strung together in long rows with two or three wires. These wired lengths of pales are used as fencing supported by posts at intervals, so the lack of contact with the soil allows them a long life. Thicker poles are used for post and rail fencing and for garden furniture. There are many other specialist uses as well, such as making the framework and handles for Sussex trugs (baskets), and formerly they were exclusively used as the poles in Hop (*Humulus lupulus*) gardens. Indeed, Hops and Sweet Chestnut have gone hand in hand for centuries. For more on the subject, we warmly recommend Chris Howkins' *Sweet Chestnut – History, Landscape, People* (2003), a superb and fascinating survey of the growth and use of *Castanea* in this country over the centuries. Unfortunately, the management of Sweet Chestnut coppice and the utilisation of its poles is labour-intensive, and much coppice has become irretrievably neglected or grubbed out, so that a valuable habitat has become greatly reduced and much flora and fauna lost. One wood I know has not been coppiced since 1961; the oak standards have all been removed for cash, and its Pearl-bordered (*Bolaria euphrosyne*) and Small Pearl-bordered (*B. selene*) fritillaries have long since gone. The neglected coppice has developed into tall, dark forest with scarcely a Bluebell to be seen.

Walnut (Fig. 12) is another southeastern European tree bearing much-valued nuts. Also like Sweet Chestnut, it is on the edge of its climatic range here and ripe nuts are produced only in the southern parts of the country, albeit further north than is possible for Sweet Chestnut. Hence we scarcely grow it commercially.

Although there is a small demand for 'wet walnuts' in September, they do not seem to store or dry well and nearly all the dried walnuts sold in the winter are imported. Apparently the custom of beating a Walnut tree was carried out both to gather the fruit and to prune the long shoots, so encouraging the production of short fruiting spurs. It gave rise to the much-quoted doggerel: 'A dog, a wife and a walnut tree / The more you beat 'em the better they be.'

The trunk of the Walnut tree has a thick, pale sapwood, but the central heartwood is of a dark, rich colour and was much prized for furniture-making, mainly from the sixteenth to eighteenth centuries, before the rise in popularity of mahogany. It still has a minority use today. The species very readily grows from nuts, which are transported and buried by Grey Squirrels (*Neosciurus carolinensis*), which greatly prize them.

The only archaeophytes other than Sweet Chestnut that can create a significant habitat are the willows (*Salix*). There are, in fact, a number of parallels between the two – both are true trees of some beauty but of limited value as timber, and both have figured in many traditional uses. Our two tallest willows, the only true trees in the genus in the British Isles, are White Willow (*S. alba*; Fig. 20) and Crack Willow. It has often been claimed that these are natives; the evidence is equivocal in the case of White Willow but we now know that Crack

FIG 20. White Willow (*Salix alba*), with its distinctive tree outline and characteristic white foliage; the timber of one variety is used to make cricket bats. (CAS)

Willow must have been introduced. Molecular (DNA) research has confirmed that most of the trees formerly known as Crack Willow are not, in fact, a distinct species, but a hybrid between White Willow and another species, native to Turkey and the Transcaucasus, described in 2009 by the Russian salicologist Irina Belyaeva as *S. euxina* and formerly known as *S. fragilis* var. *decipiens*. Only male *S. euxina* exist with us, and rather rarely at that.

White and Crack Willows and their hybrids are characteristically found along rivers and streams and in river valleys on at least seasonally wet soil. They have been much planted on riverbanks, where they perform the important job of soil stabilisation. In such places they are often pollarded, i.e. cut off at about 2–3 m, from which point slender branches arise and can be harvested when at the right size. Pollarding is effectively coppicing at a few metres above the base instead of close to it, allowing trees to be grown in grazed grassland, the new shoots being out of reach of browsing animals. Pollarded willows along lowland riverbanks form an iconic scene much painted by artists, such as those depicted by John Constable along the Suffolk/Essex River Stour. The tops of old pollards often support a number of epiphytes, including shrubs such as Elder (*Sambucus nigra*), and abundant animal life from birds to insects. The pollard stumps are often ancient and partially rotten, and provide a haven for all manner of animals and fungi. These willows have been used for rural industries such as fence- and post-making, and for supplying the thin slats in Sussex trugs. The best-known use of willow in England, however, is for making cricket bats; in fact a 'willow' is sometimes used as a synonym of the cricket bat, and 'the sound of leather on willow' is a much-hackneyed expression. The willow used for this purpose is the variety *S. alba* var. *caerulea*, known as the Cricket Bat Willow. In the Indian subcontinent, cricket bats are made from local willows as well as from those imported from England.

Almost equally famous is the use of willow in basketry and hurdle-making, taking advantage of its extreme flexibility. Willows used for this purpose are called withies or osiers, which belong to a range of species and hybrids. Probably the species most used in Britain are the two archaeophytes Almond Willow (*Salix triandra*; (Fig. 21) and Common Osier (*S. viminalis*), together with a range of hybrids involving these two species as well as the native Purple Willow (*S. purpurea*) and others. Each taxon has its particular characteristics and therefore specific uses. The withies are grown in copses or 'holts' in wet ground, where each plant is coppiced on a cycle lasting one to several years according to the thickness of stem required. Withy coppicing, like Sweet Chestnut coppicing, greatly diminished in the twentieth century, but like many rural industries it has made something of a comeback in recent years as the public have become more aware

FIG 21. Almond Willow (*Salix triandra*), like many species of the genus, has very attractive male catkins in spring. (CAS)

of the need to use sustainable materials, especially as they are so much more aesthetically pleasing than plastic in the garden setting.

Willows (and poplars) are our fastest-growing broadleaved trees, and today this attribute is being exploited to produce 'biomass'. The idea of using plant material as a renewable, carbon-neutral energy source has become fashionable as a response to climate change. Biomass is based on the concept that any living or recently alive plant material can be turned into a whole range of organic products by chemical or bacteriological means. The degree to which it is transformed before use varies greatly, from raw timber to distilled products such as certain alcohols. It is simply the case that the more dry weight per hectare that can be produced the better, and willows seem to be among the leading contenders.

OUR ANCIENT WEEDS – THE COLONISTS

Many of our archaeophytes are annual plants that are weeds in our gardens and cultivated fields, and that are rarely found in habitats not created and maintained by us. They appeared in our flora from the Neolithic period (*c.* 4000–2500 BCE) onwards (not before), some as late as Roman or even medieval times, and from the first were associated with habitations. We shall never know whether many of them are really native, perhaps once rare in open habitats such as river gravels, landslips and coastlines, and then taking advantage of the open ground provided by humans, or are aliens that arrived here with agriculture. H. C. Watson (1847–59), in *Cybele Britannica*, his pioneering survey of plant distribution in Britain, coined the term 'colonist' for this group of archaeophytes (*see* Chapter 1). They include numerous well-known and common weeds, many of them all too familiar to keen gardeners.

Cornfield weeds

The very quintessence of the archaeophyte concept is embodied in what are often described as 'traditional cornfield weeds' (Figs 2 and 22). These calendar subjects are often held in much awe, and their virtual demise much lamented, putting them on a level with our flowery meadows. However, despite their great attractiveness, they are not in the same league as flower meadows in environmental importance. Most of us are too young to remember cornfields full of flowers, but a trip to certain parts of the Continent, including the most rural parts of France and Spain, can still reveal cornfields awash with red, white, blue

FIG 22. Quintessential archaeophytes: (a) Common Poppy (*Papaver rhoeas*) and Scentless Mayweed (*Tripleurospermum inodorum*) in a neglected field; (b) Corn Marigold (*Glebionis segetum*) and Austrian Chamomile (*Anthemis austriaca*), from a planted seed mixture intended to replicate typical cornfield weeds. (CAS)

and yellow, among which a great variety of other weeds compete with the wheat or barley.

There are today many attempts to re-create these colourful cornfields in Britain, often in sectors, field heads, baulks or marginal strips (Fig. 23), but sometimes involving whole fields. They mostly succeed in reintroducing an authentic range of weed species, but in my experience the appearance of the field rarely resembles that of the original habitat. This is because the seed mixtures used seem to be thoroughly mixed, providing an intimate mosaic of species so that 2 m² shows almost the same diversity as a hectare. In the weedy cornfields of the Continent, which presumably resemble ours of times past, more often one species dominates a large area of the field, perhaps replaced by another after a few hundred metres. Hence the mosaic of the above four colours is on a much coarser scale, because the natural dispersal of one species has led to its local dominance. This large-scale patchiness is accentuated by differences in the extent and timing of cultivation in previous years, leading to strong spatial patterning in the structure of the soil seed bank.

The four colours mentioned above are achieved by the following main species: red from various species of poppy, mostly Common Poppy (*Papaver rhoeas*); white from various mayweeds or chamomiles, today mainly Scentless Mayweed (*Tripleurospermum inodorum*) but in the past often Corn Chamomile (*Anthemis arvensis*) and Stinking Chamomile (*A. cotula*); blue from Cornflower (*Centaurea cyanus*; Fig. 24); and yellow from Corn Marigold (*Glebionis segetum*) or Charlock (*Sinapis arvensis*). Many other tall plants are mixed in, vying with the corn for light, such as Corncockle (Fig. 25), Black-bindweed (*Fallopia convolvulus*),

FIG 23. Marginal strips of cornfield that have been spared the weedkiller, allowing growth of Corn Marigold (*Glebionis segetum*). (MJC)

FIG 24. The beautiful blue Cornflower (*Centaurea cyanus*), one of the much-decreased cornfield archaeophytes. (CAS)

FIG 25. Corncockle (*Agrostemma githago*), another much-decreased species, whose seeds contain toxins that contaminated flour made from the wheat with which it grew. (CAS)

Night-flowering Catchfly (*Silene noctiflora*) and several grasses. Numerous lower-growing plants occur among these, some flowering early before being too shaded by the taller plants, others flourishing later among the stubble after harvest, e.g. Pheasant's-eye (*Adonis annua*), Blue Pimpernel, Venus's Looking-glass (*Legousia hybrida*) and Weasel's-snout (*Misopates orontium*). The above examples are restricted to archaeophytes; mixed with them would be natives (e.g. Scarlet Pimpernel and Cleavers *Galium aparine*) and neophytes (e.g. Field Brome *Bromus arvensis* and Larkspur *Consolida ajacis*), all equally characteristic cornfield weeds.

Two interesting facts emerge concerning the names of these plants. First, there is the frequency of the specific epithet '*arvensis*' (or *arvense*), which signifies arable field; and, second, one notices numerous unusual English names. True, these weeds have their share of artificially coined names like Field Pansy and Corn Buttercup (though each of these have 20 or more local names that are less commonly used, such as the Sussex Kiss-me and Crow-claws, respectively), but there are also such delights as Corncockle, Charlock, Darnel, Pheasant's-eye, Thorow-wax, Fluellen, Venus's Looking-glass, Weasel's-snout and Shepherd's-needle, indicating an intimacy with these wild plants that is now largely lost; such plants required vernacular names, and the ones chosen illustrate the perceptiveness of our ancestors. Geoffrey Grigson (1955), in his superb book *The Englishman's Flora*, explored the folklore of British wild plants, including the local names given to them. Corn Marigold, for example, has about 40 alternative names, Cornflower about 30, and the (always rarer) Corncockle about 20. This amply illustrates the importance of these weeds to country folk in the past, for it was a case of 'all hands to the pumps' when hay-making or corn-harvesting came around. Despite their visual attractiveness, these plants were, of course, pests that reduced the crop yield, and in the case of Darnel (*Lolium temulentum*) and Corncockle the seeds contain toxic substances that can render the grain and flour unpalatable. Ridding the crop of such weeds before the days of selective weedkillers could be achieved by trying to ensure that the seed sown was relatively pure, or by hand-weeding. The latter method is still sometimes used to remove Wild-oat (*Avena fatua*) from an Oat (*A. sativa*) crop in a procedure known as roguing.

Several of these cornfield weeds, notably the poppies (*Papaver*), were favourite subjects of Sir Edward Salisbury in his study of 'reproductive capacity' as a means of predicting or measuring the success (abundance) of species. His detailed series of observations was summarised in his classic *Reproductive Capacity of Plants* (1942). Four species of red-flowered poppies occur with us as cornfield weeds (Fig. 26). Two of them have smooth capsules (Common Poppy *P. rhoeas* and Long-headed Poppy *P. dubium*) and two have capsules with conspicuous bristles (Prickly Poppy *P. argemone* and Rough Poppy *P. hybridum*; Fig. 27). In terms of capsule shape, two

FIG 26. The three commonest cornfield poppies (*Papaver*): (a) Common Poppy (*P. rhoeas*); (b) Long-headed Poppy (*P. dubium*); (c) Prickly Poppy (*P. argemone*). (CAS)

FIG 27. Ripe, dehisced fruits of poppies (*Papaver*) (left to right): Common Poppy (*P. rhoeas*), Long-headed Poppy (*P. dubium*), Prickly Poppy (*P. argemone*), Rough Poppy (*P. hybridum*). (CAS)

are short and broad (Common and Rough poppies), and two are elongated to at least twice as long as broad (Long-headed and Prickly poppies). A fifth species, Yellow-juiced poppy (*P. lecoqii*), closely resembles Long-headed Poppy but was not investigated in this context. Salisbury found the relative abundance (as he perceived it) of his four subjects (decreasing from the most abundant, Common Poppy, via Long-headed and then Prickly poppies, to the rarest, Rough Poppy) to be broadly correlated with their average reproductive capacities, defined as average seed output per plant × percentage seed germination.

In Table 2, the first three columns of the matrix are taken from Salisbury's results, and the other two columns (of distributional data) from the BSBI database (as at May 2012). Tetrads are 2 × 2 km grid squares, so there are 25 in each hectad (10 × 10 km), and they are a better measure of frequency. In fact, the

TABLE 2. Seed production and abundance of four species of *Papaver*.

Papaver species	Seeds per plant	Germination	Reproductive capacity	No. of hectads	No. of tetrads
P. rhoeas (Common Poppy)	170,000	64%	10,928	1,906	11,907
P. dubium (Long-headed Poppy)	13,700	42%	5,757	1,944	6,033
P. argemone (Prickly Poppy)	2,142	63%	1,347	423	775
P. hybridum (Rough Poppy)	1,680	91%	1,529	198	544

smaller the recording area, the closer to the actual frequency are the data, but tetrads are the finest division for which data are available at present. Clearly the seed productivity (seeds per plant) mirrors the frequency of the species measured as the total number of tetrads, but not as the total number of hectads. The fact that the frequency of the four poppies is loosely correlated with their reproductive capacities is perhaps rather unsurprising, but it is clear from the above results and from further data produced in more recent years by a number of independent researchers that the correlation is far from proportional.

The reason for the hectad/tetrad discrepancy is probably that the number of hectad records for Common Poppy is approaching saturation, and has been 'overtaken' (as recording continues) by the much less abundant though slightly more widespread Long-headed Poppy. Furthermore, due to the exceptionally high level of seed germination of Rough Poppy, the reproductive capacity of that species is unexpectedly high, putting it above the commoner Prickly Poppy.

Salisbury performed some experiments on the effect of air currents on seed dispersal, allowing for the differing average heights above ground level of the capsules of the four species (which increase in the same order as for the abundance of the species). He actually found very little difference between the four. Kadereit & Leins (1988) performed wind-tunnel experiments and controlled shaking actions on the ripe capsules of six species, including the above four. The most significant result was that in the two bristly-fruited species (Rough and Prickly) a lower proportion of seeds was eventually dispersed from the capsules than in the two smooth-fruited species (Common and Long-headed), and moreover in the latter two most dispersal occurred very early on, whereas in the former two it continued over a longer period. The differences are likely to be due to the larger capsule pores in the smooth-fruited species. The authors conjectured that, despite their slower and lesser seed release, and indeed lower seed production, the two bristly-fruited species might disperse over a wider area since, over a longer time span, there is a greater likelihood of exceptionally strong wind gusts being encountered.

The breeding behaviour of the species is also relevant. Prickly Poppy and Rough Poppy are completely self-fertile, and moreover are highly self-fertilised. Self-pollination, or 'selfing', is facilitated by the anthers lying adjacent to the stigmatic surface; anther dehiscence (bursting open to release pollen) takes place when the flowers are still in bud, and selfing therefore occurs before any foreign pollen could participate. Common Poppy, in contrast, is highly self-sterile, and therefore nearly 100 per cent cross-fertilised. Long-headed Poppy is self-fertile, but the actual degree of selfing varies, mostly lying around 75 per cent. Surprisingly, the percentage selfing actually increases when there is visitation by insects, which carry pollen from the edges to the centre of the stigmatic surface. Common Poppy is demonstrably more variable, both morphologically and enzymically, and probably physiologically, than the other three species, a situation attained by its outbreeding (in which the parent plants are not closely related). It is likely that this variation enables the Common Poppy to thrive in a wider range of habitat conditions than its congeners, so that it is found in sites (such as clayey soils) where the other species are not found. There are many more records for this species than for the other three species combined.

CULTIVATED PLANTS

Eighty of the archaeophytes listed in Appendix 2 are marked with only 'W' (weedy) in the final column, suggesting that the other 117 taxa have been deliberately cultivated or at least tolerated in the past as culinary, medicinal or ornamental plants. The 73 food-plants vary from some of today's major crop species to species formerly used as food-plants but now largely or totally abandoned in favour of better alternatives, or to marginal crop plants of minor use. There are 67 medicinal species, many of which are similarly no longer used and, indeed, some were probably never of any real medical benefit at all. A few retain their importance in mainstream medicine; examples are Fennel (*Foeniculum vulgare*), Greater Celandine (*Chelidonium majus*), Mugwort (*Artemisia vulgaris*) and Pot Marigold. Others are still used in homoeopathy or alternative medicine. The 19 species that became garden ornamentals amount to a paltry total, suggesting that in the Middle Ages and earlier, people did not value plants for their appearance as much as is the case today. Those that became cherished as attractive plants ideally had to earn their place in the garden by some more practical criterion, mainly as food or medicines, or they were chosen from the weeds of cultivated ground. Examples are Elecampane (*Inula helenium*), Hollyhock, White Stonecrop (*Sedum album*), Lesser Periwinkle (*Vinca minor*) and Soapwort (*Saponaria officinalis*). In fact, all of the archaeophytes in Appendix 2 marked 'O' are marked with other categories as well.

The Common Poppy can be used to exemplify the above comments. It has been used medicinally but it contains only traces of alkaloids, and its introduction as a garden ornamental is more likely to have arisen from its abundance as a weed. There is variation in the flower colours of wild plants, particularly in the presence or absence of black blotches at the base of the petals, and sometimes there are narrow borders of white beside the black areas or at the petal edge. But today's 'Shirley' poppies, named after the parish of Shirley in Surrey, where in the 1880s the local vicar bred a series of colour forms from wild plants that he collected locally, show a vastly greater range: from red to white, with all manner of pink and mauve variations, and the flowers ranging from double (*flore pleno*) to single. In the opinion of many, however, none of these can rival the red Common Poppy of the cornfields. Due to their prodigious seed-set (Table 2) and their great longevity in the soil (over 100 years), once grown in the flower garden 'Shirley' poppies are there for keeps. It has been calculated that the soil under grassland that was once used for cereals can contain as many as 280 million seeds per hectare. These lie dormant in the soil until they are exposed and then allowed to germinate, often in amazing quantity, as happened in Flanders in the First World War.

In 1894, Lady Amherst, of pheasant fame, published a manuscript poem by Jon Gardener that was written in about 1440–50 and entitled *The Feate of Gardeninge*. It was effectively a guide to gardening practice, with subtitles such as 'Of settyng and sowyng of sedys' and 'Of cuttyng and settyng of vynys'. Many of the plants named are natives, but 22 are archaeophytes and in addition there are eight species that we treat as neophytes, e.g. Hyssop (*Hyssopus officinalis*), Lavender (*Lavandula angustifolia*), Motherwort (*Leonurus cardiaca*) and Martagon Lily (*Lilium martagon*). This seems to be a good reason to reconsider these aliens as possible additional archaeophytes. Furthermore, two species were listed that are not included in our list of aliens at all: Saffron and Rue.

Saffron was formerly cultivated for the valuable spice of the same name (Saffron Walden in Essex was the English centre of cultivation in the sixteenth and seventeenth centuries) and it figured in several older British plant lists (such as those of Druce in 1928, and Dandy as late as 1958), although it does not seem to have become naturalised. Interestingly, there has recently been a resumption of its field cultivation at Saffron Walden. Rue (*Ruta graveolens*) is still commonly cultivated, mainly as an ornamental, but formerly for medicinal purposes. It does sometimes occur as seedlings on walls and the like, but it does not become established.

In addition to these main categories there are some less common but equally important uses for which archaeophytes were employed, as indicated in Appendix 2: plants grown for timber, fibre, oil, dyes, scent, soap, fodder and biomass, and as withies.

Crop plants

Hardly any of our modern crop plants are natives or are derived from natives. The best examples of native vegetables are probably Beet (*Beta vulgaris*), Cabbage (with some doubt), Carrot (*Daucus carota* subsp. *sativus*), Parsnip (*Pastinaca sativa*) and Celery. Among fruits we have Cherry (*Prunus avium*), Blackberry (*Rubus fruticosus*) and Raspberry (*R. idaeus*), and in addition there are a number of animal foods, particularly grasses such as Perennial Rye-grass (*Lolium perenne*) and legumes like Red Clover (*Trifolium pratense*) and Common Vetch (*Vicia sativa*). Actually, in the case of the above vegetables, it is unlikely that the crops evolved or were developed from our native genotypes; they were probably selected in more southern latitudes. The hundreds of species of alien plants that are now grown as crop plants have been introduced over a very protracted period, but the majority of our major food-plants have been here for a long time and are therefore archaeophytes. Most important are probably the vegetables Broad Bean, Leek, Garden Lettuce, Onion, Garden Pea, Garden Radish and Turnip; the fruits Apple, Pear (*Pyrus communis*) and Plum (*Prunus domestica* subsp. *domestica*); and the cereals Barley, Oat and Bread Wheat. The most obvious of our foods not included above are the neophytes introduced from the New World after 1500, notably Potato (*Solanum tuberosum*), Tomato (*S. lycopersicum*), Marrow (*Cucurbita pepo*), Runner Bean (*Phaseolus coccineus*) and French Bean (*Phaseolus vulgaris*), and Maize (*Zea mays*).

Broad Bean is not known as a wild plant, but it probably evolved in southwestern Asia. It was once placed in the genus *Faba*, from which the name of the legume family (Fabaceae) is derived. It does not seem to have been in existence before the Neolithic period, but by the Iron Age (c. 800 BCE–50 CE) it was in Britain, although the large-, pale-seeded varieties most used today, borne in a non-shattering pod, probably were not developed until about 500 CE. Other varieties are grown as animal feed. Today, the species is only a short-lived casual found where seeds have fallen.

Of the *Allium* crops, Onion is by far the most popular and exists as many different variants from which different parts are eaten: Onion and Shallot provide subterranean bulbs; Spring Onion has edible foliage; and Tree Onion produces bulbils in the inflorescence. Within each group a diverse range of cultivars exists. None of these, nor Leek, nor Garlic, the two other most commonly grown members of the genus, is known as a wild species, but they have close wild relatives in central and southwestern Asia. They might have been in Britain since before the Romans, and were commonly grown by the medieval period. Onion and Leek are merely occasional casuals in our area, but Garlic, by far the least grown here, is the only one of them that sometimes establishes itself, often in somewhat saline areas. Onion fields, in addition, provide a habitat for weed species.

Garden Lettuce (Fig. 28), possibly brought to England by the Romans, is usually grown in protected conditions, but sometimes as an open field crop. It exists in a wide range of cultivars, some spherical and others elongated, some of the latter growing up to 1 m tall in the Mediterranean countries. The ancestor of our most popular salad plant is most likely Prickly Lettuce (*Lactuca serriola*),

FIG 28. Garden
Lettuce (*Lactuca
sativa*), whose close
relationship to another
archaeophyte, Prickly
Lettuce (*L. serriola*),
is obvious once the
salad plant is allowed
to reach the flowering
stage. (CAS)

a common neophyte in the southern half of Britain. Traits that are desirable in
culinary terms (mainly long-delayed bolting, lack of prickles, little latex and the
tendency to form 'hearts') were developed by selection over the millennia. Like
Onion, Garden Lettuce is rather rarely found as a plant in the wild, because it is
harvested before it forms seeds (unless grown for seed production), and then only
as a casual.

We know rather little about the origin and early development of the Garden
Pea (Fig. 11) except that it has been cultivated along with barley and wheat since
the beginning of civilisation. In its native southwestern Asia it has a number of
wild or semi-wild relatives, which are variously considered separate species or

subspecies or varieties of *Pisum sativum*. One of these is var. *arvense*, the Field Pea, which is grown for animal food and occurs as a contaminant of other imported seeds. It has purple flowers and is perhaps commoner as a casual here than the white-flowered var. *sativum* (the Garden Pea).

The Garden Radish (*Raphanus sativus*), on the other hand, has a more obvious origin in the form of the Wild Radish (*R. raphanistrum*), which exists as a number of subspecies (sometimes segregated as species). One of these is the Sea Radish (subsp. *maritimus*), which is native to Britain. These wild and cultivated sorts show a range in fruit development, from long, narrow seed pods with up to 12 or so seeds in a single file (subsp. *raphanistrum*), between each of which the fruit is constricted and breaks at maturity to produce one-seeded segments that are the dispersal units, to short, broad seed pods with a number of seeds in a non-constricted, non-shattering fruit that does not reveal the number of seeds within (*R. sativus*). This development (via selection by plant breeders and growers) of non-shattering fruit is a common trait in many domestic plants that are grown for seed, but less usual in a vegetable cultivated for its root or leaves. Radishes are mostly cultivated for their roots, which are used in salads, but in Asia there are also varieties of which the leaves or young fruits are also used in salads, and in Britain fodder radish is sometimes grown: this has thin roots but a large head of leaves eaten by grazing animals or used as green manure.

An interesting feature of all these radishes is that they exhibit polymorphic (more than one form of) flower colouring. In Wild Radishes (including Sea Radish), the petals may be white or yellow; in Britain, the yellow morph is commoner in the north and west, and the white in the south and east. In Sea Radish, white flowers are very rare except in the Channel Islands, and in Garden Radish the petals are usually white or purple, although yellow ones have been reported. Studies on the effect of this variation on insect visitors have been made (Kay, 1978). In Wild Radish it was found that the yellow flowers are recessive to white flowers, but that some butterflies (*Pieris*) and hoverflies (*Eristalis*) showed strong preference for yellow flowers in their visits. On the other hand, some bees (*Apis*) showed little discrimination, while others (*Bombus*) preferred white flowers. Garden Radish is but an uncommon casual with us, sometimes arriving in birdseed.

The cultivated Turnip, in which the root is swollen into the familiar vegetable, is one of three subspecies of *Brassica rapa*, namely subsp. *rapa*. There is a nomenclatural problem here resulting from confusion between Turnip and Swede (*B. napus* subsp. *rapifera*), partly because many people do not know the difference between the two species and partly because many who know them perfectly well use the English names the other way round! The two are completely separate species with different (known) origins but showing parallel evolution under domestication, in that both include subspecies with swollen roots, rich oil-producing seeds and very leafy shoots. The vegetable Turnip usually has a greenish-white skin, white flesh and a relatively mild flavour, and is best as an addition to stews and the like. Swede is usually larger with a yellower skin, orange flesh and a distinctive strong flavour; it is more often used as a separate vegetable like potatoes or cabbage. Both can be purple-tinged. They also differ in that the Swede has a distinct 'neck' at its apex, and there are floral differences too.

Turnip is an ancient crop that has probably been used in the British Isles since before Roman times, while Swede is a much more recent development (see Chapter 9) that is thought to have been introduced to Britain from Sweden as recently as the eighteenth century. English names follow no rules and, if one regional group of people has a name for a species, who can say that is right or wrong? However, it is difficult to see any logic in applying the name Swede to *Brassica rapa*, and doing so certainly causes confusion. Further confusion comes from the fact that it is *B. napus* subsp. *oleifera*, not *B. rapa*, that is usually known as Rape in Britain. The Turnip is thought to have arisen from the Wild Turnip (*B. rapa* subsp. *campestris*), which does not have swollen roots, by human selection. Wild Turnip occurs in the wild in Britain, mainly along rivers in the south, where it is also known as Bargeman's Cabbage. It is probably a naturalised archaeophyte, whereas Turnip is an uncommon casual neophyte. The subspecies of *B. rapa* and *B. napus* that are rich in oil are much more recent developments.

The most important fruit trees in central and northern Europe are Apple, Pear and Plum, all of which are commonly found in the wild and have been present there and as crops since prehistoric times. Apple trees are frequently found in hedgerows, wood borders and within woods, bearing apples that vary from large, often coloured and edible, to small, green and very sour. The latter are called crab apples, but every intermediate seems to exist between these and the larger sorts, which appear to be cultivated Apples 'gone wild'. The true Crab Apple (*Malus sylvestris*), which is thought to be native in the British Isles, is recognisable by the leaves, pedicels and sepals being completely glabrous (hairless), and the twigs often bearing strong spines, whereas the cultivated Apple, whose correct scientific name is disputed between *M. pumila* and *M. domestica* at the time of writing, always has some pubescence (hairs, often dense) on those organs.

The apparent existence of a complete range of intermediates had been used as evidence of frequent hybridisation between these two evidently closely related species, which led to the more or less undisputed view that the wild Crab Apple figured prominently in the origin of the cultivated Apple. Since 2000, however, the use of DNA sequences and fragments and enzyme analysis, initially in

Belgium, has radically altered these views (e.g. Coart *et al.*, 2003). It is now clear that the cultivated Apple arose in Asia, probably in the mountains on the borders of Kazakhstan and China, largely or wholly from the wild species *Malus sieversii*. The gene pools of this species and the modern cultivated Apple of Europe (whence they were later taken to the southern hemisphere and the New World) are quite distinct from that of the Crab Apple, and there is little evidence that the latter has made any contribution to our modern cultivated Apple. Moreover, although the cultivated Apple and Crab Apple can hybridise quite easily, the molecular evidence does not suggest that this happens at all frequently, so that hybridisation is unlikely to be the reason for difficulties in distinguishing these two species. They are genetically distinct, but not reliably separable on morphological characters. Many of the 'cultivated apples gone wild' that one encounters are probably the stocks on which commercial varieties were grafted, the scion having died off, and were never edible apples in the first place. Others are probably seedlings from animal-sown seed that segregates into a range of genotypes, few of which would be commercially viable.

Pears are much less often grown than apples and are rarer in the wild. The identity of these trees is even more difficult than is the case with apples, because the classification of the pear is not well understood. The cultivated Pear (*Pyrus communis*) is often contrasted with a 'wild' species with small, hard, almost spherical fruits known as Wild Pear (*P. pyraster*), which is probably also an archaeophyte rather than a native with us. The cultivated Pear was selected in Asia and its journey ever westwards mirrors that of the cultivated Apple, as does its breeding programme in the past few centuries, but it has always been less important and has received less attention.

Plums form a taxonomically very complex group that, in contrast with apples and pears, contains diploids, tetraploids and hexaploids (with two, four and six sets of chromosomes, respectively). The cultivated Plum (*Prunus domestica*) is a hexaploid. Its variability is often summarised by recognising three subspecies: subsp. *domestica* (Plum), with sparsely hairy, non-spiny twigs and large fruits with strongly flattened stones; subsp. *insititia* (Bullace and Damson), with densely hairy, often spiny twigs and smaller fruits with slightly flattened stones; and subsp. *italica* (Greengage), more or less intermediate. These have, however, been so much interbred that the character correlations have been broken down and separate taxa are scarcely recognisable. Two related species, both common in the wild in this country, which probably featured in the ancestry of *P. domestica*, are the native tetraploid Sloe (*P. spinosa*) and the much-grown, mainly diploid alien Cherry Plum (*P. cerasifera*). The latter is best known as an early-flowering ornamental tree, of which pink-flowered cultivars with copper-coloured leaves adorn many a suburban street, flowering before the leaves have emerged and earlier than the cherries and almonds.

The hexaploid plums and their relatives probably arose in Asia, but further west than the apples and pears, in the Caucasus region. Both the tetraploid *Prunus spinosa* and diploid *P. cerasifera* still grow there, and it is thought that the plums evolved via the hybridisation of these two species (to produce a triploid), followed by the doubling of the chromosome number (Fig. 157). In 1936, the Russian plant breeder W. A. Rybin, who found sterile triploid hybrids between the species in the northern Caucasus, claimed to have artificially synthesised a plum-like plant by such a route, but others have disputed the evidence and prefer to believe that modern plums arose from hybrids between the various chromosome races of *P. cerasifera* alone.

More is known and more has been written about the major cereals than about any other plants. There are dozens of books and thousands of research and review papers documenting the discovery by ancient man about 10,000 years ago that they could cultivate wheat and so change their lifestyle from nomadic to one in fixed communities based on agriculture. This first occurred in the Fertile Crescent centred on Mesopotamia (in modern-day Iraq) and constituted the earliest civilisation on Earth. Today, wheat is still mankind's single most important staple food. The evolution of wheat (*Triticum*) has been a story of hybridisation between the crop species and related weeds, accompanied by the doubling of the chromosome number to form new species. This process is described in more detail in Chapter 9. The earliest wheats had 14 chromosomes (7 pairs, diploids, known as einkorns, AA or BB in Fig. 157), which increased to 28 (14 pairs, tetraploids, known as emmers, AABB in Fig. 157) and then 42 (21 pairs, hexaploids, known as spelts, AABBCC in Fig. 157), as man selected the best plants he found in his fields as the seed for sowing the following year's crop.

All these groups have been cultivated over the millennia, but the first really successful wheat crop is thought to have been the cultivated tetraploid Emmer (*Triticum dicoccon*). This gave rise to other cultivated tetraploids, two of which are still important crop species in Europe: Pasta Wheat (*T. durum*, often called Durum Wheat and used for making pasta); and Rivet Wheat (*T. turgidum*, used for animal feed (Fig. 29). Rivet Wheat was much grown several centuries ago, and has remained a major crop just across the English Channel, but it is now enjoying a revival here since the long stems of some varieties are suitable for thatching. Further hybridisation and a doubling of chromosome number gave rise to the hexaploid spelts, of which modern Bread Wheat (*T. aestivum*; Fig. 29) is by far the most important. It, or a closely related spelt, first appeared about 8,500 years ago (6500 BCE), and reached Britain about 3000 BCE, during the Neolithic period. In

FIG 29. Some common European cereals (left to right): Oat (*Avena sativa*), Two-rowed Barley (*Hordeum distichon*), Rye (*Secale cereale*), Rivet Wheat (*Triticum turgidum*), Bread Wheat (*T. aestivum*), Pasta Wheat (*T. durum*). (CAS)

recent years two of these primitive, or 'old-fashioned', wheats have been trialled again in England: Emmer and Spelt (*T. spelta*) can both be tall plants, up to 2 m high, and are being grown as thatching material, while Spelt gives a distinctively flavoured flour and bread that is becoming favoured in health-food shops. We should expect increasing records from the wild of Emmer, Spelt and Rivet Wheat if these species become more popular crops.

Barley (*Hordeum*; Fig. 29) has been cultivated for about the same length of time as wheat and also arose in the Fertile Crescent, but it has a very different evolutionary story. Both the cultivated species and its wild relative, Wild Barley (*H. spontaneum*), from which it was selected, are diploids. Modern barley exists as two major sorts – the two-rowed and the six-rowed. In both there are three florets at each node of the inflorescence, forming two opposite rows, but whereas in the six-rowed type all three florets are fertile, forming six longitudinal rows of grains in the ripe ear, in the two-rowed type only the central floret of each three is fertile, giving only two rows of grains on opposite sides of the ear. Modern Two-rowed Barley is often separated as a distinct species (*H. distichon*) from Six-rowed Barley (*H. vulgare*). *Hordeum spontaneum* is a two-rowed barley, and it is therefore thought that modern Six-rowed Barley developed from a two-rowed ancestor. Archaeological evidence suggests that this had happened by 6000 BCE, and that barley reached western Europe about 1,000 years later. At first barley was more important than wheat, but the pendulum has gradually swung so that now barley production worldwide is less than a quarter of that of wheat. Much barley today is used for animal feed or beer-making. In Britain, the vast majority of barley grown is of the two-rowed type, which is therefore the commoner type found as a casual.

Oat (*Avena sativa*; Fig. 29) tells yet another story, but one similar to that of wheat in that diploids, tetraploids and hexaploids all exist, and the commonest cultivated species, *A. sativa*, is hexaploid. The evolutionary history concerning the identity of the primitive ancestors and their hybridisations and chromosome number doublings, however, is less well known than is the case with wheat. It is thought that the modern hexaploids arose from the hexaploid Winter Wild-oat (*A. sterilis*, which is not, of course, sterile), which had earlier evolved from diploid and then tetraploid ancestors. Unlike wheat and barley, Oat did not arise in the Fertile Crescent 10,000 years ago, but much more recently and much further west, in southwestern and central Europe, so it did not feature in the early civilisation of man but became a secondary food-plant much later. It first appeared in agriculture about 1000 BCE, and was brought to Britain during the Iron Age or Roman times. In addition to the hexaploid *A. sativa*, a diploid species, Bristle Oat (*A. strigosa*), is also grown as a crop species. This is suited to poor soils in cooler climates and, in fact, is still grown to a very limited extent in some remoter parts of the 'Celtic fringe' in the British Isles.

The other cereal archaeophyte is Rye (Fig. 29), which has always been a relatively minor crop in our area and today is used primarily to make the health foods rye-bread and crispbread. It arose in or near the Fertile Crescent but later than wheat or barley, probably being selected as a crop after man encountered it as a weed of the other two cereals. Rye evolution, like that of barley, has been confined to the diploid level. It is thought to have become cultivated around 3000 BCE and to have reached Europe up to a thousand years later. It was present in Britain in the Iron Age and in Roman times, and occurs today as a fairly frequent casual.

Besides these 'major' crops there are numerous others, each of which, although 'minor' in comparison, carries its unique spectrum of chemicals, so together they are vitally important in giving us a well-balanced diet. Many of these food-plants have interesting stories to tell, not least relating to their origins and evolutionary relationship to wild species. Examples are Asparagus, Onion, Leek, Garlic, Horse-radish (*Armoracia rusticana*), Chicory (*Cichorium intybus*) and Endive (*C. endivia*).

The cultivated Horse-radish (Fig. 30a) seems to have evolved in dry regions of eastern Europe and western Asia, but has been in cultivation for only about 2,000 years. An odd feature of this crop is that it is highly (but not completely) sterile, setting virtually no seeds. The resilient rhizomes, however, provide a more than adequate means of propagation, and once established the plant is very difficult to eradicate. Its frequent occurrence on roadsides probably marks places where offending rhizomes have been discarded.

FIG 30. Two culinary archaeophytes that are now commoner on roadsides and waste places than in gardens: (a) Horse-radish (*Armoracia rusticana*), which is extremely difficult to eradicate once established; (b) Chicory (*Cichorium intybus*), very different in appearance from the vegetable product when in flower. (CAS)

Chicory (Fig. 30b) is a salad crop derived by blanching the young shoots. It looks completely different from the rather coarse, stiffly branched wild plant, which has beautiful sky-blue Dandelion-like flower heads and is common in some areas on a light base-rich soil. If the cultivated plant is left unblanched, however, it soon becomes evident that the two are one and the same, and the origin of the crop is obvious. Chicory is also sown with grass because of its ability to accumulate the element selenium, which is valuable to stock. Its close relative Endive is used as a green salad plant. It resembles a completely flattened version of a round-headed lettuce with very dissected leaves, giving rise to one of its French common names, Frisée.

Two other minor crop archaeophytes are Lentil (*Lens culinaris*) and Chick Pea (*Cicer arietinum*). These both require hot summers for a successful seed harvest, and are commonly seen in usually small fields in southern Europe and the Mediterranean region. They occur here as rather rare aliens arising from the spillage of imported seed, or from their use as birdfeed. No doubt encouraged by rash predictions of long, hot summers in future years, they are now being grown on a small scale in southern England, which might lead to increasing records of them in the wild.

Crop–weed relationships

In addition to the often intimate genetic relationships between crops and their closely related weeds, of which the best examples are found in the evolution of wheat over the past 10,000 years, obviously there has been a considerable amount of adaptation by weeds to the crops with which they grow, presumably during only the last few millennia. This is one aspect of the phenomenon known as co-evolution. The height of cornfield weeds, and the months in which they complete various parts of their life cycle, clearly need to be closely attuned to the growth and harvesting of the crop in which they grow. For example, a tall plant flowering in late summer would never set seed, and a short one flowering in midsummer might well suffer from lack of light and pollinators. Many short plants produce much of their fruit after harvest, in the stubble, and the modern habit of ploughing immediately after harvest has greatly reduced their abundance. Fool's Parsley seems to have evolved a dwarf race or ecotype (*Aethusa cynapium* subsp. *agrestis*) that is particularly suited to growing in cornfields, where at least some of the inflorescences will be left by the harvester (whether it be a man with a scythe or a combine harvester).

The relationship between cultivated Oat and wild-oats illustrates the above point. As with wheat and many other cereals, cultivated Oat is either sown in the autumn (winter oats) or spring (spring oats), the type grown depending on local conditions. By and large, in the British Isles winter oats are commoner in the south, and spring oats in the north, because of the harsher (colder and wetter) weather during winter in the latter area. The most characteristic, and in most areas the commonest, weeds of Cultivated Oat are wild-oats, of which winter and spring physiological races also occur. The two main species of wild-oats are Wild-oat (*Avena fatua*) and Winter Wild-oat (*A. sterilis*, formerly known as *A. ludoviciana*), both of which can be serious pests of cultivated Oat (*A. sativa*). Winter Wild-oat mostly germinates in the autumn and Wild-oat mostly in the spring, so it is not surprising to find that the former is more commonly found as a weed of winter oats. Winter Wild-oat would not be likely to succeed in spring oat crops, because of the soil preparation during or after winter, whereas Wild-oat could manage in either crop system. Wild-oats are a difficult weed to eradicate because they are not amenable to selective weedkillers and their disseminules (dispersal units) can mimic those of the crop. Hence wild-oat grains tend to be harvested, and replanted next season, along with those of the crop.

Physical mimicry of a crop by a weed is a widespread phenomenon, and many examples are known. The mimicry can apply to the fruits or seeds, the gross habit of the plant, or its physiological characteristics. Salisbury (1961), in *Weeds and Aliens*, commented that the seeds of Hedge Bindweed (*Calystegia sepium*) are sometimes found as contaminants of agricultural Common Vetch seed supplies. This is because the seeds of these two species are both roughly spherical and smooth, and about the same size, so if Hedge Bindweed is growing as a weed in Common Vetch (surely not a very common occurrence; Field Bindweed *Convolvulus arvensis* is much more likely to occur there) its seeds are likely to be harvested with those of the crop. In this case the crop plant is the archaeophyte. I have similarly noticed the fruiting heads (inflorescences) of Pineappleweed (*Matricaria discoidea*) contaminating loose peas – the two propagules are about the same size and shape but in this case far from being biological equivalents. The crucifer Charlock is difficult to eliminate from Rape, as are various goosefoot (*Chenopodium*) species from Sugar Beet (*Beta vulgaris* subsp. *vulgaris*), because of the overall similarity between weed and crop in each case and the impossibility of using weedkillers (unless the crops can be genetically modified to exhibit resistance to them).

Several crops are accompanied by characteristic weeds. Salisbury (1961) cited many cases, and a number of those are archaeophytes. There are, however, several reasons for these associations, only one of which is crop mimicry, and in many of Salisbury's examples it is difficult to see why a particular weed is involved other than the fact that it is very common in the region from which the crop seed was introduced. For example, the now rather uncommon archaeophyte Red Star-thistle (*Centaurea calcitrapa*) was once a characteristic weed of Lucerne (*Medicago sativa* subsp. *sativa*). The 10–20 seeds of Lucerne, each about 3 mm across, are borne in an indehiscent pod, whereas the achenes of Red Star-thistle are carried in a spiny flower head and each is about 5 mm across. It does not seem likely that a careful screening process would fail to separate these two. Red Star-thistle is no longer a weed of Lucerne but it became naturalised and is well established in a few places.

Rye Brome (*Bromus secalinus*) was once a common and characteristic weed of Rye (and also of the leguminous forage crop Sainfoin *Onobrychis viciifolia*). Rye, like wheat, is a cereal that has become free-threshing during domestication – in other words, the grains fall free from any enveloping structures (lemmas and paleas). Rye Brome is almost unique among our brome species in that the grains are broad and have inrolled lateral margins, and the lemmas become tightly wrapped round the grain, falling with it. In all but one of our other species the grains are narrower, do not have inrolled margins, and the lemmas are much less firmly attached to them. The result is that the disseminules of Rye Brome (i.e. grains, plus the lemma and palea) closely mimic Rye grains in size and shape, and in the past it was difficult to separate the two. The spikelets of Rye Brome also shatter less readily than those of most other bromes, allowing the inflorescences to reach the thresher before they have dropped to pieces. In fact, it seems that the broad grains of Rye Brome were sufficiently nutritious to allow the grass long ago to be used as a food-plant. Following the Second World War, Rye Brome became a rare plant (and Rye is no longer a common crop), but in recent years its occurrences have increased again since it is a constituent of some of the seed mixtures used in conservation schemes.

Bromus is actually a true cereal genus, as there was a species (*B. mango*) that was used as a staple food-plant in Chile. It was, however, eventually superseded by other cereals and became totally extinct in the 1850s. It resembled Rye Brome and paralleled it in evolving tardily shattering spikelets. If ancient man in the Old World had chosen differently, it might have been *Bromus* rather than *Triticum* providing our bread and ushering in our civilisation.

The annual yellow-flowered crucifer called Gold-of-pleasure (*Camelina sativa*; Fig. 31) has been much studied as it has been a major weed of Flax crops, whose growth habit it mimics very closely. Like Flax, it has actually been used in the past as both an oilseed and a fibre crop, and the possibility of its modern use as the former is at present being reassessed. Moreover, a genetically modified variant into which nutritionally valuable fish-oil genes have been introduced is about to be field-trialled. It is nowadays also being planted as cover and food for game. Gold-of-pleasure has always been far commoner and more important, both as a useful plant and as a weed, in eastern Europe and Russia, and the most detailed studies were carried out by the Russians H. B. Zinger, E. N. Sinskaia and A. A. Bestuzheva, who wrote major treatises on the species in the first half of the twentieth century. The different varieties of Flax in cultivation have been mimicked by a similar range of forms of the genus *Camelina*. The varieties of Flax used for fibre are much taller and less branched than those used for oilseed, and *Camelina*

FIG 31. Gold-of-pleasure (*Camelina sativa*), a mimic of Flax (*Linum usitatissimum*), itself formerly grown for its oil. (John Somerville)

can match both. Gold-of-pleasure is part of a complex group of species or subspecies; C. *sativa* itself occurs in a range of ruderal and agricultural habitats as well as Flax fields, but its close relative C. *alyssum* is confined to Flax fields. The latter was always much rarer in this country and has not been recorded here for more than 70 years. As might be expected from its virtual confinement to Flax fields, C. *alyssum* is the better adapted Flax mimic, being larger in all its parts than Gold-of-pleasure. In addition, the capsules of C. *alyssum* are only tardily dehiscent, mimicking those of cultivated Flax. Two other archaeophytes characteristically found in Flax crops in Europe are Garden Cress (*Lepidium sativum*) and Garden Rocket (*Eruca vesicaria* subsp. *sativa*), both of which show adaptations to Flax mimicry.

WOULD-BE NATIVES: THE DENIZENS

Apart from the colonists and the cultivated plants, the other major group of archaeophytes is comprised of the so-called denizens, another of Watson's (1847) categories. These are plants that in many respects resemble and might be treated as natives (and often have been), but about which there is some reason to believe that they are in fact alien, often because of the habitats they occupy, or their foreign distribution, or because of their absence from the pre-Neolithic fossil record. They can be contrasted with the colonists in that they occupy semi-natural habitats, rather than overtly man-made ones. Eighty-seven taxa are marked as denizens in Appendix 2. Not all of these were considered as denizens by Watson – some he classified as natives and others, which have since become commoner and have invaded closed habitats, as colonists. Indeed, many deliberately introduced plants can be envisaged as potentially passing in turn through the three states, cultivated–colonist–denizen, as they become progressively more established, and the reverse also must have occurred to some extent.

Hence denizens are plants that have achieved a good foothold in our flora, having entered closed, semi-natural communities, so that they often appear to be native. Because they have reached this status in a variety of ways, they represent rather a hotchpotch of species about which generalisations are hard to make.

Some denizens are very well ensconced in our flora and are scarcely recognisable as aliens today. Good examples of common plants in this category are Black Bent (*Agrostis gigantea*), Barren Brome (*Anisantha sterilis*), Greater Burdock, Mugwort, Hemlock, White Dead-nettle (*Lamium album*) and Hedge Mustard (*Sisymbrium officinale*). Others, although very well established, are still convincingly alien, even though some have come to occupy very characteristic habitats. Three ancient herbs much commoner in the north, found in grassy places and having gained interesting vernacular names, are Masterwort (*Imperatoria ostruthium*; Fig. 223), Good-King-Henry (*Chenopodium bonus-henricus*) and Monk's-rhubarb (*Rumex alpinus*; Fig. 32). Some have remained, or perhaps become, rare plants of particular habitats, sought after by plant hunters to augment their lists. Examples are: Grass-poly (*Lythrum hyssopifolia*), found in a few localities on seasonally wet bare mud; Purple Viper's-bugloss (*Echium plantagineum*), restricted to sandy ground in the extreme southwest; Broad-leaved Cudweed (*Filago pyramidata*), another rare plant of sandy ground; Fingered Speedwell (*Veronica triphyllos*), an East Anglian speciality; Red Star-thistle, now

FIG 32. Monk's-rhubarb (*Rumex alpinus*), unusual in the genus in having long rhizomes, has both medicinal and culinary uses; it is much commoner in the north of our area. (CAS)

in only a few rough grassy places in southern England; Field Eryngo (*Eryngium campestre*), of calcareous grassy or fairly bare places, mostly near the sea; and Rampion Bellflower (*Campanula rapunculus*) and Dwarf Elder (*Sambucus ebulus*), both very sparsely but widely scattered. A number of these rarer plants – notably Red Star-thistle, Field Eryngo and Dwarf Elder – are common in central and northern France and will soon catch the eye of the keen botanist on holiday there.

The occupation of a specific habitat, defined and limited by the natural variation in this country in soil types and climate, is the surest way that an alien can appear fully naturalised. Alexanders (*Smyrnium olusatrum*; Fig. 33) is no less characteristic of maritime habitats than many coastal natives, and along our rivers and ditches Hemlock (Fig. 34) competes with many other waterside

FIG 33. Alexanders (*Smyrnium olusatrum*) is a common sight near the sea, where it is one of the earliest plants to flower. It was once commonly eaten blanched as a vegetable but, unlike Fennel (*Foeniculum vulgare*), it has gone out of fashion. (CAS)

FIG 34. Hemlock
(*Conium maculatum*),
a deadly poisonous
plant, is very common
on roadsides
and riverbanks:
(a) welcoming visitors
to Glen Parva;
(b) portrait. (CAS)

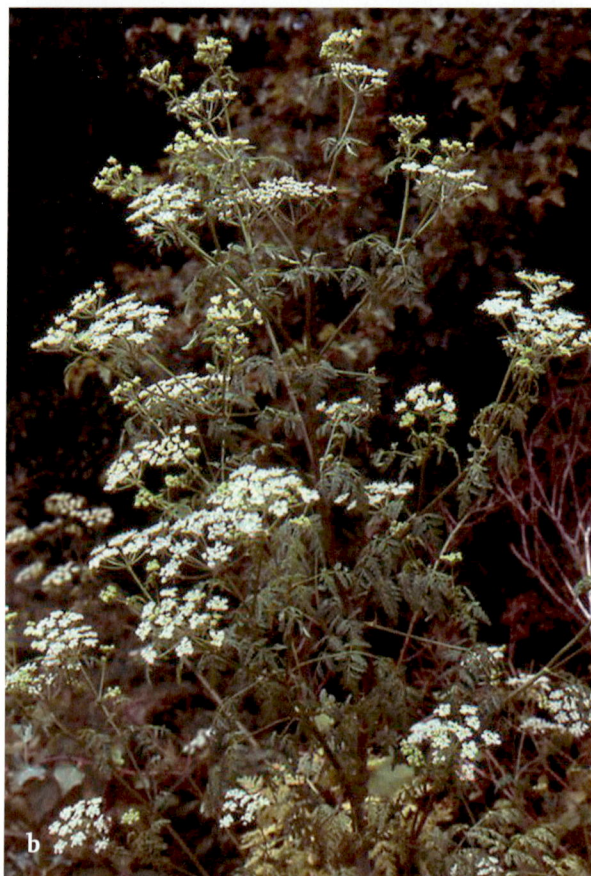

plants. The latter has also become a very characteristic roadside species, and is particularly abundant by motorways, especially if there is a drainage ditch alongside. The miles of Hemlock lining the M1 and other main roads would cause grave concern among the 'health and safety brigade' if only they knew what it was. It is an extremely poisonous plant that has been confused with various edible members of the carrot family with fatal consequences, but the tall, purple-blotched stems are distinctive enough. There is evidence that in market garden areas Hemlock can act as a reservoir for the celery mosaic virus and then reinfect adjacent crops of Celery.

Wall Barley (*Hordeum murinum*; Fig. 35) is one archaeophyte that is abundant in the southern part of the country but becomes rare further north, especially in the west; it is absent from most of Scotland and Ireland. In the 1970s, Alan Davison investigated its distribution in relation to climate, and found that it is restricted by high rainfall, low summer temperatures and high altitude, three parameters that are clearly interrelated and where one can compensate for the other two, e.g. the plant can tolerate higher rainfall in warmer areas. Wall Barley is rare, for example, in localities with an annual rainfall above 1,000 mm, or above 180 m

FIG 35. Wall Barley (*Hordeum murinum*), a grass archaeophyte that is common only in the warmer parts of the British Isles. (CAS)

altitude. This, then, is one of many alien plants that reach their environmental limits in the British Isles.

Dwarf Elder (Fig. 36) is very similar to Elder in the appearance of its flowers, fruits and leaves, but it is a rhizomatous perennial that dies down to the ground each winter. It is important to distinguish the two, as far from being delicious in pies or made into wine as is Elder, Dwarf Elder is a medicinal plant with emetic, purgative and diuretic properties. It has had many other uses in the past, from curing piles to dyeing. In *The Englishman's Flora*, Grigson (1955) treats us to five pages of its folklore, not least to the derivation of its alternative name Danewort.

Medlar (*Mespilus germanica*) is a large shrub looking rather like a Crab Apple in flower but bearing brown fruits a few centimetres across. These have a

FIG 36. Dwarf Elder (*Sambucus ebulus*) closely resembles Elder (*S. nigra*) in both flower and fruit, but is entirely herbaceous in habit and has purgative properties that make its fruits unsuitable as a pie filling: (a) flowers; (b) fruits. (CAS)

more 'open' apex than an apple, resembling a very large version of a Hawthorn (*Crataegus monogyna*) fruit, to which it is closely related. The fruits were once popular and are still enjoyed, preferably with fresh cream, by a few connoisseurs. They remain hard until affected by the frost, which renders them soft and pulpy, in which state they are eaten. This process, known as bletting, can be achieved by placing the hard fruits in a deep-freeze for a week or two. The species' close relationship to Hawthorn is illustrated by the fact that not only do sexual hybrids arise, known under the hybrid genus name × *Crataemespilus*, but graft-hybrids (chimaeras) have also been produced, going under the name + *Crataegomespilus*. The sexual hybrids are intermediate in all respects, including fruits, but the graft-hybrid is a shrub that irregularly sprouts Hawthorn, Medlar and intermediate branches with a great range of leaf shapes. The sexual hybrid is pollen- and seed-sterile, but the graft-hybrid is fertile, although it gives rise only to pure Hawthorn progeny as the inner tissues of the chimaera are derived entirely from that species.

Many other denizens have been widely used for a great range of purposes, some more effectively than others. Greater Celandine has been used to remedy various digestive-tract troubles, and its bright orange sap for topical application to warts. Caper Spurge (*Euphorbia lathyris*) is, like the rest of the genus, poisonous, and has been used as a purgative. It is also said that growing it in the garden discourages Moles (*Talpa europaea*). It gets its name from its fruits, which resemble capers, but their use as such would lead to severe regret. Fennel (Fig. 37) is used in cooking, but also has a range of medicinal applications. Both leaves and fruits of this perennial are utilised. Both the English and scientific names of Soapwort tell us that this plant can be used as a soap, the aerial parts forming a lather when crushed, but it has also been used, like so many other species, as a purgative and expectorant. Alexanders (Fig. 33) is an old vegetable no longer used but once favoured; it was recommended that the shoots be blanched, like Celery or Chicory.

The genus *Artemisia* contains about 400 species, the foliage of many of which possesses a strong scent and has been put to many uses. We have three native species and two archaeophytes. Mugwort is a tall, dull plant common in rough and marginal land and in waste places throughout the British Isles, where it has more than 20 local names. It had supposed magical properties and was used in a whole range of medical treatments. Wormwood (*A. absinthium*) is a far more attractive plant, with beautiful, finely dissected silvery leaves, a much stronger smell, and a more sinister reputation. Like Mugwort, and the not-so-distantly related Tansy, which is treated here as of uncertain status but might well be another archaeophyte, Wormwood has been used for a host of human

FIG 37. Fennel (*Foeniculum vulgare*) is still a much-favoured herb, especially used with fish, and also a blanched vegetable; it is common in rough ground, particularly near the sea. (CAS)

applications. However, it is best known for its use, along with other herbs, in the production of the original and authentic version of the alcoholic spirit absinthe, which has profound physiological and psychotic effects. Absinthe has more than 14 million hits on the Internet, so perhaps needs no further advertisement here. The disastrous results of its overindulgence in France led to the manufacture of the drink being banned.

After investigating the intimate relationships between humans and a great range of archaeophytes, one is struck by two features. First, people have striven to discover effective uses for these plants, but many if not most claims have not stood the test of time. Because of our ancestors' determination, we do not know in the majority of cases whether the plants were deliberately introduced or came as hitchhikers with people or their paraphernalia, nor whether they were deliberately cultivated or merely tolerated. Second, as we learnt the truths about our plants, uses changed and the level of favour has fluctuated. As is so often the case, one can do no better than quote Grigson (1955), either describing the rejection of a formerly valued species: 'Once in the garden, it is now outside the fence; it is now left to its own devices on the bank, or beside the ditch, or in a corner of waste land'; or, heralding a new use, 'a plant in fact with aesthetic claims to replace its discarded medicinal virtues'. These comments could be justifiably applied to many species.

CHAPTER 3

The Alien-hunters – the Discovery and Documentation of Our Alien Flora

He looketh seldom in their face, His eyes explore the ground.
Ralph Waldo Emerson, 'Manners', *May-day and Other Pieces* (1867)

T he study of the alien elements in our flora was not always considered such a worthy pursuit as it is now. The standard Floras of the British Isles from 1843 until 1952 were those of C. C. Babington (*Manual of British Botany*, 1843), G. Bentham & J. D. Hooker (*Handbook of the British Flora*, 1858), and Hooker (*The Student's Flora of the British Islands*, 1870), running to ten, eight and three editions, respectively. Of today's top 20 truly alien (i.e. neophyte) flowering plants, based on the number of 10 × 10 km grid squares (hectads) from which they have been recorded (*see* Appendix 1), fewer than half (nine, six and six, respectively) were treated in those three Floras. A study of the 'Appendix: excluded species' in Bentham & Hooker's latest edition (1930) confirms the lack of interest in aliens. A few of the 14 'missing' species are simply listed there, with such comments as 'a casual weed', 'garden escape' or 'naturalised near Edinburgh'. Clearly, escapes, casuals and naturalised plants were not considered worthy of inclusion, unless widespread. Several species not mentioned at all in 1930 were well known in this country before 1900. It seems likely that alien plants were held in such low esteem in the first half of the twentieth century that the later editions of those standard Floras were not adequately updated in that respect.

PRE-1910 ALIEN PLANT STUDIES

FIG 38. Rough Cocklebur (*Xanthium strumarium*), first reported in the British Isles in 1562 by William Turner in his *A New Herball*, is the earliest neophyte recorded here. (CAS)

Throughout the eighteenth and nineteenth centuries, and even before that, important alien data were being published. The first undoubted neophyte (i.e. excluding the alien/native uncertainties and the archaeophytes) to be recorded appears to be Rough Cocklebur (*Xanthium strumarium*; Fig. 38), which was reported in the second part of William Turner's *A New Herball*, published in 1562: 'groweth in fat grounds and in ditches that are dried up'. The next was Berry Catchfly (*Silene baccifera*; Fig. 39), which appeared in a foreign text in 1571, *Stirpium Adversaria Nova* by Pierre Pena and Mathias de Lobel, which stated that the plant occurred in damp woods in Germany, Flanders and England. The third was Procumbent Yellow-sorrel (*Oxalis corniculata*), reported as a garden weed in Somerset by William Brewer in or around 1585. Then, in 1597, John Gerarde produced

FIG 39. Berry Catchfly (*Silene baccifera*) was the second neophyte to be recorded in the British Isles, surprisingly in a foreign work, *Stirpium Adversaria Nova* by Pierre Pena and Mathias de Lobel (1571). (CAS)

his *Herball or Generall Historie of Plantes,* in which seven neophytes made their first appearance. For the next two centuries only a trickle of new arrivals was recorded, with 41 additional ones in the seventeenth century and 86 more in the eighteenth century, but from about 1800 the trickle quite rapidly changed to a torrent, as shown in Fig. 61. There are a number of very well-known early alien records, for example Sowbread (*Cyclamen hederifolium*) from Gerarde (1597, see above), Houseleek (*Sempervivum tectorum*) from 1629, London Rocket (*Sisymbrium irio*; Fig. 40) from 1650 and Birthwort (*Aristolochia clematitis*) from 1685. The 51 taxa recorded before 1700 are listed in sequence in Table 3. It is quite noticeable that many of the species recorded very early on never became common and in fact are less common today than they were in the nineteenth century. This is presumably because the uses to which they were put no longer apply; hence further introductions did not take place and the plants were no longer grown, so that the source of more wild plants dried up.

FIG 40. London Rocket (*Sisymbrium irio*), a much-discussed neophyte, was apparently commoner in the seventeenth century than it is today. (Peter Greenwood.

TABLE 3. List of the 51 neophytes recorded before 1700.

English name	Scientific name	Date
Rough Cocklebur	Xanthium strumarium	1562
Berry Catchfly	Silene baccifera	1571
Procumbent Yellow-sorrel	Oxalis corniculata	1585
Garden Angelica	Angelica archangelica	1597
Sowbread	Cyclamen hederifolium	1597
Buckwheat	Fagopyrum esculentum	1597
Motherwort	Leonurus cardiaca	1597
Honesty	Lunaria annua	1597
White Poplar	Populus alba	1597
Salsify	Tragopogon porrifolius subsp. porrifolius	1597
Cockspur	Echinochloa crus-galli	1620
Houseleek	Sempervivum tectorum	1629
Moth Mullein	Verbascum blattaria	1629
Sycamore	Acer pseudoplatanus	1632
Canary-grass	Phalaris canariensis	1632
Broad-leaved Ragwort	Senecio sarracenicus	1632
Leopard's-bane	Doronicum pardalianches	1633
Yellow Figwort	Scrophularia vernalis	1633
Bladdernut	Staphylea pinnata	1633
Lesser Canary-grass	Phalaris minor	1638
Asarabacca	Asarum europaeum	1640
Ivy-leaved Toadflax	Cymbalaria muralis subsp. muralis	1640
Goat's-rue	Galega officinalis	1640
Larkspur	Consolida ajacis	1650
Common Evening-primrose	Oenothera biennis	1650
Garden Peony	Paeonia officinalis	1650
Japanese-lantern	Physalis alkekengi	1650
London Rocket	Sisymbrium irio	1650
Greater Periwinkle	Vinca major	1650
Hollowroot	Corydalis cava	1656
Madwort	Asperugo procumbens	1660

English name	Scientific name	Date
Black Currant	Ribes nigrum	1660
Sticky Groundsel	Senecio viscosus	1660
Greater Bur-parsley	Turgenia latifolia	1660
Yellow Chamomile	Anthemis tinctoria	1661
Hairy Vetchling	Lathyrus hirsutus	1666
Reflexed Stonecrop	Sedum rupestre	1666
Rough Bristle-grass	Setaria verticillata	1666
Green Bristle-grass	Setaria viridis	1666
Sweet-flag	Acorus calamus	1668
Broad-leaved Everlasting-pea	Lathyrus latifolius	1670
Hartwort	Tordylium maximum	1670
Hairy Yellow-vetch	Vicia hybrida	1670
Birthwort	Aristolochia clematitis	1685
Sand Lucerne	Medicago sativa nothosubsp. varia	1686
Lucerne	Medicago sativa subsp. sativa	1688
Canadian Fleabane	Conyza canadensis	1690
Hairy Finger-grass	Digitaria sanguinalis	1690
Pearly Everlasting	Anaphalis margaritacea	1698
Snapdragon	Antirrhinum majus	1698
Red Valerian	Centranthus ruber	1698

From the start of the nineteenth century, most alien records were included in local Floras, which were clearly not straitjacketed by the decisions made by the authors of the national Floras. For instance, in 1816 Thomas Furley Forster had cited localities for such aliens as Borage, Lucerne, Sticky Groundsel (*Senecio viscosus*) and Greater Periwinkle (*Vinca major*) in his *Flora Tonbridgensis*; his comments clearly show that he knew that these were not natives.

Some of the earliest systematic records of aliens were made in Northumberland and Durham around the docks of Tyneside, Wearside and Teesside, e.g. at Newcastle, Sunderland and Hartlepool. Nathaniel John Winch (Fig. 41) reported many ballast aliens in his *Flora of Northumberland and Durham* (1831), some going back to 1805 and a few to 1798, mainly repeating records made in two earlier lists. Winch's first list was in an earlier local Flora entitled *An Essay on the Geographical Distribution of Plants, through the Counties of Northumberland,*

FIG 41. Nathaniel John Winch (1768–1838) was perhaps the first botanist to record aliens systematically, the earliest in 1798. (Reproduced with kind permission of the Hancock Museum, Newcastle, and the Director and Board of Trustees, Royal Botanic Gardens, Kew)

Cumberland and Durham (1819), in which the species are categorised in 11 sub-lists. His second list was an 1825 update of the first. Comparison of these lists with Winch's 1831 *Flora* shows that the latter included only one of his sub-lists: that titled 'Plants which have become naturalised' (60 species). Winch's 1819 *Essay* was a truly pioneering work. John Hogg (1867) was able to list 276 species in a paper entitled 'On the ballast flora of the coasts of Durham and Northumberland'.

Although he is not renowned for seeking or discovering alien plants, Hewett Cottrell Watson (Fig. 42) made the first steps in their scientific study in a number of works dating from 1832 and culminating in *Cybele Britannica* (1847–59) and *Topographical Botany* (1873–4). These were analytical accounts of the geographical distribution of plants in Britain and the Isle of Man (but not Ireland or the Channel

FIG 42. Hewett Cottrell Watson (1804–81) was a pioneering plant geographer who defined the categories 'denizen' and 'colonist' in his *Cybele Britannica* (1847–59). (From G. C. Druce's The *Comital Flora of the British Isles*, 1932)

Islands). In order to present his data methodically, he introduced two sets of definitions that remain with us today.

In the third volume of *Cybele Britannica* (1852), Watson devised the system of 112 vice-counties into which Britain is divided for recording purposes, these units being far less diverse in size than the administrative counties, many of which were subdivided in his scheme. Although in many respects they have been replaced by Ordnance Survey grid squares for recording purposes, vice-counties are still widely used by botanists and some zoologists, and many regional Floras are based on one or more of them. They still have the merit of pandering to chauvinistic pride ('my patch'), an important element in wildlife recording.

In the first volume of the *Cybele* (1847), Watson also classified plants according to their degree of nativeness, recognising natives, denizens, colonists, established [naturalised] aliens and casuals. These terms were defined in Chapter 1. The categories of denizen and colonist were in standard use from the time they were introduced by Watson for about a century, but were not utilised in *Flora of the British Isles* (Clapham *et al.*, 1952) and consequently became ignored by at least two generations of field botanists. For the past half-century or more, plants in these categories have been considered to be native or 'doubtfully native', but together they comprise most of today's archaeophytes. Denizens and colonists represent an important concept and it is unfortunate in our view that they were ever dropped; we accordingly resurrect them in this book (*see* Chapter 2). Their neglect for 60 years led to the obfuscation of the native–alien boundary and widespread misunderstanding of the distinction. Watson's name was commemorated in the journal of the BSBI (*Watsonia*), which ran from 1949 to 2010.

In Ireland, similar publications followed on the heels of the British ones: *Cybele Hibernica* was produced by David Moore and Alexander Goodman More in 1866, and *Irish Topographical Botany* by Robert Lloyd Praeger in 1901. The former differed from its British counterpart in not using the terms denizen and colonist, but it did include many non-natives and categorised them at three levels, as possibly, probably or certainly alien. In the second edition, however, prepared by Nathaniel Colgan and Reginald William Scully (1898), many aliens were relegated to an appendix of 'Excluded species (errors, casuals and aliens not fully naturalised)', and even the 1987 edition of *Census Catalogue of the Flora of Ireland* by Maura Scannell and Donal Synnott omitted the many species that had still not become naturalised by that date.

Workers in the second half of the nineteenth century therefore had a firm basis for their recording activities. James Brewer, in his *Flora of Surrey* (1863), provided a list of 155 aliens recorded by the River Thames at Wandsworth and

Battersea, many of which 'originated from the siftings and sweepings of corn [barley] from the Distillery' at Wandsworth. Brewer excluded these from the main text of the Flora as the majority had 'no claim whatever to be considered even British species'. He provided the list as some might become naturalised, 'when it would be of importance that their origin should be known'. If only more authors had taken this view. Henry Trimen (1866), co-author of the *Flora of Middlesex* (1869), wrote 'Exotic plants about London in 1865'; this list of 64 species from a farm at Mitcham, Surrey, owed its existence to sources similar to those of the Wandsworth species, i.e. cereal seed impurities. Another list of 21 species, not from cereals, found on the site of the South Kensington International Exhibition of 1862, was also given, compiled by Trimen's co-author, William Thiselton Dyer.

The early Floras of Liverpool (1839 and 1851, with an 1855 supplement), by contrast, contained rather few aliens, and the same is true of other Merseyside floristic works right up to *The Flora of the Liverpool District* by Conrad Theodore Green (1902). James Alfred Wheldon (1912–14) (Fig. 43), in a series of short notes (totalling 83 pages) under the title 'Some alien plants of the Mersey Province',

FIG 43. James Alfred Wheldon (1862–1924) provided the first serious contribution to the alien flora of Merseyside, in 1912–14. (From *Botanical Society and Exchange Club of the British Isles* 7(3), 1925)

recorded over 400 species, amounting to the first serious contribution to the alien flora of Merseyside. Additions were gradually made during the first half of the twentieth century, but it was left to J. P. Savidge *et al.* to publish them fully in 1963 in *Travis's Flora of South Lancashire*. Savidge included the naturalised aliens, and the casuals with more than three records (together amounting to about 315 species), in the main text, but added a supplement of about 320 casuals that had been recorded no more than three times. Even so, 635 is only a moderately high running total for the 1960s. The Cheshire aliens are less well documented, but in 1899 in *The Flora of Cheshire*, Lord de Tabley listed many within the main text. He explained that the principal source of these alien introductions was ships' ballast at Birkenhead Docks; most of the records are from there or from the sides of new roads to which the ballast had been transported.

Two other relatively early lists were provided by John Storrie, one in the *Report and Transactions of the Cardiff Naturalists' Society* (Storrie, 1876), the second as an appendix to his *Flora of Cardiff* (1886) and entitled 'Foreign plants found on the ballast heaps, near Cardiff and Penarth Docks'. The 67 species in the later list included some recorded for the first time in this country, e.g. Japanese Knotweed (*Fallopia japonica*).

Subtropical species of flowering plants and freshwater algae growing in the canals and mill-lodges (mill-pools) of Greater Manchester that were heated by the effluent from the steam engines driving the cotton mills were first noticed in 1883 (Bailey, 1884). The most notable angiosperm was Egyptian Naiad (*Najas graminea*), found in the Reddish Canal from 1883 to 1947, but some other apparently alien taxa in the genera *Utricularia* and *Zannichellia* were never further investigated. The Reddish Canal was filled in the early 1960s, and the source of warm effluent disappeared from the area in the late 1960s when the last steam engines were replaced by electric motors.

William Glasson (1890) wrote 'On the occurrence of foreign plants in West Cornwall', the plants listed actually all found within 10 miles (16 km) of Penzance. He listed 55 species, including 15 crucifers. A supplementary article in the same year listed 29 extras. Glasson waxed lyrical on the pleasures for a botanist of finding a new alien plant on an outing: 'His steps are then arrested, he is no longer indifferent, and with a feeling of strong interest he proceeds to find out all he can about his new friend'.

In 1902, J. F. Robinson included 251 aliens in *The Flora of the East Riding of Yorkshire*. He calculated that there were 114 'colonists and denizens' (naturalised aliens) and 137 'aliens' (casuals), both totals exceeding the numbers listed for the whole of Britain in the then current edition of *The London Catalogue of British Plants*. A high proportion of the latter group was recorded from Hull Docks, where Robinson said there existed 'A perfect wilderness of exotics from many lands.' Wilson (1938) included nearly 400 aliens in a paper entitled 'The adventive flora of the East Riding of Yorkshire', in which the dock aliens of Hull and surroundings featured prominently. Also highlighted were aliens from flour and oil mills and from around farms, and those of horticultural origin. The records from these two sources were summarised and supplemented by later ones by Eva Crackles in 1990 in her *Flora of the East Riding of Yorkshire*.

James Fraser (1904–14) contributed a series of important papers on alien plants found near Edinburgh. Leith Docks and Musselburgh were two well-known and productive localities in this area. South of the city he also found rich pickings on the shingle of the River Tweed between Abbotsford and Leaderfoot; this is the area later monographed by I. M. Hayward and G. C. Druce in 1919.

By 1910, Fraser claimed to have recorded about 970 species; the short lists he published after that time might have increased his total to 1,000, because they included 17 species new to Britain and a short list of Esparto-grass aliens from Musselburgh. W. and W. E. Evans (1903–04) also recorded from around Edinburgh. By 1903 they had a list of about 300 species, of which at that time half had not been noted by Fraser, and in 1904 they added about 80 more. In 1934, at the end of her *The Field-club Flora of the Lothians*, Isa H. Martin added a 'Note on alien flora'. She mentioned that over 1,000 alien species had been recorded in the Lothians, mainly from imported grain, Esparto-grass, ballast and hides, and from domestic rubbish and horticulture. It is evident that relatively few species had been added since the days of Fraser and the two Evans. Two lists of the most frequently observed species were provided by Martin: one of 103 species of casuals from Leith Docks; and the other of 141 species of aliens found more widely in the area (some species appearing on both lists). Douglas McKean, in *A Checklist of the Flowering Plants and Ferns of Midlothian* (1988), included many aliens in the main list and also provided an appendix of 'long extinct' aliens that were not mentioned by Martin. His total of 1,349 can scarcely be equalled by any other

British county; it includes, for example, 175 legumes, 159 composites, 107 crucifers and 104 grasses. In *Plant Life of Edinburgh and the Lothians* (McKean, 2002), he emphasised the use of ballast-hill material to make new roads and railways, so spreading the alien species away from the docks, and the dramatic decrease in aliens after about 1930 due to cleaner imports.

The first major work on alien plants was Stephen Troyte Dunn's *Alien Flora of Britain* (1905), for which there was a 1903 forerunner in the form of a slim volume entitled *A Preliminary List of the Alien Flora of Britain*. Dunn (Fig. 44) had provided an impressive inventory of species, and notably included many that today we would describe as archaeophytes (e.g. Shepherd's-purse) and others that are native in some parts of the country but alien elsewhere (e.g. Jacob's-ladder). Probably because of its short time (three

FIG 44. Stephen Troyte Dunn (1868–1938) wrote the first British book on alien plants, *Alien Flora of Britain* (1905). (© The Board of Trustees of the Royal Botanic Gardens, Kew; reproduced with permission)

to four years) in compilation, it does not quite justify its claim of listing 'all the presumably non-indigenous species hitherto recorded as growing spontaneously in the British Isles'; most but not all of Storrie's ballast alien records are omitted, for example. It was, however, a major advance in our documentation of aliens, although it seems that it had surprisingly little impact on British botany.

G. A. Dunlop (1908) published 'An annotated list of the alien plants of the Warrington District', covering the area 25 kilometres upstream from Liverpool/ Birkenhead. The total of more than 160 includes species that are native elsewhere in Britain.

George Claridge Druce (Fig. 45), an amateur (but far from amateurish) botanist who became prominent towards the end of the nineteenth century and was to dominate British field botany until his death in 1932, took a very keen interest in aliens for all that period. Druce overcame the very considerable problems involved in identifying alien species, often with an unknown country of origin, by consulting the very rich Oxford University herbarium, which was near his home, and by building up correspondence with a range of experts from across Europe. In this way many additional species were named, and British field botanists sent difficult specimens to him for identification by him or his contacts. Hundreds of records were published, mainly in the *Reports of the Botanical Society and Exchange Club of the British Isles* ('BEC Reports'), of which Druce was secretary. Some of the alien plants found in Britain (especially on Tweedside in the Scottish Borders) were apparently unknown to science, and were described in the *BEC Reports* as new species by two of Druce's most frequent correspondents (Josef Murr of Innsbruck and Albert Thellung of Zürich), in the goosefoots (*Chenopodium*) and pepperworts (*Lepidium*), respectively. Murr also published papers entitled 'Adventif-Flora von Gross Britannien' in the German journal *Allgemeine Botanische Zeitschrift für Systematik* in 1913 and 1914.

In 1908, Druce published his *List of British Plants*, which was effectively set up as a rival to the established *London*

FIG 45. George Claridge Druce (1850–1932), one of Britain's most distinguished and productive botanists, took a special interest in aliens and in 1908 provided the first exhaustive list of them. (From the BSBI archives, reproduced with kind permission)

Catalogue of British Plants, started in 1844 and with its tenth and penultimate edition coming out also in 1908. There are only about 240 aliens (my count) included in that edition of the *London Catalogue*, whose introduction (by F. J. Hanbury) states that 'The policy adopted in the present edition has been in the direction of reducing the number [of aliens] by expunging the names of a few species hitherto included that are of very rare or doubtful occurrence'. A glance at the species listed for many genera will confirm this policy. In goosefoots (*Chenopodium*), for example, there are 12 species in total, whereas Druce's *List* gives 18; in comfreys (*Symphytum*), the comparable figures are two and six, respectively. It is likely that this perceived lacuna in the *London Catalogue* was one reason for Druce to produce a rival. Another was that it enabled him to list many more varieties under each of the species. Druce aimed to provide a complete list of all plants ever recorded, even 'the mere ballast waif', for 'the waif of to-day may be the pest of the next decade'. He included 1,084 aliens, of which 144 he considered 'well established'. Dunn's *Alien Flora* contained about 940 non-natives (my count), but the true figure to compare with Druce's should be lower than that, for the 940 includes many archaeophytes considered by Druce to be natives. In fact, all these figures should be treated as very approximate, because they depend heavily on variations in the perceived limits of species and of their native status. In 1928, Druce issued a second edition of his list, called *British Plant List*, in which the number of aliens had virtually doubled to 1,999, including 293 classed as 'established'.

ALIEN PLANT STUDIES, 1910–52

After about 1910, evidently stimulated by the 1,084 aliens listed in Druce's *List of British Plants* and perhaps by Dunn's *Alien Flora*, increasing numbers of lists of aliens were published, mainly in the *BEC Reports* and in journals of local natural history societies, ranging from London and Cornwall to central Scotland, and in Ireland.

Greater London is today one of our richest sites for alien plants, and as our largest city and centre of trade it probably always was. Druce (1910), in a list of interesting plants from Middlesex, 'included a somewhat large number of casuals, which are especially to be found about the refuse from the street sweepings of the metropolis near the brickyards of Drayton, etc.' Shortly after, Cooper (1914) listed nearly 160 casuals from Middlesex. Interestingly, like several other lists from that period, garden escapes were omitted, 'aliens' or 'casuals' including only accidentally introduced species. Ronald Melville, later to become a leading Elm (*Ulmus*) and Rose (*Rosa*) authority, and Roy Smith, who had earlier published on Cardiff aliens (*see* below), produced a list of about 250 species largely from rubbish tips and waste ground in 'the Metropolitan area' (Melville & Smith, 1928). They gave fascinating insights into their toils on the tips of London, stressing the difficulties but concluding that they 'prefer this kind of field work to the more conventional form and have had many enjoyable excursions over rubbish heaps'. One doubts that Melville retained this preference over his long botanical career! Despite the undoubted importance of the many London Docks (falling into the vice-counties of Surrey, Middlesex and South Essex) as pathways of introduction of aliens, there is a remarkable dearth of records from them. All the other eight or so major dock areas in Britain are much better documented, and in Douglas Kent's *The Historical Flora of Middlesex* (1975), which is meticulous in including all old records (with a 47-page appendix listing records of 'over 600' casuals – 'garden escapes, birdseed, grain and soyabean aliens, etc.'), it is difficult to find any mention of dock aliens. It appears that these really did not exist in significant numbers, presumably because London was a major importing city and ships rarely entered it 'in ballast'. In Kent's chapter on the botanical exploration of Middlesex, the nearest mention one finds to dock aliens are rubbish tips on Hackney Marshes, not far from some of the old dock areas.

J. F. Rayner published articles (1924–5) on alien plants in Hampshire and the Isle of Wight. An excellent list of over 300 species, including 16 that had become naturalised over the previous century, was presented. Three non-flowering plants were included: European Larch (*Larix decidua*), Maritime Pine (*Pinus pinaster*) and Water Fern (*Azolla filiculoides*).

Further lists from the Cardiff port area were published by Arthur Wade and Roy Smith (1926) under the title 'The adventive flora of the port of Cardiff', with a supplement and a follow-up paper (Wade & Smith, 1927; Smith & Wade, 1939) They included all of John Storrie's ballast records, plus many more besides, because the authors searched more widely and recorded species from several diverse sources as well as ballast, including garden escapes and grain siftings. Their article contains interesting details of the management (especially the pattern of dumping) of the dock areas that relate to the nature of the aliens appearing there. They commented on the facts that the dumping of hot furnace slag served to provide a warmed environment for tender plants, and that many of the species had become naturalised. The first article listed about 450 species (my count) and the supplement a further 43. The 1939 publication tells of the reduced appearance of aliens in the area due to cleaner agricultural and industrial processes, and complains 'The lot of the present-day seeker after adventive plants is indeed hard'. The list of about 233 species (my count) is less than half the total from ten years before, but it includes species not in the previous list and some new to the British Isles. D. J. Pugsley (1941) provided an interesting account of the process of colonisation of coal dumps at Cwmbach, Aberdare, 30 km inland from Cardiff.

The margins of the River Clyde have long been rewarding areas for aliens, and an important paper by Robert Grierson (1931) catalogued about 400 species recorded there between 1916 and 1928, mostly on 'coups' (rubbish tips). This list actually summarised the contents of a series of notes previously published by Grierson, who echoed a sentiment frequently expressed in the middle of the twentieth century (and still so today): 'The city is burning most of its refuse now; possibly this is good for the public health, but it is ruinous for my work.' A good selection of aliens was included by J. R. Lee in *The Flora of the Clyde Area* (1933), including a number at that time not treated in the standard national Floras other than *Hayward's Botanist's Pocket-book* (last edition 1930), e.g. Indian Balsam, Russian Comfrey (*Symphytum × uplandicum*) and Rhododendron. From further south, James Fraser (1914b) had listed about 50 species, including some interesting aliens that he recorded from around the port of Stranraer.

To the north of the Firth of Forth, in Fife, many ballast aliens were being discovered, especially at the coal-exporting ports such as Charlestown and Burntisland. Many were listed by William Young (1936), but those recorded between 1820 and 1919 were more fully documented by George Ballantyne (1971), who listed 203 species, about 50 of which are plants native elsewhere in the British Isles. Ballantyne's paper also provided arguably the best account of the history of ballast aliens in this country. Later, as an appendix to his *Wild Flowers of Fife and Kinross* (2002), he listed 295 aliens from all sources, which included all those on his previous list.

Mrs Cecil Sandwith (1933), mother of the Kew botanist Noel Sandwith, published a list of 717 alien species from 'the Port of Bristol'. This included the ports of Portishead and Avonmouth, as well as many areas nearby used for dumping waste, and it incorporated the many records in James White's *The Flora of Bristol* (1912) and a list produced by Druce (1917) covering one of the waste areas, St Philip's Marsh. Sandwith gave excellent accounts of the sources of the plants listed, and of the relevant industrial practices and processes. To this day, alien records are published annually under 'Bristol Botany' in the local journal, now named *Nature in Avon*.

Other more minor ports and docks also yielded many interesting aliens. G. C. Brown (1930), in an article on the alien plants of Essex, recorded about 300 species, with Hythe Quay, Colchester, figuring prominently among the localities 'during the past 12 years'. Edgar Thurston (1929) published a list of 'alien and

British plants' found around Par and Charleston harbours, Falmouth Docks and Penzance, on the south coast of Cornwall. The British plants included are those not native in Cornwall or ones that were represented there by unusual (perhaps alien) varieties. The list included about 400 species, of which approaching one-quarter is native to the British Isles.

Dunn (1905) had stated that 'By far the most important agent of plant introduction at the present time is the importation from foreign countries of the kinds of grain which are most largely used for making flour and for distilleries'. This acknowledged the lessened impact of ballast aliens. James Fraser Robinson (1902), in *The Flora of the East Riding of Yorkshire*, recorded that 'as earthy forms of ballast have been replaced by water' [possible when iron superseded wood in hull construction] the 'ballast heap class' of aliens had disappeared, and dock aliens owed their appearance to 'the sweepings of the holds of vessels, dock sheds and railway trucks'. The demise of ballast aliens reported by Dunn, Robinson and others was, however, starkly contradicted by Thurston (1929), writing about dock aliens in Cornwall. In the absence of evidence that any of these three were mistaken, one can only assume that the Par, Charlestown, Falmouth and Penzance areas of Cornwall bucked the trend, and that in those areas the change from the use of solid to water ballast was much delayed. Thurston referred to the fact that during the First World War china clay was shipped to France, and that the vessels returned in ballast. Also, he reported comments made by L. T. Medlin and W Tresidder concerning Par and Charlestown, respectively, referring to ballast arriving in ships from various countries ranging from Scandinavia to Morocco, the ballast consisting of soil, underlying subsoil, sand, gravel and shingle.

In any case, species arriving in ballast and grain were soon numerically overtaken by plants derived from seed brought in as contaminants of wool, a source that has now, in turn, hugely diminished in importance. The best-documented area for wool aliens in the British Isles in the first half of the twentieth century was at Galashiels by the River Tweed in the Scottish Borders (Fig. 69). The plants found here were studied in detail by Ida Margaret Hayward (Fig. 46), who in collaboration with George Druce wrote *The Adventive Flora of Tweedside* (Hayward & Druce, 1919). Earlier the plants had been studied by James Fraser, but they had been around for much longer. James Britten reported in 1865 that Soft Stork's-bill (*Erodium malacoides*; Fig. 47) had been recorded as a 'shoddy heap plant' at Huddersfield, where the wool waste had been spread over adjacent fields as a manure. However, Chris Yeates of the Tolson Museum, Huddersfield, has re-examined the supporting specimen and found that it was collected in 1858 and that it was in fact a misdetermination of Eastern Stork's-bill (*E. crinitum*) (G. T. D. Wilmore, pers. comm.). In the same year, Three-lobed Stork's-bill (*E. chium*) was recorded from

FIG 46. Ida Margaret Hayward (1872–
1949) studied the wool aliens of the
Galashiels area and with George Druce
wrote *The Adventive Flora of Tweedside*
(1919). (Reproduced courtesy of the Royal
Botanic Garden, Edinburgh)

FIG 47. Soft Stork's-bill (*Erodium malacoides*),
one of the earliest recorded wool aliens, was first
noted in 1865. (CAS)

Pimlico, London, and there is an 1851 record of Eastern Stork's-bill from Yorkshire.
The latter might even be our earliest wool alien record. Hayward & Druce's book
dealt with 348 wool aliens, although a number of them were native British species.
Many were identified by Murr and Thellung, and the records of goosefoots figured
prominently in Paul Aellen's *Die Wooladventiven Chenopodien Europas* (1930). Thellung
was well practised in determining wool aliens from his experiences in Montpellier,
which resulted in his pioneering 672-page *La Flore Adventice de Montpellier* (1911–12).

Another major area yielding wool aliens was at the heart of the woollen
industry in Yorkshire, particularly around Bradford. Although this area was
worked by John Cryer, a respected hawkweed (*Hieracium*) expert, little was
published and there is scant evidence of much activity after about 1920 until
the 1950s. A shortlist of Bradford aliens published by Cryer (1920) contained
some species, e.g. Soft Stork's-bill (Fig. 47), that indicate at least an element of
wool aliens. A few other short lists from him are to be found in *The Naturalist*
between 1909 and 1916. In Cryer's obituary, Druce said that it was reputed that

he had found 500 species of aliens around Bradford, but Druce did not mention wool aliens specifically. The Yorkshire Naturalists' Union has for a long time maintained a card index of alien records, but unfortunately Cryer's records are not among them (G. T. D. Wilmore, pers. comm.). In 1952, John G. Dony (Fig. 48) collected over 40 species of wool aliens in Yorkshire.

A third area favouring wool aliens was discovered by Dony in 1946 in Bedfordshire, where wool waste (shoddy) was being used as manure by market gardeners. The aliens were found in the cultivated fields, in gravel pits and on railway sidings. In 1953, Dony listed 112 species in his *Flora of Bedfordshire*, and

LEFT: **FIG 48.** John George Dony (1899–1991) with his wife Christina. In 1946, Dony discovered the wealth of wool aliens in agricultural areas fertilised by shoddy and became a leading expert on the subject. (Mary Sheridan, reproduced courtesy of Chris Boon)

BELOW LEFT: **FIG 49.** Catherine Muriel (Kit) Rob (1906–75), a dedicated Yorkshire botanist who made significant contributions to our knowledge of aliens. (From *The Naturalist* 100 (1975), reproduced courtesy of the Yorkshire Naturalists' Union)

BELOW: **FIG 50.** Mary McCallum Webster (1906–85) was the leading alien hunter in Scotland for several decades. (William Gloyer)

by 1979, with the help of J. E. (Ted) Lousley (Fig. 51), he had identified 366 species in the county. There were many other areas using shoddy as fertiliser and yielding smaller numbers of wool aliens, such as the hop gardens around Barming in Kent, and near Portchester in Hampshire, from where P. M. Hall (1941) listed 23 species found in 1939.

One other alien-rich area that should be mentioned is around the brewery industry in Burton upon Trent, where the plants were mostly derived from 'brewery sweepings' from the maltings. Lists were supplied by Sir Roger Curtis (1931; 187 species, my count) and by Richard Burges (1946; 79 additional species, my count). Very few of these species still persist, but some became naturalised and Lesser Milk-vetch (*Astragalus odoratus*; recorded 1931) and Indian Knotgrass (*Polygonum cognatum*; recorded 1946) were still there in 2010 (Mike E. Smith, pers. comm., 2010).

FIG 51. J. E. (Ted) Lousley (1907–76) was one of the most influential alien specialists of the twentieth century. (From the BSBI archives, reproduced with kind permission)

J. P. M. (Pat) Brenan (1947), later director of the Royal Botanic Gardens, Kew, listed just over 90 alien species that he found on two visits in 1939 and 1941 to an area near Southampton Docks. Although the number is not high, more visits to other areas in the docklands would have surely raised the total significantly.

London was to assume a renewed prominence in the study of alien plants as a result of its bombing in the Second World War. The ugly and sinister bombed sites, many still existing well into the 1950s, were ideal habitats for colonisation by pioneer species, which in places turned the sites into a riot of colour. J. E. (Ted) Lousley (1944, 1946) published the results of his surveys of bombed sites in two papers, recording 27 and 112 species, respectively. Species that became notably more abundant than they previously had been in London include Rosebay Willowherb (*Chamerion angustifolium*), Butterfly-bush (*Buddleja davidii*), Oxford Ragwort (*Senecio squalidus*), Sticky Groundsel (*S. viscosus*), Canadian Fleabane (*Conyza canadensis*) and Bracken (*Pteridium aquilinum*). In a few locations Lousley found the hybrid between Oxford Ragwort and Sticky Groundsel. Believing it to be new to science, he described it in 1947 as *S. × londinensis*. Although, in 1984, it was suggested that it is the same hybrid as that previously named (as *S. × subnebrodensis*) in Hungary in 1881, and it has appeared in the British literature as this ever since, it is now known that Oxford Ragwort was not a parent of the

Hungarian hybrid, and so British botanists will now have to learn to revert to Lousley's original name for our plant.

Other authors also described the effect of the war on London's wild flora. Sir Edward Salisbury (1943) wrote on 'The flora of bombed areas' in the prestigious journal *Nature*, concentrating on the importance of an effective seed-dispersal mechanism for successful colonisation. All six species listed above are particularly well adapted for wind dispersal. Salisbury's full list of the 157 species he found on London's bombed sites was published as an appendix to Richard Fitter's *London's Natural History* (1945). There were also surveys of the bombed sites of Kew in 1942 and the City of London in 1958. It was inevitable that *London's Natural History*, written by Fitter at the end of the war as number 3 in the Collins New Naturalist series, should feature prominently rubbish tips, bombed sites and other waste ground. Parts of the book form a fascinating contemporary commentary of London's alien flora. All the alien plants of Greater London recorded up to 1956 are documented in 'A hand list of the plants of the London area' (1951–7) by Douglas Kent and Ted Lousley, published in the *London Naturalist* for the years 1950 to 1956. Bombed sites in other cities were also searched for alien plants, e.g. 20 sites in Liverpool in 1948 and 1949 (Allen, 1951). There were some notable differences between the floras in London and Liverpool: Oxford Ragwort and Canadian Fleabane were absent from the latter, and Butterfly-bush (*Buddleja*) was rare there.

Hence, in contrast to the situation 40 years previously, by 1950 we had accumulated an impressive record of our alien plants, which had, however, scarcely been assimilated into the standard Floras of the time or into any other single publication. This was soon to change.

ALIEN PLANT STUDIES, 1952–1991

Even before the middle of the twentieth century our standard Floras had become hopelessly outdated in almost every respect, because the numerous revised editions had not kept pace with developments. Many changes in taxonomic opinion, in known distributions, and in the selection of species needing inclusion, had not been adequately accommodated. In 1952, however, the publication of *Flora of the British Isles* by A. R. Clapham, T. G. Tutin and E. F. Warburg produced our first really up-to-date Flora of the twentieth century.

Apart from updating the nomenclature, incorporating the results of field and taxonomic research, and modifying the selection of native plants to be included, *Flora of the British Isles* (or 'CTW' as it became known) treated a much more representative selection of aliens, both naturalised and casual. Many more were included than in previous Floras, albeit often only as very short supplementary notes, and garden escapes in particular were given unprecedented prominence. For the first time in a standard Flora every species was marked as 'native' or 'introduced', although with an expression of doubt in many cases. Most species that we today consider as archaeophytes were labelled native or doubtfully so. Accepting the status accorded by the authors, which was often different from that given today, I counted 694 aliens in the *Flora*, a similar number to the 643 in J. E. Dandy's *List of British Vascular Plants* (1958), which omitted casuals. However, only about 450 of the almost 700 aliens in CTW were actually incorporated in the keys, which proved something of a barrier to their easy identification. *Flora of the British Isles*, with its overtly novel approach and contents, triggered a surge in taxonomic research in Britain, in the fields of both professional experimental taxonomy and detailed, largely amateur, fieldwork. The study of aliens continued apace, and a number of amateur botanists came to the fore in the second half of the twentieth century as acknowledged experts. Among the most accomplished and productive of these experts are those shown in Figures 48–59, and others are still alive and alien-hunting.

Numerous publications covering alien plants followed in the 1950s and following decades. In addition, and more significantly, the many local Floras and the annual summaries of plant records and field meeting reports being published contained an increasing number of aliens (admittedly, much to the displeasure of some old stalwarts). It was simply the case that the list of species being recorded had expanded, so that more and more aliens were being treated on an equal footing with natives.

Field recording is enormously influenced by the standard Flora being used for identification; what is included in it is recorded, while what is absent from it is not. This is particularly noticeable in genera with several similar species, only a few of which are covered. For example, CTW has only one species of pokeweed (*Phytolacca*; Fig. 52), and one of mock-orange (*Philadelphus*; Fig. 5b). Despite the fact that the species described in CTW are far from the commonest ones of those two genera that are found in the wild here (in fact, in the former case the species probably does not occur here at all), almost all records of wild plants were named as one of those two species. This is one of the few disadvantages of using any finite list of taxa as one's guide; even a full list of every species ever recorded here would not allow for the frequent occurrence of newly arrived ones. This is therefore a problem that has to be lived with; to be forewarned is to be forearmed. The advantages of the extended treatment of aliens in CTW, however, soon became apparent.

FIG 52. Indian Pokeweed (*Phytolacca acinosa*), the commonest species of the genus in the British Isles, was for long misidentified as American Pokeweed (*P. americana*) because the latter was the only member of the genus recognised in British Floras. (CAS)

It is not possible to survey comprehensively in this book the huge post-war literature on aliens, but some of the most substantial and significant items demand mention.

Lousley (1961) published an annotated list of all the 529 wool aliens by then recorded in the British Isles, from 25 vice-counties in England and two in Scotland, but none in Wales or Ireland. Later, T. B. (Bruno) Ryves (1974, 1988) reported on the grasses from a newly discovered area for wool aliens on the fruit farm belonging to Lady Anne Brewis (Fig. 53) at Blackmoor in North Hampshire. Whereas Lousley had included 187 grasses in his national survey, Ryves listed 'over 360' from one small area. Many of these were identified only after they had been grown on in a frost-free greenhouse. Both authors included very useful general data on the occurrence of wool aliens in England.

E.P. (Betty) Beattie (1962) provided what seems to be the only botanically orientated account of the use of Esparto-grass (*Macrochloa* (*Stipa*) *tenacissima*) in

FIG 53. Two leading alien hunters taking a ride on the standard public transport in Sark, Channel Islands: Florence Houseman (1909–89), on the extreme left; Anne Brewis (1911–2002), on the extreme right. Courtesy of Norah Eastwood and Geoffrey Wilmore.

the paper industry, including a list of 64 aliens that she found at a site in Fife, eastern Scotland. Those found on the south side of the Firth of Forth, in the Lothians, have apparently never been listed separately from other aliens there.

Even today, rubbish tips can be a productive source of aliens, despite most now being kept clean by burning and burying, and being fenced off from alien-hunters. The exhaustive list produced by Easy (1976) for Cambridgeshire can be usefully compared with that of Greater London by Douglas Kent and Ted Lousley from 20 years earlier.

Kent and Lousley's London list included a few aliens found in 1945–6 as a result of the importation of Soyabean (*Glycine max*) for processing, including oil-extraction, at Springwell, Middlesex. John Palmer (1977), who has probably seen more alien plants (he claims up to 4,000 taxa) growing in this country than anyone else, discovered Soyabean aliens at a processing plant at Stone, northwestern Kent, in 1973. The imported Soyabeans were contaminated with a large amount of vegetable waste from the American crop fields, and aliens were found in and around the factory as well as on various rubbish dumps in the area. Palmer outlined the processes involved, and listed 57 species, including 14 pigweeds (*Amaranthus*; Fig. 54) and 10 grasses. Soyabean itself is a common alien at such sites, in contrast to the imported Esparto-grass, which has never been found growing in this country.

Birdseed has always proved to be a rich source of aliens, and this continues to be the case today with the growing popularity of putting out seed for wild birds. Watts & Watts (1979) analysed 13 different commercial birdseed mixtures and compared them with species found growing on Harford rubbish tip, Norwich,

FIG 54. Prostrate Pigweed (*Amaranthus blitoides*), one of at least 14 species of the genus found as oilseed aliens. (CAS)

East Norfolk. Gordon Hanson & John Mason (1985) carried out a similar but much wider survey. They grew plants from many seed sources and listed those species together with the records of plants thought to be from birdseed and found on tips and in places where birdseed had been scattered. The total of 438 species included about 30 that were deliberately imported as bird food. About 90 species on the list have not been found growing in the wild (only raised from sown birdseed), leaving more than 300 that represent contaminants of the commercial product. Over 50 of these are grasses. Hanson (2000) later added 44 extra taxa.

FIG 55. Charles Edward (Ted) Shaw (1910–94) was an eccentric and inspirational alien hunter from Oldham, where he was known as 'Vicar Shaw'. (Courtesy of Patricia Francis and Gallery Oldham (incorporating Oldham Museum))

There are also a number of minor sources of alien plants that are often of considerable interest. The tanning industry formerly used oak (*Quercus*) bark and imported oak cupules ('acorn cups'), and with them some contaminating seeds also arrived. These could be found at the tanneries or in garden centres and parks where the spent material was used as mulch. Len Margetts was the first field botanist to recognise this (Grenfell, 1983), when he found 17 alien species (including 13 legumes) at two sites near Truro, Cornwall,

FIG 56. Adrian Leonard Grenfell (1939–91) was one of a younger group of alien experts, but he died tragically young before he could make his full contribution; many of his contemporaries are still hunting aliens today. (Diana Grenfell)

in 1973. By 1983, the total number of such aliens found in Cornwall was 36, including 27 legumes. Thirty additional species were recorded in 1984 (Grenfell, 1985), and a few others since. Unfortunately, Adrian Grenfell (Fig. 56) did not live to produce the full account that he promised.

By the late 1980s, our knowledge and documentation of alien plants had improved enormously from the situation in 1950, due to the efforts of the dedicated amateurs referred to above. By 1980, it was realised that the selection of aliens treated in CTW, despite its great improvement over earlier Floras, had in turn inevitably become quite outdated. Indeed, some of the aliens included in CTW in 1952 and in later editions had not been found in the British Isles since before 1930. Even in the second edition (1962), Berry Catchfly was said to be still naturalised on the Isle of Dogs in London, 110 years after it was last seen there! A successor was required for these reasons alone.

ALIEN PLANT STUDIES AFTER 1991

An entirely new work, *New Flora of the British Isles* (Stace, 1991), set out, among other aims, to treat aliens in a different way from previous Floras. First, all naturalised aliens, even if found in only one locality (as for native plants), were included, plus those casual species 'that the plant-hunter might reasonably be able to find "in the wild" in any one year'. These were all included in the diagnostic keys. Second, extinct aliens were excluded. An accurate inventory was compiled by utilising the rich pool of data held in the literature and by the many knowledgeable alien enthusiasts. The three leading experts of the time, Douglas Kent, David McClintock and Eric Clement, provided many lists and detailed opinions, and a large number of other botanists were consulted – more than 200 in 1991 and nearly 400 by the time of the third edition. In this way a uniquely accurate list of aliens was compiled and continually updated. The first edition (1991) treated fully about 1,060 naturalised aliens and 430 casual species. About 130 extra aliens were added for the second edition (1997) and about 150 for the third (2010), although some species that no longer occurred were excluded, giving a total of about 1,770 aliens. This compares with about 1,560 native species.

The *List of Vascular Plants of the British Isles* was produced by Douglas Kent (1992) (Fig. 57) to replace Dandy's 1958 *List*. Kent's *List* and the *New Flora* were compiled in parallel by their two authors, so the lists of naturalised aliens in them are identical. Unlike the *New Flora*, however, neither Dandy nor Kent included casuals. Kent's *List*, with its three supplements of 1997, 2000 and 2006, became and is still the standard list of vascular plants for the British Isles.

There are nearly twice as many aliens in the above two publications as in either Dandy's *List* or CTW, even allowing for the taxa newly omitted due to their disappearance from our flora. Moreover, for the first time, the means of identification of all these newly admitted taxa were readily available to field botanists, which inevitably led to a great increase in the number of records being made. Now (2015), after 24 years, we are reaping the benefits of this fresh approach to field botany.

In the same (post-1990) period, six other very important books on our alien plants were published. *Alien Plants of the British Isles* (Clement & Foster, 1994) is an annotated and referenced list of all alien vascular plants ever recorded from

FIG 57. Douglas Henry (Duggie) Kent (1920–98) was one of the last of the great amateurs to devote most of their lives to the study of wild British plants and their literature, and there is sadly no-one in sight to rival his knowledge. He took a special interest in documenting the records of aliens. (Courtesy of BSBI, originally from the former North Thames Gas Board)

the British Isles. It does not, however, include grasses, which were covered in a companion volume, *Alien Grasses of the British Isles* (Ryves *et al.*, 1996). Their combined total of species is 4,166 (3,586 + 580), of which the authors estimated that 945 (885 + 60) are or have been naturalised. Of those 4,166, about 1,770 are naturalised or are recurrent casuals, a further 1,686 are less common casuals or were formerly naturalised but are now extinct, and 710 have not been recorded since 1930 (the cut-off date that the authors chose to define the modern era). These two volumes together represent the first and so far only update of Druce's 1928 *List*, which included 1,999 aliens, of which 293 were considered naturalised. Actually, the total of 3,586 listed by Eric C. Clement and M. C. (Sally) Foster (1994) is something of an underestimate, because in their book there are many examples of 'groups of closely related species' being treated as a single entry, where the taxonomy was uncertain or where the records for the critical taxa could not be separated. In addition, many hybrids are treated under one of their parents, even in cases where the hybrid exists in the wild as a completely separate entity. Probably a few hundred extra taxa should therefore be added to the total.

David McClintock (Fig. 58) for many years strove to improve the information available on alien plants, particularly those originating as garden escapes. More than half a century ago he published a slim *Supplement* (1957) to *The Pocket Guide to Wild Flowers* (McClintock & Fitter, 1956), including 50 pages covering almost 450 alien species that had to be omitted from the *Pocket Guide*. Many of these had never previously appeared in any British Flora. He also planned to produce a third, companion volume to the two volumes of *A New Illustrated British Flora* (Butcher, 1961), to cover those aliens not dealt with in the two published volumes. Around 1962, McClintock drew up a list of about 930 'wanted' species that he circulated to potential collectors, and he recruited a number of artists to produce the drawings. Progress was quite rapid at first but soon slowed due to the fact that many of the aliens were rare, shy to flower and even more reluctant to set fruit. Some 350 species had been drawn by 1972, but

FIG 58. David Charles McClintock (1913–2001) was a leading amateur botanist of his day and a specialist in alien plants of all categories, but especially those of a horticultural origin. (Margaret Stace)

many of these lacked vital parts, especially fruits. By the 1980s it was clear that the original aims could not be met, and David McClintock invited some other botanists to take up the reins. Finally, in the mid-1990s, Eric Clement did so, and with his characteristic persistence he completed the task (as he had redefined it) to produce *Illustrations of Alien Plants of the British Isles* (Clement *et al.*, 2005). This work presents 444 full-page plates of aliens for which, in most cases, illustrations were difficult or impossible to find and often did not exist in our home-grown literature. A few of the species are not in Stace's *New Flora*, and indeed a number (e.g. some greenhouse weeds) are scarcely a part of our flora at all, but where an illustration was at hand it was used. Although the *Illustrations* cover only about 48 per cent of McClintock's original list, and there are no descriptions, they provide a unique collection of drawings of species that are otherwise hardly available.

New Atlas of the British & Irish Flora (Preston *et al.*, 2002) consists of maps showing the distribution of each species at the hectad (10 × 10 km square) level. The species treated were those covered in the diagnostic keys of the second edition of Stace's *New Flora* (1997). Maps of the commoner aliens were depicted in the main text, but 942 of the less common ones (many present in only one or few of the total of 3,859 hectads) were placed on an accompanying compact disc.

The final two publications are more regional in scope. Geoffrey Wilmore's (2000) *Alien Plants of Yorkshire* covers a part of England that over a long period has been more thoroughly searched for aliens than most. The main text includes about 1,300 species of alien, but there are also two supplementary lists: one of about 180 additional species for which there is inadequate documentation; and the other of about 134 marginally 'wild' species, mainly planted woody taxa. The total of more than 1,600 aliens would be hard to equal in any other similar-sized region.

FIG 59. Humphrey John Moule Bowen (1929–2001) was a professional scientist who wrote the Floras of two counties and included in them many aliens, especially garden escapes, a number of them only marginally in the wild. (Stephen Jury)

The sixth publication is *A Catalogue of Alien Plants in Ireland* by Sylvia Reynolds (2002). Previously, Ireland had been neglected regarding alien plant studies, as commented on above, but this book provides not only a complete list of species but also a summary of the ecology and the field records under their vice-counties for all but the commonest species. Many species additional to those in the *New Atlas* are included; the total is given as 920 'species and hybrids'. Decisions on native/alien status follow those in the second edition of *Census Catalogue of the Flora of Ireland* (Scannell & Synnott, 1987), so those archaeophytes treated as natives in the latter are omitted, but all extinct aliens are included. It should be remembered that a good number of Irish aliens are native in Britain. Such a pioneering work has inevitably stimulated a surge in Irish records of alien plants.

We now have an unprecedented level of information on our alien plants – a full list, means of identifying all but the rarest species, detailed distributions, and an ongoing process of documenting new records and extra species. The BSBI maintains an active and detailed survey of the distribution of all wild plants. The species treated in the third edition of the *New Flora* (2010) constitute their main working list, but all other occurrences are also documented. Most are published under 'Plant records', the origin of which can be traced back to the *BEC Report for 1911*. This feature was published in the BSBI's journal *Watsonia* from 1949 until 2010, but has now been transferred to its successor, *New Journal of Botany*. The newsletter *BSBI News* has also carried a designated section on aliens since 1975. The continually updated maps are available on the society's website. In addition, The Wildflower Society has published annually since 1960 in its *Wildflower Magazine* a section entitled 'Exotics', which deals with rare and new aliens not included in their standard list (from the third edition of *New Flora*). Appendix 1 of this book takes into account all of these sources, and represents the most up-to-date listing of our aliens available.

An important consequence of all this easily available information is that the study of aliens by both professionals and field botanists has increased in both scope and depth. Alien plants are now incorporated on an equal footing with natives in most of our publications and projects, although it is true that there are still some who prefer to ignore records of casuals and/or planted specimens, and there is still variation, both personal and regional, in the efforts being made to record aliens. Many of our very numerous local natural history societies provide a wealth of information on aliens, and Floras are more likely than ever to place them among the natives in their correct systematic sequence rather than banish them to abbreviated appendices. Geoffrey Wilmore has also maintained a privately circulated referenced annual list of additional taxa of alien plants recorded in the British literature since Clement and Foster's definitive *Alien Plants of the British Isles* (1994) and Ryves *et al.*'s *Alien Grasses of the British Isles* (1996). Up to the end of 2015, the running list totals over 500 species, although many of these are planted individuals (especially trees), and well over a half scarcely deserve a place on our standard list.

CHAPTER 4

Accidental Introductions

In fourteen hundred and ninety two, Columbus sailed the ocean blue.
Winifred Sackville Stoner Jr, 'The History of the US' (1919)

The voyage of Christopher Columbus marked a step change in the distances over which plants would be moved. Plants that had been dispersed by people before this time are defined as archaeophytes and are discussed in Chapter 2. This chapter is about the neophytes (plants introduced after 1500) that have been unintentionally introduced to the British Isles since then. The neophytes resulting from deliberate introductions are discussed in Chapter 5.

In 1488, four years before the opening up of the Americas, the first Europeans had reached the southern tip of Africa under the command of the Portuguese explorer Bartolomeu Dias. The first circumnavigation of the world was by the Magellan–Elcano expedition, which sailed from Spain in 1519 and returned in 1522 after having crossed the Atlantic, the Pacific and the Indian oceans. Sir Francis Drake passed through the Strait of Magellan into the Pacific Ocean in September 1578, in the course of his circumnavigation of the world (Cape Horn was not discovered until substantially later, on 29 January 1616).

From this period onwards, global trade by sea increased exponentially. The East India Company was set up in 1600 and their Dutch counterpart, Vereenigde Oost-Indische Compagnie (VOC, translated as 'United East India Company') was chartered two years later. VOC went on to carry out five times as much trade as its British rival over the next 200 years, sending almost a million Europeans in more than 4,700 ships to work in Asia. They traded mainly in silk, cotton, pepper, indigo dye, salt, saltpetre, tea and opium.

We begin this chapter with an overview of the big questions about our alien plants: where they came from, when they were introduced, how they were introduced and which plant families were involved. These analyses cover all aliens, after which we shall concentrate on the neophytes.

PATTERNS OF INTRODUCTION

Geographic origins

The geographic origins of our alien flora reflect the history of human exploration and the extent to which the climate of the foreign lands from which the plants came matched that of the British Isles. The archaeophytes came to the British Isles along the routes of human migration and cultural diffusion, from southwest Asia and the Mediterranean, through southern, central and northern Europe. The neophytes are pan-global in origin, but their relative numbers reflect the degree of climatic matching, the area and topographic variety of the country of origin, the richness of its flora, the amount of trade between the country and Britain and Ireland, and the attractiveness of its flora to European gardeners.

The big picture is clear (Fig. 60). Most of our aliens came from relatively close by, in Europe. Of the long-distance immigrants, most came from North America, where European settlers first put down roots, and where the climate and ecological composition of the Eastern deciduous forest biome was very like our own. Considering that the Americas received virtually the whole of the weedy flora of Europe and the Mediterranean, it is striking how few consequential invasions have taken place in the opposite direction. Charles Darwin and Asa Gray (his botanical friend and author in 1848 of *Manual of the Botany of the Northern United States*) exchanged letters on the topic. Darwin teased Gray: 'Does it not hurt your Yankee pride that we thrash you so confoundedly? I am sure Mrs Gray will stick up for your weeds. Ask her whether they are not more honest, downright good sort of weeds.' Mrs Gray wrote back that her American weeds were 'modest, woodland, retiring things: and no match for the intrusive, pretentious, self-asserting foreigners'.

Eastern Asia produced a great number of alien species, reflecting the size of its temperate flora and the passion for Chinese and Japanese plants in British and Irish gardens. South Africa, Australia and New Zealand have been the source of remarkably few British alien species, especially considering the richness of their floras and their attractiveness for horticulture. The biggest deficits, of course, are in the tropics and subtropics: the richest places on Earth for plant species. The reason the tropics produce so few aliens is simply that their plants lack the necessary frost-tolerance to see them through a British winter.

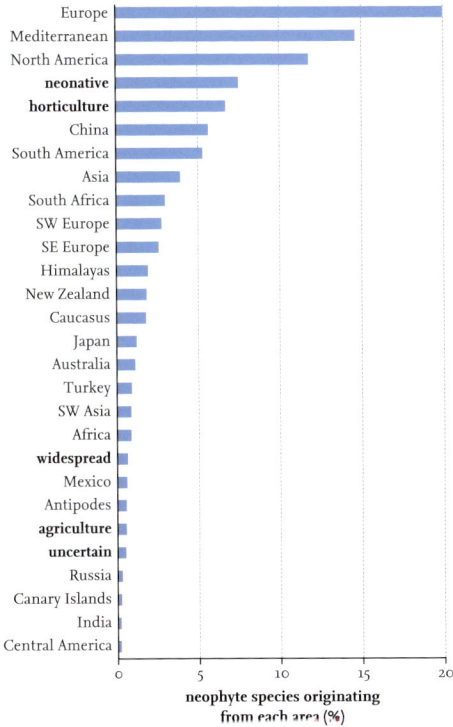

FIG 60. The geographic origins of our neophyte flora. Horizontal bars show the percentage of neophyte species originating from 28 sources. Aside from 23 geographic regions, there are five additional categories: for 'neonative' (species that arose within the British Isles by hybridisation or ploidy change involving one or more alien species; *see* Chapter 1); 'horticulture' (species that arose by hybridisation or selection within gardens); 'widespread' (species whose native distribution is wider than the other categories mentioned, such as Europe and Asia, or the northern hemisphere); 'agriculture' (crop plants with no clear wild origins); and 'uncertain' (species whose native area is unknown; often true of warm-temperate or subtropical species that now, by whatever means, encircle the globe).

It is interesting to see whether there is any pattern with invasiveness in relation to geographic origin. There are certainly examples of highly invasive plant species from origins that produced very low total numbers of aliens (New Zealand Pigmyweed and New Zealand Willowherb (*Epilobium brunnescens*) are striking examples). In addition to those, invasive species in the British Isles come from all over the world: from Europe (Rhododendron), China (Butterfly-bush), North America (Narrow-leaved Michaelmas-daisy), South America (Fuchsia), South Africa (Bermuda-buttercup), India (Indian Balsam) and Japan (Japanese Knotweed). In terms of the proportion of naturalised species that become invasive, Europe ranks relatively low at 14 per cent. The highest rate is for Asia (33 per cent), then South America (29 per cent) and North America (26 per cent). South Africa is clearly ranked low (most parts of the British Isles have few invasive South African species, although Hottentot-fig *Carpobrotus edulis* and Bermuda-buttercup *Oxalis pes-caprae* are invasive in the Channel Islands and the Isles of Scilly).

While climatic matching is clearly the most important process at work here (*see* Chapters 6 and 7), the combination of propagule pressure (the number of individual plants of an alien species that are released) and extreme selection (imposed by very high death rates) may be important.

Time course of introductions

The quality of our information about dates of plant introductions increases century by century (Fig. 61), but the broad pattern is a consistent increase in the rate of species introductions up to 1900 and then a levelling off after that date. New alien species continue to be introduced, but at a somewhat lower rate than formerly.

The time course of horticultural exploration is reflected in the dates of importation of tree species (Mitchell, 2001). The order in which different areas were explored for trees can be seen by noting the median date of introductions of trees from that region. The first area to be thoroughly explored was eastern North America (1736), then southeastern North America (1752) and the central United States (1777), with Australia substantially later (1829), followed by the Himalayas (1836), Mexico (1838), Caucasus (1841), Alaska (1842), southern South America (1847), New Zealand (1850), western North America (1851) and Tasmania (1857) in relatively short order, then finally by the far-eastern horticultural bonanza of Japan (1865), China (1899), Korea (1909) and Taiwan (1910). Very few new tree species entered commercial horticulture after the First World War, the most notable exceptions being Dawn Redwood (*Metasequoia glyptostroboides*), which was introduced from eastern Sichuan and northeastern Hubei provinces of China in 1948 (the plant had

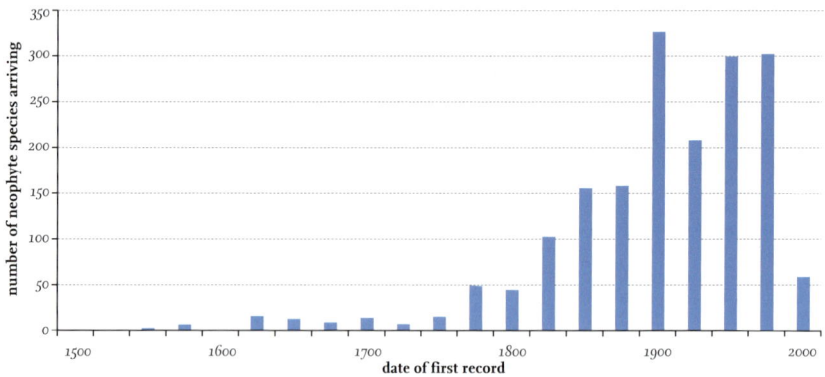

FIG 61. Pattern of arrival of neophyte species in the British Isles from 1500 to 2000 (data on new arrivals aggregated across 25-year periods within each century). There was an exponential increase in the arrival rate of alien species until the First World War, after which there was a decline in horticultural activity, with fewer large gardens and fewer people to work in them. Whether the apparent decline in the rate of new arrivals in the twenty-first century is real will not be known until 2025.

been discovered in 1941 and described in 1944), and Wollemi Pine (*Wollemia nobilis*), which was discovered in 1994 in a remote series of narrow, steep-sided sandstone gorges in a temperate rainforest wilderness area of the Wollemi National Park near Lithgow, 150 km northwest of Sydney, in New South Wales, Australia.

Taxonomic origins

Neophytes are strongly overrepresented in the British Isles compared to native species in the Asteraceae, Rosaceae, Amaryllidaceae, Iridaceae, Solanaceae and Onagraceae families. Archaeophytes, meanwhile, are strongly overrepresented in the Asteraceae, Brassicaceae, Amaranthaceae, Veronicaceae, Apiaceae, Papaveraceae and Euphorbiaceae families. There are also numerous alien plant families that have no British native representatives, e.g. Aizoaceae, Apocynaceae, Azollaceae, Berberidaceae, Juglandaceae, Lauraceae, Resedaceae and Verbenaceae.

At the other end of the spectrum, there are families that are relatively species-rich in natives but poorly represented by aliens. Indeed, there are 65 families that have British or Irish native representatives but no alien ones. Families that are relatively rich in native species, but which exhibit a relative paucity of aliens, include Juncaceae, Caryophyllaceae, Rubiaceae, Primulaceae, Ranunculaceae, Orobanchaceae, Orchidaceae and Ericaceae. Species belonging to these families appear to be poor invaders.

Mode of introduction

Intentional introductions are much more numerous than unintentional introductions and, of the intentional introductions, species introduced for horticulture are overwhelmingly the most common (Fig. 62). Of the unintentional introductions, contaminants of wool and grain provide the greatest number. Seeds introduced intentionally as food for birds make up a substantial proportion of the plant species unintentionally introduced into the environment. Neonatives are an interesting category: these are plants that originated in the British Isles from parents one or both of which evolved in different parts of the world and were brought into contact with one another here.

ALIEN PLANTS ACCIDENTALLY INTRODUCED

Ships' sweepings

Docks used to be great places to search for aliens, but our obsession with security means that access to them is all but impossible these days. From the dawn of the seafaring age, mariners had carried animals with them on their journeys:

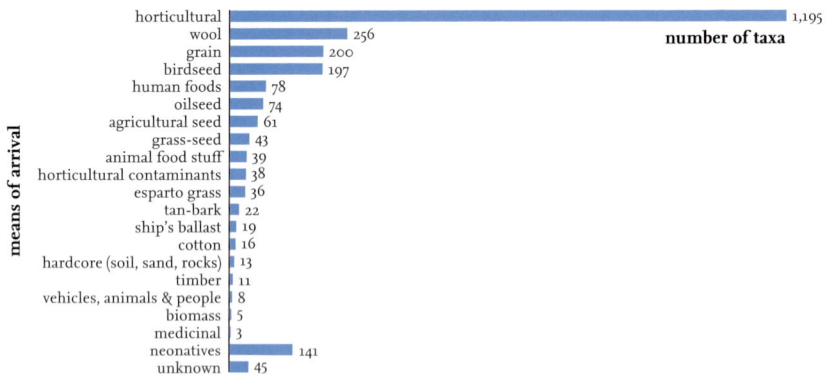

FIG 62. The means of arrival. Neonatives are taxa that arose within the British Isles from one or more alien parents (*see* Chapter 1). Much the most frequent category is of intentional introductions for horticulture. The next three groups are of unintentional introductions: wool and grain brought alien species with them as contaminants, while birdseeds were introduced intentionally as food for birds but unintentionally for planting in the wild.

cage birds for company, hens for eggs, domestic animals for milk and meat, Horses and Cattle as beasts of burden. All these animals needed feeding, and great quantities of fodder were carried, doubtless contaminated with the seeds of untold numbers of alien plants. When the ship returned to port, the hold would be swept out and heaps of dung, hay, straw and dust deposited on the quayside. The rubbish was then either sold or given away, to be spread on fields as fertiliser.

Many of the alien plants, however, were never recorded from beyond the confines of the dock itself. At Bristol, for instance, in a single year (1909) White (1912) encountered no fewer than ten alien casual species of mullein (*Verbascum*) at St Phillip's Marsh, including Moth Mullein (*V. blattaria*), Nettle-leaved Mullein (*V. chaixii*), Orange Mullein (*V. phlomoides*), Purple Mullein (*V. phoeniceum*), Wavy-leaved Mullein (*V. sinuatum*) and Hungarian Mullein (*V. speciosum*; Fig. 63).

Ships' ballast

Ships' ballast is gravel, sand, stones or any other heavy material that is placed in the hold of a ship in order to prevent it from capsizing. If the ballast was taken on board in a foreign port, it might contain viable seeds or roots of plants from that country, and some of these might survive the journey back to their home port. The first mention of ballast aliens comes from Francis Bacon (1627) in his *Sylva Sylvarum, or A Naturall Historie*: 'Earth that was brought out of the Indies and other remote countries for ballast for ships, cast upon some grounds in Italy, did put forth foreign herbs, to us in Europe not known.'

FIG 63. Species of mullein (*Verbascum*) hybridise promiscuously whenever they come into contact; 10 of our 13 species are aliens, producing 16 different hybrids in combinations between themselves and with native species. This is one of the aliens, Hungarian Mullein (*V. speciosum*), a garden biennial that is increasingly common in south-eastern England: (a) flowering plant; (b) rosette. (MJC)

Ballast aliens were most common in exporting ports like Newcastle, from which shiploads of coal had been dispatched for centuries, and back to which the dirty and dusty boats would return with loads of cheap sand, earth and stones. Importing cities like London received little ballast and had correspondingly few ballast aliens. In some ports the ballast was thrown into the sea, and the plants within it killed. But harbour regulations and the cost of dredging meant that ballast was usually taken ashore or dumped on the edges, eventually forming ballast islands or hills. The ballast hills produced a rich flora of alien plants, but most of them were casuals, disappearing as soon as the supply of ballast containing foreign seeds dried up. Only a few ballast plants were able to survive and even extend their area, Pineappleweed being the most successful example.

Ballast plants represent a category of alien that is now almost extinct. Most ships ceased to bring solid ballast back to our shores when iron hulls replaced wooden ones at the turn of the nineteenth century. Judging from the lists of ballast aliens published for Cardiff, Northumberland and Durham, and Fife, there were probably around 450 species in total, although a number of these are British natives or archaeophytes. In Appendix 1 there are only 15 species listed as originating from ballast, and only three of these do not have another vector as well.

Wallflower Cabbage (*Coincya monensis* subsp. *cheiranthos*), a subspecies of our native Isle of Man Cabbage, is scattered throughout the British Isles, but the only area with numerous records is South Wales. It is tempting to associate this concentration of records with the ports of that region, but the *Coincya* was not recorded there by Storrie (1886). The second species, Prostrate Toadflax (*Linaria supina*), is naturalised only on open sandy ground in southwestern England, particularly around Par, Cornwall; it was first recorded in Cornwall as long ago as 1848. The third species, although much rarer than the other two, has had by far the greatest impact. The American Smooth Cord-grass (*Spartina alterniflora*) was first found in Southampton Water in 1816; it spread over quite a large area of the estuary but later started to retreat and is now very restricted in distribution. The story of its hybridisation with the native Small Cord-grass (*S. maritima*), and the origin of the important and now widespread mud-binding Common Cord-grass (*S. anglica*), is told in Chapter 9.

The origin of the *Linaria* and *Coincya* species in ballast is fairly certain; both were recorded from ballast hills in County Durham in 1848 and 1972, respectively, although they are not in the early ballast alien lists from southern England. A similar history for Smooth Cord-grass is less certain, but ballast is its most likely origin in the absence of any more plausible explanations. Three other naturalised species that might owe their presence in the British Isles at least in part to ballast, from which they were recorded in the nineteenth century, but have other known vectors as well, are Annual Wall-rocket (*Diplotaxis muralis*), Hoary Cress (*Lepidium draba* subsp. *draba*; Fig. 115) and Lesser Swine-cress (*L. didymum*). Hoary Cress, also known as Curse-of-Kent, is an aggressive rhizomatous species now spread over most of the British Isles. It has also been recorded as introduced with fodder and with straw. The other two species are annuals. One additional naturalised species, Lax Viper's-bugloss (Fig. 10), might also have had an origin in ballast. It is more usually referred to in broader terms, as a dock alien, and was first recorded as such in 1927 from Barry Docks, Glamorgan, still its only locality. Possibly this species was introduced in the ballast era before it was noticed.

Soil, sand and rock aliens

Quite apart from solid ships' ballast, various forms of rock and soil have been imported for a long time, and they still are; for example, natural stone slabs sold in garden centres are frequently of Indian origin. A number of aliens have appeared from time to time associated with piles of rubble or rocks of various sorts, from sand to lumps of granite, and with no obvious other sources around, it seems likely that their origin was tied up in some way with the hard core. For example, the following rare aliens were found as casuals on piles of sand and granite chippings at Gloucester Docks in 1974: Mediterranean Chamomile

(*Chamaemelum mixtum*), American Cudweed (*Gnaphalium purpureum*), Shore Medick (*Medicago littoralis*), Yellow Serradella (*Ornithopus compressus*), Jo-jo-weed (*Soliva pterosperma*) and Nodding Clover (*Trifolium cernuum*). Most of these species are also wool aliens, but there was no evidence of that vector in this case. Shore Medick has been naturalised on a sand and gravel beach in East Kent since 2001, which is similar to its natural Mediterranean habitat. The dull-looking Bugweed (*Corispermum intermedium*), a member of the Chenopodiaceae resembling a depauperate Summer-cypress (*Bassia scoparia*), is also found on sandy soils both in its native Continental Europe and in the British Isles, where its occurrences have been mostly maritime. Perennial Pigweed (*Amaranthus deflexus*), another wool alien, has also been found on granite dumps. Two species have been tentatively linked with imported iron ore: Sand Rock-cress (*Arabidopsis arenosa*) and Ternate-leaved Cinquefoil (*Potentilla norvegica*).

It is possible that the last two species were introduced not with the ore but on imported machinery used industrially on the site. The moss *Atrichum crispum*, an American species that has been well naturalised in the British Isles since 1848, is thought to have been imported with soil.

FIG 64. Slender Rush (*Juncus tenuis*), known in the British Isles since 1795 and at first championed as an example of the 'American element' in our native flora, was probably introduced with hay. (CAS)

Fodder aliens

Two common and widely distributed aliens that have an uncertain history but are thought to be most likely contaminants of imported hay or straw are Hoary Cress (Fig. 115) and Slender Rush (*Juncus tenuis*; Fig. 64). The former was also a ballast alien (*see* above) and its origin here might additionally or alternatively be from that source. Slender Rush is discussed under 'An Irish conundrum' in Chapter 12. In this case, the alternative suggestion to its arrival in North American hay has been that it is native here. The latter suggestion is, however, nowadays abandoned by most. Slender Rush is a variable plant in America and several segregates have been described (Fig. 65), the best known of which is Dudley's Rush (*J. dudleyi*). This has been known at Crianlarich in

FIG 65. Three segregates of the *Juncus tenuis* aggregate (left to right): Slender Rush (*J. tenuis*), Lax-flowered Rush (*J. anthelatus*), Dudley's Rush (*J. dudleyi*). (CAS)

the vice-county of Mid Perthshire since 1915, and might still be there, although it has not been recorded in recent years. It has, however, been discovered in another locality. Another segregate is Lax-flowered Rush (*J. anthelatus*), which was discovered in our region much later and is still very rare. These two rare rushes have been treated as subspecies or varieties of Slender Rush by many taxonomists, but the present consensus is for their recognition as separate species. The means of arrival of these two segregates is uncertain.

Other neophytes that are thought to have arrived here as contaminants of fodder include American Fox-sedge (*Carex vulpinoidea*), which is also a wool alien from North America; Hungarian Clover (*Trifolium pannonicum*), formerly a ballast alien and also a grass-seed alien; a central European subspecies (*polyphylla*) of the native Kidney Vetch (*Anthyllis vulneraria*); and a number of grasses, including the easily overlooked Swamp Meadow-grass (*Poa palustris*) and the rare American Canary-grass (*Phalaris angusta*). In addition, some species of *Amaranthus*, a genus that has hitched a lift with nearly all possible vectors, are found as fodder and pet-food aliens, especially Purple Amaranth (*Amaranthus cruentus*) and Green Amaranth (*A. hybridus*).

Grain and seed aliens

Grain and seed aliens are one of only three categories of neophyte to exceed 100 species in Appendix 1. They are subdivided in Fig. 62 into birdseed, grain, agricultural seed and grass-seed, together amounting to about 325 species. Adding species that have not been found since 1986, the total would be in excess of 400, of which at least 60 (15 per cent) are grasses.

Seeds are imported for a great many purposes, notably cereal grain (for sowing, baking and brewing); other agricultural seeds; animal, poultry, pigeon and gamebird feed; grass-seed and wildflower mixes; and seed for wild bird food. These different uses are satisfied by seed mixes from various geographical regions, and they are used in a range of locations in the British Isles, so the long and diverse list of both deliberately imported seeds and their contaminants is fully understandable. Changes in requirements and areas of sourcing account for the appearance of new species and for the disappearance of others. Our most widespread neophyte, Pineappleweed (Fig. 127) was probably a grain alien, as were many of the species that were generally described as dock aliens ('the sweepings of the holds of vessels, dock sheds and railway trucks'; Robinson, 1902) but no longer occur. There is also a small number of species that seem to have arrived here as contaminants of seed intended for direct human consumption, including spices. Green Nightshade (*Solanum physalifolium*) and Slender Salsify (*Geropogon glaber*) are examples.

At the time that Dunn (1905) was writing the *Alien Flora of Britain*, he attributed 206 species as 'grain-sifting aliens of recent appearance and of little permanence' out of his total of 924 species. At that time, in every sack of imported grain 'countless seeds of the cornfield weeds of the country of origin come mixed with the grain. Before the grain is used, these seeds are sifted out, and are either thrown away with other rubbish on waste ground or sold for feeding domestic fowls and game' There were cereals both ancient and modern, plus a great variety of other alien Poaceae.

Today, commercial seed carries many fewer impurities than formerly, but there is a greatly increased use of seed for various 'conservation' plantings and for wild bird food (Chapter 5), and these products do not have the same requirements for purity; in fact, diversity is considered an asset. The alien species arriving in the various categories of grain and seed overlap to a considerable degree, but there are differences. The contaminants will vary according to the location of the seed harvest (which in turn depends upon the main species being collected) and according to the degree to which the disseminules are successful in mimicking those of the main species. Hence, larger-seeded contaminants are commoner than small-seeded ones in grain, and the reverse is true in grass-seed or birdseed.

Among the crucifers, Warty-cabbage (*Bunias orientalis*) is being grown on a very small scale in London allotments (and perhaps elsewhere) as 'Turkish rocket' for its mustard-flavoured flowers. The recent commercial marketing of Perennial Wall-rocket as 'wild rocket' has been mentioned in Chapter 2. Similarly, species of *Amaranthus*, notably Common Amaranth (*A. retroflexus*), Green Amaranth and

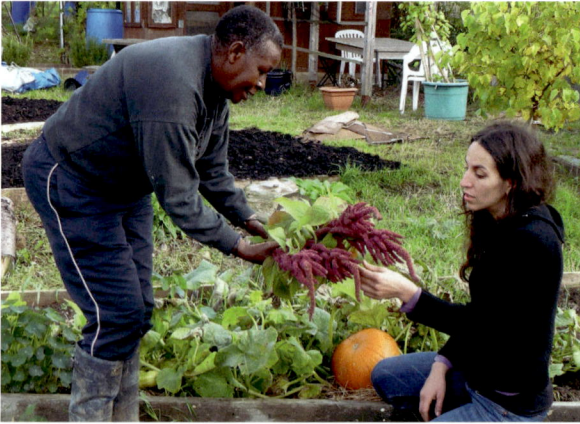

FIG 66. Fidelis Nyinakasingye showing off some of her crop of Purple Amaranth (*Amaranthus cruentus*; known locally as 'callaloo') on a London allotment. (Jeanette Sitton)

Purple Amaranth, which are both birdseed and grain aliens (as well as arriving via other vectors too), are being cultivated as a leafy vegetable under the name 'callaloo' (Fig. 66). The name and usage were probably brought into the British Isles by West Indian immigrants. (We are indebted to Brian Wurzell for sharing his knowledge of East End allotments.)

Accurate identification of a few key alien species occurring on a site can indicate the nature of the vector that brought them, and alert the recorder to other likely species. An illustration of the need for accuracy in determination is provided by the small North American grass Rough Bent (*Agrostis scabra*), a characteristic grain alien. A very similar species that has been frequently confused with it is Blown-grass (*A. avenacea*), a wool alien from the Antipodes. The distinction between these two was not fully recognised at first by British botanists, although in fact they are not especially closely related, and a number of early records of Blown-grass actually refer to Rough Bent, which also occurred in wool.

In the nineteenth century, Larkspur was a characteristic cornfield weed in Cambridgeshire, producing its attractive blue flowers either with the corn or after harvest in the stubble. According to Salisbury (1961), its seeds mimic those of Onion, and sometimes occur as contaminants of the latter. Larkspur does not seem to be part of modern cornfield seed mixes, although it would be a welcome addition because of its unique colour. Today, it is a fairly common alien, but as an escape from gardens, where it is a popular annual. Field Cow-wheat (*Melampyrum arvense*), another former cornfield weed, is now a rarity.

Austrian Yellow-cress (*Rorippa austriaca*) was originally introduced as a grain alien, but is now carried about the country by soil-moving activities and perhaps by water, near which it often grows. This change of vector, equally illustrated by Larkspur above, is also seen in the fiddlenecks (*Amsinckia*) species. These are small, harshly hairy plants of the borage family (Boraginaceae) with small yellow flowers. This American genus is taxonomically very perplexing, and many of the early records were wrongly identified. Modern revisions have shown us that the vast majority of our plants are referable to Common Fiddleneck (*A. micrantha*), which has become widespread in sandy arable land in eastern England and Scotland, where it is now dispersed by agricultural activities, probably mainly on humans and machinery, and in builder's sand. It is likely to have been first introduced with grain, grass-seed and agricultural seed.

There is a small group of species that are rather characteristic grass-seed aliens, in that they are more frequently found in newly sown lawns and grass verges than elsewhere. The contaminants vary, of course, according to the origin of the seed, which is rarely from the British Isles. The species that is currently the most discussed grass-seed alien is Austrian Chamomile (*Anthemis austriaca*), which is sold as 'Corn Chamomile' (*A. arvensis*) in many commercial wildflower mixtures (*see* Chapter 5). Long-established species that we believe owe their existence in the British Isles to introduction with grass-seed are Greater Yellow-rattle (*Rhinanthus angustifolius*) and Beaked Hawk's-beard (*Crepis vesicaria*). The former is a handsome plant differing from the common Yellow-rattle (*R. minor*) mainly in having larger flowers with a different-shaped corolla. It was formerly much commoner than now, having been recorded from only 17 hectads in the past 25 years, and rare in most of them. Beaked Hawk's-beard, which is very common on grassy roadside verges, particularly in the south, is discussed in Chapter 11. It has a number of much less common congeners, of which Bristly Hawk's-beard (*C. setosa*) and Narrow-leaved Hawk's-beard (*C. tectorum*) are also grass-seed aliens, as are at least two uncommon species of pepperwort (*Lepidium*) and two species of barley (*Hordeum*). Occurrences of Eastern Groundsel (*Senecio vernalis*) are usually in new grass verges. This species is easily confused with Groundsel (*S. vulgaris*) and the various hybrid products it forms with Oxford Ragwort (*S. squalidus*). It usually does not persist for long, but sometimes forms hybrids with Groundsel, or perhaps these hybrids are introduced with it. An unusual species that often stumps botanists at first is California Lobelia (*Downingia elegans*). This is a member of the Campanulaceae, which like *Lobelia* has bilaterally (not radially) symmetrical flowers, which are blue with a white centre. It was first recorded in the British Isles in 1978, and is usually now found in damp grassland by water or in hollows.

The closely related Patience Dock (*Rumex patientia*) and Greek Dock (*R. cristatus*) seem to be associated with clover seed, and perhaps grass-seed as well. Many of the grass-seed aliens are equally frequently derived from the contamination of crop-plant seed, in addition to those associated with grain. Landscaping associated with the installation of new mini-roundabouts at road junctions in Berkshire created a local outbreak of Greek Dock because the contractor's grass seed was contaminated with its seeds.

Birdseed aliens

Birdseed aliens now constitute the greatest number of grain and seed aliens, but this is a relatively modern phenomenon. In the past, chicken runs were a favourite hunting ground for those seeking alien plants, although today feed put out for gamebirds, either grown as crops or provided as seed, is probably a richer source. The appearance of very unfamiliar plants in garden borders or in parks is often the result of wild birds having been fed with commercial seed mixes. It is such conspicuous plants as Ragweed (*Ambrosia artemisiiflora*; Fig. 67), Lesser Star-thistle (*Centaurea diluta*), Buckwheat (Fig. 68), Niger

FIG 67. Ragweed (*Ambrosia artemisiifolia*) is a common alien from birdseed, here appearing in a garden flower bed. It is a major cause of hayfever in North America. (CAS)

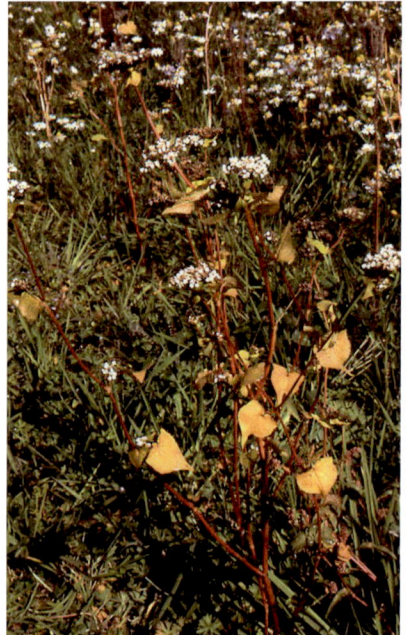

FIG 68. Buckwheat (*Fagopyrum esculentum*) is an important non-grass cereal in some parts of the world, and is now also being grown in the British Isles as a source of food for gamebirds and wild birds. (CAS)

(*Guizotia abyssinica*), Sunflower (*Helianthus annuus*; Fig. 82) and Fenugreek that draw attention to these activities, together with a number of exotic grasses that are not closely related to any native species. Grasses provide the greatest number of birdseed aliens, but several of the other larger families are prominent as well. There are about 20 crucifer genera and nearly as many legumes, but the latter outnumber the former in species because of the many representatives of the genera *Medicago*, *Trifolium* and *Vicia*. Some of the smaller annual legumes attract interest because of their weird fruit (legume) shapes, especially those resembling a bird's claw or a scorpion's tail in species like Annual Scorpion-vetch (*Coronilla scorpioides*) and Caterpillar-plant (*Scorpiurus muricatus*). The ground beneath a bird feeder may produce some highly exotic plants by late summer.

Hanson & Mason (1985) listed 32 species as being harvested for wild birdseed, excluding seed used to feed pigeons, gamebirds and poultry; all are common birdseed aliens (deliberately imported as bird food but unintentionally introduced into the wild). Even feeding wild birds is subject to changing fashions. Whereas breadcrumbs were once considered sufficient, today specially purchased seed mixes are the norm, and even these are changing over the years. For example, the sunflower 'seeds' on offer are becoming more commonly the isolated kernels rather than the whole achene, and these have fewer contaminants. In recent years it has been found that Niger seed is a favourite of Goldfinches (*Carduelis carduelis*), and special dispensers for it are commonplace in today's gardens. Gordon Hanson has grown commercial Niger samples and compiled a list of 25 contaminating species, including ten grasses. Some of these have not been recorded as aliens here, but are to be expected in the future.

Birdseed aliens are associated closely with households, and formerly a great many of them were characteristic members of council rubbish-tip floras, having come from backyard sweepings, bird-cage cleanings and kitchen waste. This habitat is no longer fruitful for the alien hunter, as rubbish, once spread over large areas, is now burnt or covered with soil or hard core very quickly, and the tips are in any case fenced off. Another similar habitat now denied the botanist on health and safety grounds is railway sidings, also securely fenced off.

Soyabean and oilseed aliens

Several plant species are imported for oil extraction. These come from different parts of the world, so the list of contaminating species (74 in Appendix 1) is very varied and does not constitute a recognisable grouping. Oil-rich species that are exploited include Soyabean, Sunflower, Flax, Oil-palm (*Elaeis guineensis*), Sesame (*Sesamum indicum*), Groundnut (*Arachis hypogaea*) and Coconut (*Cocos nucifera*), but most interest has been generated from the first of these, which along with its contaminants is mainly American in origin.

Soyabean itself is a common and distinctive feature of sites supporting Soyabean aliens, and attracts the botanist's attention. Only three of the 74 contaminating species we list do not have another known vector, and these are all rare casuals associated with Soyabean processing that have fewer than 20 post-1986 hectad records between them: Powell's Amaranth (*Amaranthus powellii*), Pinkweed (*Persicaria pensylvanica*) and Horse-nettle (*Solanum carolinense*). Two of these genera are represented by several oilseed alien species: 12 of the 21 species of *Amaranthus* and 5 of the 15 species of *Solanum* listed in Appendix 1 fall into this category. Another species of *Amaranthus* particularly characteristic of Soyabean waste is Dioecious Amaranth (*A. palmeri*).

The other well-represented family among Soyabean contaminants is the grasses, Poaceae, with non-British genera like *Digitaria, Echinochloa, Eleusine, Panicum* and *Setaria* prominent. Characteristic species that are not common elsewhere include Tropical Finger-grass (*Digitaria ciliaris*), Autumn Millet (*Panicum dichotomiflorum*) and Nodding Bristle-grass (*Setaria faberi*). These familiar alien genera do not attract particular attention at first, but once the species are identified their origin becomes clear. In the same way, particular species of other well-known alien genera are indicators of the presence of oilseed aliens. Examples are Soyabean Goosefoot (*Chenopodium bushianum*), Prickly Mallow (*Sida spinosa*), White Morning-glory (*Ipomoea lacunosa*), Giant Ragweed (*Ambrosia trifida*), Lesser Sunflower (*Helianthus petiolaris*) and a particular variant of Rough Cocklebur that is sometimes separated as *Xanthium echinatum*. In addition to Prickly Mallow, two other conspicuous plants in the Malvaceae often occur with the above: Velvetleaf (*Abutilon theophrasti*) and Bladder Ketmia (*Hibiscus trionum*). Many of the above closely resemble alien species that are commoner from other sources, notably birdseed, and careful discrimination is necessary to avoid false conclusions about the site. In addition, there are a number of other aliens occasionally accompanying the above, but too infrequent to be included in this book, e.g. Spiny Amaranth (*Amaranthus spinosus*).

Wool aliens

For centuries, it was the production of wool for export that made fortunes for British merchants. Fine cloth was manufactured elsewhere, typically in Flanders, where British wool was highly valued for its length, strength and texture. Historically, wool was the first English export to be taxed (by Edward I in 1275), and the first material to be smuggled to avoid the payment of tax (most of the illegal export was through the ports of Kent, carried out by men known as owlers). In 1662, the death penalty was introduced for anyone caught smuggling wool. It was not until much later that foreign fleeces were imported in large volumes to cater for a flourishing British-based textile industry in the late eighteenth century.

Industrial production of woollen yarn and fabrics started in valleys where water power was plentiful, in places like the Yorkshire Dales and the Southern Uplands of Scotland. With the development of coal power and the arrival of the railways, the industry shifted downstream to the coal fields, and Leeds became the capital of the Yorkshire woollen industry. The hugely increased demand for fleeces meant that wool began to be imported to the British Isles from all of the sheep-farming regions of the world: South America, South Africa, Australia, New Zealand, California and North Africa. Along with the dirty fleeces came viable seeds of many hundreds of foreign plant species. These were the wool aliens.

There were two separate routes for the introduction of wool aliens to the wild. The first was in the waste-water effluent from woollen mills resulting from washing the wool, and the second was the use of dirty wool waste as a fertiliser, applied intentionally to fields or orchards. Both routes rely on the fact the fleeces were dirty in two complementary ways. The wool itself contained seeds burred within it, tangled up in it and stuck to it. Then there were the dags: dried sheep dung that clings to the wool of the tail and hind quarters. These are a rich source of seeds of plants that were eaten by the sheep in far-distant pastures, passed through its gut unharmed and then were transported halfway round the world in faeces attached to a dirty fleece.

The Adventive Flora of Tweedside

The pioneering study of wool aliens was carried out in the south of France by Albert Thellung (1912) in his book *La Flore Adventice de Montpellier*, but a British classic followed soon after. The woollen mills of the Scottish Borders produced a cloth known as tweel, which came to be known as tweed after a London cloth merchant allegedly misread the handwriting in a letter from a Hawick woollen firm (apparently he was thinking of the local River Tweed, which is still celebrated for the excellence of its salmon fishing). George Claridge Druce introduced the book he wrote with Ida Margaret Hayward, *The Adventive Flora of Tweedside* (1919), as follows: 'The adventitious or alien flora of Tweedside is so interesting and in extent so exceptional that no apology need be made for placing on record the results of the investigations of my colleague, Miss I. M. Hayward.' This indefatigable lady scoured the riverbanks for several years, assisted by her uncles, Messrs Sanderson, who being engaged in the woollen industry were able to help Miss Hayward gain access to the mills and their grounds to look for plants. She was an intrepid collector, and Druce noted that 'she has industriously explored the pleasant banks of the Gala and Tweed, and taken hazards frequently in both these waters, with the good fortune that often attends the adventurous'. She recorded 348 species of wool aliens and a further 294 species of adventitious plants other than wool aliens. Druce continued:

Formerly, many of the effluents passed directly, with other vegetable and animal refuse, into the once pellucid Tweed and Gala (Fig. 69). In these quickly moving waters the seeds and fruits of many a strange and unexpected species were carried down to the Gala foot, where a delta like bed of shingle appears at the meeting of the waters, and forms an anchorage for many seeds from Selkirk and Galashiels. Others are carried still further down the Tweed till some eddy enables them to obtain a precarious tenure on the banks.

FIG 69. The wool mills of Tweedside were the source of scores of alien plants imported unintentionally on dirty fleeces from Australia, New Zealand, South Africa and South America. In the early days of the industry, untreated water from washing the fleeces was discharged directly into the river. (MJC)

Druce clearly understood the impermanent nature of this casual alien flora. At the start of the twentieth century he wrote:

In recent years a system of drainage has been introduced whereby these impure waters are intercepted and conducted into septic tanks, where, under drastic treatment, all or nearly all of the seeds present are destroyed. In the near future, therefore, instead of a large Alien flora appearing along the rivers only a few sporadic species can be expected to occur.

They were classic casuals, dependent entirely on repeated introduction for their presence by the river. Very few survive today.

Ida Hayward's most extraordinary find was perhaps a very rare yet beautiful needle-grass *Nassella flaccidula*, which grows as a native high up in the Andes of Bolivia on mountain slopes above La Paz at an altitude of 4,000 m. Its seeds remained in the fleece as the sheep was driven down for shearing, and stayed lodged in the wool as the fleece was hauled by bumping cart to the railhead. The mountain railway took the seeds for hundreds of kilometres down to the port at Antofagasta in Chile, and the seeds remained lodged as the fleeces were swung aboard the ship for the long sea voyage, first south to Cape Horn, then north across the Equator. We don't know the details, but it is likely that the fleece arrived at the Tweedside mill by rail. Hayward found the plant flowering on wool refuse at Selkirk in October 1913, 10,000 km from its native home. The taxon is now sunk within *N. pubiflora*.

Wool shoddy

In Yorkshire, the mill owners realised that they could make money from the dags and greasy waste that was extracted during preparation of the fleeces by selling it as fertiliser. The study of the flora that grew on the fields fertilised with this wool waste (Fig. 70), and in the railway yards (Fig. 71) where the fertiliser was handled, became the consuming passion of a dedicated band of amateur but exceptionally capable botanists, notably Ted Lousley, John Dony and David McClintock.

The word 'shoddy' is confusing because it meant three quite separate things in the industrial parts of West Yorkshire. The first shoddy was a kind of cloth made from recycled wool or cotton rags, where the fibres of the softer makes of old cloth had been torn to pieces and mixed with new wool in the appropriate proportion. The people employed in picking and tearing the rags were known as shoddy-grinders, and the district around Batley and Dewsbury was known as the Shoddy Metropolis. The second use was as a term of derision for describing goods that were cheap and nasty. The third is the use that concerns us here: 'The stuff which

FIG 70. Alien grass widows? Mrs Dorothy Lousley (left) and Mrs Christina Dony posing in a Brussels Sprouts (*Brassica oleracea*) field treated with shoddy at Maulden, Bedfordshire, in 1965. The grasses seem to be mostly *Ceratochloa* species. (Ted Lousley, reproduced courtesy of Chris Boon)

even the shoddy-making devil rejects, is packed off to the agricultural districts for use as manure'. This shoddy, the dirt and grease from a fleece when washed, called in the factories 'mouts', plus the entire substance that falls on the floor during processing, was sold at a high rate for top-dressing grassland or market gardens.

The discovery by J. G Dony immediately after the Second World War of the wealth of alien plants in wool waste used in market gardens on light soils in Bedfordshire opened up a large field for research (McClintock, 1977). The primary crop was Brussels Sprouts (*Brassica oleracea*; Fig. 70) for which the area around Biggleswade in Bedfordshire was the major UK producer

FIG 71. Rex Alan Henry Graham, the former *Mentha* expert, searching for wool aliens along the railway sidings at Biggleswade, Bedfordshire, *c.* 1950. (Ted Lousley, reproduced courtesy of Chris Boon)

(7,069 ha of local land was given over to the crop in 1938). The source of the alien plants was traced back to the wool districts of the West Riding. The wool aliens were mostly European species, often of Mediterranean origin, that had been exported in the fleeces of the Merino and other sheep taken to the Antipodes and other sheep-rearing colonies when the industry was founded in the nineteenth century, and then brought back again in dirty fleeces, an extraordinary round trip of tens of thousands of kilometres. Of course, there were some plants that had evolved in these far-flung places, in Argentina, Chile, South Africa, Australia and New Zealand, but most were returning Europeans (some of them having evolved new subspecies in their adopted foreign homes in the meantime). With such a huge potential area to search, the difficulties in finding correct names were immense. The hours dedicated to this work by Dony and Lousley represent one of the outstanding contributions to alien plant science.

At least two South African species of wool alien have persisted long after their introduction in wool and have become fully naturalised. Narrow-leaved Ragwort (*Senecio inaequidens*; Fig. 270) is now a common wind-dispersed plant on waste ground in the Greater Bradford area. Its spread in Yorkshire is quite separate from its arrival and current advance in southern England. Another species, African Bent (*Agrostis lachnantha*; Fig. 72), which was a rather rare wool alien, has recently been found naturalised in Baildon, north of Bradford, by Mike Wilcox. This is an attractive and distinctive species with long, narrow silvery inflorescences, and is well established over a considerable area of waste ground.

Another naturalised wool alien, widespread in the north temperate region, is Russian Knapweed (*Acroptilon repens*), which has been known at Hereford railway station sidings since 1950, but still (2012) occupies only a very small area. Ryves (1974) also mentioned a few grasses that persist for a number of years, but gave no evidence for their long-term establishment. Some species, however, appear

FIG 72. African Bent (*Agrostis lachnantha*), a rather rare wool alien that is seen here well established on waste ground near Bradford; it is one of the few wool aliens to have become naturalised in the British Isles. (CAS)

FIG 73. Pirri-pirri-bur (*Acaena novae-zelandiae*), probably the first and most abundantly naturalised wool alien in the British Isles, is readily carried about by people and animals thanks to the barbed spines on its fruits. (Peter Llewellyn and ukwildflowers.com)

to be well established for a period but later die out. An example is Hooked Dock *Rumex brownii*, which was said to be 'quite naturalised on Tweedside' (Hayward & Druce, 1919), but was last recorded in the British Isles in 1984. Such a fate might well await Russian Knapweed, but surely not Narrow-leaved Ragwort or the next species.

Probably the wool alien that has become the most widespread and thoroughly naturalised of all is Pirri-pirri-bur (*Acaena novae-zelandiae*; Fig. 73) from Australia and New Zealand. *Acaena* is a genus of creeping plants, woody at the base, with pinnate leaves and erect, dull flower heads that are spherical and somewhat resemble those of the related Salad Burnet (*Poterium sanguisorba*). At fruiting, however, the small, dry fruits develop barbed spines at the apex, and these are very effective at adhering to sheep's wool, dog's fur and human clothing. Another closely related species is Bronze Pirri-pirri-bur (*A. anserinifolia*), which is very similar to the above species and was confused with it at first; it probably also arrived in fleeces. Altogether we have five naturalised species of *Acaena*, and all of them are grown in rock gardens for their attractive foliage and have escaped into the wild. Once escaped they are easily spread by adhering to humans, animals and vehicles, and are now found in semi-natural vegetation in such habitats as heathland and maritime dunes. It is not possible to be certain which populations were ultimately derived from gardens as opposed to wool, and in any case most have probably resulted from secondary dispersal from the primary sites, but even today there is a concentration of records around sites that have a history of involvement with wool imports, especially in northeastern England. It is therefore reasonable to consider at least Pirri-pirri-bur as a naturalised wool alien.

Other species that might have originated from wool in some of their naturalised sites are Argentine Fleabane (*Conyza bonariensis*; Fig. 122d), Stinking Fleabane (*Dittrichia graveolens*), African Love-grass (*Eragrostis curvula*) and Small Nightshade (*Solanum triflorum*).

The vast majority of wool aliens are, or were, casuals killed by the first frosts. In Appendix 1, 256 are marked as wool aliens, but only 38 of them have not arrived here by some other means as well, and a number of those have probably not been recorded in the past decade and/or probably do have other, albeit unknown, means of arrival. Of those 38 species, 12 originated in the northern hemisphere, ten in Australia, eight in America and six in Africa, but it must be remembered that most of the first group arrived here not directly but via one of the three southern hemisphere continents. Ted Lousley (1961) listed a total of 529 species altogether; Bruno Ryves (1974, 1988) added at least 200 extra species of grass, and (1977) 20 extra *Lepidium*; and John Dony added a good number in his various publications. It appears that an updated list of our wool aliens has never been made, but the total would probably be around 800 taxa.

Many wool aliens are species that have special adaptations for dispersal by hairy animals (Fig. 74). These include spines, bristles, barbs or hooks derived from a great variety of organs, and involve all parts of the fruit or seed or of organs arising close to the fruits and dispersed with them. In a number of cases the fruits are borne and dispersed in tight clusters beset with hooks or barbs; these are known as burs and have arisen in widely different families. Judging from their frequency, the hooked fruits of the medicks (*Medicago*) and the spirally twisting fruit segments with backward-directed bristles of the stork's-bills (*Erodium*; Fig. 75) are perhaps the most successful at hitching a ride on a sheep. Illustrating the convergent evolution that has occurred, the hooks in *Medicago* and *Lappula* are spines on the fruits, in *Rumex brownii* are lateral teeth on the sepals, in *Marrubium* are the calyx lobes, in *Setaria* are barbs on the pedicel bristles, in *Tragus* are spines on the glumes, in *Cenchrus* are barbed spines on the inflorescence axis, in *Calotis* are barbed pappus-hairs, in *Acaena* and *Bidens* are barbs on the fruit

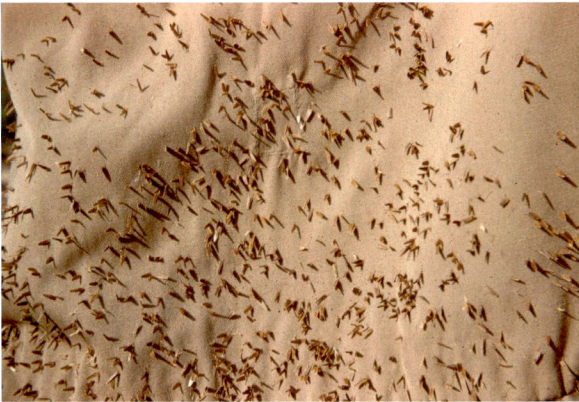

FIG 74. Barbed spines are an extremely efficient means of dispersal; these fruits are mainly from bur-marigolds (*Bidens*). (Ted Lousley, reproduced courtesy of Chris Boon)

FIG 75. Western Stork's-bill (*Erodium cygnorum*), from Western Australia, is a representative of perhaps the most characteristic and frequent genus of wool aliens. (Ted Lousley, reproduced courtesy of Chris Boon)

spines, in *Xanthium* are spines on the bracts (phyllaries) surrounding each female capitulum, and in Subterranean Clover (*Trifolium subterraneum*) are sterile, petal-less flowers mixed with the fertile ones in each inflorescence.

The cockleburs (*Xanthium*; Figs. 38 and 76) belong to the daisy family, Asteraceae, although the dull, scarcely exposed flowers do not suggest that at first sight. The male and female flowers are in separate heads (capitula); the female capitula contains only two flowers, which become surrounded by fused bracts (phyllaries), which surround each capitulum in this family. This forms a bur, which is easily detached at fruiting, and which by that time has developed strong hooked spines. These burs become very tightly entangled in wool or clothing. Three species of *Xanthium* were very characteristic wool aliens (*see* Appendix 1).

FIG 76. Spiny Cocklebur (*Xanthium spinosum*), a representative of a very frequent genus of wool aliens, develops hooked spines on its fruiting head. (CAS)

It is not surprising that the stork's-bills (*Erodium*) and *Xanthium* contribute nine of the 26 wool aliens still persisting at Flitwick (*see* below).

Some species of wool alien have seeds that can survive in the soil for many years, and in fields that were once spread with shoddy those species still reappear from time to time as the soil is turned and the seeds are exposed. Dony & Dony (1986) stated that this could happen 'as long as eight years' after the last shoddy application, but this has proved to be a considerable underestimate. Gordon Hanson told me that he recorded this phenomenon on a farm at Flitwick, Bedfordshire, which was once one of our richest areas for wool aliens (and the one on which the Donys worked), by making annual visits between 1970 and 2010 (pers. comm., 2012). The farm in question last used shoddy in 1985, but 25 years later (2010), 26 highly characteristic wool aliens were still evident.

A good number of species consistently recorded in shoddy-treated areas, and in some cases grown experimentally from shoddy, and which therefore by implication are wool aliens, are in fact natives or archaeophytes in the British Isles. These include a wide range of species, such as Fat-hen (*Chenopodium album*), Cock's-foot (*Dactylis glomerata*), Common Stork's-bill (*Erodium cicutarium*), Smooth Cat's-ear (*Hypochaeris glabra*), Fiddle Dock (*Rumex pulcher*), Black Nightshade (*Solanum nigrum*) and Hare's-foot Clover (*Trifolium arvense*). These are often species that persist in fields after the use of shoddy there, but of course it is not always known whether the persisting specimens were brought in with shoddy or existed there previously. In some cases, however, we know that the native species were exported to the former colonies long ago, possibly with sheep, and have returned here in imported wool. The best example of this is Subterranean Clover, an uncommon and inconspicuous species in the British Isles that usually grows in short turf, mainly near the sea, and has creeping stems rarely over 20 cm long. As a wool alien, however, it is a much more robust plant, forming large patches up to a metre or so across, and with larger leaves, so that at a glance it resembles a patch of White clover (*T. repens*). It is known as *T. subterraneum* var. *oxaloides*, a variant that is apparently native from southeastern Europe to the Caucasus, from where it was introduced to Australia. It was trialled as winter-green fodder at Aberystwyth in the 1920s (Chater, 2010), but as a wool alien in this country it is a casual killed by the first heavy frost. Several other species are represented as wool aliens by variants that differ noticeably from our native genotypes, e.g. Fat-hen, Common Stork's-bill and Black Nightshade.

The greatest variety of wool aliens was recorded on land treated with shoddy, because in those conditions the seeds had the greatest chance to germinate. A wide range of crops, including root crops, brassicas and Rhubarb (*Rheum* × *rhabarbarum*), as well as Hops, soft fruit and fruit trees, but not cereals, was manured with shoddy, and proved rich collecting grounds for wool-alien hunters. Certain genera or species were consistently found in areas treated with shoddy, and the field botanist was alerted to a new field of interest by spotting certain of the more conspicuous representatives. The genera *Erodium, Medicago, Trifolium* and *Xanthium*, along with various grasses, were particularly useful indicator species, with an occasional larger colourful species such as Narrow-leaved Ragwort drawing immediate attention. The eight species of medick, the ten species of stork's-bill and the 15 species of clover provided much interest as well as considerable perplexity in identification, as it is always difficult to determine aliens if their continent of origin is unknown.

Shoddy is still used today as manure in the English Rhubarb industry, which is concentrated in 70 ha of the so-called 'Rhubarb Triangle' between Wakefield, Morley and Rothwell in West Yorkshire. David Shimwell (2006) discovered a second, more recent, use of shoddy as bedding for cattle, employed in the Upper Colne Valley near Huddersfield from about 1980 onwards.

Cotton aliens

There are very few aliens that owe their presence here to the importation of cotton, in marked contrast to the situation with wool. Presumably seeds found their way into cotton by being blown into the bolls on the bushes or during the process of harvest and preparation for export. In the case of cotton, the intimate and obvious connection between the raw product and the contaminating seeds is lacking, and most species are considered merely to be 'probably cotton aliens'. The bales in which the cotton was transported were probably a bigger source of unintended alien plant introductions.

The 'classic' example of a cotton alien is the aquatic Egyptian Naiad, which occurred adjacent to a cotton mill in the Reddish Canal, Manchester, from 1883 to 1947. This and other angiosperms that might well have had a similar origin are discussed in the section on frost-sensitive plants in Chapter 12, because their survival relied on the industrially warmed water of the canal.

The only species in Appendix 1 that are considered to be cotton aliens are the very closely related Rough Bristle-grass (*Setaria verticillata*) and Adherent Bristle-grass (*S. adhaerens*), Old World grasses that are considered by many to belong to one species, and the Southern Marigold (*Tagetes minuta*), from South America. All three are more frequent as birdseed and wool aliens. Southern Marigold is usually a much taller and more branched plant than the two common garden annuals African Marigold (*T. erecta*) and French Marigold (*T. patula*), but it has unattractive tiny flower heads. The two *Setaria* species are distinctive in that the sterile bristles on the pedicels bear backward-, not forward-, directed barbs. A further species belonging to this group is Hooked Dock, from Australia, but this characteristic wool alien has not been recorded here in the post-1986 period. The only species that owes its present naturalised status to cotton imports appears to be American Pondweed (*Potamogeton epihydrus*), which has existed in unheated canals in Lancashire and Yorkshire since 1907. This species, however, is outside the scope of our book, because in 1943 it was discovered as a native in the Outer Hebrides, as one of the tiny group constituting our North American element.

Paper-making: Esparto-grass aliens

Esparto-grass (*Stipa tenacissima*; Fig. 77), often known as *Macrochloa tenacissima*, is not to be confused with *Stipa tenuissima* (now *Nassella tenuissima*), a rare wool alien increasingly recorded because of its use in garden borders. Esparto-grass is a robust, tough, densely tufted plant often over 1 m tall, common in the southern parts of the Iberian Peninsula and in northwestern Africa, whence it was imported into the British Isles, mainly Fife, for making high-quality paper. The species that accompanied it as aliens were therefore western Mediterranean plants.

Thirty-five species in Appendix 1 are noted as originating from Esparto-grass imports, but none exclusively. Birdseed, grain and wool are the main other vectors in which these species have arrived, and to which they owe their continued occurrence here. It is unlikely that any of the 64 species listed from Fife by Beattie (1962) ever became naturalised. Beattie suggested that Garden Rocket 'might even become established', but this is an archaeophyte whose presence in our flora goes back many centuries. There is a wide difference between the climates of Fife and the drier parts of the western Mediterranean, and few plants from the latter place would be expected to thrive in the former. If there had been an Esparto-grass importing business in, say, Hampshire, the list of associated aliens might have been longer (perhaps including Esparto-grass itself), with some of them becoming naturalised.

FIG 77. Esparto-grass (*Macrochloa* (*Stipa*) *tenacissima*), whose import for paper manufacture has been the source of many aliens, has never itself been recorded in the British Isles; here it is pictured in southern Spain, where it is often a dominant species on dry soils. (CAS)

Our Esparto-grass aliens not surprisingly resemble a list of species in a field botanist's notebook taken down in an area of dry lowland grassland in Andalucía or Morocco, including many common annuals, but offer few other generalisations. The plants listed by Beattie include 15 grasses, 14 composites, 9 crucifers and 8 caryophs. Perhaps more surprising is the presence of only a single legume. A few of the plants are rare European species, or even extra-European – e.g. the grasses *Vulpiella tenuis* and *Cutandia memphitica*, and the composite *Cyanopsis muricata* – and underlining the fact that most of our Esparto-grass was imported from North Africa.

Packing materials

Materials used in packing crates were a great source of unintended plant introductions, because the packaging was usually discarded only after the goods had been delivered – often a substantial distance inland from their ports of entry. Species with adhesive fruits or seeds were particularly prone to importation on all kinds of bales and other merchandise. Sacks that contain a large proportion of unwanted material such as husks, stalks, straw, soil, stones, fragments of insects, pieces of newspaper and other rubbish were another vector for untold numbers of propagules. Wooden packing materials, like crates, pallets and woodchips, would have seeds stuck to them or wedged into corners. These seeds would be unlikely to escape until the crates were broken up.

Dunnage is the name given to light material such as brushwood, mats and the like that was stowed among and beneath the cargo inside the hold of a vessel to keep it from injury by chafing with the movement of the boat and to protect it from getting wet; any lighter or less valuable articles of the cargo were also used for the same purpose, to wedge the valuable parts of the cargo into secure positions. Specialised dunnage mats used for padding the cargo would often be infested with seeds from countless voyages. These would be turned out, seeds and all, onto quays at the port of destination, and in warehouses at seaports, along railways and on other trade routes. When guano for use as phosphate fertiliser was shipped in bulk from the Chincha Islands, off the southwestern coast of Peru, the hold was lined with guano in bags, called dunnage bags, to protect the rest of the cargo and for better packing.

Sacks made of jute would be riddled with sticky or spiny seeds picked up on their journey from farm through road, rail and depot to the dock (to say nothing of their contaminated contents). Bales of cord-grass (*Spartina*) were used to cushion heavy items in the holds of ships; this may be how the American Smooth Cord-grass was introduced to the River Itchen (where it was first recorded in 1816). Seaweeds were often used as packing for transport of fresh seafood (*see* Chapter 10).

Some curious plants were used for packing. For example, the husks left over when Buckwheat was processed were once used as packaging material (Howes, 1974). An example of such use is to be found among the plant remains from the Monte Cristi shipwreck excavations. This English or Dutch trading ship, wrecked off Hispaniola in the Caribbean in the seventeenth century, was carrying a cargo that included crates of clay pipes packed in Buckwheat.

Today's packing materials are as close to seed-free as they have ever been. One could plant a great deal of expanded polystyrene, bubble wrap or corrugated paper before the first alien seedling germinated from it.

Timber aliens

Timber has been imported in large quantities from Europe and North, Central and South America, and also from as far away as Australia. Both logs and rough-sawn timber carry innumerable crevices and projections that can harbour seeds, and nine species are marked in Appendix 1 as having probably had such an origin, usually based on the coincidence of newly arrived aliens, the presence of imported timber and the absence of a more obvious vector.

The only species with no vector other than imported timber listed by Clement & Foster (1994) is Small Balsam (*Impatiens parviflora*; Fig. 78), a delicate succulent plant that is briefly discussed in the section on Indian Balsam in Chapter 11. Its natural habitat in its Asian area of distribution is damp woodland, in which it also thrives with us. It was first recorded in the wild here in 1848. Clement and Foster implicated the import of Russian timber in its introduction, but David Coombe (Coombe, 1956) believed that its 'occurrence in various timber-yards is to be attributed to transport with timber grown in local parks, rather than import with softwood from Russia'. In *The Flora of Berkshire* (1897), Druce suggested that Small Balsam might have been introduced from Asia with imported Buckwheat,

FIG 78. Small Balsam (*Impatiens parviflora*), a delicate shade-lover perhaps imported with timber or Buckwheat (*Fagopyrum esculentum*) seed from Russia after *c.* 1850. (Ruud van der Meijden, reproduced courtesy of Nelleke van der Meijden)

and this and the idea that it might be an example of dispersal with plants from nurseries are equally plausible. The species is now scattered through most of the British Isles and slowly spreading; it is locally dominant in the field layer of some damp woodland in southern England (Crawley, 2005).

The taxonomically difficult group of *Euphorbia* known as the leafy spurges are natives of Europe, mainly in the east and southeast, but are also well-naturalised weeds in North America and might have come to us from either or both areas. All except Cypress Spurge (*Euphorbia cyparissias*) are undesirable garden plants because of their quickly extending rhizomes; grain and timber are the most likely vectors. Cypress Spurge has been in the wild in the British Isles for more than 200 years and is probably a garden escape; it has been plausibly claimed as a native. In addition, we have two species (Leafy Spurge and Waldstein's Spurge *E. waldsteinii*) and three hybrids (the hybrid formed by these two species, and each of their hybrids with Cypress Spurge). They all form large, spreading patches of lime-green foliage and flowers in rough grassland by roads and similar places.

The abundant, still rapidly spreading American Willowherb (*Epilobium ciliatum*) has an uncertain origin. It might have come here from North America mixed with other more desirable plants, or as a timber alien. Two other species that we have probably gained from timber, wool and/or grain are the North American Least Pepperwort (*Lepidium virginicum*) and Rough Bent. Again, the evidence for timber transport is circumstantial.

Tanning aliens

Adrian Grenfell (1983), working with A. J. Byfield and K. L. Spurgin, discovered a rich crop of tanning aliens from Turkey during an August (probably 1983) visit to the Manor Tannery, Grampound, Cornwall. The proprietors, the Croggon brothers, explained that the cupules of Valonia Oak (*Quercus macrolepis*), imported from Turkey, were crushed and used in conjunction with Forest of Dean oak bark for the extraction of tannins used in the production of high-quality shoe leather. The spent material was used horticulturally as mulch (peat being expensive in Cornwall). The importance of this site as a source of adventives was first recognised by L. J. Margetts in the early 1970s. Nearly all the species are characteristic of Turkey, from where most acorns were imported. These aliens have usually been known as tan-bark aliens in the British literature, but as explained above, all the aliens evidently arose from the imported acorn cupules.

Although 66 species of aliens imported as contaminants of cupules have been recorded, only 22 are listed in Appendix 1, and all of these have other sources that allow their continued occurrence in our flora. They offer a contrast to the situation with Esparto-grass aliens, which proved to be a random sample of species from the native area of Esparto-grass itself, and not demonstrating any particular modes of dispersal. The tanning aliens were far from a random selection of local species because, in Grenfell's (1983) first list of 36 species, 75 per cent were legumes; no other family was represented by more than two species, and there were no grasses or composites. The notable genera involved were *Coronilla, Hymenocarpus, Medicago, Onobrychis, Ornithopus, Scorpiurus, Securigera* and *Trifolium,* including at least 19 species of *Trifolium.*

Contaminants of horticultural introductions

All gardeners will know the danger of unwittingly including pernicious weeds as contaminants of their ornamental introductions, and that care is needed in ensuring that the rootballs of woody plants and herbaceous perennials are free of the rhizomes of such undesirable plants as the native Hedge Bindweed (*Calystegia sepium*), its neophyte relative Large Bindweed (*C. silvatica*) or the archaeophyte Ground-elder, among many others. It is impossible to exclude totally unwanted seeds or other small propagules, and a number of aliens appear to have arrived in the British Isles by such a route: 41 are indicated in Appendix 1, 13 of which have no other known vector.

The presence of an alien in or around nurseries or garden centres, and in gardens in the same geographical area, is indicative of such a vector being operative. Several annual weeds come into this category; they are inconspicuous plants that are most unlikely to have been introduced intentionally. New Zealand Bitter-cress (*Cardamine corymbosa*; Fig. 79) has been with us only since 1975, but is now common in many gardens, including Buckingham Palace. It resembles our two common annual weedy native species Hairy Bitter-cress (*C. hirsuta*) and Wavy Bitter-cress (*C. flexuosa*), but differs from them in having very short, often procumbent stems with very few leaves. Although it is a more diminutive plant than either of the two natives, New Zealand Bitter-cress can be very conspicuous in its favoured habitats (trodden bare soil, cinders or cracks between paving stones) because of its broader, pure white petals.

Celandine Saxifrage (*Saxifraga cymbalaria*) is another tiny plant; it has celandine-yellow flowers that are much smaller than those of Lesser Celandine and grows in similar places to New Zealand Bitter-cress. French Speedwell (*Veronica acinifolia*) and American Speedwell (*V. peregrina*), with blue flowers no more than 3 mm across, are even more inconspicuous little weeds of flower beds. Equally uninspiring is Spotted Spurge (*Euphorbia maculata*), whose minute flowers have to be examined with a lens for one to realise that they belong to a spurge, although the trailing stems can reach 50 cm long. This species has several relatives that are weeds of greenhouses, growing in pots of other plants, but they

FIG 79. New Zealand Bitter-cress (*Cardamine corymbosa*), a relatively recent arrival in the British Isles (1975), has become common in plant nurseries, and hence gardens, especially between bricks and stones. Despite the plant's diminutive size and few flowers per cluster, the individual flowers are usually larger and the petals of a purer white than in its two common relatives Hairy Bitter-cress (*C. hirsuta*) and Wavy Bitter-cress (*C. flexuosa*). (CAS)

fall outside our scope. Pearlwort Spurrey (*Spergula morisonii*) has occurred in one locality adjacent to a nursery in East Sussex (near Tunbridge Wells) since at least 1943, and we have only one or two other records.

Some species that arrived by this route have managed to shed their garden connections. Toothed Fireweed (*Senecio minimus*), which can grow to a metre tall but has very small, unattractive flowers, has advanced onto the dunes in the Isles of Scilly, having escaped from the Abbey Gardens on Tresco. Much more common is Least Duckweed (*Lemna minuta*), which is now spreading quickly over much of the British Isles, no doubt hitching a ride with other waterweeds. Even more successful is Springbeauty (*Claytonia perfoliata*), which occurs throughout most of the British Isles, often in great abundance and far from habitation, but is rare in Ireland. A favourite habitat is in the small patches of soil retained around planted trees, where it appears to be somewhat herbicide-resistant.

Species of the genus *Oxalis*, variously called yellow- or pink-sorrels according to the colour of their flowers, are dispersed by a wide range of vegetative structures, including runners and small bulblets, as well as by seeds. At least 11 of our 14 naturalised species are dispersed along with garden ornamentals or vegetables; they vary from very common to very rare, the latter mostly being confined to Cornwall, where several species are pernicious weeds of the market gardens and bulbfields.

The willowherbs (*Epilobium*) have provided two extra aliens in the past ten years, whereas the most recent previous newcomer appeared more than 50 years ago. Both appear to have been introduced accidentally by the horticultural trade. The more interesting is Panicled Willowherb (*E. brachycarpum*), which breaks the mould of our other willowherbs in that it is an annual, the inflorescence is much branched and the stigma can be entire, as in ten of our species, or four-lobed as in our other four species. Panicled Willowherb is an unattractive weedy species, so far known only near Colchester in North Essex, but it seems likely to spread.

Transport aliens

Today, a great many aliens are carried about on vehicles. This was never truer than for our commonest neophyte, Pineappleweed, which is transported on car tyres, or for the notorious Japanese Knotweed, carried on construction vehicles. But some plants arrived here in the first instance by this method.

The species perhaps most likely to have been introduced with vehicles or on various items of baggage is Narrow-leaved Ragwort. This is a South African wool alien of long standing, but its second wave of invasion has come from northern France, perhaps involving different genotypes of this variable species. Since the opening of the Channel Tunnel in 1994, the species has become common in southeastern England and is spreading rapidly across the country. It is obviously aided, once here, by its windblown achenes. Another example might be that strange, spiny-headed, dwarf composite Jo-jo-weed, from South America, which was found in 1997 in the trampled earth of a south coast caravan site, and has since been found in Guernsey. It is well naturalised in southwestern Europe and its most likely means of arrival is on vehicles. Its bur-like capitula readily catch in clothing and the like. A second species of the genus was reported in 2012.

Other species that might have been brought in by people are three of the American species of fleabane (*Conyza*; Fig. 122). Canadian Fleabane has been here for several centuries and was probably brought direct from North America, but the other two, Guernsey Fleabane and Bilbao's Fleabane (*C. floribunda*), are South American and have reached us much more recently; they probably arrived from northern France, where they have been naturalised for far longer than here. Once arrived, they have spread far and wide, comparable with Narrow-leaved Ragwort.

More than 3 million foreign-registered vehicles enter the UK each year, replete with the seeds stuck to their tyres and embedded in the mud in their wheel arches. This represents about 1 per cent of all traffic (rising to 3 per cent in London); about half of these vehicles arrive from France, Germany or the

Netherlands, but the other half come from further afield (data from Department of Transport website).

Modern trade in the British Isles

In terms of visualising the scale of the problem of attempting to monitor imports for biological contamination (seeds, spores, roots or living individuals), it is useful to consider what we bring into the British Isles each year. We import more than 23 million tonnes of coal, 10 million tonnes of iron ore, 2 million tonnes of aggregates for construction work, 1.8 million tonnes of cement, 1.2 million tonnes of salt, and smaller quantities of countless other bulk commodities that are likely to contain viable seeds, spores or vegetative propagules. Direct importation of plant products is measured in thousands of tonnes per year (fresh weight). Of the fresh vegetables, we import more tomatoes than anything else (386,000 tonnes). Fruit imports are topped by banana (1,013 tonnes), apples (456 tonnes), oranges (275 tonnes) and grapes (246 tonnes). The challenge of estimating the numbers and specific identities of viable seeds that arrive in the British Isles stuck to the intricate and varied surfaces of this cornucopia is beyond us, but the number is most unlikely to be zero.

Horticultural imports are measured in millions of pounds of value per year, rather than by mass or number. Much the biggest market in living plants (potted plus bare-rooted) is made up of indoor plants (£121 million). This is roughly double the value of the next most important sector, which comprises forest, fruit and nut trees (£63 million) and outdoor plants (£43 million). Cut flowers are a huge business, with imported roses valued at £128 million and chrysanthemums at £115 million (these data are for 2010, from Defra website).

The unintended introductions

Three obvious points should be emphasised in conclusion. First, most species have arrived via more than one vector, sometimes concurrently but often successively in line with the changing nature of imports. Second, there are numerous examples of species arriving by one source and subsequently being dispersed within the British Isles predominantly by another. Finally, we need to admit that we have absolutely no idea how many plants have arrived in the British Isles but failed to grow big enough to be recognisable. The fact that we don't know the numbers or the identities of the failures makes it impossible to work out the traits associated with success or failure of these unintentional introductions. We should simply be grateful that so few of them became common enough to register as problems.

CHAPTER 5

Deliberate Introductions

Man, the supreme meddler, has never been quite satisfied with the world as
he found it, and as he has dabbled in rearranging it to his own design, he
has frequently created surprising and frightening situations for himself.
George Laycock, *The Alien Animals* (1966)

O nce the world was opened up to commercial shipping in the
sixteenth century, international movement of plants began in
earnest. The most conspicuous consequence for European citizens
was the appearance of a dazzling array of new foods, including potato, tomato,
maize, chilli pepper, chocolate, vanilla and pineapple, along with recreational
drugs like tobacco. Most of these newly introduced crops turned out to be benign.
They grew well enough when cossetted in gardens or in arable fields, but they
showed no tendency to escape and run rampant. Typically, it was the horticultural
introductions that led to the huge increase of alien plants in the wild.

THE PLANT HUNTERS

The great plant hunters were sent out by private collectors, commercial
horticulturalists and the directors of botanical gardens to scour the world for
valuable new species. One can read about their adventures in several popular
accounts (e.g. Musgrave *et al.*, 1998; Fry, 2012; Silvey, 2012).

It was easy to transport seeds by sea, and bulbs, corms and roots also travelled
relatively well. But plants in leaf proved to be much more of a problem, and
many, if not most, died from exposure during long sea journeys, frustrating many

commercial nurserymen and scientific and amateur botanists. The combination of salty air, lack of light (under deck), lack of fresh water and lack of sufficient care often destroyed all, or almost all, plants even in large shipments. These problems persisted for more than 200 years until a chance discovery in London in about 1829 by Nathaniel Bagshaw Ward (1791–1868), a physician with a passion for botany. He found that the ferns in his London garden in Wellclose Square, which were being poisoned by the city's air pollution, survived well if they remained enclosed in sealed glass bottles. This gave him the idea for what became known as the Wardian case, a closely fitted, glazed wooden case in which plants could be set on deck to benefit from daylight and the condensed moisture within the case, but protected from salt spray. His book, published in 1842, was entitled *On the Growth of Plants in Closely Glazed Cases*. Ward's discovery revolutionised the transport of exotic plants back to Britain.

Probably the most influential of the botanic gardens in the early days was the Chelsea Physic Garden, which was founded as a collection of medicinal plants in 1673 by the Worshipful Society of Apothecaries under the direction of Philip Miller (1722–70). The Garden is sometimes said to have been the source of several early garden escapes, like Ivy-leaved Toadflax, but there is no solid evidence for this.

The Royal Botanic Gardens at Kew evolved from modest beginnings in 1759, but were greatly extended in 1841 and they now house one of the world's largest collections of living plants, with 40,000 taxa (roughly 10 per cent of the world total of seed plants and ferns). Several species are alleged to have escaped from Kew Gardens, but again, most of these cases are fanciful. The only reasonably well-documented escapes are Gallant-soldier, which escaped in 1861 and is now a common weed of cultivated and bare ground, and California Brome (*Ceratochloa carinata*), which escaped in 1919 and is now widespread by the Thames from Chiswick to Twickenham.

The general alien flora inside botanic gardens is very similar to that of any other urban garden that has a mix of walls, beds, paths and lawns in a range of light and shade. To the standard list of plants exploiting these habitats, however, are added dozens of self-sown species that apparently never make it over the wall, and it has to be admitted that botanic gardens have not had much direct influence on the list of species becoming naturalised outside. They had enormous indirect influence, however, as they were one of the main importers of new species, and once these filtered through to the horticultural trade the route to the wild was opened.

The botanical gardens have their own internal invasions to combat. At Kew, it is Perfoliate Alexanders (*Smyrnium perfoliatum*), against which unremitting war is waged; Chelsea struggles against Honeybells (*Nothoscordum borbonicum*), Italian Lords-and-Ladies (*Arum italicum*) and Three-cornered Garlic (*Allium triquetrum*);

while Wisley (not opened until the twentieth century) battles against Few-flowered Garlic (*Allium paradoxum*), Tall Nightshade (*Solanum chenopodioides*) and Yellow Figwort (*Scrophularia vernalis*).

The Veitch Nurseries, started by John Veitch before 1808, represented the largest group of family-run plant nurseries in Europe during the nineteenth century. By the outbreak of the First World War, the firm had introduced 1,281 plants into cultivation, which were either previously unknown or newly bred varieties, but again few became naturalised in the wild.

Most people's vision of the traditional flower garden is one of borders of perennials, the so-called 'chocolate-box' image of a cottage garden. Gardens have been laid out and cherished for thousands of years, but the earliest ones of which we have any reasonable knowledge in the British Isles are from the Roman era, about 2,000 years ago. From that time onwards, alien species ('exotics') have been introduced. The first to be imported to supplement the more showy natives of the British Isles were naturally from western Europe, soon extending further afield to eastern Europe, western Asia and North Africa. Of course, any that escaped between Roman and medieval times to become part of our wild flora are what we now term archaeophytes, but none of those (herbaceous perennials or otherwise) seems to have been a solely ornamental introduction. As Stefan Buczacki in his New Naturalist *Garden Natural History* (2007) tells us, serious importation of ornamentals was commenced by 'the sailors of Tudor England' in the sixteenth century, by which time the floras of the world, notably the Americas and eastern Asia, were becoming freely available. The addition of aliens into our gardens continued at an ever-increasing pace, culminating in the 'nineteenth century flood', although it still thrives today, with an astonishing number of new species continuing to reach us. It would be easy to fill more than a chapter with these herbaceous perennials alone, but lack of space dictates illustrative examples only.

PLANTING IN THE WILD

Many plants have followed a more direct route to becoming naturalised, as they were from the start actually planted in the wild. In most cases the plantings are carried out for a justifiable purpose, such as ornamental plantings on new road verges, landscaping, nature conservation projects, or soil stabilisation schemes. Other instances can be described as innocent, albeit often controversial in the eyes of naturalists. Foremost among these are efforts at 'beautification' of the countryside. Perhaps the commonest example here is the mass plantings of daffodils, bluebells and the like by roadsides and on open spaces, especially in towns and villages, and alongside new ring roads and similar places. Quite often we can only conjecture at the reasons for 'gardening activities' in the wild; some of these acts were obviously designed simply to ascertain whether the species could survive in our conditions, or to add diversity to our flora.

There are some well-known cases of rare alien plants having been planted that have become very well established in wild places. One example is the highly exotic-looking Breckland Birthwort (*Aristolochia hirta*), from Turkey, which was discovered in 1969 in the Breckland area of Suffolk and is still thriving there; it is not known how long it has been there. Another is the strikingly beautiful Koch's Gentian (*Gentiana acaulis*), from the Alps, which exists in a few large patches on the chalk North Downs in Surrey (Fig. 80). This was originally misidentified as the related Trumpet Gentian (*G. clusii*), because in the Alps the latter is a calcicole (occurs on lime-rich soil) whereas Koch's Gentian is calcifuge (occurs on acid soil). But Tim Rich took note of its morphology rather than its ecology and detected the error. The species was discovered in 1960, but again we do not know when or by whom it was planted. On the sand dunes at Slapton in Devon, plants of Sea Daffodil (*Pancratium maritimum*; Fig. 81) have been known since 1993. This spectacular, beautifully scented bulb plant from the Mediterranean, with white flowers similar to those of a large daffodil, was probably brought back by a holidaymaker, many of whom must have been highly impressed by such a lovely plant flourishing among the sun-worshippers. It has since been found in Cornwall.

A family perhaps more prone than any other to sowing or transplantation in our countryside is the Orchidaceae. Although no orchids in Europe are more

FIG 80. Koch's Gentian (*Gentiana acaulis*) is a well-known beauty in the Alps, so makes an amazingly exotic sight on the Surrey North Downs, where it was obviously planted, at least half a century ago. (Tim Rich and the National Museum of Wales)

attractive or fascinating than the above three rare alien plants, orchids attract unparalleled interest and many examples of foreign introductions are known. The genus *Serapias* (tongue-orchids) is well represented in the Mediterranean region; its members display a range of flower structures not found in British orchids, including a virtually undivided tongue-like lower lip. In recent decades three species have been found in England: Tongue-orchid (*S. lingua*), in Guernsey in 1991 and Devon in 1998; Heart-flowered Tongue-orchid (*S. cordigera*), in Kent in 1996 and 1997; and Lesser Tongue-orchid (*S. parviflora*; Fig. 8), in Cornwall

FIG 81. Sea Daffodil (*Pancratium maritimum*) is common on the sand dunes of the Mediterranean coast, but the origin of the plants in two sites on the coasts of Devon and Cornwall remains a mystery. (CAS)

in 1989. The first two of these did not persist, but the last possibly survives to this day although, unfortunately, it is not known whether this has been unaided. Because of the extremely small size of the Cornish population, material was taken and propagated in cultivation, and then replanted to augment the wild plants. A survey in 2008 found only one plant in flower, but the precise origin of that plant is unknown. There have been no reports of flowers since then (up to the end of 2014). Orchid seeds are very light and easily transported by the wind. Suitable old industrial sites in the British Isles are often quickly colonised by various orchids whose nearest locality is several tens of kilometres away. But transport from the Mediterranean seems less likely, and the first two species mentioned above were surely introduced by man. The same is likely true of the single plants each of Bertoloni's Bee-orchid (*Ophrys bertolonii*) and Sawfly Orchid (*O. tenthredinifera*) that were found in Dorset in 1976 and 2014, respectively. But Lesser Tongue-orchid is a less certain case. The flowers are small and inconspicuous, scarcely 'exotic' in the popular sense, and apparently the species has recently been detected in Brittany, much closer than the Mediterranean.

THE PLANTS

Crop plants

Almost all of the 80 or so crop plant species in Sally Francis's *British Field Crops* (2009) are on our list of wild alien plants of the British Isles listed in Appendix 1, as would be expected of any crop widely grown and transported as seed or seed-containing produce. A small number are native species but most of the more commonly grown species are archaeophytes, as mentioned in Chapter 2, having been grown here long before 1500. Species from the Americas, however, and many of today's minor crops, are neophytes. Those few that are not yet on our list of

wild aliens in the Appendix are recently adopted or trialled species, which should
be expected in the wild in the future. Crop species are found as aliens in fields
where they have been grown (where they are known as 'volunteers'), around farm
buildings and railway sidings and, increasingly, along roadsides.

Six of the most common neophyte crops come from the New World. French
Bean and Runner Bean are popular vegetables, grown on a field scale in southern
Britain, but they are not commonly found in the wild, and then only as casuals
that survive for just a year or two. The former is very variable, with a wide range
of cultivars, and, although it is often known as a dwarf bean, there are tall
climbing varieties. Since some varieties of Runner Bean are white-flowered, the
more obvious distinctions between the two disappear. However, Runner Bean is
a perennial with tuberous roots (although usually not surviving our winter) and
has cotyledons that remain subterranean (hypogeal germination), whereas French
Bean is an annual with above-ground cotyledons (epigeal germination). It is rare
to find hypogeal and epigeal species in the same genus.

Perhaps two of our most popular vegetables are Potato and Tomato. Like the
last two species, both were introduced from the Americas about 400 years ago, but
the two *Solanum* species take to our climate much more readily. Neither species
can normally survive our winters as plants, but potato tubers are very persistent
where they have been grown or have fallen, and tomatoes readily regenerate from
seed. Germination of tomato seeds seems to be enhanced by travelling through
the gut, and they readily sprout if allowed the space on sewage farms – there are
many tales of workers at such sites harvesting the ripe fruit for consumption.
The old argument as to whether a tomato is a fruit or a vegetable is quite futile,
since, of course, it is both, like French and Runner Beans. An increasing range of
varieties of both solanums is now being grown here, but virtually all appear to fall
into the above two species.

Sunflower (Fig. 82) is a popular garden ornamental. It is also grown as a crop,
mainly as food for poultry, gamebirds, cage birds and wild birds, but also for use
in various foods for human consumption. The species is potentially the tallest
annual in our flora, despite similar claims made for Indian Balsam. Sunflower
oil is produced in only very small quantities in the British Isles, but the breeding
of more cold-tolerant varieties might change that. At present, even its growth
here as a commercial seed source is somewhat marginal. In more recent years it
has been used in mixed plantings for game and wild-bird cover and food, and its
popularity is rising for all these uses.

Jerusalem Artichoke (*Helianthus tuberosus*) is a perennial species in the same
genus as Sunflower and a minor vegetable. The tubers are harvested like potatoes
and, also like potatoes, they are very persistent in the ground, naturalised

FIG 82. Sunflower (*Helianthus annuus*) must be one of our best-known flowers, and is a valuable source of oil for cooking. It is now planted in the British Isles for this purpose and as a food source for wild birds and gamebirds: (a) commercial crop in Suffolk; (b) well-grown plant in a sewage works in Lancashire. (CAS)

colonies being common on the sites of old gardens and allotments. This species is also used in gamebird and wild-bird plantings, which will lead to more cases of naturalisation. When used as a game planting, it can be allowed its natural perennial habit.

Maize, known (confusingly to us) as corn in the USA, is much less common as a wild plant than its extent of cultivation might predict, but it is mostly grown as a vegetative forage. The grain is used in poultry food but can be produced successfully only in southeastern England, where a small amount is also used for corn on the cob (sweet corn), as it is in many sheltered private gardens. It is also being used today for game cover.

A number of other minor crops are also being utilised for the last purpose, and the list of species involved is growing. Most of these are being recruited from birdseed crops, when their use as gamebird or wild-bird cover benefits from the provision of both food and physical protection. Grasses in this category include Triticale (× *Triticosecale rimpaui*), Common Millet, Pearl or Bulrush Millet (*Pennisetum glaucum*), Canary-grass (*Phalaris canariensis*), Bulbous Canary-grass (*P. aquatica*), Great Millet (*Sorghum bicolor*), and Japanese Millet (*Echinochloa esculenta*), White Millet (*E. frumentacea*) and Shama Millet (*E. colona*). All of these except Bulbous Canary-grass are annuals, and all can produce seed, at least in favourable summers, and hence are likely to become commoner than Maize as aliens.

Species other than grasses used for game cover are also increasing in diversity. Apart from those mentioned previously, among the most frequent species found are

Buckwheat, Quinoa (*Chenopodium quinoa*), amaranths (*Amaranthus*), melilots (*Melilotus*, known as sweet clovers in the US), Lucerne (Alfalfa in the US), and Ethiopian Rape or Texsel Greens (*Brassica carinata*). These all also set seed and we should expect increasing numbers of records from the wild. All except Lucerne are annuals.

Quinoa (Fig. 83) is not only grown for game but also for human consumption. It is a Mexican and South American species that was cultivated and eaten as a cereal there from about 4,000 years ago, as one of the many species that were informally 'trialled' by people as staple food-plants before maize, wheat and rice assumed almost complete dominance. Other species of *Chenopodium*, including the native Fat-hen, were also eaten in the past and used for animal feed in Europe (including the British Isles) and America. Quinoa, however, has seen a rise in popularity in this country in the past few decades as a 'health food', mainly imported from South America. The United Nations declared 2013 as the International Year of Quinoa.

The most important fruit crop that occurs as a neophyte is Garden Strawberry (*Fragaria ananassa*). It has a complex evolutionary history involving the hybridisation of American species (the European strawberries, such as our Wild Strawberry *F. vesca*, do not figure in its ancestry at all). Garden Strawberries are abundantly fertile and very attractive to birds, and most wild occurrences probably originate from bird-sown seeds. Once established, the plants can soon develop into large patches by means of their rampant stolons.

Besides strawberries, there is an ever-increasing range of fruits being grown in the British Isles, often heralded as important health foods and equally often soon falling out of favour again. An example is the scrambling shrub Duke of Argyll's Teaplant (*Lycium barbarum*; Fig. 84), which is widespread as a naturalised plant in hedges over most of the British Isles except northwestern Scotland. Recently, its small, elongated red fruits have been promoted under the name goji berries as a health food based on their high concentration of carotenoids. The plant

FIG 83. Quinoa (*Chenopodium quinoa*), an ancient staple in parts of South America, is now being sold in health-food shops; here it is being grown as food and cover for wild birds and gamebirds. (CAS)

FIG 84. Duke of Argyll's Teaplant (*Lycium barbarum*) is a woody hedgerow scrambler from China whose red berries are sold as 'goji berries' to promote healthy eating. (CAS)

is becoming widely grown here in gardens and allotments, but is not grown commercially outside China, whence our supermarket supplies are imported.

There is a constant drive to find more and better oilseed crops. It is partly a question of which species are most favoured by our environment, and partly one of their oil chemistry, which determines the uses to which the crop is best applied. Further breeding programmes can change both. Oilseed Rape (*Brassica napus* subsp. *oleifera*) is the prime example of a crop that was hardly known here a few decades ago, but whose vivid yellow flowers now cover large areas of the countryside in spring. Its origin is discussed in Chapter 9. Conspicuous areas of escaped brassicas have been a familiar sight in yards and by roads and tracks for centuries, but, whereas 50 years ago they would most likely be Cabbage or Turnip (*B. rapa* subsp. *rapa*), today Oilseed Rape is by far the most frequently encountered. Another species of *Brassica* used as a minor oilseed crop in the British Isles is Chinese Mustard (*B. juncea*).

Oil-producing archaeophytes have been discussed in Chapter 2. In the 1990s, Abyssinian Kale (*Crambe hispanica* subsp. *abyssinica*) was introduced as an oilseed crop, and has been grown commercially since 2001; by 2004, the first records were made from the wild. Sunflower and Soyabean are two other oilseed crops that are being developed, but in these cases their seeds have long been imported to the British Isles for oil extraction or other purposes, and wild records date from a century or so ago. More recently, and perhaps more surprisingly, Honesty is being evaluated for its oil content. It is a plant that is frequently seen naturalised by roads and similar habitats, having escaped from or been dumped from gardens. If it succeeds as an oilseed crop it could become one of our commoner aliens. Despite its specific epithet *annua*, Honesty is a rare example of a biennial crop that is harvested in its second year; unsurprisingly, the development of annual varieties is a major aim for breeders. The plant's oil has specialist lubricant and medicinal applications.

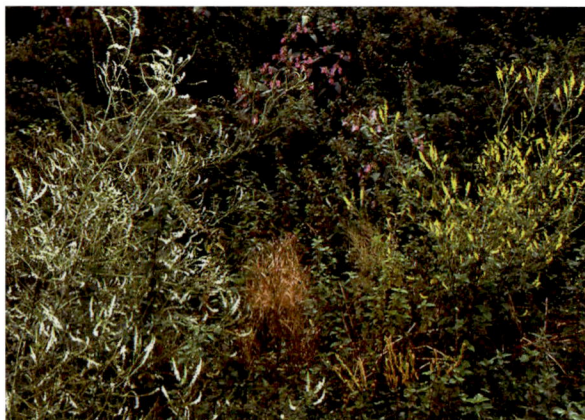

FIG 85. Ribbed Melilot (*Melilotus officinalis*) and White Melilot (*M. albus*), two common alien legumes, growing on waste ground with Indian Balsam (*Impatiens glandulifera*). (CAS)

In the past, many leguminous crops used as forage (silage or hay) or as green manure have been grown, but the range seems far narrower today. Crops now seen rather rarely include Lucerne; Ribbed Melilot (*Melilotus officinalis*), the archaeophyte Tall Melilot (*M. altissimus*) and White Melilot (*M. albus*; Fig. 85); Crimson Clover (*Trifolium incarnatum*) and Alsike Clover (*T. hybridum*); Sainfoin (*Onobrychis viciifolia*, possibly a native or an archaeophyte); and even the archaeophyte Common Vetch. The native clovers – Red Clover and White Clover – are more commonly grown, which contrasts with the situation in Europe, even as near as central France, where a wide range of species is still utilised. In the British Isles, Narrow-leaved or Blue Lupin (*Lupinus angustifolius*) and White Lupin (*L. albus*, which can be white- or blue-flowered) might become more common. Seeds of both species, as well as those of Yellow Lupin (*L. luteus*), are grown and imported as animal food pulses. Most of our grass crops are also mainly native species, the notable exception being Italian Rye-grass (*Lolium multiflorum*) and its hybrids with Perennial Rye-grass (*L. perenne*).

A specialist class of crop grown in the British Isles for only the past 30 years or so is that used for biomass, as discussed in Chapter 2. A very large range of species has been, and is being, tried for this purpose, varying from woody plants to grasses and other herbaceous perennials. Among the neophytes, most effort is being made here with silver-grasses (*Miscanthus*; Fig. 86). These are giant East Asian grasses that can grow to 3 m tall; they have strong rhizomes that are left in the ground year after year, the stems being allowed to die, desiccate and brown before harvesting takes place in late winter. The most favoured taxon is Giant Silver-grass (*M. × giganteus*), the hybrid between Chinese Silver-grass (*M. sinensis*), which is grown here as a garden ornamental, and Amur Silver-grass

FIG 86. Chinese Silver-grass (*Miscanthus sinensis*) being grown en masse as a biomass crop. (MJC)

(*M. sacchariflorus*). It is often known colloquially as 'elephant-grass', but this name is properly applied to tall grasses of the African plains. The area under *Miscanthus* is likely to grow steeply in the next decades, but it does not set seed here and its potential for naturalisation other than by persistent rhizomes is very limited.

One further specialist crop is worth mentioning because the species involved, Red Buffalo-bur (*Solanum sisymbriifolium*), is a well-known though rare casual from various sources. It is a spiny annual, with white to purple flowers, red fruits and deeply lobed leaves, which is being planted in small areas near potato crops as a nematode (eelworm) trap. This species selectively attracts potato cyst nematodes, which, however, cannot complete their life cycle in that host plant. If this trial proves successful, Red Buffalo-bur is likely to become a more common alien.

Apart from the above categories, there are many other minor crops grown on a small scale that cannot be called field crops. These include vegetables such as Salsify (*Tragopogon porrifolius*) and Sweet-potato (*Ipomoea batatas*), salad plants such as Cucumber (*Cucumis sativus*) and Coriander (an archaeophyte), herbs such as Lovage (*Levisticum officinale*) and Costmary (*Tanacetum balsamita*), spices such as Ajowan (*Trachyspermum ammi*) and Fenugreek (an archaeophyte), fruits such as Blueberry (*Vaccinium corymbosum*) and Large-flowered Tomatillo (*Physalis philadelphica*), and health and beauty products such as Clary (*Salvia sclarea*) and evening-primroses (*Oenothera*). They all contribute to our alien flora; there are very few species grown for any of these purposes that are not on our alien list, and a story could be told about each one.

Mints (*Mentha*) probably represent the culinary herb best known to the British public. This is a taxonomically complex genus, often posing problems in identification. The culinary favourites are a mixture of archaeophytes (e.g. Spear Mint *M. spicata*) and neophytes (e.g. Apple Mint *M.* × *villosa* (Fig. 87), Peppermint

FIG 87. Apple Mint (*Mentha × villosa*), a herb grown for its culinary use. Due to its rampant rhizomatous growth, it is often thrown out of gardens, when it can quickly become naturalised. (CAS)

M. × piperita and Eau de Cologne Mint *M. × piperita* var. *citrata*).

The Fig (Fig. 88) carries as much mystique about its cultivation and usage as Asparagus. Figs often grow in the wild here on inaccessible walls, railway sidings and riverbanks. The pips or 'seeds' contained within their so-called fruits are, in fact, the actual fruits, but they appear not to ripen in the British Isles to a stage when they can germinate. Hence the trees found in the wild here are the offspring of fruits imported from the Mediterranean and sold in greengrocers, not the offspring of locally grown Fig trees. Many of the localities for wild Figs are on riverbanks, often shady ones, which is in stark contrast to their arid Mediterranean habitat. This is explained by their origin: wild Figs mostly germinate from human sewage, which is deposited on city riverbanks in times of spate. The Figs seen on walls of railway cuttings and derelict buildings must result from remains discarded directly by humans or further dispersed by birds. Some of our wild Fig trees are very old – certainly more than 50 and probably over 100 years.

The promotion of healthy eating and living is leading to an ever-increasing range of minor crop species, sometimes aided by specific projects. For example,

FIG 88. Fig trees (*Ficus carica*) are often found on riverbanks, where they probably arrived in sewage outlets, but are also frequent on old walls, where they are more likely bird-sown from discarded human food; they do not set seed in the British Isles. (CAS)

citizens of Leicester of Asian origin have been encouraged to adopt allotments for growing Asiatic vegetables, herbs and spices that do not travel well from the Indian subcontinent but can be grown here with a little care. In the London area especially there are many schemes encouraging the cultivation of a wide range of plants in 'community allotments' and the like. For example, 'winter salad leaves' or 'Japanese greens' are mixtures of brassicas that are sown in autumn and eaten as salads in winter when the leaves are young. The mixtures typically consist of two or more of Borecole (*Brassica oleracea* Acephala Group), Mizuna (*Brassica rapa* var. *nipposinica*), Chinese Mustard and Garden Rocket.

Horticultural origins

For more than 100 years horticulture has been the main source of aliens in our wild flora. Since the demise or near demise of fresh aliens arriving in ballast and wool, the preponderance of aliens of horticultural origin – mostly simply garden escapes or throw-outs – has become overwhelming. Nearly two-thirds (64.3 per cent) of the neophytes listed in Appendix 1 fall into this category. If we considered only the new arrivals in, say, the past half-century, that percentage would be significantly higher. A considerable proportion of the huge wealth of species grown in ornamental gardens has found its way into the wild and, once there, has behaved in varying ways from fleeting casuals to well-naturalised and spreading perennials. The range is bewildering, but can be discussed under the headings most often employed by gardeners. At present, there are roughly 70,000 different kinds (species, hybrids and cultivars) of plants for sale in the UK, and these are listed in the Royal Horticultural Society's online Find a Plant database (available at www.rhs.org.uk), compiled from the catalogues of more than 540 nurseries.

Trees

Trees have probably the greatest effect on our landscape and wildlife of any group of plants. We do not have a large number of native species (about 35 angiosperms, excluding the microspecies of whitebeams *Sorbus*), but about 110 alien tree angiosperms are listed in Appendix 1. The conifers are covered in Chapter 10. There is also a large number of other trees that have been planted but not with sufficient prominence to be regularly recorded by field botanists. The resources for recruiting new species into our wild alien flora are already in place, and this recruitment is bound to increase in the coming years.

Tree-planting is a favourite way of commemorating national events, such as the important jubilees of the current monarch, either as individual specimen trees or as groups, avenues or even forests. Trees chosen to be in the public eye are usually spectacular and attractive, and often widely spreading in habit. In the second half of the twentieth century there developed a much keener awareness of the environment and the need for greenery, and in parallel with the Clean Air Acts much roadside planting took place, both in towns and on country trunk roads. Plantings in such areas demand narrower, less spreading trees, often fastigiate (narrow in outline) cultivars. As planting increased, so did the range of species utilised.

The New Towns Act 1946 signalled the development of several new towns, usually involving the massive expansion of small communities. Anyone who has visited one of these 23 towns, such as Milton Keynes and Telford, cannot have failed to be impressed by the many kilometres of ring roads and feeder roads that have been constructed, and by the enormous amount of tree-planting that has taken place on their verges, often on banks in cuttings that have been made to lessen the local impact of traffic noise and pollution. John Kelcey, former ecologist and linear park manager for the Milton Keynes Development Corporation, says that about a million trees were planted there each year from 1972 to 1991 – some 20 million in total (pers. comm., 2012). About 215 different species were involved, many as very numerous cultivars, and huge numbers of shrubs were also incorporated (probably about as many again). Planting was very dense, and over time natural competition caused much thinning, but these areas still act as informal arboreta, providing a great opportunity to hone one's identification skills.

Following a policy document issued by the Countryside Commission in 1987, the National Forest was commenced in 1990. This project aims to create a patchwork of forest covering more than 500 km², stretching from Needwood Forest in Staffordshire to Charnwood Forest in Leicestershire, and incorporating the towns of Burton upon Trent and Ashby de la Zouche, as well as many smaller communities. The forest is combining commercial forestry, public recreation and wildlife conservation, aiming for 33 per cent cover of the whole area through the planting of 15 million trees. By 2009, 7.8 million trees had attained a cover of 18 per cent, and it is reckoned that completion of the original aim will be reached in about 30 years' time.

Such developments have a profound effect on our environment. In the National Forest, there is about an 87:13 split in the percentage of broadleaved to coniferous species. As in virtually all of these schemes, many exotic species have been planted; Corsican Pine (*Pinus nigra* subsp. *laricio*) is one of the two commonest conifers. Moreover, there has been no attempt to use British stocks of native species such as Pedunculate Oak, and very often unfamiliar forms of our natives are evident. All the material planted at Milton Keynes was purchased from the Netherlands.

New trees being planted include extra species of genera already here – often natives – as well as novel varieties of species that we already have. For example, ash trees (*Fraxinus*) planted today are as likely to be either Narrow-leaved Ash (*F. angustifolia*; Fig. 89a) or Manna Ash (*F. ornus*; Fig. 89b), both from southern Europe, as our native Ash (*F. excelsior*). Both are susceptible to the recent outbreak of ash dieback. In recent years, Indian Horse-chestnut (*Aesculus indica*) has become very popular as a substitute for the common Horse-chestnut, because unlike the latter, it is not susceptible to the disfiguring leaf-miner moth *Cameraria* (*see* Chapter 14). Many of the birches (*Betula*) now being planted are Himalayan Birch (*B. utilis* var. *jacquemontii*), which has a much whiter bark than our native Silver Birch (*B. pendula*) and is usually also referred to as 'silver birch' for obvious reasons. It is, however, a much less graceful tree than our native species. Limes (*Tilia*) are now being used in a much wider range than our two native species and their hybrid, the Common Lime (*T.* × *europaea* = *T. platyphyllos* × *T. cordata*); the leaves of Silver-lime (*T. tomentosa*) have attractive white undersides. This general trend can be seen in many other genera, e.g. *Acer, Alnus, Malus, Populus, Prunus, Quercus, Salix* and *Sorbus*, some of which are discussed in other chapters.

Oak (*Quercus*) is a large genus containing deciduous and evergreen species, and ranging from tall forest trees to short shrubs. The commonest alien is Turkey Oak (*Q. cerris*), dealt with in Chapter 11. The two other commonly naturalised species are the North American deciduous Red Oak (*Q. rubra*) and the Mediterranean Evergreen

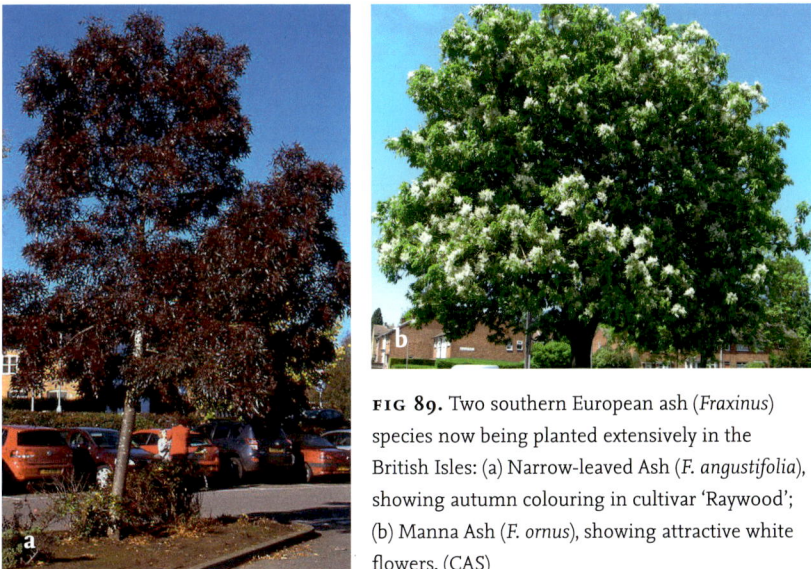

FIG 89. Two southern European ash (*Fraxinus*) species now being planted extensively in the British Isles: (a) Narrow-leaved Ash (*F. angustifolia*), showing autumn colouring in cultivar 'Raywood'; (b) Manna Ash (*F. ornus*), showing attractive white flowers. (CAS)

Oak, or Holm Oak (*Q. ilex*), both of which regenerate freely in some areas. Red Oak performs better on sandy soils and often occurs in the same places as Turkey Oak, e.g. Sherwood Forest in Nottinghamshire and the New Forest in Hampshire. Evergreen Oak is much more markedly southern in its distribution, thriving on well-drained soils whether limestone or sandstone. Notably, it has colonised St Boniface Down, a large area of chalk downs above Ventnor on the Isle of Wight, since the Second World War, the acorns having been dispersed by Rooks (*Corvus frugilegus*) and Jackdaws (*C. monedula*) into the former chalk grassland. Another area of abundance is the northern part of East Anglia, where Evergreen Oak has formed dense thickets under pines on the coastal sand dunes, especially around Holkham.

Apart from our native Alder (*Alnus glutinosa*), two other members of the genus, Grey Alder (*A. incana*) and Italian Alder (*A. cordata*), have been much planted in towns, along main roads and in small plantations. They both regenerate, but Grey Alder also produces abundant suckers, making it unsuitable for anything other than a large garden. It flowers earlier than the other two species, in January and often even before Christmas, and when it is in full catkin (especially the orange-twigged 'Aurea' cultivar) is very attractive in the winter sun. Like our native species, it prefers wet ground, and is also tolerant of exposed conditions; it is often planted in the uplands, where it suckers freely. Hybrids have been found in the wild in all three combinations of the above species.

Whitebeam (*Sorbus*) is a difficult genus in taxonomic terms, and is briefly discussed in Chapter 8. There are ten alien taxa present in the British Isles, of which Swedish Whitebeam (*S. intermedia*), with strongly pinnately lobed leaves and reddish-orange fruits, is the most common. It is well naturalised in many places throughout our region as an understorey species in light woodland on commons and the like, often in similar places to *Amelanchier* (*see* below).

In addition, many hitherto unfamiliar genera are being utilised in tree-plantings. Examples are Maidenhair-tree (*Ginkgo biloba*), Tulip-tree (*Liriodendron tulipifera*), Judas-tree (*Cercis siliquastrum*), gums (*Eucalyptus*), Foxglove-tree (*Paulownia tomentosa*; Fig. 90a), Caucasian Wingnut (*Pterocarya fraxinifolia*; Fig. 90b) and others that have not yet made it into Appendix 1.

However, there are several alien species that have been here a long time and become thoroughly naturalised. Sycamore, Walnut, Sweet Chestnut and Horse Chestnut are dealt with elsewhere in this book; probably the majority of the general public would in fact be surprised to learn that they are not natives. Some other non-native species are significant members of our wild flora and deserve some mention.

False-acacia (*Robinia pseudoacacia*; Fig. 91) is a North American tree that is much planted and well naturalised in some areas of the south. Like *Eucalyptus*, it is a reminder to us that our cool climate can be to our advantage, as in warmer

FIG 90. Two exotic trees that are becoming naturalised, mainly in urban areas: (a) Foxglove-tree (*Paulownia tomentosa*), here self-sown at the back of a London council shrub border; (b) Caucasian Wingnut (*Pterocarya fraxinifolia*), its leaves and male catkins showing its close relationship to Walnut (*Juglans regia*). (CAS)

FIG 91. The beautiful flowers of False-acacia or Locust Tree (*Robinia pseudoacacia*) provide nectar for bees and honey for us. Luckily, it does not become an invasive pest in the British Isles as it does in southern Europe. (CAS)

climates, such as the Mediterranean, members of these two genera thrive and are becoming much too common. In parts of Europe, honey from False-acacia is much prized; many will have seen the numerous advertisements for *miel d'acacia* along French roads. Despite its scientific, English and French names, 'acacia' is a misnomer, as *Robinia* is not closely related to the true *Acacia* and it is difficult to see any similarity. In America, the species is known as Locust-tree.

Members of the true *Acacia* genus, which number more than a thousand and are found mainly in Africa and Australia (although the species of those two continents have recently been split into different genera), have large sprays of little pom-poms of yellow flowers; one of these, *A. dealbata*, is particularly popular in florists' shops. Some of the species of drier regions are very spiny, but many are not and one, Australian Blackwood (*A. melanoxylon*), is locally naturalised by the sea in southwestern England. Confusingly, both the English and French common name for the genus *Acacia* is mimosa, which is the scientific name for a related but completely different genus.

London Plane (*Platanus × hispanica*) is one of our best-known trees, at least to town-dwellers, and many magnificent specimens are to be seen in our parks and squares, some more than 40 m tall. It was a seventeenth-century introduction to England but its origin, probably in France around 1650, is somewhat obscure. It has usually been considered as the hybrid between the American Plane (*P. occidentalis*) and the eastern Mediterranean Oriental Plane (*P. orientalis*), and this might well be true. However, molecular (DNA) studies have not always confirmed this, some supporting the supposition but others contradicting it. Morphologically (and molecularly), London Plane is closer to Oriental Plane, and it could be a mutation of that species. It is fertile, and does self-sow, but it is not well naturalised.

The southern beech genus (*Nothofagus*) is a southern hemisphere relative of the Fagaceae (Nothofagaceae). Two species are grown in the British Isles in plantations as timber trees, especially in the southwest. These two species, Roble (*N. obliqua*) and Rauli (*N. alpina*), both come from Chile and western Argentina, and both there and in the British Isles they have been found to hybridise. The hybrid was not described until 2004, as *N. × dodecaphleps*; the type specimen came from a tree still growing in the Forestry Commission grounds at Alice Holt Lodge, Surrey.

Mulberries (*Morus*) are commonly found in parks and large gardens, and have been grown in the British Isles since the sixteenth century. The well-known children's nursery rhyme 'Here We Go Round the Mulberry Bush' was, however, a nineteenth-century composition. The only species to thrive here is Black Mulberry (*M. nigra*), which sets abundant fruits that turn from red to purplish black when fully ripe, when they can be used in jams and jellies or eaten raw. This species is rare outside southern and central England and Wales, and does not often reproduce even there, but there are bird-sown plants naturalised by the River Thames in Middlesex. Mulberry is probably best known as the food-plant of silkworms, but the ideal species for this purpose is White Mulberry (*M. alba*), which does not thrive in England but is much grown in southern Europe as a shade tree (as is Black Mulberry) and to supply local silk industries. Attempts to utilise Black Mulberry as silkworm food led in the past to failures in silk-producing enterprises in England, but the most recent one (at Lullingstone Castle in Kent) used White Mulberry and was able to produce the silk gown for Queen Elizabeth II's wedding dress in 1947 and Coronation gown in 1953.

Laburnum is a very well-known garden ornamental, much grown despite the fact that most of the plant – especially the unripe fruits, seeds and bark – are very poisonous. Although the genus has English names like Golden Rain and Golden Chain, it is most often known by its scientific name. By far the commonest *Laburnum* in gardens and planted or self-sown in the wild is Laburnum (*L. anagyroides*), which is native to hilly country in southern central Europe. The only other species in the genus is Scottish Laburnum (*L. alpinum*), which grows in the same general area of Europe and sometimes in close proximity, and is also grown in gardens here. It has longer racemes that bear a larger number of smaller flowers, has much less hairy inflorescences, and the fruits are different

in morphology. It also flowers rather later in the season. Both in the species' native range and in cultivation, hybrids (*L. × watereri*) occur, the most desirable of which combine the larger flowers of one species with the longer racemes of the other. The hybrid has a much-reduced fertility, and both it and Scottish Laburnum are rarely found self-sowing.

Juneberry (*Amelanchier lamarckii*; Fig. 92) is related to *Sorbus* and, as a small tree with attractive large white flowers, it is justifiably popular in gardens. It is particularly valued because it flowers very early in the season, usually at the beginning of April, and also produces its fruits before almost any other tree with succulent fruits, giving rise to its English name. The small red 'fruits', botanically akin to a very small apple or haw, are much sought after by birds at a time when there are few berry-like fruits available. They commonly self-sow on acid, sandy soils and, in such places as the commons of southeastern England, regeneration takes place on a grand scale. The genus is complex; the species are not easily distinguished, and the ones that have become naturalised in the British Isles

FIG 92. Juneberry (*Amelanchier lamarckii*) is frequently bird-sown in the wild: (a) in full flower on southern heathland in April; (b) fruiting in June. (a, MJC; b, CAS)

received at least five different names until it was shown in 1970 by Fred-Günter Schroeder that they are all referable to the North American *A. lamarckii*.

Shrubs

Shrubs tell a very similar story to trees, and there are many more of them. They have been used in great numbers for amenity planting in council schemes, especially in parks and along roads, and are a striking feature of landscaping around out-of-town shopping areas. Many of them produce succulent fruit and are sought by birds in winter – there have been numerous examples of rare vagrant birds attracting large numbers of 'twitchers' to supermarket car parks. For the same reason, shrubs are often planted for wild-bird or game cover. Their attractive fruits and the ability of many to produce vigorous rhizomes, stolons or suckers means that a large number of them have become well naturalised.

An example of a more specialised use of shrubs is those planted as tall evergreen hedges, particularly pittosporums (*Pittosporum*) and daisy-bushes (*Olearia*), both from New Zealand (Fig. 93a), which are used to shelter bulbfields in the Isles of Scilly. These two quite unrelated genera display parallel evolution in a number of their vegetative structures, so that in the absence of flowers misidentifications can be easily made by careless recording. In particular, Karo (*P. crassifolium*) can be mistaken for Akeake (*O. traversii*), and Kohuhu (*P. tenuifolium*) for Akiraho (*O. paniculata*). All four are commonly used as hedging, and all are somewhat frost-sensitive and suffer badly in the rare unusually cold winters that are experienced in the Isles of Scilly. They do, however, recover. The same is true of the evergreen Late Cotoneaster (*Cotoneaster lacteus*), which is used as hedging in parts of East Anglia, for which it was promoted by Max Walters of the University Botanic Garden, Cambridge, because it provides cover and food for birds and small mammals in winter. Several other hedging plants are used in the Isles of Scilly and other maritime areas because of their salt-tolerance, e.g. New Zealand Holly (*O. macrodonta*; Fig. 93b), Hedge Hebe (*Veronica × franciscana*), Evergreen Spindle (*Euonymus japonicus*; Fig. 258), New Zealand Broadleaf (*Griselinia littoralis*), Shrub Ragwort (*Brachyglottis × jubar*) and Escallonia (*Escallonia rubra*).

There are about 76 genera that include alien species with succulent fruits; more than 50 of these are trees or shrubs, a few of which deserve further attention.

Cotoneaster (Fig. 155), which must be one of the largest and commonest genera without a well-known vernacular name, contains about 400 species, of which over 100 are cultivated in our gardens. Most of these are now known in the wild, a tribute to bird-mediated dispersal, but the abundance and variety of those species has risen very rapidly in quite recent times. In the first edition of Clapham *et al.*'s *Flora of the British Isles* (1952), just four alien species were treated, and by the time

FIG 93. Species of *Olearia* and *Pittosporum* are New Zealand evergreen shrubs that are not fully hardy in most of the British Isles but thrive by the sea in the extreme west and southwest: (a) *Olearia–Pittosporum* hedges protecting bulb fields in the Isles of Scilly; (b) attractive flowers and holly-like leaves of New Zealand Holly (*O. macrodonta*) used as hedging on the Isle of Skye. (CAS)

of the third edition (1987) this had risen to only nine. But in the first edition of Stace's *New Flora of the British Isles* (1991), 44 species were included, rising to 67 in the second (1997) and 85 in the third edition (2010). This steep rise is partly due to the increased numbers of species grown in gardens and by councils, partly to increased recording of garden escapes by field botanists, and partly to a decrease in grazing, allowing grassland colonisation in many areas. This latter tendency threatens several important sites of natural history interest. Of the 85 alien cotoneasters in Appendix 1, 27 are considered to be naturalised and the rest are survivors, i.e. species that have been (presumably bird-) sown into the wild but are not yet reproducing there. *Cotoneaster* is a very diverse genus, varying from dwarf creeping shrubs to trees up to 18 m tall, and with a wide range of branching patterns. They therefore fit into many different planting schemes and wild habitats. The 'berry' colour (they are not true berries, but effectively tiny apples) ranges from red to orange and yellow, and black to purple, and the fruiting season from July to late winter. *Cotoneaster* species occur in all sorts of rough ground, especially by roads and railways, in quarries and other 'brown-field' sites, and in hedges and on walls. Their abundance and variety is bound to continue to rise.

Other genera with a range of naturalised 'berry'-bearing species are the barberries (*Berberis*), dogwoods (*Cornus*), honeysuckles (*Lonicera*), roses (*Rosa*) and brambles (*Rubus*). The first two are commonly planted en masse by councils, forming dense thickets that are important to wildlife and provide abundant fruits for dispersal.

Lonicera is a variable genus, with both climbing and stiffly upright species, deciduous and evergreen leaves, and berries that can be red, orange, black or violet in colour. Himalayan Honeysuckle (*Leycesteria formosa*; Fig. 264), also known as Pheasant-berry, is not a honeysuckle but is in the same family (Caprifoliaceae). It is a soft-wooded shrub up to 2 m tall with pendent inflorescences. The purplish-black berries hanging in purple-bracted clusters are very juicy and extremely attractive to birds, for which it is planted on estates. Any garden nearby will soon display bird-sown examples, and because of its unusual appearance this species is one of those most often taken to botanists for identification by members of the general public, often fearing it to be poisonous.

The British Isles has a good range of native roses, but one alien that stands out is Japanese Rose (*Rosa rugosa*; Fig. 94). This species has deliciously scented red or white flowers up to 9 cm across and has very weak spines, making it our most user-friendly rose. It is much planted by councils along roads and in town shrub borders, although we do not yet have the many miles of main road verges and motorway central reservations mass-planted with it as is so common on the Continent. The hips are massive globose, bright red organs up to 2.5 cm across, and in addition the species spreads vigorously by rhizomes. It prefers well-drained soil, and is characteristic of sand dunes where, once introduced, it spreads very rapidly, potentially becoming a pest in some places. The brambles (*Rubus*) form another varied genus, some species of which are unarmed. One such is Chinese Bramble (*R. tricolor*), a dwarf evergreen shrub with bristly stems and creeping and arching stolons. This is often used as mass-planting ground cover, and its abundant orange fruits ensure that it gets dispersed by birds. At the other

FIG 94. Japanese Rose (*Rosa rugosa*) is one of the most spectacular and sweetly scented roses. It is much planted by councils and transport authorities, and becomes very well naturalised on dunes and other sandy ground from its fruits and rhizomes. (CAS)

FIG 95. Billard's Bridewort (*Spiraea × billardii*), a hybrid between Steeple-bush (*S. douglasii*) and Pale Bridewort (*S. alba*). It spreads quickly by its rhizomes to form extensive patches, often causing it to be thrown out of gardens. (CAS)

end of the growth-habit scale is White-stemmed Bramble (*R. cockburnianus*), which has spiny, white-bloomed stems that are up to 5 m long and black fruits, and is planted for winter colour. Where established, this plant scrambles aggressively over other vegetation. In 2009, a hybrid between this and Raspberry (*R. idaeus*) was found in East Sussex, the first time it had been seen anywhere in the wild.

But there are many shrubs that rarely or never self-sow, yet have become thoroughly naturalised by vegetative propagation. Two examples may be quoted. The brideworts (*Spiraea*) are shrubby Rosaceae, of which the most commonly naturalised are the three often confused and closely related species Bridewort (*S. salicifolia*), Steeple-bush (*S. douglasii*) and Pale Bridewort (*S. alba*), along with their three possible hybrid combinations (Fig. 95). The European Bridewort has been grown here since the seventeenth century and was first recorded in the wild in 1805, but the other two species are North American, and they and their garden hybrids were not introduced into gardens until later; their earliest dates from the wild here are in the second half of the nineteenth century. Nevertheless, field botanists were very slow to realise that Bridewort had been largely replaced in gardens, and therefore in the wild, by the two American species. In the first edition of CTW, only Bridewort was mentioned, yet by then it was far from the most commonly encountered taxon. Today, the species is known for certain only from Scotland, but many erroneous records persist in the archives. All these taxa are vigorously rhizomatous and can form dense continuous thickets. The two most frequently encountered taxa outside gardens are *S. × pseudosalicifolia* in England and *S. × rosalba* in Scotland

Stag's-horn Sumach (*Rhus typhina*), from North America, is also rhizomatous but puts up isolated suckers at various distances from the parent stem. It has become particularly characteristic of railway embankments and cuttings, perhaps because it escapes eradication in such places. The dense purple inflorescences, held high on the plant, coupled with its brilliant autumn coloration, make it a favourite garden plant that all too soon becomes too oppressive, at which point it gets thrown over the garden fence.

Some conspicuous shrubs that demand attention and are very persistent scarcely reproduce at all, so are classed as survivors. Tamarisk (*Tamarix gallica*) is one such species. Despite its lack of reproduction, it has potential as part of our wild flora; for example, the alien Tamarisk Plume moth (*Agdistis tamaricis*) was discovered on a Tamarisk tree in Jersey in 2007, and five species of fungi are confined to the species. Tamarisk grows best in conditions that mimic its southwestern European habitats: sandy ground near the sea. It is much planted in seaside towns, especially along the south coast of England. The 1919 novel *Tamarisk Town* by Sheila Kaye-Smith is based in Hastings, East Sussex.

Herbaceous perennials
Most of the genera from western or central Europe that were introduced in order to enrich our gardens have native representatives here, but the foreign species are either more showy or offer a wider range of habit or flower colour. Prominent examples are anemones (*Anemone*), bellflowers (*Campanula*), knapweeds (*Centaurea*), blue-sow-thistles (*Cicerbita*), eryngos (*Eryngium*), crane's-bills (*Geranium*), hellebores (*Helleborus*), peas (*Lathyrus*), oxeye daisies (*Leucanthemum*), mallows (*Malva*), sages (*Salvia*), scabiouses (*Scabiosa*), campions (*Silene*), meadow-rues (*Thalictrum*) and valerians (*Valeriana*).

Campanula and *Geranium* are two large genera with a very wide range of species, and surely few reasonably sized gardens are without at least one representative of each. Both have several well-naturalised species in varied habitats throughout the British Isles. *Campanula* offers many growth forms adapted to different situations. There are tall, erect species such as Peach-leaved Bellflower (*C. persicifolia*), which was so well naturalised more than a century ago that some considered it to be a native, and Milky Bellflower (*C. lactiflora*), which prefers a moist climate and is commonest in Scotland, where it can reach 2 m tall. At the other extreme are creeping (or hanging) species that are at home on walls and rocky places, such as the gardener's so-called 'port and posh', Trailing Bellflower (*C. poscharskyana*) and Adria Bellflower (*C. portenschlagiana*), both from the Adriatic coast. *Geranium* also exhibits a wide range of habitat requirements. Some plants do best in full sun, such as the truly magnificent, albeit ineptly named, Purple Crane's-bill (*G. × magnificum*; Fig. 96), which is often naturalised in roadside verges with tall grasses. Others are shorter, such as French Crane's-bill (*G. endressii*) and Pencilled Crane's-bill (*G. versicolor*). As common as both of these species in gardens and in the wild is their garden hybrid, *G. × oxonianum*.

FIG 96. Purple Crane's-bill (*Geranium × magnificum*), with an inept English but highly appropriate scientific name, is a very popular garden hybrid that has become well naturalised on grassy roadsides in many places. (CAS)

Cambridge also has its eponymous hybrid, *G. × cantabrigiense*, a garden cross between Rock Crane's-bill (*G. macrorrhizum*) and Dalmatian Crane's-bill (*G. dalmaticum*).

Several herbaceous perennial imports come from the European mountain ranges. For instance, the beautiful but sickly scented Perennial Cornflower (*Centaurea montana*), the late-winter-flowering White Butterbur (*Petasites albus*) and the tall French Meadow-rue (*Thalictrum aquilegiifolium*, which despite its lack of petals has pink to lilac plumes of flowers owing to its coloured stamens) are natives of the Alps; Pyrenean Valerian (*Valeriana pyrenaica*) and the above-mentioned French Crane's-bill are Pyrenean; and Fox-and-cubs (*Pilosella aurantiaca*) is from the Carpathians. Others are from Mediterranean islands, such as Corsican Hellebore (*Helleborus argutifolius*) from Corsica, Purple Toadflax (*Linaria purpurea*) from Sicily and Mossy Sandwort (*Arenaria balearica*) from Mallorca. Despite their varied origins, all the above thrive in our lowland Atlantic conditions.

Genera not native in the British Isles have also been introduced from the same areas, including buck's-beard (*Aruncus*), astrantia (*Astrantia*), blue cupidone (*Catananche*), goat's-rue (*Galega*), dame's-violet (*Hesperis*), cotton thistles (*Onopordum*), peonies (*Paeonia*), green alkanet (*Pentaglottis*), sages (*Phlomis*), crown vetches (*Securigera*) and periwinkles (*Vinca*). Some members of these, notably Goat's-rue (*G. officinalis*), Dame's-violet (*H. matronalis*) and Green Alkanet (*Pentaglottis sempervirens*), thrive and are as well established as any native species; in fact, the latter two are in our top 52 neophytes (Chapter 11). Masterwort (*Astrantia major*) is much less common than these, but nevertheless was once considered as possibly native; its compact, broadly bracted heads make it an extremely distinctive umbellifer. Peony (*Paeonia mascula*) has been well established on Steep Holm, an island in the Bristol Channel, since at least 1803, possibly dating from many centuries earlier when a priory existed there, as the plant was used medicinally. Apart from a small patch on nearby Flat Holm (dating from 1982), there are no other confirmed sites for it, but it is one of those plants over-recorded because in older Floras it was the only species of its genus that was covered. As far as we know, all other records are for Garden Peony (*Paeonia officinalis*), the well-known garden plant.

Moving further afield, new species or genera were introduced from southeastern Europe and southwestern Asia up to the Caucasus. The following are examples of alien species in British genera: tall, bright yellow-flowered Fern-leaf Yarrow (*Achillea filipendulina*); Great Forget-me-not (*Brunnera macrophylla*), a short, clump-forming perennial with heart-shaped leaves and very bright blue forget-me-not-like flowers; the majestic Greater Sea-kale (*Crambe cordifolia*), growing up to 2 m and looking nothing like our native maritime Sea-kale (*C. maritima*); the aptly named Giant Hogweed (*Heracleum mantegazzianum*; Fig. 97), which can grow to 6 m and smother surrounding vegetation with its huge leaves, and which can be a danger to the public if its ability to hypersensitise the skin to sunlight, causing severe blistering, is not respected; the popular Honesty, with striking flowers and fruits; the yellow-flowered Dotted Loosestrife (*Lysimachia punctata*; (Fig. 253); the giant-flowered Oriental Poppy (*Papaver pseudoorientale*); several comfreys (*Symphytum*), including Creeping Comfrey (*S. grandiflorum*), which soon invades a garden border and is frequently thrown out, only to take over its new roadside site, and White Comfrey (*S. orientale*; Fig. 98), surely the most beautiful

FIG 97. Giant Hogweed (*Heracleum mantegazzianum*) is an undesirable alien because of its severely sun-sensitising sap. (CAS)

FIG 98. White Comfrey (*Symphytum orientale*), surely the most attractive member of the genus, is well naturalised due to its persistent roots and fertile nutlets. (CAS)

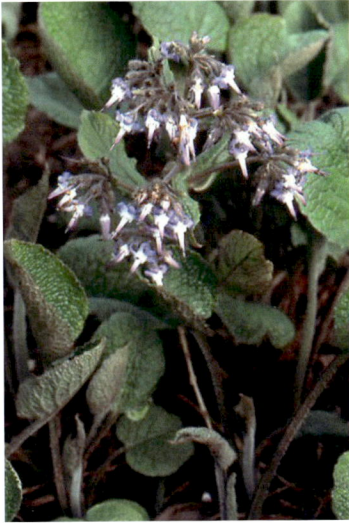

FIG 99. Abraham-Isaac-Jacob (*Trachystemon orientalis*) is a rhizomatous shade-lover whose name is derived from the various shades of blue, purple and white combined in each flower. (CAS)

of its genus, especially as it flowers in very early spring; and Abraham-Isaac-Jacob (*Trachystemon orientalis*; Fig. 99), which must be one of the most effective ground covers available and additionally has conspicuous blue flowers in late winter. The distinctive vernacular name of the last, also used in Germany, has been applied to other members of the Boraginaceae, such as comfreys and lungworts whose flowers tend to change colour as they mature. In the case of *Trachystemon*, the strange flowers, with turned-back petals and long, extended stamens, present a mixture of dark and light blue, white, lilac and violet all at the same time.

A whole new range of plants was introduced to gardens when eastern Asia – especially Japan and China – was scoured for worthy garden plants from the mid-eighteenth century onwards. Well-loved garden perennials that have escaped into the wild include false-buck's-beards (*Astilbe*, in Saxifragaceae), often confused, as its name suggests, with the buck's-beard (*Aruncus*, in Rosaceae); elephant-ears (*Bergenia*), covering the ground with large glossy leaves and displaying welcome pink flowers in winter; Giant Meadowsweet (*Filipendula camtschatica*), living up to its name and with some other unknown species spawning the red-flowered *F. × purpurea*; two species of leopardplant (*Ligularia*), giant relatives of the ragworts (*Senecio*); the tall, handsome plume-poppies (*Macleaya*), so called because they belong to the Papaveraceae, but have no petals and virtually no resemblance to any poppy; and Giant Butterbur (*Petasites japonicus*; Fig. 139), the largest species in its genus, with creamy flowers. Weeds came too. The vector for Chinese Mugwort (*Artemisia verlotiorum*) is unknown, but it is a dull plant that does not flower until late autumn and so cannot have been introduced for its appearance. Unlike its relatives the archaeophytes Mugwort and Wormwood, it is rhizomatous and spreads vigorously, though not far. The dreaded Japanese Knotweed and Giant Knotweed (*Fallopia sachalinensis*) are fully covered in Chapter 9.

South African plants are very poorly represented among our naturalised dicotyledons, presumably because of our unsuitable climate, although quite the reverse is true of the showy-petalled monocotyledons (*see* below). Among the former, two daisy-like composites stand out: Treasureflower (*Gazania rigens*) and Cape Daisy (*Osteospermum jucundum*). Both open their flowers fully only in strong sun, of which more is available in South Africa than in the British Isles. They are mostly naturalised here in the extreme south.

Australia is similarly poorly represented, and for the same reasons. One garden favourite is Strawflower (*Xerochrysum bracteatum*), often known as Everlastingflower, but this is scarcely naturalised with us. There are similarly few species from New Zealand, despite its climate being more similar to ours. However, with no disrespect intended, there are relatively few very showy herbaceous perennials in that flora, but the many New Zealand shrubs that are grown here and have become naturalised make up for this, as do some ornamental grasses and sedges (*see* below). One dicotyledon from New Zealand that is established here but in only one spot in Shetland is New Zealand Everlastingflower (*Helichrysum bellidioides*). As its name suggests, it resembles our common daisy, but that is misleading as the white rays are in fact bracts. It is actually technically a shrub, as its thin, creeping stems are woody. Another is Toothed Fireweed (*Senecio minimus*) a weedy-looking plant growing up to 1 m but with very small flower heads, which is becoming naturalised on dunes in the Isles of Scilly. It was originally a wool alien, but it probably owes its existence in the Isles of Scilly to its contamination of imported horticultural specimens. There are, on the other hand, many European species that are very successful aliens in New Zealand.

In recent years more plants from these southern hemisphere countries have been grown in the mild southwestern parts of the British Isles, but it is to be expected that only a relatively small proportion will become naturalised other than very locally. The same is true of plants from the Canary Islands, such as Cineraria (*Pericallis hybrida*) and Giant Viper's-bugloss (*Echium pininana*).

In contrast to plants from South Africa and the Antipodes, American species have made a huge impact on both our garden scene and our naturalised flora. This increasingly applies to South as well as North America, because both include a much greater extent of temperate climate than either of the other regions. One only has to consider the range of perennials in the average garden, or survey the vast areas of waste ground, often alongside railways, covered in Michaelmas-daisies and goldenrods, to appreciate the North American influence. Composites (Asteraceae) are particularly well represented by, among others, pearly everlastings (*Anaphalis*), Michaelmas-daisies (*Aster*), tickseeds (*Coreopsis*), dahlias (*Dahlia*), fleabanes (*Erigeron*), gumplants (*Grindelia*), sneezeweeds (*Helenium*), sunflowers (*Helianthus*), coneflowers (*Rudbeckia*; Fig. 100a) and goldenrods (*Solidago*; Fig. 100b).

FIG 100. Two tall yellow-flowered North American composites that are common garden plants and have become part of the alien flora of the British Isles: (a) Coneflower (*Rudbeckia laciniata*); (b) Early Goldenrod (*Solidago gigantea*). (CAS)

At least seven American species of *Aster* are naturalised in the British Isles, ranging from the spectacular insect-attracting Hairy Michaelmas-daisy (*A. novae-angliae*) to the decidedly dull-looking Delicate Michaelmas-daisy (*A. concinnus*), the reasons for whose importation must remain a puzzle. Coastal Gumplant (*Grindelia stricta*), from the coast of northwestern USA, has been well naturalised since 1961 on the sea cliffs at Whitby, Yorkshire, where it has been recorded by many people as at least three different species. Its determination as the species *G. stricta* was made by Meredith Lane, the American specialist on the genus, in 1990, which should settle any doubts.

Apart from the composites, several other very important showy garden perennials originated in America. Pride of place must go to the phloxes (*Phlox*) and lupins (*Lupinus*), the former rarely and only marginally naturalised here, and the latter very well established in many places (*see* Chapter 9). In addition, one might mention bleeding-hearts (*Dicentra*), Coralbells (*Heuchera sanguinea*), evening-primroses (*Oenothera*, which are mostly biennials), and White Burnet (*Sanguisorba canadensis*). Two close relatives of Coralbells are Fringe-cups (*Tellima grandiflora*) and Pick-a-back-plant (*Tolmiea menziesii*). These do not draw attention from a distance, but close up their weird flowers are fascinating. The latter sprouts new plantlets at the junction of petiole and leaf, by which the ground can become densely carpeted by one clone. It does not set seed, at least in my garden, whereas Fringe-cups does so abundantly.

Naturalised South American imports include the nasturtiums (*Tropaeolum*), with both annuals and perennials; yellow-sorrels and pink-sorrels (*Oxalis*); several vervains (*Verbena*), especially the now extremely popular and increasingly naturalised Argentine Vervain (*V. bonariensis*); and Magellan Ragwort (*Senecio smithii*; Fig. 234), a beautiful white-rayed species from the extreme south, now well established in the extreme north of Scotland.

Annuals

Many plants grown as annuals in the British Isles are, in fact, perennials that either cannot survive our winters or do not perform reliably well in the second year. Wallflower (*Erysimum cheiri*), Snapdragon (*Antirrhinum majus*) and California Poppy (*Eschscholzia californica*; Fig. 101) are just three of the many in this category. Some of these species when naturalised can last for many years before they die, and sometimes become large and floriferous in the process. Strict annuals by definition can reproduce only by seed, and they often quickly produce large quantities of it. For this reason, annual aliens are commonly found on pavements or rough verges around flower beds where they have been cultivated, and on rubbish tips or roadsides where the plants have been discarded.

Like their perennial counterparts, annual horticultural aliens in the British Isles are mainly natives of Europe or America. European examples include such traditional favourites as larkspurs (*Consolida*), candytufts (*Iberis*), peas (*Lathyrus*), Sweet Alison (*Lobularia maritima*), Virginia Stock (*Malcolmia maritima*), Night-

FIG 101. California Poppy (*Eschscholzia californica*), a popular garden plant, is usually found as a casual, but in the extreme south this perennial is well naturalised. (CAS)

scented Stock (*Matthiola longipetala*), Love-in-a-mist (*Nigella damascena*) and various campions or catchflies (*Silene*). Some of these genera – e.g. *Iberis, Lathyrus, Matthiola* and *Silene* – also contain equally popular perennial representatives. The white-flowered Sweet Alison was once a feature of virtually every annual display, usually at the front of a border and alternating with blue Garden Lobelia (*Lobelia erinus*; *see* below). It is a common casual from this source, but over the past two centuries it has become well naturalised on sandy ground near the sea in southern England, where it finds a match for its native Mediterranean habitat.

Probably the most popular annuals from North America are Clarkia (*Clarkia unguiculata*) and Godetia (*Clarkia amoena*), Mexican Aster (*Cosmos bipinnatus*), California Poppy (Fig. 101), Meadow-foam (*Limnanthes douglasii*, often called Poached-egg Plant because of its pale petals that are deep yellow towards their bases), Baby-blue-eyes (*Nemophila menziesii*), various sweet tobaccos (*Nicotiana*) and Phacelia (*Phacelia tanacetifolia*; Fig. 102). Like the European examples, these plants offer a wide variety of habit and colour, covering the whole range of reds, blues, yellows and white, and providing colourful displays in gardens and sometimes on tips. Phacelia is now grown in fields as a nectar source and has become a frequent arable weed in some areas.

Mexican genera include Flossflower (*Ageratum houstonianum*) and the misleadingly named African Marigold and French Marigold, all extremely

FIG 102. Phacelia (*Phacelia tanacetifolia*) is now often found as an alien because of its increasing cultivation for a variety of uses, and as a garden ornamental that is very attractive to bees. (CAS)

popular in gardens but rather uncommon casuals in the wild. From South America come Petunia (*Petunia × hybrida*), a garden hybrid of two tropical species, despite which it is frequently persistent in sheltered spots over here; and Nasturtium, a species of a more complex hybrid origin from Peruvian parents.

Some much-grown annuals frequently found as casuals come from elsewhere: the multi-coloured Annual Toadflax (*Linaria maroccana*), from Morocco; Nemesia (*Nemesia strumosa*) and the common bedding plant Garden Lobelia (*Lobelia erinus*), the latter often self-sowing from year to year, from South Africa; and China Aster (*Callistephus chinensis*), from China.

In recent decades, annual species, or perennials treated as annuals, have become particularly popular grown in windowboxes and hanging baskets, probably encouraged by television gardening programmes and because more people now live in houses or flats with small or no gardens. These plants have a long flowering period over the summer, and often set copious seed that falls to the ground and grows up on grass verges, and in gutters and pavement cracks. At first, the species so grown were the above bedding annuals, often as specially bred varieties with suitable attributes (such as a hanging rather than upright posture), for example, Garden Lobelia, Nemesia, Baby-blue-eyes, Nasturtium and, particularly, Petunia. But in recent years a wider range of genera has been imported specifically for such containers, and several of these are now being recorded as casual escapes. Foremost among these are Annual Marguerite (*Mauranthemum paludosum*), with large white and yellow daisy-like flowers, from southeastern Europe; Fern-leaved Beggarticks, from Mexico, with finely divided leaves and bright yellow flower heads bearing large ray florets, unlike the usual forms of our two native bur-marigolds; Swan River Daisy (*Brachyscome iberidifolia* – often misspelt '*Brachycome*' and in gardening circles usually going by that name rather than the Australian vernacular; Fig. 103a), from Australia, with finely divided leaves and daisy-like flower heads bearing usually blue to purple ray florets; and three South African genera. These are Silver-bush Everlastingflower (*Helichrysum petiolare*), grown for its silver-haired leaves and long, spreading stems, and usually not flowering before it is frosted; Kingfisher Daisy (*Felicia bergeriana*), so called because of its bright blue ray florets and rich yellow disc florets, and with leaves that are much less dissected than in *Brachyscome*; and Bacopa (*Sutera cordata*; Fig. 103b), with stiffly trailing stems and abundant flowers with five white to pink or mauve lobes. Several of the above are potentially perennials, and the *Helichrysum* and *Sutera* are woody shrubs. All are normally killed by our first frosts, but Silver-bush Everlastingflower has become naturalised in sheltered places in the Isles of Scilly and the Channel Islands, and is now becoming a pest in the former (Rosemary Parslow, pers. comm., 2015).

FIG 103. Two ornamentals now very commonly grown, particularly in hanging baskets, from which viable seed is scattered: (a) Swan River Daisy (*Brachyscome iberidifolia*), from Australia, more usually known as Brachycome; (b) Bacopa (*Sutera cordata*), from South Africa. (CAS)

Petaloid monocotyledons

The term 'petaloid monocotyledons' refers to those herbaceous monocotyledons that possess showy, obvious perianth segments (usually three sepals and three similar petals, together called tepals) and are placed in the families Liliaceae (in the broadest sense), Iridaceae and Orchidaceae. The Liliaceae have now been subdivided into many smaller families, five of which contain aliens in our flora that together with members of the Iridaceae are the subjects of this section (Lesser Tongue-orchid, discussed above, is the only orchid that qualifies – marginally – for inclusion in this book). Many of the species are extremely well-known garden plants. A good number are cultivated on a field scale in market gardens for the cut-flower trade, and others are planted in large-scale schemes to brighten up parks, roadsides and so on. A minority of the species reproduce readily by seed, and can become locally dispersed in that way, but the majority rely on vegetative growth and have very limited ability to spread, except by people's activities. They are often merely survivors, but can remain vigorous for a long time. In market gardens in southwestern England, for example, many taxa persist in field corners or neglected fields long after their deliberate cultivation has ceased. We can conveniently consider the main genera concerned in three groups according to their subterranean organs.

Bulb-bearing genera are purchased in enormous numbers every year, often to be planted out on a large scale, and 19 genera are represented in our alien flora. The best known and most thoroughly naturalised are the European genera onions (*Allium*), snowdrops (*Galanthus*), bluebells (*Hyacinthoides*), lilies (*Lilium*), grape-hyacinths (*Muscari*), daffodils (*Narcissus*), squills (*Scilla*) and tulips (*Tulipa*).

Bluebells (*see* Chapter 9) set abundant seed and are thus effectively spread over short distances by a primitive censer mechanism, whereby the stiff, dry stems bearing opened capsules flick the seeds out when brushed by people or animals. This can lead to dense colonisation of an area, but it takes time: it is reputed that between five and 20 years elapse between germination and flowering. Species in the related genus *Scilla* (which now includes *Chionodoxa*) also set seed readily and flowering occurs after a few years only. Another blue-flowered genus is *Muscari*, whose members self-sow prodigiously and become invasive pests in some gardens; they are often thrown out and are consequently frequently found on rubbish dumps and along roadsides. The most frequently naturalised species is Garden Grape-hyacinth (*Muscari armeniacum*), in which the upper flowers are actually sterile, and serve merely as an attractant. In another naturalised species, Tassel Hyacinth (*M. comosum*), the sterile and fertile flowers are starkly different in appearance, the upper sterile ones being bright blue and obviously much reduced in structure, while the lower ones are brown, yet are the ones that are fertile and offer nectar to pollinators. Tassel Hyacinth is rarely naturalised here, but in the Mediterranean it is one of the commonest weeds of arable land and is not destroyed by annual ploughing. It is also common a long way north in France.

Galanthus species are often abundantly naturalised in woodland, so much so that the commonest one, Snowdrop (*G. nivalis*), was considered as 'probably native' in British Floras right up until 1962. *Galanthus* spread by seed and multiplying bulbs, but some varieties of Snowdrop are doubles (*flore pleno*) and have no reproductive parts in their flowers. Before 1990, only that species was included in our Floras, but we now know that five species and three hybrids are naturalised in the British Isles. One of them, Queen Olga's Snowdrop (*G. reginae-olgae*), usually comes into flower in autumn, giving rise almost annually to letters to newspapers suggesting that spring has come early.

Exactly two-thirds of our 21 species of *Allium* are aliens, although some are extremely well established. All of them smell of some variation of onion or garlic when freshly crushed. The most distinctive and commonest is the white-flowered Three-cornered Garlic (Figs 238 and 242), which is extremely abundant in parts of the southwest. It occurs in cultivated fields, where, as with Tassel Hyacinth in the Mediterranean region, ploughing propagates and disperses the bulbs rather than suppresses them. It has spread into field borders and hedgerows, so much so that it is now a major feature of the spring flora. It closely resembles our native Ramsons (*A. ursinum*) from a distance, but differs in having much narrower leaves that are not narrowed to a petiole at the base, and sharply three-angled flowering stems. The other three species that can be described as locally common differ from Three-cornered Garlic in that the flower head often consists of a mixture of flowers and small bulblets ('bulbils'), which are effective agents of dispersal. All three can have either a mixed head of flowers and bulbils, or one of flowers alone, and two of the species can lack flowers altogether and have only compact heads of bulbils. In the case of Few-flowered Garlic (Fig. 104), the commonest type has heads with just one white flower mixed in with the bulbils; this species can be a difficult pest of shady flower borders, and thrives in the wild. Although most species of the genus have white to pink flowers, Yellow Garlic (*A. moly*) is an attractive yellow-flowered plant without any bulbils and with broad leaves.

Two species of *Lilium* are very well naturalised; both fall into the 'Turk's-cap' group, in which the pendulous flowers have divergent tepals that are rolled back towards the flower stalk. Martagon Lily is a handsome plant up to 1.5 m tall, with a long inflorescence of sickly scented purple flowers. It is abundant in the European mountains, and with us it is so well established in woods that in

FIG 104. Few-flowered Garlic (*Allium paradoxum*) typically bears only one or few flowers mixed with many bulbils per inflorescence; dispersal of the bulbils can cause this species to become a pest of flower borders and woodland. (MJC)

the past some considered it to be native. Pyrenean Lily (*L. pyrenaicum*; Fig. 237) is a shorter but equally spectacular plant with larger, deep yellow flowers and an even sweeter and sicklier scent. It prefers moister conditions and is commoner in the north.

Tulips form another popular garden genus. The European Wild Tulip (*Tulipa sylvestris*) is a rather sombre species with pale yellow flowers; it used to be much commoner than it is now. It has been known in the wild since the eighteenth century and, like the Martagon Lily, was considered in the nineteenth century to be native. It differs from our common Garden Tulip (*T. gesneriana*), whose origin is uncertain, in that when still in bud the flowers are pendent, straightening up as they come into flower.

The change in posture shown by Wild Tulip occurs in reverse in one of our most popular garden genera of all, *Narcissus*, in which the flowers are erect in bud but turn over as they open, when for some reason they become susceptible to the pull of gravity. We have one native species, the much-feted Daffodil (*N. pseudonarcissus* subsp. *pseudonarcissus*), but about 25 other taxa (15 species and 10 hybrids) have become naturalised because they have been so extensively planted in semi-wild places. The genus *Narcissus* is centred on the western Mediterranean (primarily the Iberian Peninsula and Morocco) and shows a big range in variation: dwarfs a few centimetres high to tall plants reaching above 50 cm; single to many flowers on a stem; and white, yellow or orange flowers with or without contrasting tepal and trumpet colours. The whole of this range occurs in our wild flora. Hybridisation seems to be able to occur in virtually every species combination, and an enormous amount of breeding to produce new cultivars has taken place, so that more than 26,000 taxa are listed in the Royal Horticultural Society's International Daffodil Register database (available at www.rhs.org.uk). This, coupled with oversplitting of wild daffodils into too many species, has resulted in a somewhat chaotic taxonomic situation.

The most frequently naturalised *Narcissus* taxa in the British Isles are Spanish Daffodil (*N. hispanicus*, sometimes known as *N. pseudonarcissus* subsp. *major*), which resembles a robust version of our native Daffodil in which the tepals are as deeply coloured as the corona (not distinctly paler than it); Pheasant's-eye Daffodil (*N. poeticus*), with white tepals and a very much shorter, red-rimmed trumpet; and Nonsuch Daffodil (*N.* × *incomparabilis*), the hybrid between Pheasant's-eye Daffodil and our own Daffodil, with a trumpet of intermediate size and pale yellow tepals. Of all the other naturalised taxa, Tenby Daffodil (*N. obvallaris*), is of special interest. This species (sometimes classed as a variety or subspecies of *N. pseudonarcissus*) was formerly very common in Pembrokeshire, especially around Tenby, hence its English name. Although nearly all the truly wild examples have been taken by gardeners or ploughed up, it still exists in hedgerows and rough places, and particularly in graveyards, in southwestern Wales and rarely elsewhere. Tenby Daffodil was formally described in 1796 from Pembrokeshire, and, until recently, when it was reported from one area of Spain, was unknown elsewhere. It is most similar to Spanish Daffodil, but has smaller flowers and the tepals are not twisted at the base as they are in the latter. It was once considered a likely native of Wales, which is possible, but is more likely to be an escape from early cultivation.

Corm-bearing genera have an organ of perennation that is as compact and easily stored as the bulb; the British Isles have nine cormous genera but only one native species, the Wild Gladiolus (*Gladiolus illyricus*) of the New Forest, Hampshire. By far the largest number of our naturalised species (13) belongs to *Crocus*, some of which produce large amounts of viable seed. Six of the species, all but one purple-flowered, flower in the autumn, and seven, both purple- and yellow- and sometimes white-flowered, flower in the spring. The commonest species and one known here in the wild since the eighteenth century is the common purple- or white-flowered Spring Crocus (*C. vernus*). The best-known site for this species is at Inkpen in Berkshire, where it has been recorded since about 1800, although the origin of the 400,000-plus plants there is uncertain. A fairly close relative, Early Crocus (*C. tommasinianus*; Fig. 105), has been recorded in the wild in the British Isles for only the last 50 years or so, but in places it has become extremely well established, probably because it sets seed more abundantly than any of our other species. It differs from Spring Crocus in having slenderer flowers with a white centre, and narrower leaves. Our commonest autumn-flowering species, Autumn Crocus (*C. nudiflorus*, so called because the

FIG 105. Early Crocus (*Crocus tommasinianus*) is one of the earliest species to flower in spring and probably the most successful at regenerating from seed. (CAS)

leaves do not appear until the following spring), from southwestern Europe, has been known in the wild for the longest time, since 1738. It occurs over most of the British Isles but is particularly common in south Lancashire and Cheshire; it is characteristic of the lawns of several of the magnificent Tudor houses on the southern outskirts of Manchester. By contrast, the yellow-flowered species, including the common garden Yellow Crocus (*C. × luteus*), are much more reluctant to set seed and are less well established.

Montbretia (*Crocosmia × crocosmiiflora*; Fig. 203) is an extremely popular garden plant producing masses of orange flowers in late summer. It has corms that multiply very quickly, but that also produce horizontal rhizomes that can colonise new areas much faster than corms alone. It quickly outgrows its allotted space and gets thrown onto waste ground or into hedgerows, where it is very resilient. Although the genus *Crocosmia* is South African, Montbretia was raised artificially in France in 1879 by the well-known plant breeder Victor Lemoine (1823–1911), who crossed *C. pottsii* with *C. aurea*. In the wild in the British Isles, it prefers an Atlantic climate and hence is much commoner in the west, particularly in western Ireland, where it is extremely common, being one of the characteristic hedgerow plants there, along with another late-flowering alien, Fuchsia (*Fuchsia magellanica*; Fig. 243). In recent decades, *Crocosmia* has gained in popularity and several other species are grown in gardens, three having become naturalised. More hybrids have been produced, notably by Alan Bloom, and a number of these are becoming naturalised. Some of the modern *Crocosmia* are tall (up to 1.5 m) and have very bright red flowers that some would call 'hot' and others 'gaudy' or even 'garish'. Montbretia itself is partially fertile and can form hybrids with other species.

Many petaloid monocotyledons produce neither bulbs nor corms; some are strongly rhizomatous while others remain as tight clumps. We have seven genera in this category, including African lilies (*Agapanthus*; Fig. 106), Peruvian lilies (*Alstroemeria*), day-lilies (*Hemerocallis*), irises (*Iris*) and red-hot-pokers (*Kniphofia*; Fig. 107).

Red-hot-pokers represent a truly exotic-looking South African genus that one might guess would rarely become naturalised in the wild here for any length of time. Such a guess is widely off target, because there are some extraordinary colonies around the British Isles that are positively thriving, especially in coastal areas. Perhaps the best is on sand dunes at the Point of Ayr, at the mouth of the River Dee in Flintshire. Here, there is a very colourful and extensive colony that is evidently self-propagating. The Flintshire specimens seem to come under Greater Red-hot-poker (*Kniphofia × praecox*), a complex hybrid of uncertain parentage.

FIG 106. African Lily (*Agapanthus praecox*), a beautiful South African monocot: (a) inflorescences; (b) spreading and well naturalised in the Isles of Scilly on sand dunes. (CAS)

Iris constitutes an unusual genus in that it includes both rhizomatous and bulbous species; in fact, bulbs are exceptional organs to find anywhere in the family Iridaceae. There are 12 naturalised *Iris* taxa (or 13 if Snake's-head Iris *Hermodactylus tuberosus* is included, as is now mooted), which cover a good range of vegetative and flower morphology, with flowers of varying shades and combinations of blue and yellow. The leaves of the common Bearded Iris (*I. germanica*) represent a type very common in the genus (vertical, flat, with two identical faces, successive leaves intimately overlapping at the base). Nine of our alien species have leaves of this sort, which are found also in 11 other genera

FIG 107. Greater Red-hot-poker (*Kniphofia* × *praecox*) is an unmistakable garden plant that will spread rapidly by rhizomes when in a suitable habitat, as here on the sand dunes at the Point of Ayr, Flintshire. (Goronwy Wynne)

of our wild Iridaceae, including *Gladiolus* and *Crocosmia*. Three of our irises, however, have long, narrow, cylindrical to four-angled leaves. Such leaves are characteristic of Dutch Iris (*I.* × *hollandica*), which is one of our most popular cut-flower subjects and persists in the bulbfields of the southwest. Snake's-head Iris, with its distinctive solitary flowers bearing a mixture of yellow, green, brown, purple and black pigments, also differs in lacking rhizomes or bulbs (it possesses tuberous roots), and its leaves are flat with a clear upper and lower surface. None of the iris species is common, although several are widely distributed and some of long standing, e.g. Blue Iris (*I. spuria*) has been recorded in Lincolnshire fen ditches since 1836.

In addition to the above families, the Araceae provide six genera of naturalised plants, which always attract attention because of their weird flower structures. The main organ of attraction is not the perianth but the spathe, a large bract that variously envelops the inflorescence of tiny flowers. The large white Altar-lily (*Zantedeschia aethiopica*; Fig. 238) is commonly sold as a cut flower, for which it is grown on a field scale in the Isles of Scilly. There it has escaped into hedges and field borders, especially where water tends to lie, forming a very conspicuous element in the wild flora. Another genus of large plants is the skunk-cabbages (*Lysichiton*), whose members favour swampy ground, where they spread by means of rhizomes as well as by seed. American Skunk-cabbage (*L. americanus*; Fig. 266), from western North America, has yellow spathes up to 35 cm long, enveloping the foul-smelling flowers within.

Grasses, sedges and rushes

Grasses, sedges and rushes have always had a place in the garden, though not until recent times have some been popular as border plants. In the past, aliens from Europe such as Various-leaved Fescue (*Festuca heterophylla*), Broad-leaved Meadow-grass (*Poa chaixii*), Slender Cock's-foot (*Dactylis polygama*) and White Woodrush (*Luzula luzuloides*) have been used in informal ('natural') woodland settings, along with some natives such as Wood Meadow-grass (*Poa nemoralis*), Wood Millet (*Milium effusum*), Great Woodrush (*Luzula sylvatica*) and various native sedges (*Carex*), and they have become naturalised. Ronse & Braithwaite (2012) found that the three commonest grass-like plants recorded in places in southern Scotland that suggested that they had been planted as shade-tolerant species were the native Wood Meadow-grass and the aliens Broad-leaved Meadow-grass and White Woodrush.

Other more vigorous species have been planted in or by water, and eventually form large patches through their rhizomatous growth, e.g. Rattlesnake-grass (*Glyceria canadensis*) and Manchurian Rice-grass (*Zizania latifolia*; Fig. 136). A third niche for grasses in gardens is as a large commanding specimen in a prominent position. Pride of place here goes to the pampas-grasses (*Cortaderia*; Fig. 137), which are discussed in Chapter 8 and have become well naturalised. Much more recently, the silver-grasses (*Miscanthus*) have been similarly employed. The species most often planted is Chinese Silver-grass (Fig. 86), which is 1–2 m tall with handsome silver and purple terminal inflorescences in the form of a tuft of long, densely hairy racemes.

Smaller ornamental grasses have for long been rather minor components of the herbaceous border. They include annuals like Greater Quaking-grass (*Briza maxima*; Fig. 227) and Hare's-tail (*Lagurus ovatus*), which both set seed readily and can easily escape outside the garden. The latter is so well established and locally dominant on the maritime dunes in St Ouen's Bay, Jersey, that it is surprising to learn that it was introduced there in the 1870s. Its abundance has led to its fluffy flower heads being 'much used in the *Battle of Flowers* floats since the 1940s' (Frances Le Sueur in her *Flora of Jersey*, 1984). In the 1959 Battle of Flowers procession, Hare's-tails were used to construct a group of full-sized sheep (Fig. 108).

FIG 108. Hare's-tail (*Lagurus ovatus*), a small ornamental Mediterranean annual, is extensively naturalised on the dunes at St Ouen's, Jersey. Here it is seen collected in quantity to model life-sized sheep on a float in the island's 1959 Battle of Flowers. (CAS)

Among the perennial small grasses are the densely tufted Glaucous Fescue (*Festuca glauca*) and the mat-forming Spiky Fescue (*F. gautieri*). The former is a name often applied to various native fescues that have bluish-grey foliage, but the true plant is native only to one part of the south of France (Roussillon) and has so far not appeared in any Flora of Britain or Ireland. It is, however, cultivated a great deal both in gardens and in council planting schemes, and very recently records have been made of its escape into surrounding areas by self-sowing. Spiky Fescue comes from the Pyrenees and is not closely related to any other north European fescue. It is becoming increasingly available in garden centres, but so far has not escaped from gardens. It is, however, well established in a limestone quarry in northwestern Yorkshire, where it was probably deliberately planted along with two other Pyrenean aliens – Fairy Foxglove (*Erinus alpinus*) and Round-leaved St John's-wort (*Hypericum nummularium*) – probably over a hundred years ago. It is a self-incompatible species and does not set seed in Yorkshire, suggesting that the colony, measuring about 4.5 m across, is a single clone.

Two other species of fescue that provide many problems in identification are sown in thousands of tonnes to create grassy places during landscaping or road-building. One is Hard Fescue (*Festuca brevipila*), which was previously known under a variety of names and is one of those species confused with Glaucous Fescue. It is a dwarf, densely tufted plant belonging to the Sheep's Fescue group, but due to its relatively robust habit it is often misidentified as a member of the Red Fescue group. It prefers well-drained soils and is especially common in East Anglia. The other is our largest representative of the Red Fescue group, *F. rubra* subsp. *megastachys*, sometimes separated as *F. diffusa*. In the grass-seed trade this is known as Strong Creeping Red Fescue; it thrives on rich soils, including clays, when it can grow to 1 m tall and extend at least as far laterally in one year by means of its rhizomes.

In recent times, following a fashion largely dictated by celebrity gardeners, a more exotic range of grasses has been introduced to the garden border and council display, especially the so-called needle-grasses of and related to the widespread genus *Stipa*. The commonest of these is Pheasant's-tail (*Anemanthele lessoniana*), from New Zealand, once known as *Stipa arundinacea* and still sometimes so labelled. This is a densely tufted, wispy-looking plant whose leaves become an attractive reddish brown and fray at the ends as the season advances. It sets a huge amount of seed and soon infests the surrounding area, and records of it in the wild are rapidly increasing. Probably the next most common is Argentine Needle-grass (*Nassella* (formerly *Stipa*) *tenuissima*), from South America, but so far this has not proved as effective at self-sowing.

In addition to grasses, there has been an increase in the number of sedges introduced as border or woodland plants. The one that has been most often recorded as an escape is Silver-spiked Sedge (*Carex buchananii*), from New Zealand. This has attractive red-tinged stems and leaves, and silvery ripe inflorescences. It is to be expected that other sedges will in due course become commoner.

Rockery plants

Rockeries have been a feature of gardens for a long time; many of the most magnificent ones still to be admired today are Victorian, constructed at a time of cheap labour. Virtually any plant can be added to a rockery, but the most characteristic are low-growing plants that hang or creep over the surface and into the intricate crannies of the rocks. Such species can soon take hold and are difficult to extricate, and if they are rampant plants they soon become pests that are thrown out and readily become naturalised. Two notorious species that were introduced as rock-garden plants and have since become common wild aliens are New Zealand Willowherb (Fig. 224) and Slender Speedwell (*Veronica filiformis*; Fig. 254), both covered in Chapter 11. Few gardeners would give them space today.

Among the most common other rock-garden plants with a similar invasive habit are Garden Arabis (*Arabis caucasica*), Aubretia (*Aubrieta deltoidea*), Corsican Mint (*Mentha requienii*; Fig. 109), Mind-your-own-business (*Soleirolia soleirolii*; Fig. 110) and Ivy-leaved Toadflax (Fig. 111). Given the right opportunity these all quickly take to the wild. Garden Arabis is as much at home hanging off the tall limestone cliffs of the Derbyshire Dales as it is in the garden. Ivy-leaved Toadflax rapidly extends its influence by both stolons and seed, and is in fact our seventh most frequent neophyte (Chapter 11). It is equally at home on walls, where it is

FIG 109. Corsican Mint (*Mentha requienii*) is a tiny prostrate plant that roots along the length of its stems and possesses a distinctive pleasant scent; it is naturalised on relatively bare ground in damp places. (CAS)

FIG 110. Mind-your-own-business (*Soleirolia soleirolii*), an inconspicuous dwarf creeper from the nettle family, is naturalised on damp walls, rocks and banks: (a) on a mortared wall; (b) growing over the cheeks, forehead and nose of one of the sculptures (*The Giant's Head*, with Montbretia *Crocosmia* × *crocosmiiflora* as its hair) at the Lost Gardens of Heligan, Cornwall. (CAS)

FIG 111. Ivy-leaved Toadflax (*Cymbalaria muralis*), known in the British Isles for nearly four centuries, is a familiar sight sprawling over walls and rocks. (CAS)

often found with another proficient self-sower, Yellow Corydalis (*Pseudofumaria lutea*); both come from southern central Europe.

Stonecrops (*Sedum*) constitute a large genus, of which 17 species are naturalised in the British Isles. They are succulent species, meaning that their leaves are swollen and 'fleshy', containing water-storing tissue, which often lends them a cold feel when touched and gives rise to the common name of 'ice-plant', which is equally applied to many other succulent genera. Another characteristic of many of the species is that leaves broken off a plant can take root, so the species have a very efficient method of vegetative propagation in addition to reproduction by seed. They can therefore become aggressive in a garden rockery, and can soon become naturalised in the wild. The growth habit of the genus is diverse, ranging (in our species) from upright, short shrubs to long-creeping or mat-forming species. The leaves also vary from more or less flat to greatly swollen and almost spherical. By far the commonest alien *Sedum* is White Stonecrop, which is an archaeophyte with greatly swollen, albeit small, leaves and flat heads of white flowers. The next most common species are Reflexed Stonecrop (*S. rupestre*), a more robust plant with a flat head of bright yellow flowers, and Caucasian-stonecrop (*S. spurium*; Fig. 112), which has flat leaves and a flat head of pinkish-purple flowers.

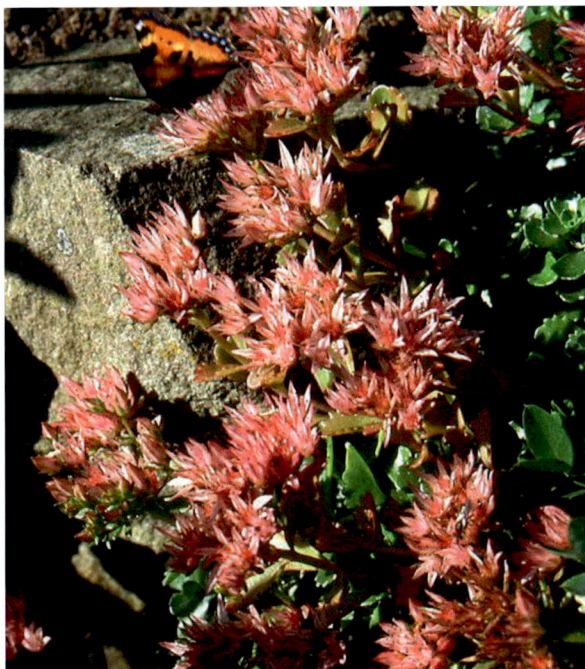

FIG 112. Caucasian-stonecrop (*Sedum spurium*), one of the commoner garden representatives of this genus, readily spreads by its creeping stolons and becomes naturalised in open dry places. (CAS)

Aquatics

Aquatic plants, whether they are submerged, floating, emergent or even marginal, are especially favoured because of the ready availability of a constant supply of water and its solutes, and because the water to some extent provides a buffer against extremes of temperature. Once established, therefore, they more often than not thrive, and if they are vigorous species they can soon dominate, often to the detriment of rare native species in the site. Many aquatic plants owe their existence in the wild to their being discarded by aquarists when they become too prolific – they are thrown into wild ponds and streams instead of onto the compost heap. Some of our most pernicious weeds arrived in this way, and are now causing alarm in the conservation world because of their threat to a fragile and diminishing habitat that is refuge to rare native species. The most notorious examples are probably New Zealand Pigmyweed (Fig. 113), Parrot's-feather (*Myriophyllum aquaticum*) and Floating Pennywort (*Hydrocotyle ranunculoides*; Fig. 134). The British government banned the sale of these three species, plus Water Fern (*Azolla filiculoides*) and Water-primrose (*Ludwigia hexapetala*), from April 2014. In the nineteenth century, Canadian Waterweed (*Elodea canadensis*) caused as much concern, but towards the end of that century it decreased in abundance and in fact was never as aggressive as some of the species mentioned above.

Two true water-lilies have become naturalised following their deliberate planting either in the wild or in tended ponds that have become neglected. Water-lilies have strong, thick rhizomes and, once established, they are difficult to eradicate other than by draining the pond. Marliac's Water-lily (*Nymphaea × marliacea*) resembles our White Water-lily (*Nymphaea alba*) but has pink or pale yellow flowers. The North American Spatter-dock (*Nuphar advena*) is similar to our Yellow Water-lily (*Nuphar lutea*) but the flowers are raised on their stalks above the water's surface.

We do not have many alien emergent species. Manchurian Rice-grass (Fig. 136), Rattlesnake-grass and the blue-flowered Pickerelweed (*Pontederia cordata*) are mentioned elsewhere. Apart from our native Arrowhead (*S. sagittifolia*), there are three naturalised species of *Sagittaria* in the British Isles, all found in only a few places but well established there. The commonest is the North American Duck-potato (*S. latifolia*), which resembles Arrowhead but lacks the latter's purple base to the petals and has yellow, not purple, anthers. Both have arrow-shaped leaves, whereas the other two alien species do not. Sweet-flag (Fig. 249) is one of the weirdest plants in our flora. The leaves arise from rhizomes and resemble those of the native Yellow Iris (*Iris pseudacorus*), with which it often grows, but many of them are transversely wrinkled and when crushed they give off a pleasant sweet,

FIG 113. New
Zealand Pigmyweed
(*Crassula helmsii*), one
of the aquatic aliens
causing great concern
at their rapid and all-
enveloping spread
in ponds, lakes and
ditches: (a) choking a
ditch; (b) close-up to
show the leaves and
flowers; (c) warning
sign indicating one of
the damaging effects
of this plant, at Pett
Level, Sussex. (CAS)

spicy smell similar to that of Bog-myrtle (*Myrica gale*). The inflorescence is a dense cylindrical spike of many tiny yellowish-green flowers, projecting laterally from the stem. Fruits are not formed in the British Isles, and probably not anywhere in Europe. The genus was once placed in the Araceae, differing in the absence of the characteristic spathe, but molecular (DNA) work has shown that *Acorus* was the first genus to split from the monocotyledonous line, i.e. it is a sister group to the rest of the monocots. It is now placed in its own family, Acoraceae, which in systematic sequence is the first family of the monocots.

There are many more marginal (waterside or marsh) aliens, including several in the Araceae. The large plants *Zantedeschia* and *Lysichiton* were mentioned earlier. Bog Arum (*Calla palustris*) is a much smaller European plant with a flat white to greenish-white spathe that is not wrapped around the flowers. Golden-club (*Orontium aquaticum*), from North America, is so far known in only one locality in the British Isles; it has a golden-yellow flower spike on top of a white stem, and a small greenish spathe that soon withers. The so-called Fish-plant (*Houttuynia cordata*), also very rare, superficially resembles a member of the same family, but it is in the remotely related Saururaceae, one of the few families of angiosperms that were (like the water-lilies) differentiated before the dicots and monocots diverged. This Asian plant has heart-shaped leaves, which in one of the commonest cultivars have red and cream blotches, and a dense terminal spike of very small flowers, at the base of which are four large white petal-like bracts. Much larger are two strongly rhizomatous members of the Saxifragaceae, with large leaves borne at the tops of long, stout petioles, and many-flowered inflorescences also raised high by their stalks, both organs arising direct from the rhizomes. Rodgersia (*Rodgersia podophylla*), from eastern Asia, has palmately compound leaves and yellowish-white flowers, and the North American Indian-rhubarb (*Darmera peltata*) has lobed leaves with the petiole joining in the centre, not at the edge, and pink flowers. Both are scattered over the British Isles.

Much larger still are Giant-rhubarb (*Gunnera tinctoria*; Fig. 251) and Brazilian Giant-rhubarb (*G. manicata*). These are South American, but the genus contains at least 40 species found throughout the southern hemisphere. They are rhizomatous plants with leaves and dense compound inflorescences arising from the ground. Some are truly gigantic, with petioles up to 5 m tall and leaves as far across, while others are tiny plants with leaves only about 1 cm across on petioles no more than 5 cm long. Our species generally have petioles 1–2 m long, bearing leaves 1–2.5 m across, and form dense thickets attracting much attention in the gardens of stately homes and in places like Kew Gardens. The plants do not usually present a problem, but in parts of Ireland, particularly in County Mayo and County Sligo, Giant-rhubarb has become seriously invasive, 'taking over hedges, ditches, roadside verges, whole paddocks up to 0.5 ha or more and open hillsides' (Jarvis, 2011).

CONFUSING LOOKALIKES

It has become popular to plant colourful annuals as a cheap and cheerful way of sprucing up parts of run-down landscapes, especially in urban areas. Commercial wildflower mixtures have a highly stereotyped composition, but interesting contaminants are quite frequent. The mixtures almost inevitably contain a set of crowd-pleasing archaeophytes (Fig. 2), such as Corncockle, Austrian Chamomile (typically sold as 'Corn Chamomile' – *see* below), Cornflower, Corn Marigold, Wild Carrot (*Daucus carota* subsp. *carota*) and Common Poppy, with a smattering of native species like Viper's-bugloss (*Echium vulgare*), Oxeye Daisy (*Leucanthemum vulgare*) and Musk-mallow (*Malva moschata*) as the principal colourful elements. The most frequent imposters are Beaked Hawk's-beard, Hoary Mustard (*Hirschfeldia incana*; Fig. 114), Wall Barley (Fig. 35), Hoary Cress (Fig. 115), Bristly Oxtongue (*Helminthotheca echioides*) and Common Field-speedwell (*Veronica persica*; Fig. 200). These mixtures often include cheap agricultural seed from eastern Europe of the same species but different genotypes from the intended British counterpart: the most frequently noticed of these are Fodder Burnet (*Poterium sanguisorba* subsp. *balearicum*), *Hordeum murinum* subsp. *leporinum* and *Lotus corniculatus* var. *sativus* (*see* below).

There is some confusion about what to call these annual wildflower mixtures. They are often presented as 'wildflower meadows' but this is clearly wrong.

FIG 114. Hoary Mustard (*Hirschfeldia incana*) was in 1958 convincingly naturalised only in the Channel Islands; it is now a common plant of waste ground and is still spreading. (CAS)

Meadows are species-rich grasslands dominated by perennial species, grown for hay in spring and summer, with the aftermath grazed by cattle in autumn. Historically, the species in most of the annual wildflower mixtures were cornfield weeds, but apparently it is not politically correct (or commercially wise) to refer to intentionally planted wildflowers as 'weeds'. They are called 'cornfield mixes' by some firms.

There is an increasing number of alien neophytes that closely resemble well-known native species or archaeophytes, and which are therefore very easily overlooked and misrecorded. As mentioned in Chapter 2 and elsewhere, it is particularly important that alien plants are determined with great care, because field botanists can be tricked into

FIG 115. Hoary Cress or Curse-of-Kent (*Lepidium draba*) is an aggressively rhizomatous species thought to have been introduced to the British Isles with fodder or straw. (CAS)

a false sense of confidence when confronted with apparently familiar plants. Two archaeophytes that are especially noted as annual weeds of agriculture, Thorow-wax and Corn Chamomile, are susceptible to over-recording because of the presence of lookalikes, which in each case are now more common than the former weeds.

Thorow-wax is strange in both name and appearance (Fig. 13). It is surely strange, too, that it is completely missing from Geoffrey Grigson's *The Englishman's Flora* (1955). It is most unusual among the umbellifers (Apiaceae) in having simple entire leaves that completely encircle the stem from which they arise, making it a most distinctive plant. The species' common name, according to Gerarde (1597), is derived from the fact that 'the stalke waxeth throwe the leaves', but Salisbury (1961) also noted that the leaf surface has a waxy appearance. It formerly grew in arable land, mainly in southern England, but it has probably not been found in a cornfield since the early 1960s. During the 1960s, its close relative False Thorow-wax (*Bupleurum subovatum*) became common as a birdseed alien, and still is so, and was then often misidentified as Thorow-wax. According to the *New Atlas* (2002), Thorow-wax was recorded from 273 hectads up to 1969, from only two between 1970 and 1986, and from 13 since 1987 (mostly as deliberate plantings).

Corn Chamomile is one of the iconic cornfield weeds of past years that have greatly decreased in the second half of the twentieth century. It does still occur in cornfields, but often as a result of sowings of 'cornfield mixes', along with poppies, cornflowers and so on (*see* above). It is also said to occur as a grass-seed alien, but the extent to which this is true is uncertain because of the occurrence of the very similar Austrian Chamomile (Fig. 22b) in newly sown grass verges and other plantings. In this case it seems that the confusion might be more deep-seated. In 2012, Brian Wurzell and I came across an area of parkland in East London that had been sown with a cornfield weed mix, producing bright swathes of Corn Marigold interspersed with Cornflower, Corncockle, poppies and (apparently) Corn Chamomile. The last, however, proved to be Austrian Chamomile, indicating that the latter has entered the commercial seed business masquerading as Corn Chamomile. It has also been found that some mixes sown in cornfield edge strips in Carmarthenshire and Hampshire have been contaminated in this way, and it is probably more widespread than we realise. Austrian Chamomile has evidently arrived here both as an accidental grass-seed contaminant and as a deliberate introduction confused with one of our archaeophytes. In parts of central Europe both are cornfield weeds but are rarely found growing together, with Austrian Chamomile favouring basic and Corn Chamomile acid soils (František Kruhalec, pers. comm., 2012), but the habitats of these two species in the British Isles (cornfields and new grass) are no longer of any use in distinguishing them. They both have a similar sweet scent when crushed, but differ mainly in the shape of the receptacular scales found mixed with the florets in each flower head.

Lookalikes might have arrived as contaminants, or have been deliberately planted as superior agricultural crops, as roadside vegetation or in various conservation schemes, and sometimes were accidentally imported under the illusion that they belonged to a British species. The mimics and the natives differ to varying degrees, leading to their recognition as different species or only different subspecies or varieties, or not being accorded taxonomic status at all despite their distinctive appearance. Critical taxonomic groups such as the Common Knapweed (*Centaurea nigra*) aggregate and the Oxeye Daisy (*Leucanthemum vulgare*) aggregate are two of the most obvious examples that have become even more complicated in this way, as are several legumes, especially in the genera *Trifolium, Onobrychis, Vicia* and *Lotus*.

Common Bird's-foot Trefoil (*Lotus corniculatus*) is a colourful legume that is less abundant than formerly due to the loss of so much 'unimproved' grassland. It is important as fodder, as the food-plant of several insects (including the blue butterflies), in nitrogen-fixing and for foraging bees, and often features in seed mixes on new roadside verges. One genotype being planted is starkly different in appearance from our familiar plant, being far more robust, and stiffly upright in growth habit. Moreover it often has hollow stems, a character lacking in our native plants but present in the related native Greater Bird's-foot Trefoil (*L. pedunculatus*). This character, often used in keys to distinguish the two native species, is clearly no longer reliable. The robust genotype is known as var. *sativus*; it comes from central Europe and is very conspicuous on the verges of many of our new motorways and bypasses.

In the past few decades, 'conservation plantings' have become very widespread, aided by various government subsidies to landowners aimed at encouraging a more wildlife-friendly countryside. The creation of hedgerows and field-head copses and scrub has been prominent, but unfortunately most of the plants utilised for such development have been sourced on the Continent. In many cases it is visually obvious that the genotypes now being found in these schemes are foreign, because the species are not entirely characteristic of our native material. Peter Sell (2007) drew attention to confusing introduced lookalikes that he had found in Cambridgeshire plantings, particularly of roadside and hedgerow trees and shrubs. Several, if not all, of his observations apply equally elsewhere in England and Wales, if not more widely. One taxonomically difficult genus involved is the hawthorns (*Crataegus*). Even allowing for the considerable natural variation shown by our two native species, Hawthorn and Midland Hawthorn (*C. laevigata*), more and more hawthorns are now appearing in the countryside showing a range of variation beyond that to which we are accustomed. Much of this is due to the planting of shrubs from central Europe that come under Large-sepalled Hawthorn (*C. rhipidophylla*, formerly known as *C. curvisepala*). This species differs from Hawthorn in minor characters of the leaves and fruits that are difficult to describe but easy enough to detect by eye. It is now common to find hawthorn hedges planted within the past 20 years that show along their length a great range of (particularly) leaf morphology. There are also intermediates between the newcomer and our two native species. These are probably fertile hybrids that have been introduced as such along with Large-sepalled Hawthorn, and if it has not happened already we should expect new hybrids to arise here soon.

Other genera discussed by Sell (2007) include the shrubs *Cornus* and *Viburnum*, and, although he did not mention it, hazels (*Corylus*) are in exactly the same category. Newly planted Dogwood (*Cornus sanguinea*) hedges, such as are found in car parks at supermarkets, medical centres and the like, are more likely to consist of the eastern European *Cornus australis*, which is probably best treated as a subspecies of our native species because the two are geographically separated and differ mainly by minute hair characters of their leaves. Similarly, most planted hazels have different shaped nuts and surrounding cupules from those of the native Hazel (*Corylus avellana*) and also come from further east; at least some of their nuts closely resemble the imported dried varieties sold here at Christmas. In *Viburnum* there are lookalikes that mimic both our two, very different, native species, Guelder-rose (*V. opulus*) and Wayfaring-tree (*V. lantana*). Some of these arise not as deliberately planted shrubs but as stocks of more exotic garden species (e.g. Fragrant Snowball *V.* × *carlcephalum*) that have been allowed to grow up unchecked. Because of these deliberate introductions of foreign genotypes of British species or their close relatives, the identification of wild plants will steadily become more difficult over future years, and, if we do not keep a careful note of what is being planted or sown, it might become impossible.

LONG-LASTING ALIENS

Most aliens do not spread far from their original point of introduction, and many of these offspring soon die out. This is true by definition, of course, of all casual species, but here we are discussing species that have the means of survival for many years, and perhaps the potential to spread either by seed or vegetatively, yet remain extremely localised. There are many plausible reasons for this, but, as noted elsewhere, the lack of a means of sexual reproduction is certainly not one of them, as we have many examples of sterile aliens that have spread enormously and even become major pests. Japanese Knotweed, Winter Heliotrope (*Petasites fragrans*), Canadian Waterweed and New Zealand Pigmyweed are notorious examples.

Trees and shrubs are prime examples of plants that 'just sit there', without spreading or showing any signs of disappearing, and many individuals could be cited that have been well known for centuries. For instance, we do not know whether Service-tree is a native, an archaeophyte or a neophyte. We do know, however, that a single specimen of it grew in Wyre Forest, Worcestershire, where it was first noticed in 1677 under its local name Witty Pear, having been there for a good number of years before then. It would almost certainly still exist today if it had not been maliciously burnt down in 1858. However, saplings had previously been raised from the tree, and one of them was planted in the same spot, much later, where it still grows. Some naturalised aliens that remain in their place of introduction for a long time without spreading are known on the Continent as *Stinzenplanten* (Netherlands) or *Stinsenpflanzen* (Germany), and have been discussed by several authors (see, for example, Bakker & Boeve, 1985). The term is mainly applied to geophytes (spring-flowering species that completely die down above ground before high summer, such as crocuses, snowdrops and anemones) that are frequently planted in shady places by large houses. *Stins* specifically refers to stone-built houses in northern Friesland (Piet Bakker, pers. comm., 2013).

It is not at all surprising that individuals of woody plants have survived *in situ* for a very long time, but more remarkable when comparable examples are found among herbaceous plants. Silver Lady's-mantle (*Alchemilla conjuncta*) has been known for about 200 years at the location where it was originally planted in Scotland; Auricula (Fig. 116) and Pyrenean Columbine (*Aquilegia pyrenaica*) were planted in Caenlochan Glen 115 and 130 years ago, respectively; and the annual Sand Toadflax (*Linaria arenaria*) was sown in Devon 120 years ago. Other notable examples of long-standing introductions still to be found in their original localities after well over a century are: Salmonberry (*Rubus spectabilis*), at Hythe, Kent, present since the 1850s; Sweet Scabious (*Scabiosa atropurpurea*), at Folkestone, Kent, present since 1862; Garden Chervil (*Anthriscus cerefolium*),

at Ross-on-Wye, Herefordshire, present since 1867; Malling Toadflax (*Chaenorhinum origanifolium*), at West Malling, Kent, present since 1880; and Sand Toadflax (*Linaria arenaria*), present on the sand dunes at Braunton Burrows, Devon, since 1893. All but Salmonberry are rare plants thriving in their classic localities but scarcely known elsewhere. The list will undoubtedly grow. No botanist who visits Fetlar, that fantastic treeless bird island in the Shetland Islands, lying further north than Oslo, Stockholm or St Petersburg, can fail to wonder at the small but healthy patches of Giant Knotweed and Lovage that seem to show no signs of decline after 82 and 54 years, respectively.

FIG 116. Auricula (*Primula auricula*) is grown in gardens in a huge variety of colours, but the colony planted in Caenlochan Glen, Angus, around 1880 is coloured like the native plants in the Alps (shown here) and probably came from there. (Peter Hall)

Other examples of 'just sit there' species are given in the section on frost-sensitive neophytes (Chapter 12), particularly the giant monocotyledons such as *Agave, Cordyline, Fascicularia* (Fig. 117b), *Furcraea, Ochagavia* (Fig. 117a), *Phormium, Trachycarpus* and *Yucca*. These all flower well, but either do not set seed or the seed fails to germinate or to find a niche suitable for development.

One other group in this category deserves mention: the bamboos. These giant grasses were and still are much planted in (often unsuitably small) gardens or on larger estates, where they can be handsome subjects in shrubberies, as hedges or as isolated specimens, according to habit. About 11 Asian species are naturalised in the British Isles, ranging in height from less than 1 m to 8 m. Some of the species form clumps, with very short rhizomes, while others spread much further by virtue of long, extensive rhizomes. Bamboos flower spasmodically and apparently irregularly, but there seems to be a great deal of synchronisation between different plants of one species and even between different species. It is often said that all the individuals of a species across the world flower at the

FIG 117. Two members of the pineapple family make an extraordinary sight on the Cornish coast: (a) Tresco Rhodostachys (*Ochagavia carnea*), forming mounds on the maritime heath (old dunes) on Tresco, Isles of Scilly; (b) Rhodostachys (*Fascicularia bicolor*) in close view on another Scillonian island, showing the inflorescences and spiny leaves. (CAS)

FIG 118. Broad-leaved Bamboo (*Sasa palmata*), the largest-leaved bamboo in the British Isles and the one that flowered most prolifically in the 1960s, here competing with nettles and encroaching scrub. (CAS)

same time, and after flowering they die. Each statement is probably true of a few species, but neither strictly applies to any of our commonly grown or naturalised species. Despite that, the pattern of flowering in our naturalised bamboos is certainly remarkable. Hardly anyone had seen bamboos flowering in the British Isles during the first half of the twentieth century, but from 1961, and especially after 1964, flowers started to appear abundantly all over these islands in many of our species, including the two commonest, Broad-leaved Bamboo (*Sasa palmata*; Fig. 118) and Arrow Bamboo (*Pseudosasa japonica*). This period of heavy flowering lasted for about 15 years, with some lingering on for longer, but once again flowering specimens are now rarely encountered. In Broad-leaved Bamboo, flowering was almost ubiquitous, with virtually every clump producing flowers, and no other species matched this. Some bamboo species regularly produced fruits in that reproductive period, and in some cases this could be germinated. The sober truth is that we have no precise idea of the triggers for flower production in this group of grasses.

BOTANICAL FRAUDS

Introducing alien plants into natural habitats was a clandestine sport indulged by certain nineteenth- and twentieth-century botanists. Whether it was done out of a sense of mischief, in order to confound the experts who came upon the plant subsequently, or in the hope of future self-aggrandisement when they themselves could 'discover' the plant to universal acclaim, we shall probably never know. Sometimes, no doubt, they simply wanted to enrich an area or find whether the species could grow there. In the past, plants introduced in this way were known quaintly as 'false wildings' (White, 1912). The first record of this behaviour involved the great John Gerarde. When Thomas Johnson prepared a new edition of Gerarde's *Herball: or Generall Historie of Plantes* in 1633, he passed on the malicious gossip about Gerarde's surprising record of Peony, which he purportedly found on a Rabbit warren in Southfleet: 'I have been told that our author himselfe planted that Peionie there, and afterwards seemed to find it there by accident: and I do beleeve it was so, because none before or since have ever seen or heard of it growing wild since in any part of this Kingdome.'

Several of the more notorious cases of deliberate plantings in the wild have involved montane plants, perhaps because our mountain flora is so poor compared with that of much of the Continent. Sikkim Cowslip (*Primula sikkimensis*) and Pyrenean-violet (*Ramonda myconi*) were both planted in Snowdonia in 1921 and are still there. The most famous of the false wildings of the mountains are known as 'the Angus rarities'. The renowned Scottish gardener George Don Sr (1764–1814) created a garden at Forfar said to be 'scarcely surpassed in Britain' in the number of species grown. A high proportion of the plants were 'alpines', many of them non-British. He was also a highly accomplished and energetic field botanist, and during that early period was responsible for the first records of many of our native Scottish rarities, such as Mountain Sandwort (*Minuartia rubella*) and Alpine Forget-me-not (*Myosotis alpestris*). Some of the plants he claimed to have found, however, are usually not accepted as British natives. These include Alpine Buttercup (*Ranunculus alpestris*), Alpine Catchfly (*Silene alpestris*) and Three-toothed Cinquefoil (*Potentilla tridentata*), as well as a good number that are now known to be misidentifications of commoner species, often unusual variants of them.

Don's herbarium specimens of several of these species still exist, but his wild localities for them have never been found. Forced to choose between fraudulent claims and a genuine mix-up between his wild collections and garden specimens, most botanists have opted for the latter. George Don was the first to report Purple Colt's-foot (*Homogyne alpina*; Fig. 119) from the British Isles in 1813 'growing on rocks by the side of rivulets in the high mountains of Clova, as on a rock called Garry-barns'. A general air of scepticism, coupled to Don's reputation for unreliability, meant that his genuine herbarium specimens of Purple Colt's-foot at Kew were put down as European plants that had been grown in Don's famous garden. However, the Glen Clova plant was refound in 1951 by Alf Slack, presumably in Don's original locality. The dilemma had therefore changed in nature; the plant does occur in the wild, so the choice is between a native colony and a deliberate planting. On European distributional grounds the species would

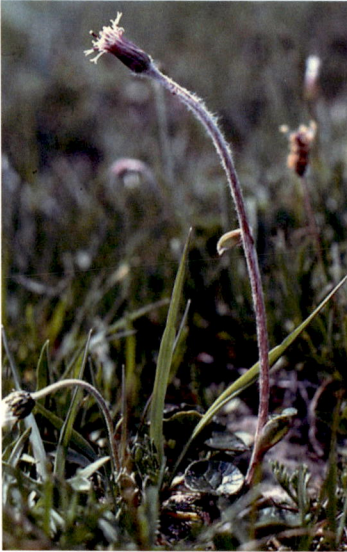

FIG 119. Purple Colt's-foot (*Homogyne alpina*) was first found in the Scottish Highlands in 1813, but its status – native or planted – remains uncertain. It was not seen for well over a century until its rediscovery in 1951. (CAS)

not be expected in the British Isles; we have it on our native/alien uncertain list.

Another of Don's discoveries was Silver Lady's-mantle, which he found in Glen Clova before 1812. This locality was lost to knowledge, refound in 1853 by A. O. Black, then lost again, and refound for a second time by R. Corstorphine in 1917. It was accepted as native by Druce (1918), and as late as the second edition of *Flora of the British Isles* (Clapham *et al.*, 1962) was considered 'possibly native'. Its European distribution argues strongly against its being native here, and it is today not accepted as such by anyone. Unlike Purple Colt's-foot it is commonly grown in gardens and frequently naturalised, but it must have been deliberately planted somewhere as remote as Clova (and in other Scottish mountain sites).

Two Scottish rarities associated with the name of William Brack Boyd (1831–1913), *Salix* × *boydii* and *Sagina boydii*, are shrouded in mystery. Both taxa are frequent these days as garden plants, popular in rockeries; the *Sagina* forms a neat, tight cushion and the willow is somewhat like a compact version of the celebrated native Scottish rarity Woolly Willow (*Salix lanata*). They have never been recorded in the wild by anyone else (in Britain or elsewhere), and it has been suggested that either Boyd made the herbarium sheets using plants growing in his garden, or he planted them in the wild, then 'discovered them', and the planted individuals died without leaving any offspring. Alternatively, it is plausible that he found them in the wild but removed the only individual for cultivation in his garden. *Sagina boydii*, for long treated as a separate species, is now thought to be a variant of the common Procumbent Pearlwort (*Sagina procumbens*); in DNA characters it falls within the range of that species.

The flora of Caenlochan, that breathtaking botanical hotspot in Forfarshire, has been augmented by at least four deliberately introduced species, including Alpine Forget-me-not, Fairy Foxglove, Pyrenean Columbine and Auricula (Fig. 116). The last two were planted in 1895 and 1880, respectively, at about

900 m, and are still there. The Auricula might well be genuine material from the Alps, as Florence Houseman told me in 1988 that its flowers are dull yellow, as is usual in Alpine examples, rather than the brighter varied colours of garden cultivars.

The 'Inchnadamph Zoo' is on a limestone outcrop near Glac Mhor above Loch Assynt in Sutherland, where five intentionally introduced species survive in a single site. We have no idea who planted them, or when, or how many species were introduced but failed to survive. The 'zoo' was first described by G. M. Richards and A. E. White in 1992 and included Fairy's-thimble (*Campanula cochleariifolia*; first recorded in 1992), Fairy Foxglove (1992), Spring Gentian (*Gentiana verna*; 1997), Oxford Rampion (*Phyteuma scheuchzeri*; 1992, the most successful of the five species at this site) and Alpine Catchfly (1997). None of the planted species has spread to nearby limestone crags outside the 'zoo' (Evans *et al.*, 2002).

The famous plant hunter Reginald Farrer (1880–1920) was well known as an eccentric. In one infamous incident, he loaded a shotgun with seeds collected on his foreign travels and fired them high into a limestone cliff in a gorge near his garden at Clapham, near Ingleborough in West Yorkshire, where he grew, and for a short time sold, alpine plants, many of which he had collected in the Himalayas.

Among the more attractive false wildings are Sea Stock (*Matthiola sinuata*), deliberately introduced on sand dunes at Broughton Burrows in Wales; and Large-flowered Butterwort (*Pinguicula grandiflora*), native with us only in southwestern Ireland, which was planted on moors in Cornwall and Devon in the late nineteenth century – a 'well-meant but misleading endeavour to extend the range of this beautiful plant' (Scully, 1916). White (1912) told us about Honey Garlic (*Nectaroscordum siculum*) on St Vincent's Rocks, Bristol, where 'a very few plants which I have no doubt were planted by some misguided enthusiast... I find no other instance recorded of its introduction as a false wilding'. There is a small but vigorous patch of the species deep in a Leicestershire wood, where it was also obviously planted. In a similar category is the colony of Pick-a-back-plant, which is thriving in remote dense woodland in a Derbyshire National Nature Reserve, jostling for space with the likes of natives Herb-Paris (*Paris quadrifolia*), Giant Bellflower (*Campanula latifolia*) and Narrow-leaved Bitter-cress (*Cardamine impatiens*).

The carnivorous Pitcherplant (*Sarracenia purpurea*) is a native of *Sphagnum* bogs and peaty barrens in northeastern USA. Pitcherplants were introduced into *Sphagnum* bogs west of Termonbarry in Roscommon in central Ireland in 1906, where the species now thrives, becoming locally abundant in 'great colonies'. These plants were the source of several subsequent deliberate introductions (Walker, 2014). Two other pitcher plants, Yellow Pitcherplant (*S. flava*) and White

Pitcherplant (*S. drummondii*), were introduced there at the same time but did not flourish (Praeger, 1934). The same species was planted on Burnt Hill, Chobham Common, in Surrey, among *Sphagnum* close to the railway, where they still grew in 2014. The history and current status of Pitcherplant in the British Isles have been summarised by Walker (2014).

In north Wales, Snowdon suffered from a 'contamination of its native flora. As the result of the sowing, on a steep slope under the Parson's Nose, of the seeds of no less [*sic*] than two hundred and forty species of alpine plants, in 1937–39 there were still "thousands of seedlings" of three alien species of *Primula*, but "no flowering plants"' (Blakelock, 1953, quoted in Raven & Walters, 1956). The story is told in detail by McLean (1997).

In 1948, a Second World War bomb crater at Brockham Hill, Surrey, was found to support an amazing selection of alien species, including four species of foxglove, *Digitalis lutea*, *D. grandiflora*, *D. lanata* and *D. ferruginea*. An obvious attempt to mislead others into thinking that the aliens might have been seed contaminants of the original bomb seems simply amusing in hindsight, but at the time it caused some debate (Lousley, 1949b). In fact, the trick was not well planned, for, although the first two foxglove species do occur near the English Channel in France and Belgium, the second two are found naturally no nearer than Italy. The *D. lutea* persisted 'for a good many years' there reported Lousley.

The most celebrated of the botanical frauds involved Professor John Heslop Harrison, FRS (1881–1967), of Newcastle University, who ran an annual expedition to Rum in the Inner Hebrides. Indeed, there is a whole book written about it by Karl Sabbagh (1999), entitled *A Rum Affair: How Botany's Piltdown Man was Unmasked*. Harrison is accused of fraudulently introducing the non-British *Carex bicolor*, *C. capitata*, *C. glacialis*, *Epilobium lactiflorum* and *Erigeron uniflorus* to the island during the 1940s. They were accepted (with reservations) by many at the time and all except the *Epilobium* were included (with caveats) in Dandy's *List of British Vascular Plants* (1958). They were all undoubtedly planted, but nevertheless they did exist for a while as aliens in our flora. Only *Carex bicolor* survived until 1950 (in fact, it lasted until 1961), and therefore none is strictly relevant to the present book. One can read the botanical details of the affair in Preston (2004) and Pearman *et al.* (2008). The incident was certainly much less scandalous than is suggested by Sabbagh, and there was no organised campaign to deceive as there had been with Piltdown Man. What remains as a mystery is why a scientist, so distinguished and academically accomplished as Harrison, should behave so foolishly.

CHAPTER 6

The Ecology of Establishment

We are seeing one of the great historical convulsions of the world's fauna and flora.
Charles Elton, *The Ecology of Invasions by Animals and Plants* (1958)

In order to understand the behaviour of a plant as an alien, it is a good idea to see it growing in its native environment. Let's go back sufficiently far in time so that humans have no measurable impact on its ecology. Our plant species is surrounded by other plants with which it competes for space, and for access to its essential resources of water, light and nutrients. It is attacked by range of viral, fungal and bacterial pathogens, and by numerous herbivores, both mammalian and invertebrate. It employs the services of mutualist organisms for pollination (insects, perhaps), fruit dispersal (birds) and nutrient capture (mycorrhizal fungi).

It is not obvious how much of our plant's evolutionary history has been spent in the company of this precise set of native species, because the ecosystem in which it currently lives has been dismantled and reassembled on numerous occasions as ice ages have come and gone. It is most likely that the different species went their own independent ways on each occasion, reassembling in different combinations during each interglacial period (Davis, 1987). The essential point, however, is that our species is able to persist in this pristine ecosystem. It can obtain all of the nutrients it requires, it receives all of the services it needs from its mutualists, and it can survive the depredations of its army of natural enemies. It can deal with the extremes of climate that it experiences. The prevailing disturbance regime creates opportunities for recruitment, and it has the means to survive between successive opportunities for regeneration. Whatever the vicissitudes of the environment, it takes them in its stride.

In particular, our plant exhibits the ability to recover from setbacks in its population size. This is an essential feature of all species in their native environments. A species that did not exhibit the ability to increase when rare would soon be driven to extinction as each successive wave of mortality drove its population further and further down. This central ecological principle is embodied in *the invasion criterion*: population size increases when population density is low. Let's assume that in this pristine environment, there are 1,500 native vascular plant species, and that they show 1,500 different combinations of traits. They all differ in their ecology, but they all pass the invasion criterion. The ability to increase when rare is clearly *not* the prerogative of a limited set of traits.

People now intervene, and move our plant from the place where it evolved into an alien environment where it is surrounded by a brand-new set of competitors, enemies and mutualists, with which it has shared no recent evolutionary history. Will it survive, and if it does, will it increase? Answering these questions is the object of this chapter. The essential issues involved are beautifully encapsulated in the Bible's Parable of the Sower (Matthew 13: 4):

> *Some seeds fell by the wayside, and the fowls came and devoured them up:*
> *Some fell upon stony places, where they had not much earth:*
> *And forthwith they sprung up, because they had no deepness of earth:*
> *And when the sun was up they were scorched;*
> *And because they had no root they withered away.*
> *And some fell among thorns; and the thorns sprung up and choked them:*
> *But others fell into good ground, and brought forth fruit,*
> *Some an hundred fold, some sixtyfold, some thirtyfold.*

Note that as well as cataloguing the hazards facing the seed (predation, unsuitable microsites and plant competition), this biblical reference provides the first published estimates of the intrinsic rate of increase of an annual plant. In this literary example, the net multiplication rate, traditionally denoted by λ (lambda), is between 30-fold and 100-fold (so $30 < \lambda < 100$). In terms of lambda, the invasion criterion states simply that $\lambda > 1$. Each of our invading aliens needs to leave, on average, more than one descendant in the next generation if it is to naturalise.

The minimum requirements for an alien species to establish in its new environment are an appropriate soil and climate, the resources it requires to grow (light, water, nutrients, CO_2), reasonable levels of protection from resident herbivores and pathogens (enemy-free space), a regeneration niche to allow seedling recruitment or vegetative spread, reasonable levels of protection from resident plants (competitor-free space) and a supply of services from resident mutualists sufficient for its needs for pollination (birds or insects), fruit dispersal (rodents, perhaps) and nutrient capture (mycorrhizal fungi and nodulating bacteria). From this list it is obvious that the perfect alien, if such a species were to exist, would be hardy, fertile, competitive, well defended, reproductively self-reliant and catholic (or flexible) in its choice of substrate and nutrient supply.

We start by investigating the different kinds of vascular plant species, the different kinds of environments they experienced and the different ecological circumstances in which they evolved as natives.

LIFE FORM

Life history theory tries to explain how evolution designs organisms to achieve reproductive success. The design is a solution to an ecological problem posed by the environment and subject to constraints intrinsic to the organism. Why are plants small or large? Why do they mature early or late? Why do they have few or many offspring? Why do they have a short or a long life? Why must they grow old and die (Stearns, 2000)?

The most conspicuous plant trait is its life form. Of all the many classifications of life form, the one that has stood the test of time is the system proposed by the Danish plant ecologist Christen Christensen Raunkiær (1860–1938) in his classic work *The Life Forms of Plants and Statistical Plant Geography*, published in 1934. The classification is based on the location in which the plant's buds are held during the unfavourable season (Table 4).

In the British Isles, a broad span of life forms is found among the most successful alien species, illustrating the fact that there is no single optimum life history when it comes to invasiveness (Fig. 120). Of the alien species that are most widespread within the British Isles, three are annuals (Pineappleweed, Common Field-speedwell, Italian Rye-grass), two are trees (Sycamore, Horse-chestnut), two are hemicryptophytes (Slender Speedwell, American Willowherb), two are shrubs (Gooseberry *Ribes uva-crispa*, Snowberry *Symphoricarpos albus*), one is a chamaephyte (Ivy-leaved Toadflax) and one is a rhizomatous geophyte (Japanese Knotweed). Not all of them produce copious seeds, although most do, but Japanese Knotweed and Slender Speedwell spread entirely by vegetative fragments, and Snowberry seldom recruits from seed in the British Isles.

194 · ALIEN PLANTS

TABLE 4. Definition of life-form categories according to Raunkiaer's (1934) classification, with examples and references to illustrations. The abbreviations of the categories as used in Appendices 1 and 2 are also provided. We find the distinction to be arbitrary between helophytes (marsh plants) and rooted hydrophytes on the one hand, and hemicryptophytes growing in wet ground on the other, so we have not recognised the helophyte category. Where a taxon exists in more than one life form, we have given the one we believe is commonest. For biennials or short-lived perennials, we have categorised the overwintering stage.

Term	Definition	Examples
Phanerophytes	Trees and shrubs; woody plants with buds >0.25 m above soil	Sycamore (*Acer pseudoplatanus*; Fig. 199) Japanese Rose (*Rosa rugosa*; Fig. 94)
Chamaephytes	Buds just above soil level ≤25 cm above soil; some are woody	Ivy-leaved Toadflax (*Cymbalaria muralis*; Fig. 111) Tresco Rhodostachys (*Ochagavia carnea*; Fig. 117a)
Hemicryptophytes	Herbaceous perennials and biennials, buds ± at soil level	Japanese Knotweed (*Fallopia japonica*; Fig. 273) Japanese Cowslip (*Primula japonica*; Fig. 146)
Geophytes	Herbaceous perennials, buds below soil level	Sea Daffodil (*Pancratium maritimum*; Fig. 81) Winter Heliotrope (*Petasites fragrans*; Fig. 210)
Hydrophytes	Water plants, either rooted or free-floating	Sweet-flag (*Acorus calamus*; Fig. 249) Large-flowered Waterweed (*Egeria densa*; Fig. 141)
Therophytes	Annuals; unfavourable season spent as seeds	Canadian Fleabane (*Conyza canadensis*; Fig. 122a) Sunflower (*Helianthus annuus*; Fig. 82)

The same genetic individual can produce potentially extensive patches of shoots by means of branching rhizomes, roots or stolons (e.g. alien trees like False-acacia or Tree-of-heaven *Ailanthus altissima*). One of the great advantages of clonal growth in highly disturbed habitats is that small clonal fragments can establish and grow in such a way that the disturbance itself acts as a means of multiplication and dispersal.

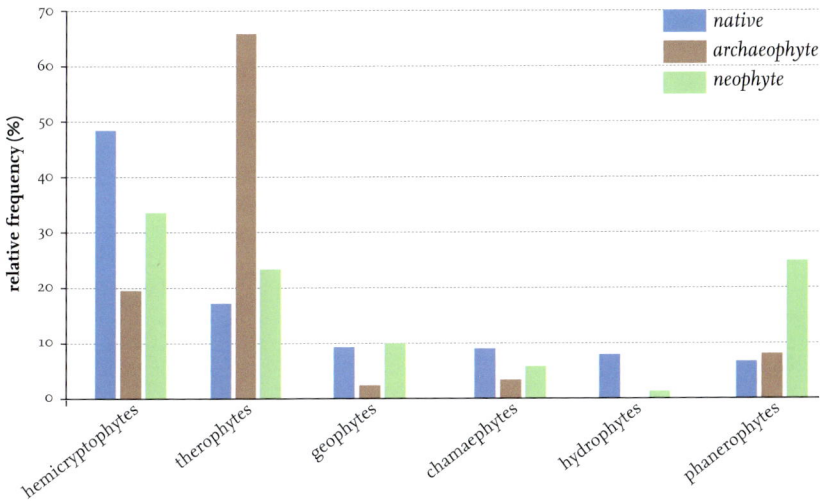

FIG 120. The relative frequencies (percentages) of native and alien plant species exhibiting different life forms (definitions in Table 4). The life forms are ranked left to right, on the basis of the percentage of the native flora they comprise (blue bars). Nearly half the native species are hemicryptophytes. Archaeophytes (brown bars) show a preponderance of therophytes (winter annuals and spring annuals), while neophytes (green bars) are relatively strongly represented by phanerophytes. (Species identities from Stace (2010) (natives), Appendix 1 (neophytes) and Appendix 2 (archaeophytes); life forms assigned by CAS and MJC, selecting the most frequent life form in cases where a species exhibits two or more forms)

REGENERATION NICHE AND SEEDLING ESTABLISHMENT

Plant species may differ in the details of the sets of circumstances that permit regeneration from seed. Peter Grubb (1977) called this the *regeneration niche*, drawing attention to the fact that all vascular plants share the same, relatively small number of resources (water, nitrogen, phosphorus), which apparently limits the opportunities for trophic niche specialisation, but there are no such obvious limits to regeneration conditions. The regeneration niche is a multi-dimensional concept embracing the physical conditions required for germination (dormancy-breaking stimuli, the correct temperature, the right soil moisture conditions, the appropriate levels of oxygen in the soil air, and the appropriate light intensity and the right colour of light – some seeds that will germinate in white light will not germinate in green light, because this indicates competition from established vegetation), the right kind of soil surface disturbance (e.g. the creation of a seedbed), freedom from competition, and protection from the enemies of seeds and seedlings.

Species may need different kinds of years for regeneration from seed (dry summers followed by wet autumns or frost-free winters, for instance). Or they may need very particular kinds of places for seedlings to become established: these are called *recruitment microsites* (or just *microsites* for short). These microsites need to provide all the necessary stimuli for germination (the right levels of moisture, temperature, light and oxygen), they must provide protection from extremes of the abiotic environment (not too hot, too cold, too wet or too dry) and, crucially, they must protect the seedling from native herbivores and pathogens. But perhaps the most important criterion is that the microsite provides sufficiently low levels of interspecific competition from established vegetation that the seedling can grow big enough to stand on its own. Gardeners know all about creating microsites, and what hard work it can be (cultivation, weeding, watering, prevention of damping-off, slug and pigeon control, and so on).

The tricky issue is how do we know what the regeneration niche of our alien species will be in its new environment? A simple, if slightly underhand, way of answering this question is to do an experiment. We sow the seeds and see what happens. If sturdy young plants appear, then microsites were present. If they don't, then something else was happening. In the first case, we say that recruitment was seed-limited: more seeds mean more plants. In the second case, we say that recruitment was not seed-limited. We need to do more work if we want to know *why* recruitment was not seed-limited. There are two obvious possibilities: either there were no microsites; or the seeds we sowed were eaten by animals or killed by fungal or bacterial pathogens. The first case is referred to as *microsite-limitation*, and the second as *predator-limitation* (Fig. 121).

At low rates of seed input, seed predators may consume all of the seeds, with the result that recruitment is prevented. This is the region of predator-limited recruitment. At intermediate seed densities, recruitment will be proportional to seed input. It is in this region of seed-limited recruitment that herbivores whose feeding reduces seed production might be expected to influence plant dynamics. At high rates of seed input, recruitment is limited by microsite availability, and substantial reductions in seed production caused by predators might have no measurable impact on recruitment. In some habitats in some years there may be no microsites at all. Under these circumstances there will be no seedling recruitment, however much seed is introduced. We can use these simple ideas to create an operational definition of the regeneration niche, as *the necessary and sufficient set of conditions under which recruitment is seed-limited.*

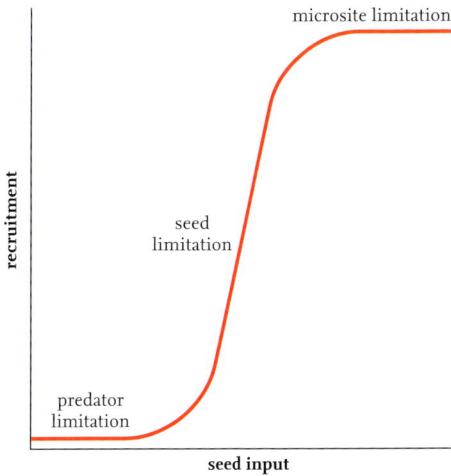

FIG 121. Recruitment limitation in plants. There are three distinct circumstances: predator-limitation; seed-limitation; and microsite-limitation (*see* text for details). Note that there may be substantial ranges of sowing density where increasing seed supply does *not* lead to increased plant numbers.

In grasslands at Silwood Park in Berkshire, experimentally determined regeneration niches were rather similar for two alien species of Asteraceae: the annual Guernsey Fleabane (Fig. 122b) and the perennial Narrow-leaved Ragwort (Fig. 270). There was no recruitment by either species when seed was sown into intact grassland, presumably as a consequence of intense competition from established perennial plants. When competition was reduced by cultivating the ground, grazing by Rabbits (*Oryctolagus cuniculus*) effectively precluded recruitment by Guernsey Fleabane and significantly reduced it for Narrow-leaved Ragwort. However, when the cultivated ground was fenced against Rabbits, both species were capable of regeneration from seed, and the rate of recruitment was increased when slugs and snails were excluded using chemical molluscicide (but exclusion of insects had no measurable impact on plant recruitment in either case). Thus for these two alien species in this particular habitat, the experiment allows us to rank the importance of the various ecological factors affecting invasion: interspecific competition from established native plants was much the most important barrier, then herbivory from resident generalist herbivores, with Rabbits more important than molluscs, and molluscs more important than insects. In the longer term, Guernsey Fleabane failed to naturalise in the vicinity of these experiments, but Narrow-leaved Ragwort spread to bare roadsides and open waste ground nearby. Also, once established following cultivation and protection from herbivores, individual Narrow-leaved Ragwort plants were able to survive for a decade or more, despite the return of intense competition in the regenerated grassland (Scherber *et al.*, 2003).

FIG 122. Fleabanes. There have been successive waves of invasion by species of fleabane (*Conyza*), all weedy plants with small, dull-coloured flower heads: (a) Canadian Fleabane (*C. canadensis*); (b) Guernsey Fleabane (*C. sumatrensis*); (c) Bilbao's Fleabane (*C. floribunda*); (d) Argentine Fleabane (*C. bonariensis*). (a–c, CAS; d, MJC)

The fact that recruitment is not seed-limited all the time, even in habitats that are clearly suitable on all other criteria like climate, soil and resources, is one of the main reasons that repeated introduction of propagules is so crucial for initial establishment of alien plants. In studies of alien establishments in New Zealand, for instance, Clare Veltman found that the number of years, the number of sites per year and the number of individuals per site per year were the most important factors influencing establishment, and far more important than any of the biological traits of the introduced species or the recipient habitats (Veltman *et al.*, 1996).

TRAITS

A species is a constellation of interacting traits. Traits are so obviously important that it is frustrating that they predict so little. We know that a species can increase when rare in its native habitat, so all of the combinations of traits that we find bundled together as extant species are capable of invasion, so long as the circumstances are right. Those circumstances of climate, microclimate, substrate, resources, competitors, natural enemies and mutualists are bound to be different in the alien range than they were in the native range, if only to the extent that the cast of characters is different (Sun *et al.*, 2013), but we have to admit that all species are potential invaders, just as all communities are in principle invasible. Because all trait combinations possessed by extant plant species are capable of exponential increase when rare in at least one set of environmental circumstances, it is impossible, even in principle, to predict the success of a plant in an alien environment on the basis of traits alone. This is because the new ecological context is unknown, and indeed unknowable without the experience of the introduction itself.

As students of ecology and evolution have drummed into them repeatedly, it's not the genotype (genetic make-up), and it's not the environment, but the genotype-by-environment interaction that matters. The plant is not free to maximise its fitness one trait at a time. Natural selection works on the entire phenotype (observable characteristics), and this is the environment-dependent integration of all of its traits. Trade-offs between traits are inescapable and absolutely universal.

Seed size
It seems fairly obvious that seed size would be important for alien plants, because in head-to-head competition within the same microsite, then surely the larger seed is likely to produce the winning seedling? Well, possibly. Seed size is

involved in the most profound of all the botanical trade-offs: plants that produce large seeds must necessarily produce fewer of them. This seed size vs. seed number trade-off means that during an invasion, small-seeded species are likely to find their way into many more microsites than are large-seeded species, and as a result of this, are more likely to find themselves in microsites where there is no superior competitor present. The aliens Guernsey Fleabane and Bilbao's Fleabane (Fig. 122b, c) are fine examples of highly successful invasive species with copious but tiny wind-dispersed seeds. In contrast, some of the most successful alien trees, like Turkey Oak, Horse Chestnut and Sweet Chestnut, have massive seeds.

Across species, seed mass is positively correlated with plant size. Plants with bigger canopy area, plant height, stem diameter, plant mass and canopy volume have significantly heavier seeds. There are also positive correlations between seed mass and time to first reproduction, plant lifespan and reproductive lifespan. Although small-seeded species produce more seeds per unit canopy area per year than do large-seeded species, the latter tend to have larger canopies and to reproduce for more years, with the upshot that seed mass is not significantly related to the total number of seeds produced by an individual plant throughout its lifetime. To understand the forces shaping the evolution of seed mass, we need to consider plant size and longevity, as well as seedling survival rates and the number of seeds that can be produced for a given amount of energy (Moles *et al.*, 2004).

Phenology

Phenology is the science of the timing of things: the dates of bud burst, flowering, anthesis, seed ripening, leaf coloration or leaf fall, and the extent to which these timings are affected by day length (as affected by latitude) or by annually varying factors like the earliness of spring (as determined by accumulated heat). Robert Marsham (1708–97) is the founding father of modern phenological recording. He was a wealthy landowner who kept systematic records of 'indications of spring' on his estate at Stratton Strawless, Norfolk, from 1736. These were in the form of dates of the first occurrence of events such as flowering, bud burst, emergence or flight of an insect. Consistent records of the same events were maintained by generations of the same family over unprecedentedly long periods of time, eventually ending with the death of Mary Marsham in 1958, so that trends can be observed and related to long-term climate records. For example, over the past 250 years, the first leafing date of oak appears to have advanced by about eight days, corresponding to overall warming of the order of 1.5° C in the same period (Fitter & Fitter, 2002).

Phenology can be vitally important for alien species, particularly if dispersal involves a rapid shift of hemisphere so that summer suddenly becomes winter.

Once acclimatised, there might be vacant niches for species that grow out of peak season, for instance. Species that can respond opportunistically to favourable growing conditions might be expected to prosper, particularly if this gave them a head start over the native species (e.g. after rains or fires, or at the beginning of spring). Shifts in flowering time might allow coexistence with pollen-compatible native species. Evolutionary modifications to reproductive systems could promote the colonising ability of invading populations. Reproductive timing is an important target of selection during range expansion (Barrett *et al.*, 2008).

It is a conspicuous feature of plant-hunting in the British Isles that relatively more alien species than natives flower very early (e.g. snowdrops *Galanthus*, Winter Heliotrope, Winter Aconite *Eranthis hyemalis*) or they only start to flower very late in the year (e.g. Sowbread, mugworts *Artemisia*). For 43 native and 30 alien shrub and liana species common to the understorey of deciduous forests in the eastern United States, Jason Fridley (2012) showed that extended autumn leaf phenology was a common attribute of eastern USA forest invasions, where alien species are extending the autumn growing season by an average of four weeks compared with natives. Fridley found that most alien species capture a significant proportion of their annual carbon assimilate after canopy leaf fall, a behaviour consistent across five phylogenetic groups and virtually absent in natives.

Leaf traits

There is enormous variation in leaf size, shape, thickness, degree of dissection, toughness, nitrogen content and maximum photosynthetic rates across species, and it is inevitable that leaf traits will be important for colonising plants. Leaves need to intercept light and allow CO_2 into the interior of leaf, while minimising water loss and respiratory costs and protecting tissues against consumption by herbivores. These conflicting demands impose challenging design constraints. Some groups of adaptations are correlated: for example, tough, leathery leaves that are adaptations for drought resistance are often effective anti-herbivore defences. The density of the stomata and their location (on both surfaces or just one surface, protected from airflow by a dense covering of hairs or not) are key determinants of water-use efficiency (e.g. contrast the flimsy leaves of Orange Balsam *Impatiens capensis* with the tough leaves of Blue Globe-thistle *Echinops bannaticus*). Variations in leaf size are among the most extreme, from giants like the Brazilian Giant-rhubarb, with leaves more than 2 m across, to miniatures like Mind-your-own-business, with leaves less than 6 mm across.

Shade tolerance is involved in a classic trade-off with growth rate. Traits that allow a plant to maintain a low death rate in shade (e.g. well-defended evergreen leaves with high construction costs) are bought at the price of reduced growth rate in full sun. Leaf traits in our alien plants vary from long-lived, leathery and shade-tolerant in Rhododendron and Cherry Laurel (*Prunus laurocerasus*), to short-lived, flimsy and light demanding in Hemlock.

Relative growth rate

A trait that summarises many of the most important morphological and physiological features of a plant is its relative growth rate (RGR). This is the dry weight gain per unit time, and is usually measured on a logarithmic scale so that it is dimensionless (RGR = (log(dry weight at the end) – log(dry weight at the start)) / time between the start and the end). Experience has shown that the shorter the period over which the weight gain is measured, the more useful is RGR in distinguishing species (Grime *et al.*, 1988). Also, RGR early in life has been found to be a more discriminating feature than RGR measured once the plant is mature. This has led several investigators to propose that seedling RGR should be used as a predictor for invasive behaviour in alien plants. For pines, high RGR, relatively small seed masses and short generation times characterise the species that are successful invaders in disturbed habitats. While net assimilation rate, leaf mass ratio and specific leaf area were all found to contribute positively to RGR, specific leaf area was found to be the main component responsible for differences in RGR between invasive and non-invasive pines. Analysing phylogenetically independent contrasts of known invasive and non-invasive alien woody horticultural species showed that the RGRs of invasive species were significantly higher, suggesting that this may serve as a useful screening tool for invasiveness for woody angiosperms (Grotkopp & Rejmanek, 2007).

Are there consistent traits of weediness?

There has been a long debate about whether or not there are predictable traits of weediness. A paper by Herbert Baker (1974) was extremely influential, and weedy traits are often referred to in aggregate as 'the Baker list' (Table 5). Plants that have none or a very few of the characteristics are unlikely to be weedy, but weeds certainly do not need all of the traits to be pestilential. Rapid growth through the vegetative phase to the flowering condition is a relatively consistent feature of weeds, and enables them to flourish in environments that are favourable for only a short period of time (as well as to produce more than one generation each year when the favourable season is prolonged). Rapid vegetative growth also provides the necessary basis for phenotypic plasticity (*see* below).

The main problem with Baker's analysis is that it focuses on weeds of arable agriculture; that is to say on plants from early successional and nutrient-rich communities. It is not at all obvious that patterns exhibited by weeds under these

TABLE 5. Baker's list of the attributes of an ideal weed. For Baker, a plant is a weed 'if, in any specified geographical area, its populations grow entirely or predominantly in situations markedly disturbed by man (without, of course, being deliberately cultivated plants)' (Baker, 1974).

1	Germination requirements fulfilled in many environments.
2	Discontinuous germination (internally controlled) and great longevity of seed.
3	Rapid growth through vegetative phase to flowering.
4	Continuous seed production for as long as growing conditions permit.
5	Self-compatible but not completely autogamous or apomictic.
6	When cross-pollinated, unspecialised visitors or wind utilised.
7	Very high seed output in favourable environmental circumstances.
8	Produces some seed in wide range of environmental conditions; tolerant and plastic.
9	Has adaptations for short- and long-distance dispersal.
10	If a perennial, has vigorous vegetative reproduction or regeneration from fragments.
11	If a perennial, has brittleness, so not easily drawn from ground.
12	Can compete interspecifically by special means (rosette, choking growth, allelochemicals).

circumstances would match the traits of invasive alien plants from, say, nutrient-poor late-successional communities.

Plasticity

Phenotypic plasticity is the ability of the same plant genotype to take on different sizes, shapes or flowering behaviour under different environmental conditions. Because of the modular construction of most vascular plants (buds can produce shoots that support more buds), the same genotype can often exhibit a wide range of phenotypes when grown under different environmental conditions, simply by producing more or fewer modules, or by varying the size of the modules. Some plant parts typically show very low levels of phenotypic plasticity (like flower size or seed size), while others show much higher levels of plasticity (like seed number per fruit, leaf length, root length, hairiness, etc.).

Phenotypic plasticity is commonly considered as a trait associated with invasiveness in alien plants because it may enhance the ability of plants to occupy a wide range of environments (Davidson *et al.*, 2011). But do invasive plant species really have greater phenotypic plasticity than non-invasive species? And if so, how does this affect their fitness relative to native non-invasive species? What role might this play in plant invasions?

Assuming that plasticity enhances ecological niche breadth and therefore confers a fitness advantage, recent studies have posed two main hypotheses: invasive species are more plastic than non-invasive or native ones; and populations in the introduced range of an invasive species have evolved greater plasticity than populations in the native range. A successful invader may benefit from plasticity as either a jack of all trades, better able to maintain fitness in unfavourable environments; or as a master of some, better able to increase fitness in favourable environments; or as a jack and master, combining some level of both abilities (Richards *et al.*, 2006). For the flora of the British Isles, there is no compelling evidence that invasive species are more plastic than non-invasive or native species.

Evolvability

The capacity of a lineage to evolve has been termed its evolvability – its ability to generate heritable, selectable phenotypic variation. This capacity may have two components: to reduce the potential lethality of mutations; and to reduce the number of mutations needed to produce phenotypically novel traits (Kirschner & Gerhart, 1998). Evolvability may involve reducing the constraints on change and allowing the accumulation of non-lethal variation. It is plausible that this may have come about in the course of selection for robust, flexible processes that were suitable for complex development and physiology, and that it might have been specifically selected in lineages undergoing repeated radiations. The central problem is that almost all novel variation is maladaptive. We need to know whether certain properties can bias the kind and amount of phenotypic variation that is produced in response to random mutation, in such a way that more favourable and non-lethal kinds of variation are available on which natural selection can act. Models confirm in principle that rules for generating phenotypic variation can affect the evolvability of a system (Kirschner & Gerhart, 1998). Recent research suggests that transposable elements (jumping genes) could play an important role in enhancing the evolvability of alien species. Current understanding of the activity dynamics of transposable elements in genomes is that periods of relative dormancy are followed by bursts of activity, often induced by biotic and abiotic stress, such as exposure to novel habitats. Frequent transposition leads to genomic rearrangements, thus producing new genetic variants that might allow rapid adaptation of invasive species to novel environments (Schrader *et al.*, 2014).

Theoretical results suggest that disturbance in the form of fluctuating environments might select for organismal flexibility, or alternatively, the evolution of evolvability. Once genetic variance is generated via mutations, temporally fluctuating selection across generations might promote the accumulation and maintenance of genetic variation. We need to remember, too, that natural selection works with the best traits available at the time, not the best possible. Recent studies suggest that the invasion success of many species might depend more heavily on their ability to respond to natural selection than on broad physiological tolerance or plasticity (Kirschner & Gerhart, 1998).

Breeding systems

We need to know whether different breeding systems (*see* Chapter 8) are associated with greater or lesser success in plant invasions. Breeding system matters in a study of alien species, because one would imagine that obligate outbreeders would be more likely to suffer the ill effects of low population density (Allee effects) than self-compatible species because of the low probability of receiving suitable pollen. Whether to get involved with sex at all, and whether the costs of inbreeding are great enough to militate against self-fertilisation, are important issues for invading plant species. There is a continuum from obligate outbreeders (self-incompatible plants), through facultative outbreeders, to selfers and then apomicts (individuals that produce seeds without sex, so that every embryo has the exact same genotype as its mother). The apomicts might be expected to do best when environmental conditions in the next generation are expected to be the same as in the present generation, and obligate outbreeders when conditions are unpredictable. Perhaps facultative outbreeders get the best of both worlds, and hence might make the most successful invaders? Within the common alien plants of the British Isles, we find the complete spectrum, from outbreeders like Oxford Ragwort, to self-compatible species like Common Field-speedwell, to species like Japanese Knotweed and the archaeophyte Horse-radish, which never set seed at all.

Genetics

Having exactly the right genotype might make all the difference as to whether or not an alien species becomes established. To this end, the more introductions that are made, the more propagules per introduction and the wider the range of genotypes in each sample, the better. If there are genetic gradients (clines) in the native environment, then material collected from many different locations on the cline will increase the chance that at least one of them is adapted to the new environment.

The classical view is that invasive species are genetically depauperate because of founder effects. The founder effect is the loss of genetic variation that occurs when a new population is established by a small number of alien individuals from a larger native population. The new alien population is often very small and therefore shows increased sensitivity to genetic drift, an increase in inbreeding and relatively low genetic variation. As a result of the

206 · ALIEN PLANTS

loss of genetic variation, the new population may be distinctively different, both genetically and phenotypically, from the parent population from which it is derived, and in extreme cases, the founder effect is thought to lead to the speciation and subsequent evolution of new species. Overall, there tend to be significant losses of both allelic richness and heterozygosity in introduced populations, and large gains in diversity are rare, but populations with diminished diversity may still evolve rapidly (Dlugosch & Parker, 2008). However, evidence from numerous studies of alien plant population genetics suggests that any negative effects of genetic bottlenecks have been largely overcome by multiple introductions involving samples from different parts of the native range (Huttanus *et al.*, 2011).

Just as a species has no chance of successful invasion if it happens to be highly palatable to the resident herbivores, so an outbreeding species will struggle if it happens to be sexually compatible with a closely related dominant native plant. Suppose that an acorn of an alien white oak (*Quercus*, Section *Quercus*) is planted in southern England. It grows happily for 25 years up to reproductive maturity, then attempts to set its first acorn. It is a racing certainty that the stigma will be pollinated by the native Pedunculate Oak. The acorn may produce a plant that survives to reproductive maturity, but it has only 50 per cent of the alien genes. When this hybrid flowers after 25 years or so, the likelihood is that its female flowers will again be pollinated by the dominant native tree, so the alien genes decline to 25 per cent, and so on. This process means that, *other things being equal*, it will be much easier for self-compatible species to invade than for obligate outbreeders, and that obligate outbreeders are more likely to invade an ecosystem that does not contain any close relatives with which they happen to be pollen-compatible.

If, on the other hand, the results of hybridisation of the first generation of alien stigmas with pollen from the native dominant was immediately associated with hybrid vigour, and this was coupled with mutations or recombinations that created self-compatibility, or apomixis, then a super-invader might have been created. Under these circumstances, the abundant pollen of the native dominant is no longer a problem. Fortunately, we have few recorded cases of this behaviour, but the origin of Common Cord-grass is an object lesson in what can happen (*see* Chapter 9).

So far, we have dealt with the traits of a species that evolved in its native range. It remains for us to consider the ecological circumstances under which it lived in its native range. In particular, we need to understand the successional stage to which it is adapted, the resource supply within that successional stage, and the severity of the abiotic conditions that prevail.

SUCCESSION

Ecological succession is the process by which plant communities replace one another in an orderly sequence over the course of decades or centuries, as bare ground eventually becomes mature forest. If the process starts from bare rock, after glaciation, say, or following a lava flow, or from bare sand or estuarine mud, then it is referred to as *primary succession*. If it starts with a fully formed soil after tree-fall, flood or cultivation, then it is referred to as *secondary succession*. Physical disturbance, fire and herbivory each has the ability to destroy the existing biomass and to reset the successional clock.

A primary succession involves gradual change from a situation with abundant light but very low soil-nutrient availability, to a climax forest with a relatively nitrogen-rich soil but deep shade at the soil surface. Primary successions are characterised by relay-floristics (facilitation) in which species B cannot invade until species A has lived there long enough to create the conditions under which species B can establish. In due course, species B outcompetes species A. Species B then creates the conditions in which species C can establish, and the relay goes on until a stable equilibrium is developed in which the dominant plants are capable of replacing themselves. This community is the end point of the succession, and is known as the climax community. Primary successions are often little affected by people, at least in their early stages, and consequently alien plants are typically unimportant. There are some important exceptions in other parts of the world (e.g. the nitrogen-fixing shrub Fire Tree *Morella faya* (Myricaceae) short-circuits primary succession on volcanic lava in Hawai'i (Vitousek & Walker, 1989)), and the neonative Common Cord-grass (Fig. 167) is important in driving primary succession on tidal mudflats in the British Isles and elsewhere.

Because the soil is already formed at the beginning of a secondary succession, vascular plants can grow from the very start, and facilitation is much less important than it is in primary succession. The structure of the plant community in the early stages of a secondary plant succession is strongly influenced by chance events: which species set abundant seed that year, the particular conditions affecting seed immigration, and the size and composition of the soil seed bank all influence the botanical composition of the community in the years immediately after disturbance. A key ecological difference between primary and secondary successions is in the importance of inhibitory processes. One can think of these as the opposite of facilitation: now species B cannot invade until the competitive ability of species A has been impaired (as might happen as the result of a build-up of specialist root-feeding herbivores and fungal pathogens in the soil beneath it (van der Putten *et al.*, 2013)). These days, however, alien species are often conspicuous during the early stages of secondary succession. On abandoned arable soils in southern England, for instance, one finds abundant alien annuals like Canadian Fleabane, while alien shrubs like Butterfly-bush are frequent in mid-successional communities, and late successions are characterised by alien trees like Sweet Chestnut.

For both primary and secondary cases, succession proceeds because the dominant life forms pollute their environment, making it unsuitable for their own continued existence and inviting invasion from other species. In real rather than textbook successions all three processes (facilitation, inhibition and initial floristic composition) take place simultaneously, with each individual species reacting in its own individual way. What is clear is that the resources for which plants compete change completely: there is plenty of sunlight at ground level in the beginning and very little at the end of succession, but competition for microsites may be equally intense throughout.

The most important determinant of successional dynamics is the *disturbance regime*. In the pristine habitat this was driven by natural environmental processes, but these days the disturbance regime is determined more or less directly by people. The most useful definition of disturbance is by Philip Grime (1979): 'the mechanisms which limit the plant biomass by causing its partial or total destruction'. The disturbance regime comprises the temporal pattern of biomass destruction, its frequency, its intensity and its predictability. In the context of invasion biology, the word 'disturbance' is typically used in the context of disturbance caused by people: soil cultivation, drainage, construction work, eutrophication, earthworks, soil compaction, recreational pressures, and so forth.

It is hard to produce a simple classification of disturbance regimes, because there are at least two essential components: disturbances that cause destruction of existing biomass, leading to reduced competition; and disturbances of the soil surface that create a seedbed. Like so much else in invasion biology, the notion that invasion requires disturbance is often based on circular reasoning: pristine ecosystems are uninvaded, therefore pristine ecosystems are uninvasible. But clearly this is not true. Pristine ecosystems are pristine because they experience zero propagule pressure. They are uninvaded because the alien species simply have not been introduced. As we shall see in Chapter 13, it is clearly true that the most disturbed habitats are those that have the most alien species in them, and that alien species reach the highest levels of abundance in these disturbed habitats. But we need to be careful to distinguish between cause and correlation. One of the most important determinants of alien plant species richness and abundance is propagule pressure, and propagule pressure is greatest in habitats most closely associated with people. These same habitats are often defined as being the most disturbed.

Fire regime

Fire is a much less important ecological factor in the British Isles than in many other parts of the world. There are two reasons for burning. The first, on upland heaths, is as part of grouse management to maintain a patchwork of heather of different ages to optimise the balance of tall plants for cover and short plants for high-quality food production (Hudson, 1992). The second is on lowland heaths, to prevent domination of the heather by invasive native species like Scots Pine, Silver Birch and Downy Birch (*Betula pubescens*). As explained in Chapter 13, both dry heath and wet heath have very low levels of alien vascular plant representation, as measured either by number of species or abundance and impact, and burned grouse moors have neither higher nor lower levels of invasion than unburned moors. There is a suggestion, however, that the alien Heath Star Moss (*Campylopus introflexus*; Fig. 247) is more invasive on burned than on unburned heath (Clement & Touffet, 1990).

The r-K continuum

In a textbook model of population growth, plant numbers start out low and increase exponentially at a rate traditionally denoted by *r* (known to ecologists as 'little r'), which is the per capita difference between birth and death rates, measured at low density. Eventually numbers become high enough for competition to begin to be influential, so population growth slows. As total population size goes up, the death rate increases and the birth rate declines, until eventually the two rates converge and the population comes to an equilibrium. This population density is traditionally denoted by *K* (the 'carrying capacity'). The two parameters, *r* and *K*, provide a convenient way of summarising one of the most important trait axes in ecology, describing successional change and reflecting the time that has passed since the last major disturbance.

Immediately after a disturbance, *r*-selected species exhibit traits associated with rapid population growth under conditions where competition is relatively unimportant (fast growth, early reproduction, small size, short life, small seeds, self-pollination, wide dispersal, long-lived seed banks), while *K*-selected species exhibit traits associated with slow (or no) population growth, where competition is relatively important (large size, long life, protracted pre-reproductive development, large seeds, exotic pollination and fruit-dispersal mechanisms). At the *K* end of the spectrum, competition is predictably intense and opportunities for establishment are scarce. *K*-selected individuals are therefore expected to reproduce repeatedly over a long period (polycarpy or iteroparity), and to put proportionally more energy into survival than reproduction (examples include trees like Lawson's Cypress *Chamaecyparis lawsoniana* and herbs like Summer

Snowflake). In contrast, *r*-selected individuals live in unpredictable, often early successional (and hence ephemeral) habitats. They may well put all of their energy into reproduction and dispersal, so that they die immediately after seed production (monocarpy or semelparity), as illustrated by annuals like Sticky Groundsel or biennials like Orange Mullein.

The S-C continuum

The other major trait axis in plant ecology has to do with resource supply. Habitats differ in the supply rates of light, water, nitrogen and phosphorus. These four factors can be compressed into a single axis, the two ends of which Peter Grime (1979) christened *stress tolerators* (*S*) and *competitors* (*C*). When resources are scarce, we find plants that grow slowly, hold onto their hard-won resources tenaciously, defend themselves against herbivores and use the scarcest resources with great efficiency. When resources are plentiful, we find plants that grow quickly, are profligate with their tissues, attain large final size and are relatively palatable to herbivores. At the *S* end of the axis we find species that are conservative, with small, tough leaves that have a low specific surface area and low nutrient content as an adaptation for long-term resource conservation. Good examples are Houseleek and Thick-leaved Stonecrop (*Sedum dasyphyllum*), which grow on essentially soil-free substrates like roofs and walls. At the *C* end of the axis we find species that are ruthlessly acquisitive, with large, tender leaves that have a high specific surface area as an adaptation for the rapid acquisition of resources. The fast-growing annual balsams (*Impatiens*) are good examples of this strategy.

There are certainly fewer alien species in environments that are traditionally regarded as harsh (very dry, very nutrient-poor or very exposed habitats; *see* Chapter 13), and the hypothesis has been proposed that stressful environments are less invasible than benign environments. The most obvious criticism is that experiments have not been carried out to test this, because there has been no control over propagule pressure. Perhaps it is simply that people are less likely to introduce stress-adapted species into stressful environments? There is also the problem of what we mean by a stressful environment. John Harper (1925–2009) felt that all environments are equally stressful; it is just that they are stressful in different ways. His tongue-in-cheek definition of stress was 'what I imagine I shouldn't like if I were a buttercup'. When water and nutrients are scarce, then plants that are good competitors for water and nutrients do best. But when water and nutrients are plentiful, plants that are good competitors for light do best. Neither circumstance is obviously more or less stressful than the other, and competitive exclusion of less well adapted species occurs in both habitats.

ESTABLISHMENT IN THE ALIEN ENVIRONMENT

So much for the behaviour of species in their native environment. We now move the plant to the British Isles. In thinking about the establishment of alien species in their new environment, it will be useful to distinguish between intentional and unintentional introductions (i.e. contrasting Chapters 4 and 5). The reason is that intentional introductions are likely to be cossetted from the outset. The more that people value the species, the harder they are likely to try to get it established. They will sow plenty of seeds, they will weed them and water them, and they will protect them from pests. Contrast this with an unintended stowaway. It arrives on foreign shores in small numbers, perhaps in the wrong habitat, with no friends and many enemies. There are lots of ways for an invasion to fail, and the order in which we deal with the various barriers facing an invading species can only reflect an approximate ranking of their importance in any particular case.

It goes without saying that a source of propagules is the essential prerequisite. Propagules are like botanical lottery tickets: the more lottery tickets allocated to a species, the greater its probability of successful establishment. The importance of chance and timing in determining the success of an invasion is reduced if there are multiple introductions, in different places and at different times of year, each of which is made up of a mixture of different genotypes. Usually, propagule pressure will decline rapidly as one moves away from human settlements, factories, canals, ports, roads, railways, airports and other communication routes into semi-natural habitats. A useful definition of a pristine habitat is one where the propagule pressure of alien plants is zero. Needless to say, there are few pristine habitats left on Earth.

For intentional introductions, the greater the effort, the greater the likelihood of naturalisation. This is reflected in the fact that for horticultural species, both planting frequency in European gardens and time since introduction significantly increases naturalisation success, but the effect of planting frequency on naturalisation success is much stronger for non-trees than for trees (Bucharova & van Kleunen, 2009).

Climatic matching

A great many species could germinate and grow happily if they arrived in the new environment during well-watered conditions in spring. The crunch comes with the onset of the first unfavourable season. This might be a protracted summer drought, or exposure to very low temperatures or icy winds in winter. The probability of persistence through the first year is correlated with the fit between the fundamental niche in the native environment and the weather conditions experienced in the new one. This fit is known as climatic matching.

The beach-head population

Intentional introductions are likely to be taken directly to their target habitat, and are more likely than not to find this to be relatively convivial. Some unintended introductions are also unintentionally cosseted, as are the species that arrive as contaminants of crop seeds or horticultural specimens. Most unintentional introductions, however, are likely to arrive on a dock, on a riverbank, in a rubbish dump or in a railway yard. The probability of establishment therefore depends on their pre-adaptation to these randomly imposed circumstances. They may be a stepping stone to the habitat in which they are destined ultimately to become naturalised, or this may be as far as they ever get into the ecology of their new country.

Pre-adapted competitive superiority

Of the purely ecological hypotheses about successful alien plant establishment, perhaps the most straightforward is that, of all the alien plants species that were introduced, the ones that succeed are simply those that just happened to exhibit competitive superiority over the native plants. The evolutionary play in their native environment led to the creation of what turned out to be a superior phenotype in an alien environment, simply because the ecological theatre was different: there was a different cast of characters (competitors, natural enemies and mutualists) and the species were interacting in a different abiotic setting.

It is also plausible that the native species carry excess genetic baggage and that this could put them at a competitive disadvantage compared with the aliens. The native species have to be capable of withstanding the worst that the native environment can throw at them: the most protracted droughts, the latest spring frosts, the most violent July gales. These may be once-in-a-millennium events, meaning that the natives have substantial spare capacity in normal years, as a result of natural selection to which they were exposed in the past. As long as none of these extreme events occurs, the alien may be at an unfair advantage because it does not carry this excess baggage. Of course, this is a short-term state of affairs, and the alien will suffer if and when these extreme events do occur. However, we have been importing neophytes only since 1500, and 500 years is the merest blink of an eye in terms of plant evolution.

Native species richness and invasibility

It is important to know whether species-rich environments are more or less invasible by alien plant species. Both Charles Elton and Charles Darwin clearly thought that species-rich communities were *less* invasible. They assumed that the more native species there were, the fewer opportunities there would be for alien species to invade. More species meant more biotic resistance, so diverse native communities should be less invasible than species-poor communities, as a result of more intense competition and higher levels of herbivory.

They also believed that there was more likely to be a vacant niche in a species-poor community than a species-rich one. This is presumably because they thought that all habitats had roughly the same number of niches, and that in species-poor communities rather few of them were filled. But why should habitats that differ in resource supply, climate and soil have the same number of niches? This doesn't make any sense as a fundamental assumption, and it has found little support from recent field studies. Where carefully controlled analyses have been carried out, they tend to show the opposite pattern, with communities that are rich in native species being rich in alien species too (Fridley *et al.*, 2007). With hindsight, it is plausible to argue that the presence of lots of native species demonstrates that many species are capable of coexisting under those ecological circumstances, and under those circumstances it should be easier, not harder, for alien species to establish.

Phylogenetic relatedness and invasibility

Darwin speculated that introduced plant species will be less likely to establish a self-sustaining wild population in places where the native species are close relatives. This has come to be known as *Darwin's naturalisation hypothesis* (Thuiller *et al.*, 2010). There are main two ideas operating here. It is assumed that closely related plant species are likely to be better competitors than more distantly related plants. Also, closely related native species are likely to support herbivores and pathogens that are assumed to be more likely to attack alien invaders than natural enemies that are adapted to more distantly related plants.

We should note, however, that both of these ideas are more often assumed than they are demonstrated. There is some support, however, from recent experimental work using microcosms involving native and alien bacterial species. Invader abundance was best explained by phylogenetic distance between the invader and its nearest resident relative (Jiang *et al.*, 2010). But field studies on vascular plants tend to show much larger effects for ecology than phylogeny in predicting invasiveness. At small spatial scales, where neighbouring individuals interact, competition is likely to be most important and one might predict that distantly related species would be more likely to establish. At larger, geographic scales, then adaptation to the abiotic environment, especially the prevailing climate, will be most important and one might predict that close relatives would be more likely to establish because near relatives are more likely to be adapted to the prevailing abiotic conditions. If environmental filters drive trait convergence of native and non-native species, then we might expect a positive effect of relatedness on invasibility at large spatial scales. If competition drives trait divergence then we might expect to see a negative effect of relatedness at small spatial scales. It would be fair to say that no consistent pattern has yet emerged from data on real invaded ecosystems (Ness *et al.*, 2011).

People plants

Some species are pre-adapted to live with humans and their domestic livestock. They enjoy the nutrient-rich, high-pH substrates that people typically produce, they relish the disturbance regimes that people impose, and they tolerate the attentions of domestic livestock. These 'people plants' evolved with Europeans over thousands of years and have come to specialise on the habitats in which people live and work. The most obvious people plants are those aliens that were pre-adapted to live in completely novel habitats that have been created by people (railway ballast, rubbish dumps, mortared brick walls, pitched roofs, canals, etc.).

In the British Isles, our semi-natural vegetation is all more or less unnatural: it is cut-over, grazed, drained, and both fertilised and acidified by atmospheric inputs. This characteristic suite of intentional and unintentional habitat management practices has been imposed wherever Europeans have settled. These human-influenced habitats are characteristically rich in alien plant species. But is this because the disturbance regime renders these habitats more invasible, or is it due to pre-adaptation of a set of people plants as the result of centuries of natural selection in the company of humans? The dominance of invasive species is often assumed to reflect their competitive superiority over the displaced native species, but it may be nothing more than pre-adaptation to the management regime and associated disturbance involved in living close to people.

Invasibility and insularity

There are a number of hypotheses about pre-adapted competitive superiority that might boil down to nothing more than differences in the relative importance of people plants. For instance, New World ecosystems are supposed to be more invasible than Old World ecosystems, temperate ecosystems are supposed to be more invasible than tropical ecosystems, and islands are supposed to be more invasible than similar continental areas. In each case it could simply be that people plants evolved in temperate continental conditions in the Old World, and when people took them to the New World, to the tropics or to islands, they could exhibit their pre-adaptation to living with people.

Species that evolved on continents are supposed to be superior competitors to species that evolved on remote islands, and remote islands typically have small floras with high rates of endemism and highly depauperate faunas. The situation in relation to alien plants on remote islands is confounded by the habit of the ancient mariners of guaranteeing a future supply of fresh meat by introducing pigs, goats and sheep to the islands they visited. The depredations of these alien herbivores had profound negative impacts on most island floras, rendering them much more invasible by alien plants than would otherwise have been the case.

The experiments of introducing island endemics to continents, or tropical plants from one location to another within the tropics, have never been properly controlled for propagule pressure. It is plausible that pre-adapted competitive superiority might account for success, but it is just as likely that propagule pressure is responsible for the difference.

Population dynamics and invasion

If the alien is to naturalise, then it must leave more than one descendent in the next generation. The number of effectively dispersed seeds needs to be discounted by mortality suffered at every stage: germination, seedling emergence and growth to reproductive maturity. Likewise, the plant must grow big enough to produce sufficient flowers and have the resources to ripen sufficient seeds to satiate all of the granivorous animals, and still have seeds left over to account for all of the inevitable failures involved in searching out suitable microsites. The accounting can be done at any stage in the life cycle, but a convenient time is adulthood: our measure of success is the number of adult plants in the next generation, divided by the number of adult plants in this generation. We call this lambda, λ, and the species has passed the invasion criterion when $\lambda > 1$. This definition works equally well for alien plants like Japanese Knotweed that do not produce seeds but spread by fragmentation of rhizomes or stems. Species for which lambda is consistently less than one would decline inexorably to extinction.

Persistence despite local extinction

Species can never guarantee that they will be replaced in their exact location by another individual of their own kind. This risk is the fundamental reason why all plants must be capable of effective dispersal. Every death of a sedentary organism is a small-scale local extinction. Our species needs to be able to cope with such local extinctions, so that the larger-scale population, the metapopulation, survives despite local extinctions among its component patches.

The key point is that the patches are linked by dispersal. Occupied patches produce migrants that disperse over the landscape as a whole. Migrants landing in an unoccupied patch have a chance of establishing a new population. Occupied patches have a certain probability of going extinct each year, and the average patch occupancy time is the reciprocal of this probability. So if the per year extinction risk is 0.05, then the average patch occupancy time is 1/0.05 = 20 years. Despite the extinction of patches every year, the metapopulation as a whole is stable and persistent so long as the rate of establishment of new patches by dispersal from occupied patches is sufficiently great. The important message is that instability and local extinction is completely compatible with stability and persistence at a larger spatial scale.

Species that are good colonisers (for instance, those producing many small windborne seeds) are typically poor competitors (because they are likely to lose out in competition for the same microsite to a large-seeded species). Likewise, species that are good competitors are typically poor colonisers (e.g. because they produce fewer, larger seeds). This fundamental principle is the basis of the *colonisation / competition trade-off*. Potentially, this is an important mechanism for species coexistence, because the inferior competitor obtains a refuge from competition in patches of the metapopulation that do not currently contain the superior competitor. If there is a perfect negative correlation between competitive ability and dispersal ability, then there is no limit to the number of plant species that can coexist by this mechanism (Tilman, 1994). Note that the good competitor in such a system is bound to be seed-limited, so we should be able to test for coexistence by a colonisation/competition trade-off by sowing seeds of the superior competitor.

Competitive exclusion

Much of the early and continuing interest in competition has centred on how so many competing species coexist. Hutchinson (1957, 1965) posed the 'paradox of the plankton', asking how 30 or more species of algae could coexist in a few millilitres of lake or ocean water, when there were only a few limiting resources, and when the open waters of lakes and oceans were so homogeneous because of wind-driven mixing. At that time, theory predicted that no more species could coexist than there were limiting factors or resources. The same paradox occurred for terrestrial plants and animals. The Earth's 270,000 or so species of vascular plants compete for the same few limiting factors (usually one or two from the shortlist of water, light, nitrogen, phosphorus, potassium and calcium).

The competitive exclusion principle plays a central role in theoretical ecology. It has a history dating back to Charles Darwin, Charles Elton, Georgii Gause, George Evelyn Hutchinson and Robert MacArthur, but in its modern manifestation it can be stated like this: in an environment that is spatially uniform, and that is temporally constant from year to year, then, given long enough so that any transient dynamics have damped away, one species will exclude all others. Competitive exclusion means that you cannot have more species than there are distinct niches.

As soon as one begins to think about each of the component assumptions, it becomes clear how species coexistence might occur. Real environments, for example, are spatially heterogeneous, so different species could persist in different kinds of patches. Real environments experience different kinds of years, through variation in weather and variation in the sizes of populations of competitors, enemies and mutualists, so different species could do best in different kinds of years, so long as they were capable of surviving through the inevitable runs of poor years; this is called a storage effect (Chesson, 1985). Real environments are subject to disturbances that reset the successional clock, so the ecosystem may not have been stable long enough for competitive exclusion to have happened.

As a theoretical challenge, Hutchinson's paradox of diversity has now been resolved. It turns out that there are four classes of mechanisms that allow many species of plant to coexist in their native habitat despite niche differences: spatial heterogeneity (places differ), temporal heterogeneity (years differ), competitive-colonisation trade-offs (inferior competitors may be able to escape from superior competitors), and frequency-dependent or density-dependent natural enemies (recruitment may be difficult close to long-lived parent plants because of herbivore and pathogen build-up). A fifth possibility is that nothing is happening other than neutral drift in the relative abundances of essentially identical species. Again, the five mechanisms need not work in isolation; several may be operating simultaneously, and interactions between them are likely to promote coexistence even further.

Fundamental niche

The fundamental niche defines the set of conditions under which our species could be expected to prosper *in the absence of competitors and natural enemies*. Obviously, there could be many such factors and they could interact in complicated ways, so the fundamental niche is likely to be multi-dimensional. In a famous address to the Cold Spring Harbor Symposium, the eminent United States ecologist George Evelyn Hutchison was being only slightly tongue-in-cheek when he defined the fundamental niche 'an *n*-dimensional hypervolume' (Hutchinson, 1957).

The *n*-dimensional hypervolume may sound complicated, but the essential point is simple. Inside the fundamental niche, our plant would be expected to prosper (i.e. it would increase when rare), and outside the fundamental niche the plant would be expected to decline to extinction. To get a picture of the issues involved, we need pick only two abiotic factors; let's say temperature and soil pH. The climate goes from too cold at one extreme to too hot at the other, with 'just right' somewhere between. Likewise, the soil goes from too acid at one extreme to too calcareous at the other, again with 'just right' somewhere between. The trick is that we draw these two environmental factors as the axes of a graph, and put them at right angles to one another (in the jargon, the axes are said to be orthogonal, with the implication that they act independently on the plant).

The fundamental niche is shown in the left-hand panel of Fig. 123. In the grey area, the species fails the invasion criterion; one might find plants growing under these conditions, but they would leave less than one offspring on average in the next generation. In the red area of the fundamental niche, the species passes the invasion criterion, by which we mean that, in the absence of competitors and natural enemies (in a fenced and weeded garden, for instance), an average individual of this species would leave more than one individual in the next generation. Notice that the species in this example tolerates a wider range of temperatures on soils that are more calcareous (the niche shape does not need to be symmetrical). The red area defines the extent of the fundament niche of this species in these two dimensions, and it is predicted that it would naturalise throughout the red area in the alien environment if it had left behind its important competitors and natural enemies in the native range.

Realised niche

Now we expose our species to the full rigours of competition (Darwin's 'entangled bank') and to the depredations of its full suite of natural enemies. These will include herbivorous animals, both specialist and generalist, and the gamut of plant pathogens, fungal, bacterial and viral. The herbivores reduce the plant's competitive ability and increase the death rates of its seeds and seedlings. Competing plants reduce the availability of light, water and nutrients, which slows the plant's shoot growth and root growth, reduces its flower and seed production, and increases the risk of premature death. These combined insults mean that the plant now fails the invasion criterion in substantial regions of niche space where it would be able to prosper in the absence of competitors and natural enemies (the red regions of the lower panel of Fig. 123). Only in the green region of Fig. 123 is the plant capable of forming self-replacing populations when confronted with the full array of herbivores, pathogens and competitors. The realised niche shows that our plant is apparently rather specialised, confined to cold environments at high soil pH.

The German plant ecologist Heinz Ellenberg (1913–97) made an important contribution to niche theory by pointing out that plants are not found in the

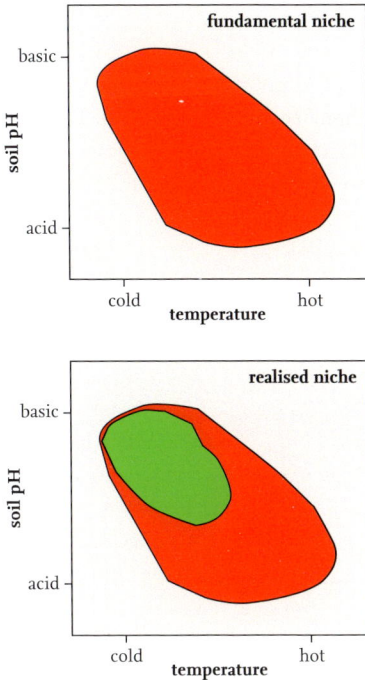

FIG 123. The fundamental and realised niches compared. This example has only two limiting factors: air temperature and soil pH. In the grey area, the species fails the invasion criterion and would leave less than one offspring on average in the next generation. The red area above is the fundamental niche, where the species would naturalise in the absence of competitors and natural enemies. The realised niche (green area below) could be much smaller than the fundamental niche if competitors and natural enemies are influential.

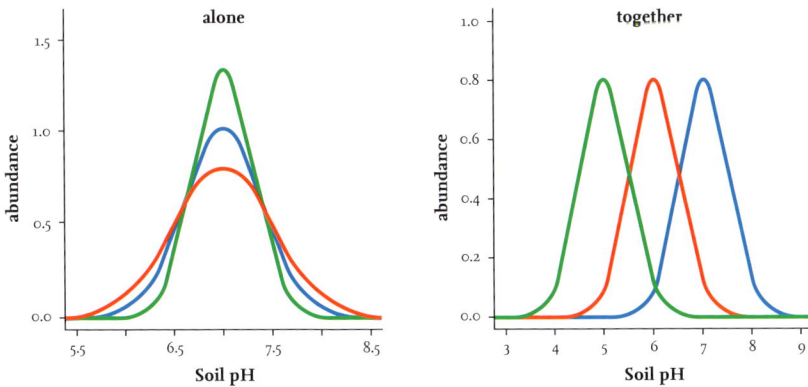

FIG 124. Ellenberg's rule: plants do not grow in the field where things are best for them (left panel), but where their competitors and natural enemies allow them to grow (right panel). Grown on their own, all three species (green, red and blue) grow best at a soil pH of 7 (left panel). Grown in competition with one another, the three species peak in performance at different soil pHs. Now the green species looks as if it is a calcifuge (lime avoider) and the blue species looks as if it is a calcicole (lime lover; right panel).

field where 'things are best for them'. On the contrary, *plants are found where their competitors and natural enemies allow them to grow*. This insight turned conventional wisdom on its head, and Ellenberg backed it up with a series of elegant experiments. He grew plants in the greenhouse along experimentally controlled gradients of soil water, nutrient supply, light and soil pH. He discovered that, growing on their own, most species thrived under exactly the same conditions – a fact that many an expert gardener would recognise. However, when Ellenberg grew the plants in competition with one another on the same experimental gradients, he found that they peaked in performance under *different* conditions (Ellenberg, 1953). Their realised niches were different from their fundamental niches (Fig. 125).

The key point here is that the realised niche is only a part of the fundamental niche (potentially a rather small part). And in the native range, that part is determined by the identities of the plant's native competitors and native natural enemies. It is by no means obvious, therefore, how moving a plant to a new, alien range will affect the size of its realised niche, because the identities of its competitors and natural enemies will be completely different. A niche model based on a description of the realised niche as a native is likely to be useful in predicting the realised niche in the alien range, but only to the extent that competitors and natural enemies have similar effects in both native and alien ranges. The attributes of the native range are likely to predict the alien range best in those cases where the native range is determined almost exclusively by abiotic conditions like climate and soil, and where biotic interactions like competition and predation have little impact on geographic distribution (although, of course, they may have major effects on abundance within the native range).

This is important, because if competitors and natural enemies *do* restrict the realised niche to a small fraction of the fundamental niche in the native range, then climate and other range models are likely to underestimate greatly the potential for spread in the alien range where the competitors and natural enemies will be different (or missing altogether in the most extreme cases). Only if the realised niche is roughly the same size as the fundamental niche in the native range can data from the native range be used confidently to predict likely distributional limits in the new alien environment.

There are some tricky operational issues, too. The fact that a combination of ecological factors lies outside the fundamental niche does not mean that we would never find plants growing under those circumstances. What it means is that if we were able to dedicate the necessary resources to studying those plants over the long term, we would discover that they left fewer than one surviving offspring on average (i.e. they failed the invasion criterion). But if we are carrying out a study on geographic range and the environmental factors associated with that range, then we would almost certainly include that plant, and those ecological circumstances, in our data set even though, in fact, they should lie *outside* the fundamental niche. This means that definitions based on geographic survey of adult plants are bound to overestimate the extent of the fundamental niche.

Similarly, because competitors and natural enemies have populations that fluctuate in abundance over time, then the boundaries of the realised niche will change through time and over space. In addition, they may not be consistently related to the abiotic factors that determine the boundary of the fundamental niche, and that we would like to use to predict range expansions or contractions under climate change.

Vacant niche

Clearly, vacant niches do exist, the most obvious examples being in habitats created by people. There was a vacant niche for a shrub that could live on brickwork: Butterfly-bush filled that niche. There was a vacant niche for a herb that lived on railway ballast: Oxford Ragwort filled that niche (which resembles the volcanic lava that supports its ancestors). The point is that there are few species of the native flora pre-adapted to take advantage of these new and unfamiliar substrates.

The concept of vacant niches in semi-natural habitats is more nebulous. A big part of the problem is that we cannot test for the existence of a vacant niche other than by doing the introduction experiment, and seeing if the alien species establishes successfully. If it does, then clearly there was a vacant niche. If it doesn't, then there wasn't. This tautology is not particularly illuminating.

The old idea of a vacant niche is predicated on the notion that adding a new alien species will increase species richness without consequence for native species. For this to work, there must be one or more resources that are permanently or temporarily unused by the resident community (Fig. 125). For example, it must be the case that there is a vacant niche for invasion by an alien species that is more drought tolerant than species A in Figure 125, and another species that is more tolerant of waterlogging than species F, and this is true, in principle, for any niche axis. This is sometimes known as the *Goldilocks principle*: there must always be room at the ecological extremes that are too hot or too cold for the natives, too wet or too dry, and so on. So there must be at least ten vacant niches in every ecosystem via this mechanism if we allow the existence of niche axes for water, temperature, light, nitrogen and phosphorus. It is by no means obvious, however, that such species exist, given the trade-offs that must

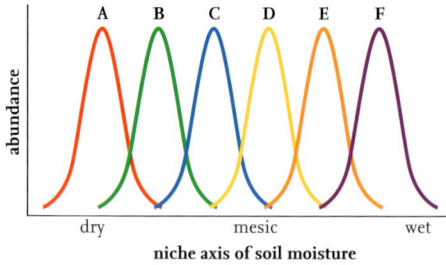

FIG 125. Here, six native species (A to F) are displayed along a niche axis of soil moisture. The height of the curve indicates the abundance of the species at this point on the niche axis, and the spread of the curve illustrates the niche width of the species (all species in this example have the same niche width). The location of the species on the axis is determined by the interaction of their soil-water preferences and the outcome of interspecific plant competition.

inevitably be involved between these extreme traits and other essential traits that constrain the species' biology.

Now we ask what happens if an invading species succeeds in establishing between species C and D in Figure 125? Is it inevitable, for instance, that the abundance distribution of species C must change? If it does change, does this mean a contraction in niche width, or a shift of the whole distribution to the left? If this happens, is it inevitable that species B will suffer? And so on. There is no theory to help us with this (although there is much in the literature on 'limiting similarity', which deals with the distance between the peaks of the abundance distributions of each of the native species as a function of the widths of their distributions). One can imagine four kinds of invaders. The first squeezes in between the existing residents with negligible impacts on their abundance. The second pushes C to the left and D to the right. The third replaces C with no effect on D (or vice versa). The fourth eventually replaces both C and D. In the first two cases, invasion increases species richness, in the third it has no effect on richness, and in the fourth case invasion reduces species richness. There clearly can be no single answer to the question of how invasion affects species richness.

Good examples of locally abundant aliens that squeeze in between the native species with negligible impacts on their abundance are the monkeyflowers (*Mimulus*) inhabiting gravelly riverbanks and rocky upland streams, and New Zealand Willowherb on damp bare ground in the uplands. The opposite effect is illustrated by Green Alkanet (Fig. 126), which can form extensive monocultures on banks and beneath hedges, and clearly has strong negative impacts on most of the native species that formerly occupied this habitat.

There are examples where one abundant alien species appears to be replaced by another (e.g. Canadian Fleabane by Guernsey Fleabane by Bilbao's Fleabane, the moss *Orthodontium gracile* by *O. lineare*, or Green Field-speedwell *Veronica agrestis* by Grey Field-speedwell *Veronica polita* by Common Field-speedwell) but

FIG 126. Green or Evergreen Alkanet (*Pentaglottis sempervirens*) has attractive, bright blue flowers produced early in the year and well into high summer, but it becomes very untidy, spreads its seeds around prodigiously and can exclude native vegetation: (a) dominating the floor of birch woodland in winter; (b) forming a monoculture on a grassy bank. Both habitats in Silwood Park, Berkshire. (MJC)

in no case are we sure about the mechanisms involved (e.g. whether it was direct interspecific competition, a shared pathogen, the newcomer exploiting the niche that is for some reason vacated by the previous occupant, or just coincidence).

There are no examples from the flora of the British Isles of an alien plant causing the extinction of a native species. It goes without saying, however, that for sedentary organisms like vascular plants, every established alien takes up space that might otherwise be occupied by a native.

Novel weapons hypothesis

An alien species might obtain an unfair advantage over the natives by exposing them to 'weapons' that they have never previously experienced. This argument, called the *novel weapons hypothesis*, is most often used in the context of interference competition (rather than exploitation competition), and the best documented cases of this have to do with the use of secondary plant chemicals (this is known as *allelopathy*: the deleterious process by which one plant influences others nearby through the escape or release of toxic or inhibitory substances into the environment).

The test of the novel weapons hypothesis is that the alien's competitors in its new environment suffer more from the allelopathic chemicals than do the equivalent competing species from its original environment, thereby affording the alien species a competitive advantage. A convincing experiment would require growing the alien plant with its competitors from its original range, with and without the allelochemicals, then again with the competitors from its new range, with and without the allelochemicals, and demonstrating that the alien plant performed disproportionately better with the competitors from its new range only in the presence of the allelopathic chemicals.

Diffuse Knapweed (*Centaurea diffusa*) is one of the most destructive invasive alien weeds in the western USA and allelopathy appears to contribute to its invasiveness (Callaway & Aschehoug, 2000). The allelopathic agent involved, 8-hydroxyquinoline, is at least three times more concentrated in soils invaded by Diffuse Knapweed in North American than it is in the weed's native Eurasian soils. The chemical also has stronger phytotoxic effects on grass species from North America than on grass species from Eurasia. Soil microbes are also an important part of this story: Eurasian soil microbes from the native environment might have evolved natural resistance to 8-hydroxyquinoline, while North American microbes have not (Vivanco *et al.*, 2004).

With another invasive alien member of the same genus, Spotted Knapweed (*Centaurea maculosa*), soil microbes from the home range have stronger inhibitory effects on its growth than soil microbes from invaded parts of North America. It seems that *Centaurea* species and soil microbes participate in different plant–soil feedback processes at home and away. In native European soils, the presence of *Centaurea* promotes a soil biota that has increasingly *negative* effects on the plant's growth, possibly leading to its control. But in soils from North America, *Centaurea* cultivates a soil biota that has increasingly *positive* effects on itself, which may contribute to the success of this exotic species there (Callaway *et al.*, 2004). Another effect of phytochemicals is on mycorrhizal fungi. Allelochemicals that are benign to resistant mycorrhizal symbionts in the home range may be lethal to native mutualists in the introduced range and indirectly suppress the plants that rely on them, affording competitor release to the alien (Callaway *et al.*, 2008).

There are plenty of stories about allelopathy in alien plants in the British Isles (Tree-of-heaven, Black Walnut *Juglans nigra* and Rhododendron are prime examples), but none of them has been convincingly demonstrated by critical experiments in which other potentially confounding processes like resource competition or herbivory have been ruled out.

Enemy release hypothesis

Alien plants may do better in their new environments because they have left behind their specialist herbivores and pathogens in the native environment. This is known as the *enemy release hypothesis* (Keane & Crawley, 2002). In the new habitat, natural enemies that are strict specialists on the native flora cannot switch to feed on the aliens. So instead of having to deal with both generalists and specialists as in its native environment, the plant has only to deal with generalist natural enemies in the alien environment, and this gives it a competitive edge.

If enemy release is associated with successful invasion, then its converse, enemy acquisition, is likely to be a common cause of failure. Imagine the fate of a species that evolved on a remote island where there were no mammalian herbivores. It is introduced to the British Isles, where it now faces the combined onslaught of voles, rabbits, deer and a host of domestic animals. In the absence of evolved defences against these herbivores, the alien plant species is most unlikely to remain competitive with native vegetation, and the invasion will fail. Similarly, our native slugs, snails and fungal pathogens are likely to make short work of the seedlings of any alien plant species that lacked the kinds of defences and tolerances exhibited by seedlings of the native flora.

The persistence of our species in its native habitat is proof that it can deal with the continual insults of the resident natural enemies, both specialist and generalist. It achieves this through a complex set of compromises between rarity and abundance, shelter and exposure, tolerance and avoidance, damage and compensation. Plant defences are often classified as either *resistance* traits or *tolerance* traits. Resistance traits reduce herbivore damage by making the plant less attractive to the herbivores, by dint of being too tough (physical defences) or too distasteful (biochemical defence, through secondary plant compounds such as poisons, digestibility-reducing compounds like tannins and chemical deterrents). Tolerance traits limit the negative fitness consequences of herbivore damage through various mechanisms of plant compensation. Classic plant defence theory predicts that there is a trade-off between growth rate and investment in chemical and physical defences, so that in the absence of herbivores, the undefended palatable genotypes would outcompete the expensively defended plants. The *resource availability hypothesis* (Endara & Coley, 2011) predicts that plant populations evolve to optimise individual plant growth rates according to the availability of resources in the environment, which then constrains allocation to plant defences, affecting the types and levels of optimal defence investment. The simplest predicted pattern from this model is a negative relationship between plant growth rate and investment in constitutively expressed defences.

The best evidence for the importance of enemy release comes from work on fungal pathogens and viruses. On average, 84 per cent fewer fungal species and 24 per cent fewer virus species infect each plant species in its naturalised range than in its native range. In addition, invasive plant species that are more completely released from pathogens are more widely reported as harmful invaders of both agricultural and natural ecosystems (Mitchell & Power, 2003). The longer-term evolutionary response to enemy release might be that the resources used for defence in the native environment could now be diverted towards increased competitive ability in the alien environment, but experiments to test this have not produced compelling results.

Enemy release may be a short-term benefit, as natural enemies in the new range evolve or adapt behaviourally to include the alien species in their host range. For 124 European plants introduced to North America, fungal and viral pathogen species richness has increased with time since introduction. Alien plants introduced 400 years ago now support six times more pathogens than those introduced 40 years ago (Mitchell *et al.*, 2010).

It is plausible that the most problematic alien plants in the British Isles (e.g. Indian Balsam and Butterfly-bush) are beneficiaries of enemy release, since many of them appear simultaneously to benefit from very low levels of attack by British generalist natural enemies and to have brought with them few, if any, specialist enemies from their native ranges. If so, this makes them potential targets for biological control, should this ever be deemed to make economic sense (Japanese Knotweed and Indian Balsam are currently the target of classical biocontrol programmes in England; see Chapter 15). Rhododendron has already recruited an abundant alien herbivore (the highly colourful scarlet and green Rhododendron Leafhopper *Graphocephala fennahi* from the USA), but this has had a negligible effect on its growth rate or fecundity.

When an alien plant does recruit a specialist herbivore, the result can be spectacular. The most conspicuous example in recent years has been the arrival and spread of the Horse-chestnut Leaf Miner (*Cameraria ohridella*), which arrived in 2002 after its rapid spread across Europe (Gilbert *et al.*, 2004), as discussed in Chapter 14.

Plant size at home and away

Two of the most frequently repeated assertions about established aliens are that the individual plants are larger than typical individuals in the native habitat, and that the biomass of the species is greater than in the native range. It turns out that neither of these is generally true.

It is a fact that some plants are much bigger in their alien range. No one who has ever seen Pedunculate Oak growing in South Africa or New Zealand can fail to be impressed by both the size of the mature tree and the size of its acorns when freed from its army of European pests and pathogens. Equally, travellers in Hawai'i who see Gorse (*Ulex europaeus*) growing in thickets 4 m tall are likely to believe in enemy release. Perhaps the most extreme case is Monterey Pine (*Pinus radiata*). When one sees this as a twisted and parasite-infested shrub on the Monterey Peninsula in California, it is hard to imagine that once released from its natural enemies it would go on to become the most frequently planted commercial conifer in the world (Critchfield & Little, 1966).

On the other hand, many alien plants are the same size in both ranges. Where European grassland species have been compared growing at home and away in a wide range of temperate environments all over the world, the clear pattern is that they are neither bigger nor smaller as aliens (Firn *et al.*, 2011). So while there are clear examples of plants that are significantly bigger as aliens, this is certainly not a pattern that is repeated across all successful alien species.

In most discussions of the 'plants are bigger as aliens' hypothesis, the clear but usually implicit assumption is that they are bigger because of enemy release. In fact, the same result could equally well come about because the plants benefit from competitor release in the alien environment. The distinction is very seldom tested by carrying out manipulative experiments in the native environment, in order to demonstrate that when the herbivores and pathogens are excluded (e.g. using a combination of fences, insecticides and fungicides), the plants achieve the same large size as observed in the alien range. One outstanding exception that demonstrated the plausibility of enemy release was the work done on Broom (*Cytisus scoparius*), which grows to great size as an alien in California and New Zealand. Controlled experiments showed highly significant increases in size, fecundity and survival of Broom when sprayed with insecticides in its native environment in southeastern England. It transpired that the key natural enemies in the native environment were the Pea Aphid (*Acyrthosiphon pisum spartii*) and a chrysomelid beetle, *Phytodecta olivacea* (Waloff & Richards, 1977).

Sufficient resident mutualists

Assuming that the alien plant has found itself with both competitor-free and herbivore-free space, it now has to grow. If it needs to form mycorrhizal associations in order to forage effectively for soil resources, then it must be capable of recruiting the local generalist mycorrhizal fungi to this task. The requirement for a specific mycorrhizal fungus that is found only in the native environment would doom the invasion to failure, unless the seed was wise enough to bring the fungus along with it, as spores in the seed coat or as a systemic infection. We have no idea how common this cause of failure is in practice, but many intentional introductions have been assisted by the simultaneous intentional introduction of mycorrhizal fungi, especially in commercial forestry (Richardson *et al.*, 2000).

Should the alien plant grow large enough to flower, then it must make do with the local generalist pollinators, and in the event that it sets seed, it must use the local generalist seed dispersers. Of course, the ideal invading species would have no need for any mutualists all; it might be a non-mycorrhizal, self-pollinating species with wind-dispersed seeds, like Hoary Cress (Fig. 115).

Evolution after invasion

If there is sufficient time before an ill-adapted alien is driven to extinction, then perhaps it may be possible for it to respond to the new regime of natural selection and evolve to become a better competitor. It needs the appropriate mutations, and the heritability of these mutations, to turn around a situation where it was failing the invasion criterion ($\lambda < 1$) to one where it now passes the invasion criterion ($\lambda > 1$). This might be made more likely if the plant was also benefiting from enemy release, because it could divert resources, which previously had to be used for defence against natural enemies in the native habitat, into increased competitive ability in the alien environment (growing taller perhaps, producing a more extensive root system, or recruiting a more effective mycorrhizal partner).

Perhaps more likely, however, is a process that might be mistaken for evolution, where the passage of time simply allows for the importation and release of more genotypes from the native range, one of which turns out to be better pre-adapted to the alien environment, thereby giving the alien species a competitive edge over the residents.

We are now in a position to separate our alien species into three groups: the failures that never showed their faces; the casuals that grew big enough to be recognisable, but which failed the invasion criterion; and the naturalised species that formed self-replacing populations. In the next chapter we shall consider why most of the naturalised species are so uncommon, and why a handful of them become pests.

CHAPTER 7

The Ecology of Abundance

The wanderings of plants about the world have existed ever since their first evolution, and are still continuing.
Henry Nicholas Ridley, *The Dispersal of Plants Throughout the World* (1930)

W hy do some naturalised species become abundant, but most remain uncommon? After all, every one of our naturalised aliens has passed the invasion criterion, so they all have the ability to increase exponentially in their new environments. So why do most alien plant species, just like most native ones, stop increasing at such low population densities? And what is it about the few species that keep on increasing until they attain high and occasionally damaging populations?

These are questions about abundance. They concern the population density of alien plant species and the processes that regulate population density. These are the mechanisms that slow down, and then eventually stop, the tendency of the naturalised population to increase exponentially. In aggregate, they are referred to as the *density-dependent process* because, for a mechanism to regulate population size, the intensity of its operation must increase as population size increases. We saw in Chapter 6 that, for naturalisation, it is essential that our invading population benefits from some form of *rare-species advantage*. In a sense, this chapter is about the opposite of that. Population regulation is about *common-species disadvantage*. As we shall see, the population density at which an invading species eventually settles down is not determined by one factor or another, but by the *interaction* of several factors, including density-independent factors like weather and soil conditions, and potentially density-dependent processes like interspecific competition with native plants, intraspecific

competition with other members of the invading species, herbivory and
pollination success.

Because distribution is so much easier to measure than abundance, there
is a temptation to use distribution data as a surrogate for abundance. The
problem is that distribution data are so highly scale-dependent; it is impossible
to distinguish the presence of a single individual within a 10 km square from a
continuous monoculture of the plant. The issue is nicely illustrated by the case
of Pineappleweed (Fig. 127), the most widespread neophyte in the British Isles,
recorded from 3,530 of the 3,795 hectads. Within Berkshire, it is present in all
the hectads, but in only 50 per cent of the 1 km squares. Within Silwood Park,
it occurs in just 8 per cent of the 110 ha, and was not recorded at all in random
samples at spatial scales smaller than this. In the few places where Pineappleweed
does grow in Silwood Park, the population density varies between one and ten
plants per square metre, and no patch containing the plant extends to more
than 25 m². This means that the peak density where the plant grows (10 m^{-2})
overestimates the average density in Silwood Park more than 7,500-fold.

The only objective way to estimate plant abundance over large areas is to use
replicated randomised samples. In some cases it may be possible to count the

FIG 127. The
most common and
widespread neophyte
in the British Isles:
Pineappleweed
(*Matricaria discoidea*).
(CAS)

plants in a specified area (e.g. woody perennials like Turkey Oak) but it will often be impossible to recognise individuals (e.g. with patch-forming perennials like Ground-elder), in which case abundance will have to be quantified by measuring cover or biomass (the latter, of course, is destructive and hence inappropriate in many circumstances).

Variation in abundance from place to place (so-called spatial heterogeneity) is often just as important as average abundance, and this, too, needs to be established from replicated randomised samples: one useful measure of spatial heterogeneity is the variance:mean ratio. The key point is that these data are so expensive and labour-intensive to obtain that we know distressingly little about the abundance, and even less about long-term changes in abundance, for most of our alien flora.

POPULATION REGULATION

The birth rate of our alien population is likely to go down as population size increases; individual plants will be smaller because of more intense competition, and consequently each plant is likely to produce fewer seeds. The opportunities for seedling establishment might also go down as more and more sites are occupied by adult plants. The risk of death of established individuals is likely to go up with population density, most obviously because competition reduces plant size, and small plants are much more likely to die than large ones. The risk of death from natural enemy attack is also likely to go up with population size, as specialist enemy populations build up from year to year, and generalist enemies are attracted to switch to what is now a more abundant food source. Whatever the details, the alien population will continue to increase in abundance as long as the birth rate is greater than the death rate.

Sooner or later, however, there will come a point where the declining birth rate meets the increasing death rate. The population density at which the birth rate and the death rate are equal is called the *equilibrium population size*. For reasons that we shall discuss shortly, this happens at low densities for most alien plant species. The few that continue increasing until they reach high population density are the aliens that we refer to as *invasive species*.

Recruitment limitation

The simplest form of population regulation is recruitment limitation. If there are only a few microsites available for seedling establishment, then this imposes the most straightforward kind of density-dependent regulation. Recruitment is not seed-limited, so the higher the number of seeds produced, the smaller the proportion of them that will find their way into suitable microsites. Suppose that there are 100 microsites capable of supporting seedling establishment. In the first year there are 200 seeds competing for access to the microsites. Assume that all of the microsites are taken. The seeds that do not find microsites die, so the death rate this year is (200 – 100)/200, or 50 per cent. In another year, there might be 1,000 seeds competing for access to the 100 microsites, so in this year the death rate is (1,000 – 100)/1,000, or 90 per cent. The death rate increases with the number of seeds, and we say that the death rate is density-dependent. This is an example of the simplest kind of intraspecific competition. It is called *contest competition* because there are winners and losers. Some seeds get access to a recruitment microsite and prosper, and the others die.

One can picture this ecological process as being played out in a carpet of perennial native vegetation. For most species, there is no possibility of establishment from seed in the body of the carpet because the intensity of competition from the resident plants that make up the fabric of the carpet is so great. The only possibility for recruitment is when holes appear in the carpet. The holes might be caused by small-scale disturbance by animals, or by the death of individual native plants. So long as there are sufficient seeds, the number of holes in the carpet (the number of microsites) determines the maximum number of alien plants in the next generation, and the plant will be rare when the number of holes is small. The invading plant is not in charge of its own density, to the extent that if it produces more seeds it will not increase its total population size.

Of course, an individual plant producing more seeds than its conspecifics may increase in fitness relative to the other invading individuals, if by producing more seeds it gains access to a greater fraction of the microsites than its competitors, and if there was a heritable basis to its increase in seed production. There is a constant tension between the good of the individual (Darwinian fitness) and the abundance of the species, and natural selection typically operates via individual selection rather than group selection.

Seed limitation

When microsites are plentiful, then recruitment might be seed-limited. The more seeds that are produced in one generation, the larger will be the alien plant population in the next generation. Under these circumstances, population density will be determined by the factors that regulate seed production. These are likely to include competition from native plant species, herbivore and pathogen attack by native natural enemies, pollination and seed-dispersal services and, if our alien becomes abundant, competition from individuals of its own kind (intraspecific competition). We shall consider each of these in turn, bearing in mind that it is likely to be the interaction between these factors that determines the density at which the death rate first exceeds the birth rate.

Competition from native species
Once our species is established as a seedling, its subsequent performance is strongly dependent on its growth rate, the length of time it continues to grow and the final size it attains at maturity. Established neighbouring plants are a major determinant of this performance. They compete with our alien plant in two contrasting ways. Competition for light is essentially a contest: the plant that grows tallest wins its place in the sun, while the shorter plant gets to live in the shade. For soil resources like water and nutrients, the competition is more like a scramble. All the plants take up resources and, in so doing, reduce the supply for all the other plants in the vicinity. In a pure scramble, all of the plants would die once the resources run out. This does not happen in practice because the plant species are not identical.

A factor is defined as being a *resource* for a species if increases and decreases in the factor lead to increases and decreases in the specific growth rate of the species. We assume that a species continues to increase in abundance until the resource is reduced to the point at which the supply rate is insufficient to support any further population growth. The key insight from this model is that the species that reduces the resource supply to the lowest level will competitively exclude all the other species (Titman, 1976). It is said to be the 'best competitor for this resource' because it can carry on increasing in abundance at resource supply rates that are too low for any of the other species. With this way of looking at interspecific competition, plants are said to be good competitors for a resource when they can thrive at a low supply rate of that resource. So a good competitor for nitrogen might prosper on nitrogen-poor soils by hosting a nitrogen-fixing bacterium in nodules on its roots, or by trapping and digesting invertebrate animals in sticky traps. Alternatively, the good competitor might have evolved to show exceptionally high efficiency in using that particular resource, or have evolved to require less of it by dint of very low tissue concentrations of that nutrient.

Why would an alien be a better competitor than a native? After all, the native has evolved here, so it ought to be better adapted to local conditions. The simplest explanation is that the different evolutionary history of the alien has pre-adapted it to the point where it is the superior competitor. In its native environment it had to 'try harder' as it were, so that it arrives with an ability to grow at resource supply rates that are lower than those that could support the native species.

Another plausible explanation is that the alien is what we might call a 'people plant'. It evolved in close proximity to human beings and is therefore adapted to the habitats we create, the domestic animals we keep and the disturbance regimes we impose. The native species might have evolved with much lower human population densities and has not had any opportunity to respond to the selection pressures that a high human population imposes. This is just another kind of lucky pre-adaptation.

Attack by native herbivores

One of the most popular explanations for the competitive superiority of successful invasive alien plants is the *enemy release hypothesis*: plants are supposed to do well in the alien range because they have left behind their specialist herbivores and pathogens in the native range (Chapter 6). But whether or not our alien left behind an influential suite of natural enemies in its former home, it must be capable of dealing with the native herbivores and pathogens that live in its new environment. Because we are assuming that our species has passed the invasion criterion and is naturalised, it is clear that it can cope with the native natural enemies, at least when it is rare. The question here is the extent to which herbivores in the alien range make the difference between our alien becoming abundant or remaining at low densities.

It is clear that the effects of native herbivores could lie anywhere along a continuum from no effect at all (because the alien plant is totally unpalatable to them), to complete elimination (because the alien plant is so palatable that it is consumed entirely on first encounter). Between these extremes, it is possible that native herbivores regulate the alien plant population at stable low densities. This might come about because the alien plant is reasonably palatable, but the herbivores switch to feeding on it only after it passes a threshold level of abundance.

It might be that the alien plant passed the invasion criterion because it has a refuge where it is safe from the depredations of herbivores, but it is so attractive that the herbivores eat it up whenever they encounter it outside of its refuge. In this case, the population of the alien plant would be determined by the size of the refuge.

To understand the impact of the native herbivores on the abundance of the alien plant, we need to know what regulates herbivore numbers. Are they regulated at high densities by the availability of large quantities of native food-plants? Or are they regulated at low densities by their own predators, parasites and diseases? It is crucial to know whether the native herbivores will consume our alien plant on sight, or whether it is relatively unpalatable by dint of its physical or chemical defences. Finally, we need to understand how the alien plant responds to attack by these native herbivores. Can it compensate by increasing its photosynthetic rate or by mobilising stored reserves for regrowth?

Availability of mutualists

If it is not plant competitors or natural enemies that are keeping our naturalised alien scarce, then perhaps it is a shortage of mutualists. As with our thinking about herbivores, we need to bear in mind that the alien has passed the invasion criterion in becoming naturalised, so there is clearly no evidence for an absolute shortage of essential mutualists for pollination, fruit dispersal, resource capture (nodulating bacteria or mycorrhizal fungi) or defence (e.g. protective ants). We would not expect to see naturalisation in alien plant species that had obligate dependencies on species that were likely to be left behind. So if there are enough generalist pollinators, seed dispersers and mycorrhizal fungi to pass the invasion criterion, how likely is it for a shortage of mutualists to be the cause of population regulation at low densities?

We should recall that one of the most successful of our aliens, Oxford Ragwort, is typically an outbreeder, so clearly in this case at least the generalist native pollinators (hoverflies, honey bees and solitary bees) perform an adequate if not a perfect service. When seed production is experimentally compared between open-pollinated and hand-pollinated plants of Oxford Ragwort, seed-set is rarely found to be pollen-limited (Simon Hiscock, pers. comm., 2012), but this is not to say that it cannot happen.

THE LAG

A recurrent theme in writings about biological invasions is the existence of a more or less protracted lag phase, during which the alien species remains at low density. Then, many decades or even centuries later, the species takes off and increases to high densities. The implicit assumption in most of these cases is that the species is waiting, as it were, for something to happen that will tip the balance and allow it to switch from being a casual (where $\lambda < 1$) to being naturalised ($\lambda > 1$). There are several examples of lag-phase studies (Daehler, 2009; Aikio *et al.*, 2010) showing that most species that are going to become invasive do so reasonably quickly, but that long periods of seemingly consistent behaviour can be poor predictors of what invaders will do in the future.

Nothing's happening

The simplest explanation is that there is no lag at all. The lag phase might be nothing more than a reflection of the way that people perceive the process of exponential growth. Suppose that the alien population has passed the invasion criterion, and is increasing each year by a factor of 1.2. It started out, say, with a population density of just 1. If we do not perceive anything as having happened until the population exceeds some threshold value like 5 million individuals, then it looks as if there has been a lag phase of 80 years. In fact, of course, the population has been changing at exactly the same constant rate, increasing by 20 per cent year after year. It might not look like it from Fig. 128a, but the constant increase would have been clear if we had plotted the logarithm of population size rather than population size. When there is no lag, as here, the graph of log(numbers) against time is a straight line (Fig. 128b).

Alternatively, the lag might reflect a detectability threshold, as understood in diffusion models of plant dispersal. Under the diffusion model, a very small number of individuals disperse a very great distance indeed, so their population density (number of individuals per unit area of ground) would be so low that their probability of detection would be effectively zero (Shigesada & Kawasaki, 1997). Of course, the detectability threshold would depend on the kind of plant, so that trees would have much lower detectability thresholds than tiny annuals.

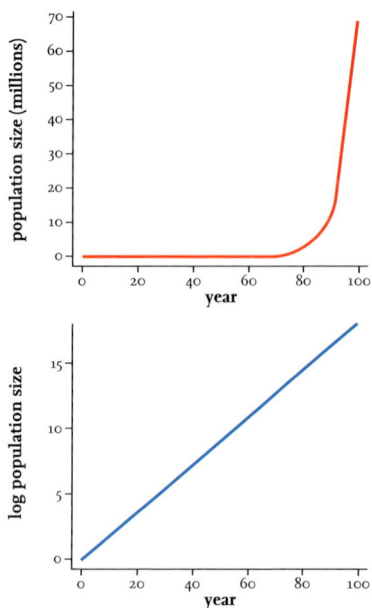

FIG 128. (a) Population size (in millions, red line, above) as a function of time (years since introduction). It looks as if there is a time lag up to year 80, after which the population erupts. (b) This impression is false, as can be seen by plotting the logarithm of population size (blue line, below) against time. Exponential growth (here an annual increase of 20 per cent) occurs because next year's population is a constant multiple of this year's population ($\lambda = 1.2$ in this example).

Evolution

The thinking here is that the species is waiting for the right genetic mutation to arise, and that this will tip the balance. This is a low probability event, so it could take a long time to happen (if indeed it ever happens). A good example of post-introduction evolution is afforded by Oxford Ragwort, which since its introduction to the British Isles has undergone homoploid speciation to produce our plant, which until recently was endemic to us. It might well be the case that it did not escape properly until it had evolved into *Senecio squalidus*.

Multiple introductions

A much quicker way to get the right genotype is through multiple introductions from different parts of the native range, or from different habitats. The number of genotypes in each introduction is likely to be a small random sample from a potentially large pool of genotypes, so it will take time to find a good fit between genotype and environment. The lag phase represents the time required to get the right genotype imported, rather than evolution by natural selection on the ground in the new alien range. We don't know for sure, but a good example of this process might be American Willowherb (Fig. 129a), which first established in England in the Midlands before 1891 but didn't spread. Another introduction into southeastern England in the 1920s spread rapidly throughout the British Isles (Fig. 129b), overrunning the original Midlands population in due course (Preston, 1988). American Willowherb is the alien plant species with the fastest-known spread in the British Isles so far. Although we can't be certain, the failure of the nineteenth-century introduction to spread might be plausibly ascribed to genotype. Another possible example of this process is Narrow-leaved Ragwort (Fig. 270), first known as a rare wool alien but now a locally common weed of waste ground in southeastern England.

Climate change

If the environment changes, then the wrong genotype can become the right genotype. Some of the plants that will prosper in future climates are already naturalised, but are present at low densities. A change in winter temperatures, for instance, or a reduction in summer rains, may tip the competitive balance in favour of the previously disadvantaged alien species, allowing it to increase in range and abundance. At this point, the population takes off and the lag phase is over.

Invasion debt

This is another name for *sleeper weeds*: we have introduced a lot of aliens in the past that are not problems now, but may become problems in the future

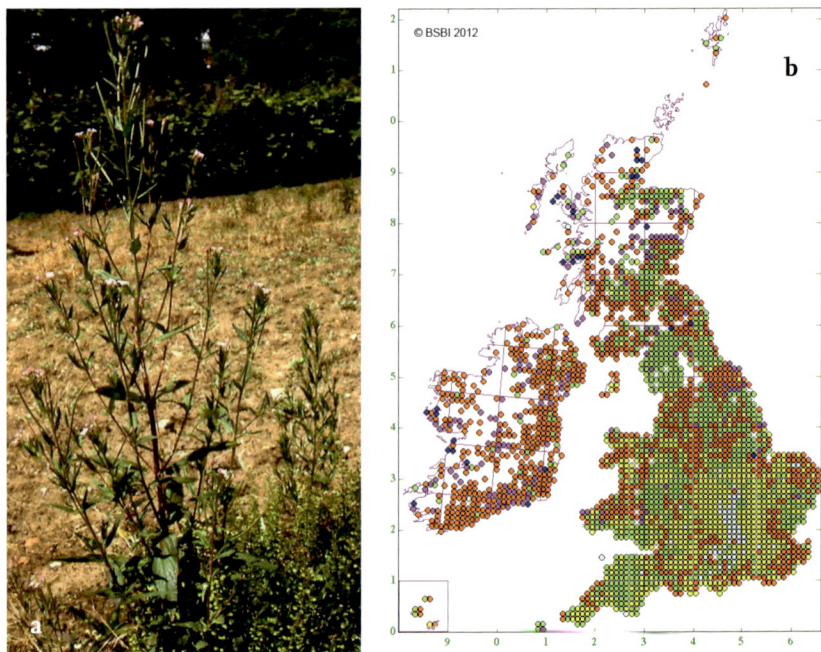

FIG 129. American Willowherb (*Epilobium ciliatum*): (a) although not common in the region until the mid-twentieth century, it is now the commonest willowherb in central and southern Britain, where it hybridises with all our other eight lowland species; (b) the range expansion of American Willowherb from its epicentre to the west of London in 1900 to nationwide by 2000. Grey dots pre-1930; yellow 1930–69; green 1970–86; red 1987–99; purple 2000–09; blue 2010 onwards. The pattern is approximately radial, reflecting both the spread of the plant and the spread of the ability to identify it. (a, CAS)

(Dullinger *et al.*, 2013). These species may be quietly increasing exponentially at low densities, or they may be waiting for environmental change or for the appearance of the right genotype. The problem is that we don't know which species they are, and we are unlikely to want to spend the money to try to find out.

Invasional melt down

This is the ecological nightmare where native vegetation is replaced wholesale by a rag-bag of alien plants. Under this scenario, the more invaded the vegetation is, the more invasible it becomes (Green *et al.*, 2011). There are depressingly many examples of this process on tropical islands (see Chapter 15 for examples), but fortunately we have no examples (yet) from the British Isles. Invasional melt down is an example of positive feedback where invasibility increases with the degree of invasion.

GEOGRAPHIC RANGE: DISPERSAL AND SPREAD

The species is naturalised (Chapter 6) and has increased in abundance to its limit at the site of introduction (as explained above). This section is about the way that the species extends its alien geographic range (Fig. 130). There are two contrasting models to describe. The first, 'ripples on a pond' model is essentially ecological and is based on the reproductive rate and dispersal biology of the plant (Fig. 131). The second, 'starburst' model involves longer-distance movement, where the range extension is caused intentionally or unintentionally by movement of the plant, typically by people but also potentially by birds, flooding or rivers (Fig. 132).

Most investigations of alien plant range extension are single-species studies, where the only variables considered to be important are the reproductive rate of the alien plant and the distance over which its propagules are typically dispersed. Technically, this is a probability density called the *dispersal kernel*, which shows the likelihood of a propagule being deposited at different distances from its parent. These single-species models are thoroughly described by Shigesada & Kawasaki (1997). The implicit assumption is that recruitment is seed-limited and that competition with native vegetation is unimportant in determining the rate of spread. Likewise, the nature of the landscape through which the range is extending gets short shrift, and no account is taken of soil types, patch size, patch type, fragmentation, the existence of dispersal corridors or the effects of herbivore feeding. Most tellingly, however, is the absence of any role for long-distance dispersal of propagules by people or other agencies.

Figure 131 illustrates the *ripples on a pond* model of range expansion for an invading species in a spatially uniform habitat. At the start (top left), there is a single individual in the centre of the pond. This individual produces four offspring, then dies ($\lambda = 4$ at low density). Each offspring moves one step, either north, east, south or west (top right). Next, these individuals reproduce and disperse in the same manner, then die (now $\lambda = 9/4 = 2.25$; bottom left). And so on (bottom right, where $\lambda = 16/9 = 1.78$). In the interior of the invaded area, we assume that if there is more than one individual per location, then there is contest competition, and one individual takes control of that cell and all the other individuals in that cell die. This is why λ decreases with the passage of time,

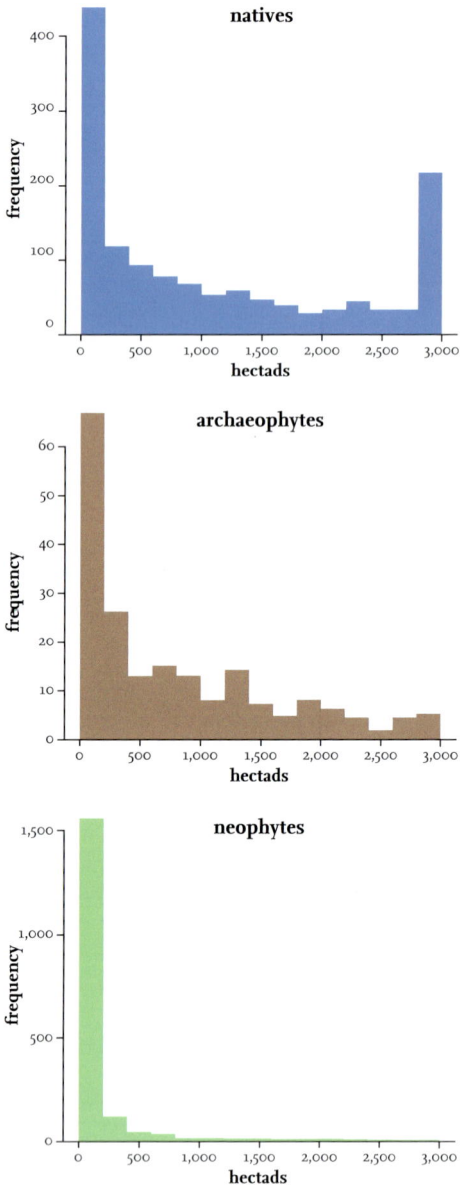

FIG 130. Geographical distributions of natives (blue), archaeophytes (brown) and neophytes (green) in the British Isles, showing the frequency with which species were represented across 14 categories of range size (measured as the number of hectads (10 × 10 km squares) occupied in 2000). Note the large differences in scale across the three histograms. Most neophytes show extremely limited spatial distributions, and very few neophytes are ubiquitous (the L-shaped pattern in the bottom panel). Archaeophytes are intermediate with several widespread species (middle panel). Natives show a contrasting pattern with a large class of close-to-ubiquitous species (top panel). (Data from BSBI)

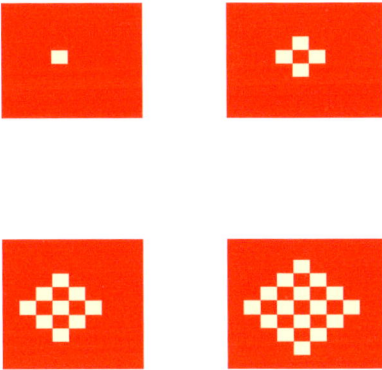

FIG 131. Range expansion in an alien plant driven solely by reproduction and dispersal is known as the 'ripples on a pond' model. The invasion starts with a single individual (top left) and spreads stepwise, with increases in abundance (to bottom right). (*See* text for details.)

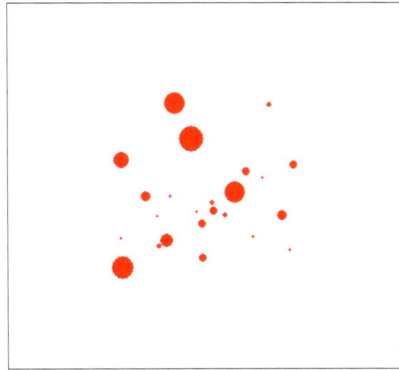

FIG 132. Range expansion in an alien plant mediated by people is known as the 'star-burst' model. New appearances are concentrated where propagule pressure is greatest: in towns and near communication routes. (*See* text for details.)

as intraspecific competition becomes more intense in the interior of the range. There is no competition on the leading edge of the invasion front.

Note that along a given radius (say, westward from the initial individual) the invasion front moves at a constant velocity of one square per generation. Notice also that the spread is determined entirely by the reproductive and dispersal behaviour of the individuals on the front of the wave. The assumptions that we make about dynamics inside the infestation (contest competition in this case) are not consequential, because we do not allow leap-frogging dispersal where individuals produced in the heart of the invasion could overtake the individuals on the invasion front.

This simple model is representative of a class of more complex models based on diffusive movement (Shigesada & Kawasaki, 1997). The radius of the invasion increases linearly with time, so the infested area increases with the square of time. This is why the square root of the infested area is often plotted against time since introduction; the plot of the square root of area against time is often roughly linear, e.g. Skellam's (1951) celebrated analysis of the spread of Muskrat (*Ondatra zibethicus*) in Europe, and the spread of American Willowherb in England, illustrated in Figure 129b.

The other extreme model of dispersal has much less to do with small-scale dispersal kernels and more to do with essentially unpredictable, long-distance dispersal effects. This long-range dispersal may be natural (the ducks' feet model) but more often will be brought about by people, intentionally or unintentionally. We can call this the *star-burst model*, after the image of the sky during a firework display. Local dispersal occurs subsequently, from a mass of widely separated multiple foci of infection.

In the star-burst model of alien plant dispersal and range expansion (Fig. 132), individuals are dispersed over long distances, typically by people, and then begin expanding like ripples on a pond from these foci of infection. Differences between the sizes of the infected patches reflect differences in time since infection, and differences in local ecological conditions affect both the rates of population growth and dispersal. After four generations or so, the fastest-growing patches are about to coalesce with later-established smaller patches, while in each generation, new infestations arise by immigration in locations determined by propagule pressure (i.e. not by dispersal from existing neighbouring patches, although these could obviously be a source of propagules).

As with the ripples on a pond model, these star-burst models play down the importance of biological interactions with competitors or natural enemies. In real systems, a variety of interspecific interactions might be expected to influence the rate and extent of range expansion during an invasion: resource competition might lead to the exclusion of all species other than the species with the lowest requirement for the rate of resource supply; interference competition might exclude less aggressive taxa; priority effects might mean that the first species to a site excludes the others; and shared natural enemies might lead to apparent competition, slowing the rate of range expansion in what would otherwise be the superior competitor.

The two components to spread, reproduction and movement, can compensate for one another to a degree, so that a rapidly reproducing species that moved only a short distance in each generation might spread at the same effective rate as a species that had a low reproductive rate but high dispersal distance. To begin with, we can think about a single trait, combining parental reproduction and movement of the offspring away from the parent that characterises the rate of spread in a new environment. This will be measured as distance per unit time along a radius away from the point of introduction. What can we expect of this measure? Will it be constant, for instance? Will it tend to speed up as the invasion gathers momentum? Or will it tend to decline as the suitable habitat is used up? Given long enough, of course, both models will fill the entire area so long as the net reproductive rate is greater than 1. The spatial patterns they exhibit between arrival and the achievement of equilibrium are therefore essentially transient.

A case study with American Willowherb

North America was on the receiving end of virtually the entire weed flora of Europe, but North American plants do not do particularly well as invasives in Europe. A classic exception to this rule is American Willowherb (Fig. 129a). Within a century, this species went from recent arrival to the most widespread and abundant willowherb in the land. The date of the first record (1891) might be somewhat late, because the plant could have been mistaken for a native species for some years before the first voucher specimen was preserved. It is a classic example of a people plant, growing in a broad range of habitats near houses, roads, railways, factories, walls and waste ground, as well as on the disturbed edges of semi-natural habitats. Its spread throughout the British Isles is better described by the 'ripples on a pond' model than almost any other widespread alien species.

THE WIDESPREAD AND ABUNDANT ALIENS

The abundant alien species are a subset, and often a very small subset, of the most frequently naturalised species (Table 6). They have rather little in common with one another, except that like the people who introduced them, they are sociable. There are trees like Sycamore and shrubs like Butterfly-bush. There are annual herbs like Pineappleweed and perennial herbs like Ground-elder.

High-impact alien species

Amongst the 1,809 neophytes listed in Appendix 1, 1,138 are naturalised, 342 are casual and 329 are survivors. Of the naturalised neophytes, 5 per cent are invasive of semi-natural vegetation and at least 6 per cent are locally abundant in man-made and disturbed habitats. For the terrestrial habitats, the one trait that unites the high-impact aliens is that they are thicket-forming (like Rhododendron, Cherry Laurel (Fig. 133), Japanese Knotweed, Japanese Rose, Snowberry and brideworts *Spiraea*). These species form dense monocultures to the exclusion of virtually all other vascular plants. The individuals are highly tolerant of intraspecific competition and exert very strong interspecific competition on native vegetation. The thickets cast such dense shade that they are capable of preventing recruitment by seed and their competitive root systems can resist ingress of other species by vegetative means. High-impact herbaceous plants that rely on recruitment from seed include Indian Balsam and Giant Hogweed; these tend to reach their peak biomass in naturally disturbed waterside habitats. The other high-impact terrestrial species are those that thrive in man-made substrates

TABLE 6. The top 30 most frequent alien plant species in London, Berkshire and East Sutherland, based on replicated, fixed-effort, randomised sampling across all habitats. Species common to all three lists are shown in pink; species found in the top 30 in only one of the three locations are shown in green. Species capable of achieving biomass at which they are likely to influence vegetation structure are marked with an asterisk. Planted species not reproducing by seed or spreading vegetatively are shown by (P). The habitats of these alien species are described in Chapter 13.

Rank	London		Berkshire		East Sutherland	
1	Butterfly-bush *Buddleja davidii*	*	Sycamore *Acer pseudoplatanus*	*	Ground-elder *Aegopodium podagraria*	*
2	Guernsey Fleabane *Conyza sumatrensis*	*	White Dead-nettle *Lamium album*		Sycamore *Acer pseudoplatanus*	*
3	Sycamore *Acer pseudoplatanus*	*	Barren Brome *Anisantha sterilis*		Soft Lady's-mantle *Alchemilla mollis*	
4	Petty Spurge *Euphorbia peplus*		Butterfly-bush *Buddleja davidii*	*	Montbretia *Crocosmia × crocosmiiflora*	*
5	Shepherd's-purse *Capsella bursa-pastoris*		Mugwort *Artemisia vulgaris*	*	Russell Lupin *Lupinus × regalis*	*
6	Green Alkanet *Pentaglottis sempervirens*	*	Green Alkanet *Pentaglottis sempervirens*	*	Sitka Spruce *Picea sitchensis*	*
7	American Willowherb *Epilobium ciliatum*		Cherry Laurel *Prunus laurocerasus*	*	Flowering Currant *Ribes sanguineum*	
8	Bristly Oxtongue *Helminthotheca echioides*		Ground-elder *Aegopodium podagraria*	*	Butterfly-bush *Buddleja davidii*	*
9	Common Mallow *Malva sylvestris*		Shepherd's-purse *Capsella bursa-pastoris*		Himalayan Cotoneaster *Cotoneaster simonsii*	*
10	Cherry Laurel *Prunus laurocerasus*	*	American Willowherb *Epilobium ciliatum*		Intermediate Bridewort *Spiraea × rosalba*	*
11	Mugwort *Artemisia vulgaris*	*	Horse-chestnut *Aesculus hippocastanum*		Feverfew *Tanacetum parthenium*	
12	Red Dead-nettle *Lamium purpureum*		Hybrid Black Poplar *Populus × canadensis* (P)		Lodgepole Pine *Pinus contorta*	*
13	Wall Barley *Hordeum murinum*		Snowdrop *Galanthus nivalis*		Snow-in-summer *Cerastium tomentosum*	

Rank	London	Berkshire	East Sutherland
14	Lavender _Lavandula angustifolia_	Common Field-speedwell _Veronica persica_	Japanese Rose _Rosa rugosa_ *
15	Cabbage-palm _Cordyline australis_ (P)	Petty Spurge _Euphorbia peplus_	Welsh Poppy _Meconopsis cambrica_
16	White Dead-nettle _Lamium album_	Privet _Ligustrum ovalifolium_	American Willowherb _Epilobium ciliatum_
17	Firethorn _Pyracantha coccinea_	Darwin's Barberry _Berberis darwinii_	Aubretia _Aubrieta deltoidea_
18	Barren Brome _Anisantha sterilis_	Greater Periwinkle _Vinca major_	Oriental Poppy _Papaver pseudoorientale_
19	Trailing Bellflower _Campanula poscharskyana_	Lavender _Lavandula angustifolia_	Norway Spruce _Picea abies_
20	Yellow Corydalis _Pseudofumaria lutea_	Sweet Chestnut _Castanea sativa_	* Lilac _Syringa vulgaris_ (P)
21	Horse-chestnut _Aesculus hippocastanum_	Guernsey Fleabane _Conyza sumatrensis_	* Snowberry _Symphoricarpos albus_ *
22	Annual Mercury _Mercurialis annua_	Purple Toadflax _Linaria purpurea_	Dotted Loosestrife _Lysimachia punctata_
23	Michaelmas-daisy _Aster × salignus_	* Crack Willow _Salix × fragilis_	* Pineappleweed _Matricaria discoidea_
24	Procumbent Yellow-sorrel _Oxalis corniculata_	Aubretia _Aubrieta deltoidea_	London Pride _Saxifraga × urbium_
25	Red Valerian _Centranthus ruber_	* Bristly Oxtongue _Helminthotheca echioides_	* Hybrid Bluebell _Hyacinthoides × massartiana_
26	Privet _Ligustrum ovalifolium_	Rhododendron _Rhododendron ponticum_	* Garden Peony _Paeonia officinalis_
27	Large Bindweed _Calystegia silvatica_	Wilson's Honeysuckle _Lonicera nitida_	Shepherd's-purse _Capsella bursa-pastoris_
28	Prickly Lettuce _Lactuca serriola_	Garden Grape-hyacinth _Muscari armeniacum_	New Zealand Willowherb _Epilobium brunnescens_
29	Oxford Ragwort _Senecio squalidus_	Firethorn _Pyracantha coccinea_	Shasta Daisy _Leucanthemum × superbum_
30	Ground-elder _Aegopodium podagraria_	* Lilac _Syringa vulgaris_ (P)	Darwin's Barberry _Berberis darwinii_

FIG 133. Cherry Laurel (*Prunus laurocerasus*) is one of the commonest evergreen shrubs in British gardens. Its handsome spires of white flowers appear early in spring and are often browned by frosts, but it spreads both by seed and by layering of the lower branches. (CAS)

like brickwork, where the dry, low-nutrient substrates have not recruited an influential native flora (e.g. Butterfly-bush or Red Valerian *Centranthus ruber*). All these plants achieve dominance through a combination of large size, high growth rate and freedom from influential natural enemies.

In aquatic systems, where spatial uniformity of the substrate is more likely to be the norm, dominance by alien plants is frequent even if this dominance is seasonal or relatively short-lived – e.g. Water Fern, Least Duckweed, Floating Pennywort (Fig. 134) or New Zealand Pigmyweed (Fig. 113). As a result of rapid growth followed by vegetative division, these species are capable of forming dense colonies that blanket the entire water surface, reducing light availability to submerged aquatic plants.

Other alien species become abundant without ever causing any measurable harm to native vegetation, largely as a result of their inhabiting low-value habitats such as urban waste ground. The most obvious examples of these are American Willowherb (Fig. 129a), probably our most frequent willowherb, the fleabanes (*see* below) and the soldiers, Gallant-soldier and Shaggy-soldier (Fig. 6).

FIG 134. Floating Pennywort (*Hydrocotyle ranunculoides*), recently arrived (1990) in the British Isles from North America, is threatening to disrupt canal traffic and smother other wildlife, as here in Tottenham. (CAS)

The Fleabanes

The four alien fleabanes of the genus *Conyza* (Fig. 122) illustrate the issues involved in the determination of abundance. Field experiments at Silwood Park in Berkshire have shown that none of these species is capable of regeneration in shade or in closed vegetation, presumably as a result of competition from established vegetation. This means that their overall abundance is limited by the availability of open ground in full sun. Within this habitat, however, they show contrasting behaviour.

Argentine Fleabane has always been local and rare in the British Isles, seldom more than casual, yet it is an abundant garden weed in New Zealand and Japan, where it is spread as a weed of plant pots. The plant is much less abundant with us than in these other temperate islands, but we don't know why.

Canadian Fleabane was the first of the four to naturalise and is geographically the most widely distributed within the British Isles. It has shown interesting changes in abundance, declining as a waste-ground species but remaining as a locally abundant weed of cultivated ground and fallow arable land.

Guernsey Fleabane has increased steadily in abundance on open ground in southern England, to the point where it is often the most conspicuous alien on railway ballast (on London's railways, for example, it is now more numerous than Oxford Ragwort).

Finally, Bilbao's Fleabane is still a rarity in many of its British locations, but at Silwood Park it has increased to dominance on brick paving, outcompeting the former dominant Guernsey Fleabane in the process.

Both of the last two species are capable of forming dense monocultures, so clearly they have the ability to tolerate high densities of conspecifics in their new alien environment, perhaps as a result of the lack of natural enemies. It will be interesting to see if the propensity of Bilbao's Fleabane to outcompete Guernsey Fleabane is repeated in other parts of the British Isles.

Feral Oilseed Rape

Oilseed Rape is a locally abundant alien whose populations are largely transient. In many of its habitats it is a casual species relying on repeated introduction ($\lambda < 1$), but it can be a visually spectacular feature of motorway verges in early spring (Fig. 135). Crawley & Brown (1995, 2004) studied the population dynamics of feral Oilseed Rape for ten years between 1993 and 2002 in 3,658 adjacent permanent 100 m quadrats in both verges of the M25 motorway around London. The aim was to determine the relative importance of different factors affecting the observed temporal patterns of population dynamics and their spatial correlations. A wide range of population dynamics was observed in different locations (including downward and upward trends, cycles, local extinctions and

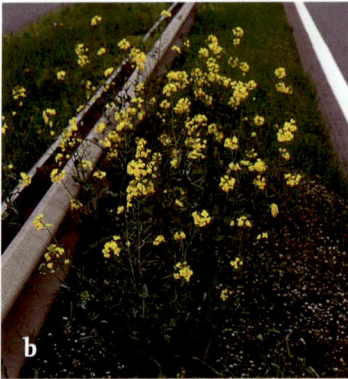

FIG 135. Oilseed Rape (*Brassica napus* subsp. *oleifera*): (a) monocultures, now a familiar sight; (b) plants colonising a roadside verge from spilled seed. (CAS)

recolonisations), but overall the populations were not self-replacing ($\lambda < 1$). Many quadrats remained unoccupied throughout the ten-year study period, but a few were occupied at high densities for all ten years. Most quadrats showed transient Oilseed Rape populations that typically lasted one to four years. There were strong spatial patterns in mean population density, associated with soil conditions and with the successional age of the plant community dominating the verge, and these large-scale spatial patterns were highly consistent from year to year.

The importance of seed spilled from trucks in transit from the farm to the oil-processing plant at Erith in Kent is illustrated by the fact that Oilseed Rape populations were roughly threefold higher on the 'to Erith' verge, where the passing trucks were full of rape seed, than on the 'from Erith' verge, when the trucks were returning empty. Quadrats in which $\lambda > 1$ were much more frequent in the 'to Erith' verge, indicating that seed immigration can create the spurious impression of self-replacing population dynamics. Where there was spatial synchrony in this system, it appears to be driven by at least three processes: small-scale local forcing caused by soil disturbance like roadworks; intermediate-scale forcing as a result of seed input; and large-scale climatic forcing (e.g. winter rainfall) that affects the motorway as a whole (Crawley & Brown, 2004). This is a good example of a case where the species is naturalised, but many of its local populations are casual.

WHY ARE MOST ALIEN PLANT SPECIES UNCOMMON?

The balance of evidence suggests that the factors determining population size in plants can be ranked in order of importance, with competition from other species as the most important, followed by attack by natural enemies (herbivores and pathogens), and with shortage of mutualists (pollinators and mycorrhizae) as the least important. Individual cases, of course, can depart from this scheme. The spatial extent of the population within a suitable habitat will be determined by the availability of regeneration niches, and this in turn is likely to depend upon the disturbance regime, both natural and human-imposed.

There are seven ways of being uncommon but only one way of being ubiquitous (Table 7). As to spatial distribution, we have seen that climatic niche models parameterised from the native range are reasonably successful in predicting distribution as an alien within the British Isles. This suggests that, in the native range, the realised niche is not substantially smaller than the fundamental niche. This, in turn, suggests that competitors and natural enemies have a bigger role to play in determining population density than spatial distribution. This is an important and far-reaching insight that was possible only after studying alien plants in both their native and alien ranges.

So what is it that makes the difference between commonness and rarity? The first thing to recall is that left to their own devices, and protected from their competitors and natural enemies, nearly all species are capable of becoming abundant. The second thing is that context is all important. It is not the traits of the species that determine what happens, but the interaction between those traits and the environment in which the alien plant finds itself. And, crucially, that environment contains competitors and natural enemies with which the plant has no shared long-term evolutionary history.

It turns out that the answer to the question 'what is it that makes the difference between commonness and rarity' is both technical and quite subtle.

TABLE 7. Seven forms of rarity as described by Rabinowitz (1981). The table shows that there are seven ways of being uncommon but only one way of being ubiquitous. The rows show geographic range within the British Isles (wide or restricted). The outer columns show population density (high or low) and the inner columns show the species' habitat range (many or few). The 'real rarities' are in the bottom right cell: they occur at low density in few habitats and have a restricted geographic range.

	High population density		Low population density	
	Many habitats	*Few habitats*	*Many habitats*	*Few habitats*
Wide range	Pineappleweed (*Matricaria discoidea*) Sycamore (*Acer pseudoplatanus*)	Monkeyflower (*Mimulus guttatus*) Common Field-speedwell (*Veronica persica*)	Feverfew (*Tanacetum parthenium*)	Large-flowered Evening-primrose (*Oenothera glazioviana*)
Restricted range	Three-cornered Garlic (*Allium triquetrum*)	Fritillary (*Fritillaria meleagris*) Hottentot-fig (*Carpobrotus edulis*)	Fly Honey-suckle (*Lonicera xylosteum*)	Argentine Fleabane (*Conyza bonariensis*) Limestone Woundwort (*Stachys alpina*) Magellan Ragwort (*Senecio smithii*)

It has to do with *the nature of the density dependence* experienced by the alien plant in its new environment. For most naturalised alien plant species, the population density at which the death rate first exceeds the birth rate is low, and that it why the plants are scarce. It is only a very small number of naturalised alien plant species that are capable of continuing to increase in abundance despite the fact that their numbers are high already.

We know the details in very few cases, because it takes carefully controlled experiments carried out under field conditions to discover the answers. One suspects that the commonest cause of rarity in alien plants is microsite limitation; there are simply rather few places in which a seed has a decent chance of becoming an adult plant. Since soil disturbance is the classic way of relieving microsite limitation, this would account for the oft-reported positive correlation between disturbance and alien plant abundance.

In short, the general scarcity of alien plants is due to the interaction between scarce microsites, voracious native herbivores and native competitions. There is little to suggest an important role for missing mutualists but, of course, we know only about the successfully naturalised aliens.

CHAPTER 8

Alien Genetics and Breeding Systems

Nought may endure but mutability!
Percy Bysshe Shelley, 'Mutability' (1816)

When an alien plant is introduced it can have effects on the native flora at two different levels. First, it acts as an organism, with possible repercussions on the ecology of the region, ranging from highly advantageous (e.g. providing vegetation cover to bare areas, or new food-plants for humans and other animals, especially invertebrates) to extremely harmful (e.g. invading natural habitats and ousting native species, or causing physical damage, notably to waterways). Second, it acts as a vector for its DNA, as a sort of parcel containing its genetic information. The organism is the physical guardian of its DNA, and the degree of impact its DNA will have in the new region initially depends on the ecological success of the alien plant. If the plant is not very successful, and cannot gain a foothold (e.g. if the plant is frost-sensitive and dies out each winter, and is hence a casual), its DNA might die out without leaving a trace. It is possible, however, for the DNA of even the most unsuccessful alien plant to survive in its new environment if the plant can hybridise with natives or already established aliens to form hybrids that are more successful than the new alien. Effectively, the new alien DNA will have 'jumped ship' and found refuge in a more successful vector.

It is not, however, just the ecological and physiological performance of the alien taxa that determines the fate of their DNA; their breeding behaviour (mode of reproduction) also has a major impact. The most important aspects of this are the percentage of outbreeding (i.e. how the genetics of the plant and its environmental interactions determine the degree to which it mates with other plants), and the extent and success of interspecific hybridisation. This chapter will consider the first of these two major topics; the next chapter the second. It will become apparent that the aliens in our flora show just as much variation in these characteristics as do our native plants.

BREEDING BEHAVIOUR

The reproductive behaviour of an alien plant is first of all defined by whether or not it can reproduce sexually. Many alien arrivals in the British Isles, finding themselves in a suboptimal environment, never produce seed or even flowers in this country. These are mostly plants from warmer countries, which do not encounter sufficiently long or hot summers here. Very often they are 'short-day plants', i.e. those that flower towards the end of the season, when the nights get longer, but then fail to do so here before the bad weather sets in. Field botanists who hunt for aliens in such places as rubbish tips or shoddy fields often find it necessary to grow their collections on in a greenhouse in order to obtain flowers or fruits, only then enabling identification. Many well-known examples come from the Amaranthaceae, especially goosefoots (*Chenopodium*) and pigweeds (*Amaranthus*), in which fruits and seeds are often diagnostic. Others are biennials or perennials that do not flower in their first year, and are not frost-hardy, and hence never reach their second year.

Plants that do not flower either remain as casuals, needing reintroduction in successive years, or they might become established and spread by purely vegetative means. A notable example of the latter is Manchurian Rice-grass, which is well established in Sussex and Surrey, where it forms extensive patches by ponds and lakes (Fig. 136), but which has never been seen to flower there. It had been known at one of these ponds for more than half a century before its eventual determination and report in 2000, but its permanently vegetative state had defied previous attempts to name it. Examples of non-hardy perennials that never get a chance to flower are numerous; Date Palm is frequent on rubbish tips but never gets past the seedling stage.

Dioecious and gynodioecious species
Another reason for reproductive failure is shown by dioecious plants (i.e. those in which the two sexes occur on separate plants) of which only one sex occurs in the British Isles or in any one locality. Table 8 lists the 35 genera in which dioecious alien flowering plants occur in our flora. There is, however, no

FIG 136. Manchurian Rice-grass (*Zizania latifolia*) has never been found with flowers in the British Isles but thrives in the vegetative state by a few lakes in southeastern England; here it forms the dense stand in the foreground. (Michael Hollings)

absolute distinction between dioecious and non-dioecious species. For a start, many normally dioecious species occasionally bear flowers or flower parts of the opposite sex; willows (*Salix*, especially weeping willows), nettles (*Urtica*) and dog's mercury (*Mercurialis*) are well-known examples. In other cases the species are what can be called subdioecious – the plants are very predominantly either male or female, but some or all bear a few flowers of the other sex, e.g. spindles (*Euonymus*) and buckthorns (*Rhamnus*). Similar to these are the so-called polygamous species, where the sexes are variably distributed, a plant bearing any combination of male and female parts, e.g. ashes (*Fraxinus*). Whatever way the term is defined, there will be marginal cases.

In a rather different category are gynodioecious species, in which there are separate female and hermaphrodite plants. This is a widespread phenomenon. Delannay (1978) usefully listed 223 gynodioecious species in the European flora, dispersed in 25 families, more than 40 of which are aliens in the British Isles. Delannay wisely cautioned that the degree of expression of gynodioecy in a species can vary at the level of the individual, the population or the species, and that many species have not been examined critically, so that his list is likely to be incomplete. Gynodioecious plants are not included in Table 8, but *Cortaderia* and *Fallopia* are discussed here as examples of this condition.

TABLE 8. Alien dioecious flowering plants in the British Isles.

Genus	Species	Genus	Species
Acer	A. negundo	Populus	P. alba
Amaranthus	A. palmeri		P. balsamifera
Anaphalis	A. margaritacea		P. × berolinensis
Aruncus	A. dioicus		P. × canadensis
Asparagus	A. officinalis		P. × canescens
Aucuba	A. japonica		P. × generosa
Baccharis	B. halimifolia		P. × hastata
Cannabis	C. sativa		P. × jackii
Coprosma	C. repens		P. trichocarpa
Cotinus	C. coggygria	Rhamnus	R. alaternus
Cotula	C. dioica	Rhus	R. copallina
	C. squalida		R. coriaria
Egeria	E. densa		R. typhina
Elodea	E. callitrichoides	Rumex	R. acetosa subsp. ambiguus
	E. canadensis		R. scutatus
	E. nuttallii	Ruscus	R. hypoglossum
Euonymus	E. japonicus	Salix	S. acutifolia
	E. latifolius		S. alba
Fraxinus	F. angustifolia		S. daphnoides
	F. ornus		S. × ehrhartiana
Griselinia	G. littoralis		S. elaeagnos
Ilex	I. × altaclerensis		S. eriocephala
Lagarosiphon	L. major		S. euxina
Laurus	L. nobilis		S. × forbyana
Mercurialis	M. annua		S. × fragilis
Muehlenbeckia	M. complexa		S. × mollissima
Myrica	M. pensylvanica		S. × pendulina
Myriophyllum	M. aquaticum		S. × sepulcralis
Oemleria	O. cerasiformis		S. triandra
Petasites	P. albus		S. udensis
	P. fragrans		S. viminalis
	P. japonicus	Silene	S. latifolia
		Spinacia	S. oleracea
		Stratiotes	S. aloides
		Vallisneria	V. spiralis

The tall, clump-forming garden plants known as pampas-grasses (Fig. 137), belonging to the southern hemisphere genus *Cortaderia*, are normally gynodioecious. The female plants are more attractive than the hermaphrodite ones, and form their distinctive plumes even if not fertilised, so they are much the commoner in gardens. Hermaphrodite plants, however, also occur. Two species are involved: Pampas-grass (*C. selloana*), from South America; and Early Pampas-grass (*C. richardii*), from the South Island of New Zealand. They occur in the wild on waste ground, embankments of roads and railways, and maritime cliffs and sand dunes. Although Early Pampas-grass is decidedly less common, it sets much more seed than Pampas-grass. The reason for this is explained by Edgar & Connor (2010) in the *Flora of New Zealand*; Pampas-grass is an alien in New Zealand and Early Pampas-grass is one of five endemic *Cortaderia* species there. Wild populations of the

FIG 137. The ever-popular Pampas-grass (*Cortaderia selloana*) is gynodioecious, with female and hermaphrodite plants, although the hermaphrodites are self-sterile and so effectively behave as male plants. (CAS)

former consist on average of 51 per cent hermaphrodites and 49 per cent females, while the latter have 62 per cent and 38 per cent, respectively. The hermaphrodite plants of Pampas-grass, however, are self-sterile, so they set little seed and are effectively males, whereas the hermaphrodite plants of Early Pampas-grass are self-fertile and set plenty of seed. It is not known to what extent the plants recorded as naturalised in the British Isles are actually garden throw-outs, have arisen from blown seed, or started as the first and multiplied by the second route. Certainly some seeding occurs in naturalised colonies, but the relative abundance of the two sexual forms (morphs) of these species in the wild here appears not to have been investigated.

The genus *Fallopia* contains the most notorious gynodioecious species in our flora: Japanese Knotweed. This vigorous plant, now common throughout the British Isles, is gynodioecious but only the female plant exists here, so its spread is entirely vegetative, by means of its strong, rapidly growing and invasive rhizomes. This was enormously aided initially by our liking for it as an attractive garden plant, and latterly by our prodigious earth-moving activities. In fact, it has been found that all the plants here, as well as most if not all the rest in Europe, North America, Australia and New Zealand, belong to a single clone, which John Bailey has claimed as the largest female organism on earth. This plant amply proves that sexual reproduction is not necessary in order to achieve enormous, almost unparalleled ecological success, and there is no sign of the very limited genetic base possessed by this plant having any detrimental effect on its dominance.

Simple observations on our alien dioecious species would add usefully to our knowledge. For example, the shrub Osoberry (*Oemleria cerasiformis*) is usually grown as the male plant because its flowers are in longer pendent clusters (Fig. 138) and it is therefore more strongly scented (in early spring). The female is grown in some gardens and, if males are close by, it sets abundant distinctive black fruits. But I am not aware that the sex of wild plants has been noted. If they are all males, then it is likely that they arose as relics or throw-outs and have become naturalised by suckering. But if both sexes occur, they are more likely to have arisen from seed. Another case is Pearly Everlasting. Reports of its sexuality vary from dioecious to hermaphrodite with a great predominance of one sex

FIG 138. Osoberry (*Oemleria cerasiformis*) is a dioecious shrub grown for its scented early flowers. The male plants are more attractive than the females, so most cultivated and escaped plants are of this sex: (a) male inflorescence; (b) the strange fruits – each flower produces a cluster of between one and five drupes, which turn black when ripe. (CAS)

or the other on each plant. Careful study of the flower heads of different clones would reveal the true situation and tell us to what extent, if at all, this species is truly dioecious.

Kay & Stevens (1986) calculated that 4.3 per cent of native angiosperms in the British Isles are dioecious (4.4 per cent if the conifers are included). In some territories the comparable figure is much higher – for example, it stands at about 25 per cent in New Zealand. Table 8 lists 67 dioecious aliens in our flora, which gives a percentage (c. 3.7 per cent) of a similar order as that relating to our native flora.

The majority of the species in Table 8 occur as both sexes in close proximity, at least on most occasions. More relevant to us are those species that regularly appear in a locality, or even in the whole of the British Isles, as a single sex, so preventing sexual reproduction. Several on the list are shrubs or trees that are often present in very small numbers, and in those cases only one sex might be present in any particular locality. Buck's-beard (*Aruncus dioicus*) is an example of a herbaceous perennial. Few dioecious plants are annuals, thereby running the risk of never reproducing, but Spinach and Annual Mercury (*Mercurialis annua*) are examples.

Some genera are worth closer scrutiny. Hemp is found in the British Isles mostly as a birdseed alien, but it is also increasingly being grown on a field scale for fibre, paper and biomass. Many commercial varieties are available, most specially suited to the production of one commodity, although there are varieties that are grown both for fibre and seed from the same crop. For seed production, obviously solely the female plant is required, but there are varieties available that are monoecious, with male and female flowers on the same plant, and others that are dioecious but in which there is a great predominance of female plants. Since the halluginogenic drug is produced in greatest quantity by the bracts surrounding female flowers, cultivation for this purpose also demands female plants. Websites explaining how to grow Hemp for illicit purposes are quick to warn readers to 'get rid of those male plants ASAP', not only because the males are useless, but also because once the female plants are fertilised, drug production by them is reduced. Hemp is a short-day plant, not flowering until the longer nights at the end of summer, so it is often the case that plants found in the wild have not reached flowering. As with the pampas-grasses, it is not known what proportions of casual wild plants are male, female or hermaphrodite.

Asparagus is usually strictly dioecious. There is a mystique associated with the cultivation (as well as with the cooking and eating) of the alien Garden Asparagus. It has been grown since the times of the ancient Greeks (c. 2000 BCE) and has been championed as a medicinal as well as a culinary plant. In the wild here, it is now much commoner than our native Wild Asparagus (A. *prostratus*), and it often occurs in the same sorts of habitats, i.e. maritime dunes and cliffs, as well as generally on rough or waste ground. From a culinary point of view, male plants are preferred, because they produce more shoots (spears), but these are thinner than those of the female plants. The reasons usually given for the preference of males are that the production of fruits by the females weakens the plants, that productivity declines with age, and that the female plants do not live as long. Today, all-male cultivars are available and are now preferred by growers. Much of the Garden Asparagus found in the wild, however, originated from bird-sown seed when both sexes were cultivated.

The genus *Petasites*, the butterburs, contains one native species, Butterbur (*P. hybridus*), and three fairly frequent alien species. All are functionally strictly dioecious, although female plants of at least some species have one or two apparently male flowers in the centre of each flower-head (capitulum). These, however, must be sterile, because after very many years of closely observing butterbur colonies I have never seen a seed produced on an isolated female clone, and no pollen can be found in them. These neuter, male-resembling flowers secrete nectar (unlike the female flowers), and their function is probably to attract insects to female plants.

Male plants of the native Butterbur are found throughout most of the British Isles, but female plants have a much more restricted distribution and probably define the native area of the species (from the central Midlands of England to the Scottish borders). Formerly, male Butterbur was planted as an early pollen and nectar source for bees, and probably this is the origin of the plant in most, if not all, of the southern half of England. Not only were the male plants preferred by bee-keepers and their bees, but they are also more attractive visually and are preferentially planted in large gardens.

The latter is equally true of the three alien *Petasites* species. White Butterbur, from the lower mountains of Europe, is the only one with pure white flowers and the only one suitable for normal-sized gardens, as it has relatively small leaves and rhizomes of limited extent. Giant Butterbur, from Japan, has very large leaves and cream-coloured flower heads (Fig. 139), and is suitable for only very large gardens. Both species are scattered throughout much of Britain, but entirely (or almost so) as male plants. Females occur in a few botanic or other large gardens, being grown for interest or by accident from sown seed. The third species, Winter Heliotrope, is a rampantly rhizomatous species that was introduced for its early pollen and nectar, and for its very attractive appearance (Fig. 210), with neat, small leaves in summer. It has relatively slender flower heads produced from December to February and is beautifully scented like

FIG 139. Giant Butterbur (*Petasites japonicus*) is one of three naturalised species in the genus; in all the male plants are more showy than the females and greatly outnumber them both in gardens and in the wild here. (CAS)

almond or vanilla. Tony O'Mahony, in his *Wildflowers of Cork City and County* (2009), however, described it as 'the greatest single alien plant threat to wildlife habitats in Ireland', and further referred to it as 'a veritable ecological cancer' and 'an abomination of an introduction'. The female flower heads do not have the distinctive pleasant aroma of the males, but as far as we are aware, female plants do not occur anywhere in Europe, either wild or cultivated. It seems highly likely that Winter Heliotrope is a native of North Africa (from Algeria to western Libya) and an alien in the whole of Europe.

All eight genera of the aquatic family Hydrocharitaceae in our flora contain dioecious species. Five of these genera are aliens, or probably so, including Water-soldier (*Stratiotes aloides*; Fig. 140a), a most distinctive plant that is much cultivated in garden ponds and lakes. It forms robust floating rosettes of long, narrow leaves with saw-like edges (looking a bit like the top of a pineapple), and has copious white roots hanging down into the water, sometimes reaching the mud below. The rosettes are evergreen, but in autumn they sink below the water surface, possibly to escape the frost and ice, rising to the surface again in spring. There has been considerable conjecture about the mechanism facilitating this rise and fall. After undertaking simple experiments, Cook & Urmi-König (1983), who monographed the genus, concluded that the rise is due to the development of new leaves with gas-filled intercellular spaces, while the sinking is caused by the plants becoming weighed down by dead, waterlogged leaves. When the dead leaves were removed from the sinking plants in the autumn, the plants rose to the surface and stayed there all winter.

Water-soldier spreads quickly by means of thin rhizomes that produce new plants as offsets, and in small ponds reduction in numbers is constantly in need

FIG 140. Water-soldier (*Stratiotes aloides*) is a floating aquatic that will anchor in the mud if its roots reach down far enough. It has often been claimed as a native species, but all plants in the British Isles are females: (a) female plant showing leaves and flower; (b) rosettes showing only the tips of the leaves in the Ashton Canal at Droylsdon, Manchester. (CAS)

of review. Plants are often discarded, or are thrown in ponds or canals (Fig. 140b) to 'beautify' them, and probably most of the records have such an origin. We are not certain whether this species is a native or an alien in the British Isles. All our plants are female, but female flowers contain staminodes (vestigial sterile stamens) and occasionally the latter do produce some pollen, although its fertility seems not to have been tested. Nevertheless, fruits with viable seeds have never been found in the British Isles.

The four other genera in the Hydrocharitaceae are all what we would call submerged pondweeds or waterweeds, and three of them are *Elodea*-type plants. *Elodea* itself is, or was, a very familiar plant to biology students, for it was used as material demonstrating the release of oxygen by green plants in sunshine through photosynthesis. If a water-filled test tube is inverted over a plant under water and in full sun, bubbles of oxygen can be seen rising up the tube and displacing water at the top. This rise of bubbles can also be seen by lying down and peering into a garden pond or local duck pond on a hot, sunny day.

When I was a student, the species used in experiments was Canadian Waterweed (*Elodea canadensis*), an alien from North America that was first recorded in Ireland in 1836, Scotland in 1842 and England in 1847, although all of these dates are disputed. In 1848, Cambridge University's Professor Charles Babington, believing it to be a newly discovered native plant, gave it the new name *Anacharis alsinastrum*. The vegetative parts of Canadian Waterweed are entirely submerged, and after hot summers when the water is warm, the tiny, inconspicuous flowers are produced on long, thin pedicels that reach the water's surface. Only female plants occur with us, and this has always been the case from the first introduction, except that male plants were discovered in one pond near Edinburgh in 1879, where they persisted only until 1903.

We have no idea how many times Canadian Waterweed has been introduced to the British Isles, but evidently it is more than once, and in a few cases the appearance of the plant was linked to the importation of aquatic plants from North America. Once it became common, it is likely that many new wild populations were derived from material discarded from infested ponds. One of the early areas of abundance was East Anglia, where the plant was said to have originated from Babington's studies in Cambridge in the 1840s. The rapid spread of Canadian Waterweed in the British Isles has therefore been entirely by vegetative means. By 1880, it had colonised most of the region and had become a pest of waterways with still water. From that date, however, a decline was noted. In the early twentieth century, its reduction in previously densely colonised areas was marked, but it was still spreading in the remoter parts of our islands where it had then only recently reached.

In 1966, another North American species of *Elodea*, Nuttall's Waterweed (*E. nuttallii*), was discovered in Oxfordshire. In many places in England this species has taken over from Canadian Waterweed as the commonest *Elodea*, and in many individual ponds it has actually replaced it. We briefly discussed this phenomenon of replacement in Chapter 6, but it should be emphasised that, whatever the reasons for it in this and other cases, the *general* marked decline of Canadian Waterweed was well underway long before Nuttall's Waterweed arrived here. Since its arrival, Nuttall's Waterweed has spread rapidly and within 20 years it had occupied most parts of England and Wales. It has now reached Ireland and Scotland, but in those two countries it is still much less common than Canadian Waterweed. Without detailed examination, one would never realise that one *Elodea* had succeeded another. Whenever plants of Nuttall's Waterweed have flowered they have proved to be females.

Curly Waterweed (*Lagarosiphon major*) is often known to aquarists as *Elodea crispa*, but it differs from all elodeas in that its leaves are not regularly arranged in pairs or whorls, but are variously whorled to spiral up the stem. They are also more strongly curled back. Its native area, South Africa, also differs from that of the elodeas, but it similarly originated here (in 1944) as an aquarist's throw-out and has spread over much of the region except the northern half of Scotland. Curly Waterweed seems to grow closer to the water surface than *Elodea* species, so the pedicels are shorter, and flowering is much more regularly observed. As in *Elodea*, only females have been found here.

The third *Elodea*-like genus is the South American *Egeria*, which was discovered in 1950 in warm-water effluent from mills in Greater Manchester (Fig. 141a). Large-flowered Waterweed (*Egeria densa*) is another aquarist's throw-out, and in fact in Manchester the warm canals and lodges (mill-pools) were used by enthusiasts as a reservoir for the plant, which they often called *Elodea densa*. Vegetatively, it closely resembles *Elodea* species, but the white flowers are much larger – up to 2 cm across (Fig. 141b) – and resemble those of the native Frogbit (*Hydrocharis morsus-ranae*), which also grows in those canals. Now that the water is no longer heated (electric motors having replaced steam engines), flowers are not produced. The flowering plants examined have all turned out to be male. Whereas the pollen of *Elodea* and *Lagarosiphon* is dispersed by surface water currents,

FIG 141. Large-flowered Waterweed (*Egeria densa*) occurs mainly in warmed water and flowers only in those conditions: (a) typical conditions in a Manchester mill-lodge, with Large-flowered Waterweed in the foreground and steamy water showing in the background; (b) close-up showing flowering shoot. (a, CAS; b, courtesy of California Department of Food and Agriculture, 2001).

Egeria is an insect-pollinated species, visited by flies, etc. Although artificially warmed water must favour its flowering, plants have since been found in flower in unheated water, but late in the year after the water has been warmed by the summer sun. In California, this species is a major weed of waterways and it is illegal to sell it: 'it is in almost every low-elevation pond or lake in the state' (Fred Hrusa, pers. comm., 2012).

The final member of the family is quite different in appearance. Tapegrass (*Vallisneria spiralis*) consists of a rosette of long, narrow, ribbon-like leaves (up to 80 cm long) rooted in the muddy gravel under water up to more than a metre deep. The earliest record is from Worcestershire in 1868. Most of the records of Tapegrass are from warmed water, e.g. from the Manchester mills, but apart from that area it has always been a scarce plant except in the River Lea Navigation Canal, which runs from Hertfordshire to Middlesex and Essex, where it was discovered in 1961. Here the water is not artificially warmed other than by the generally higher temperatures of Greater London and perhaps by the high level of pollution. By the early 1970s, Tapegrass grew along about 9 km of the canal and was locally dominant. It is still established in canals in the area, but more usually it is recorded as short-term appearances resulting from aquarists' discarded material, which has probably always been its origin.

Each rosette of Tapegrass is unisexual. The rosettes produce thin runners at substratum level, and these form many 'daughter' rosettes, eventually developing a dense sward. The difficulty of tracing which rosettes are vegetatively interconnected is probably the origin of statements that the species is monoecious rather than dioecious; perhaps in some cases it is. The solitary female flowers are borne on long pedicels that reach the water's surface. The male flowers, which occur in groups, break off their pedicels and float to the water's surface, where they are carried around by wind and water currents, eventually bumping into a female flower and pollinating it. The actual flowers are tiny and lack petals. After fertilisation, the pedicels of the female flowers become spiralled, so pulling the growing fruit down into the water. John Swindells tells me that Tapegrass produces female flowers and fruits in the River Lea, and Weiss & Murray (1909) noted the 'characteristic female flowers' produced by plants that had been introduced to the artificially warm Reddish Canal in Manchester 'nearly 40 years ago'. It is not certain, however, whether the fruits reach maturity.

The water-milfoils (*Myriophyllum*) are another genus of aquatic plants but quite unrelated to the last. Their finely pinnately divided leaves, carried three to five in a whorl on branching stems, are submerged, but the small, inconspicuous flowers occur on erect spikes above the water's surface. Our three native species are hermaphrodite, bearing male, female and hermaphrodite flowers on one shoot, but the introduced Parrot's-feather, from South America, is dioecious. Only the female has been found here, so this is another example of a plant relying on vegetative spread.

The poplars (*Populus* species and hybrids), together with the willows (*Salix* species and hybrids), form the almost entirely dioecious family Salicaceae. The willows are discussed in Chapters 2 and 9. The vast majority of poplars in the British Isles are planted aliens; regeneration has been detected in extremely few localities. The species most frequently planted are White Poplar (*P. alba*) from Europe, and Western Balsam-poplar (*P. trichocarpa*) and Eastern Balsam-poplar (*P. balsamifera*) from North America. Planted trees of these species might be expected to be represented by roughly equal numbers of each sex, but this is not necessarily so. Female White Poplars are vastly commoner than the male, and Western Balsam-poplar is rather commoner as a male. This is due to the method of propagation, almost entirely by striking woody cuttings, and if a particular genotype is selected (or used by chance) this will be the one that becomes common. When named cultivars, which normally consist of one clone, are involved, the proliferation of one sex is inevitable.

White Poplar, so called because of the white-felted lower leaf surface and the whitish bark, is tolerant of salty conditions and is therefore much planted by the sea. Fastigiate varieties are available. It produces abundant suckers and soon forms thickets, and can colonise and stabilise sand dunes, but its prodigious vegetative reproduction makes it unsuitable for gardens and formal parks. Its hybrid with our native Aspen (*Populus tremula*) is the Grey Poplar (*P.* × *canescens*). This was formerly claimed to be a native species, but that was at a time when it was considered a separate species rather than a hybrid. In any case, contrary to the situation with White Poplar, male trees are much commoner than females. Grey Poplar does produce suckers, but strangely these are not as abundant as they are in either of its parents, making it a more acceptable specimen tree.

By far the commonest poplars in our towns and cities are members of the 'Euroamerican' hybrid known as Hybrid Black Poplar or Black Italian Poplar (*Populus* × *canadensis*, sometimes erroneously called *P.* × *euramericana*). This is a hybrid between our native Black Poplar (*P. nigra*) and the North American Eastern Cottonwood (*P. deltoides*), which is uncommon here as a planted tree. The hybrid was first produced in France about 260 years ago, but is now represented by a large number of cultivars, each one a single clone and therefore wholly male or female (Fig. 215). Up to about a century ago, there were only four cultivars that were at all frequent: the males 'Serotina' and 'Eugenei', and the females 'Marilandica' and 'Regenerata'. These were much planted for amenity, in parks and by roads and railways, and are still common features of many areas, recognisable at a distance after practice by their different distinctive outlines. All are large trees, and 'Serotina' and 'Regenerata' have a widely spreading habit, making them less suitable in today's traffic-dominated highways. 'Eugenei' is a narrow tree, probably having Lombardy Poplar (*Populus nigra* 'Italica') as its male parent (and 'Marilandica' as the female). During the past century, however, Hybrid Black Poplars (among other poplars) have also been commonly grown in plantations for paper and for matchsticks, and much further breeding has taken place, and is still pursued. As a result, many more cultivars are now available, not only for plantations but also for roadsides and similar areas. Most of the newer cultivars are relatively narrow in growth form, making them more suitable for both purposes. Apart from habit and growth form, the most important features considered desirable in new hybrid poplars are resistance to bacterial canker, *Melampsora* leaf rust and the leaf fungus *Marssonina*.

FIG 142. Perhaps the most commonly planted poplar today is the hybrid balsam-poplar known as 'Balsam Spire' *Populus × hastata*; it has a very rapid growth rate but its poor-quality timber and brittle branches confine its use to screening. (CAS)

Four other hybrids are fairly frequent as planted trees. One is a hybrid between the above-mentioned Western and Eastern balsam-poplars. Balsam-poplars form section *Tacamahaca* of the genus *Populus* and are not native in Europe. Their leaves produce a strong balsam scent when they are unfolding. The above two species are largely separated in North America, as their common names suggest, but they overlap along the Northern Rocky Mountain Axis from Alaska to Wyoming, where they hybridise. These hybrids are known as *P. × hastata*. The poplar most planted today in Britain, with the cultivar name 'Balsam Spire', is a hybrid between the above two species that was deliberately produced at Harvard in the 1930s, and made available for sale over here in 1957. However, it does not closely resemble the wild North American *P. × hastata* hybrids as it is narrowly fastigiate (Fig. 142). 'Balsam Spire' is a female clone now much planted at field margins and along roads. It has an extremely rapid growth rate (exceeding

1.5 m per year for the first 15 years) and can attain a height of 20 m in 13 years (Jobling, 1990). It is useful as a windbreak and as an amenity tree as it grows well on 'unfavourable' soils such as heavy clays, but the wood is very soft and the branches extremely brittle. It comes into leaf early in the season and soon expands its large, scented leaves, which along with its fastigiate growth make it very easily recognised.

The other three common hybrid poplars are pairings of black poplars with balsam-poplars. In these, the distinctive characters of the balsam-poplars are discernible but occur in a 'diluted' form. The best known is Balm-of-Gilead, which has been much misunderstood in the past, having been cultivated in Britain for nearly 250 years. For a long time it was thought to be a balsam-poplar, because the characters of that section are much more dominant than in other balsam-poplar/black poplar hybrids. American taxonomists have found that it matches the wild hybrid *Populus deltoides* × *P. balsamifera*, and should be known as *P. × jackii*. Balm-of-Gilead is a female clone that is best distinguished as cultivar 'Gileadensis'. It produces abundant suckers, sometimes forming thickets in damp places, and is highly susceptible to bacterial canker. Another cultivar, 'Aurora', is commonly planted in parks. It is easily recognised by its very large variegated leaves with cream-coloured patches of chlorophyll-free tissue. One strongly suckering patch of 'Gileadensis' growing by a canal in Leicester has sported some branches with 'Aurora' variegation.

It is evident that dioecious plants run the risk of not being able to reproduce sexually, and this is particularly true of our alien species that owe their presence here to chance introductions. Such plants cannot become naturalised unless they have means of spreading vegetatively, as in the case of Canadian Waterweed, Japanese Knotweed and Balm-of-Gilead.

Self-incompatibility

Hermaphrodite plants (i.e. bearing both male and female sex organs) can be self-fertile (self-compatible) or self-sterile (self-incompatible). Two plants of a self-incompatible hermaphrodite species might be able to mate successfully (the two said to be of different 'mating strains'), or not (both being of the same 'mating strain'). Self-incompatibility with no compatible mating strain being available is another reason that plants, both native and alien, can fail to produce viable seeds. As with dioecious species, this lack of a suitable partner is more likely when the individuals are spread thinly on the ground, as is true of many aliens. The proportion of species that have some sort of self-incompatibility mechanism is unknown, but it is likely to be at least 50 per cent (A. J. Richards, pers. comm., 2012).

An incompatibility mechanism in flowering plants is mediated by a biochemical reaction at the style–pollen or stigma–pollen interface (or rarely at the interface between the pollen and some other, more deep-seated part of the pistil). For this reason it is restricted to angiosperms, whose ovules are protected within an ovary bearing a receptive stigma, with or without a style. Pteridophytes and gymnosperms hence do not possess incompatibility mechanisms, and nor do bryophytes, whereas in some algae a distinctly different system operates. The incompatibility mechanism has been demonstrated in many representatives of the so-called primitive angiosperms or pre-dicots, for instance in members of the families Magnoliaceae (tulip-trees), Lauraceae (bays) and Nymphaeaceae (water-lilies), families that diverged from the main line of angiosperm evolution before that line split into the dicotyledons and monocotyledons. The fact that self-incompatibility occurred so early in the angiosperm phylogeny is not at variance with the possibility that it evolved only once in the whole of the flowering plants, and that it has been lost, and regained, many times during angiosperm evolution. It is possible that, in plants that have lost effective self-incompatibility, the genetic mechanism still exists in outline but has been rendered inoperative, so that its reinstatement is nothing like such a complex evolutionary process as was its initial development. Plants that lack self-incompatibility are mostly those whose absolute need to produce seed overrides the desirability of outbreeding – notably annuals that do not have another chance to breed in the following year. In many genera we find that the perennial species exhibit self-incompatibility, whereas the annuals do not. All other things being equal, self-compatibility might prove to be an advantage in an alien plant that is in the process of colonising new areas in which it will at first tend to be few in number. But other things are not equal, and many aliens are self-incompatible, including some that are very successful colonisers, such as Oxford Ragwort.

The genetic mechanisms controlling the biochemical reaction at the stigma–pollen or style–pollen interface in self-incompatible plants are various. There are two main categories of mechanism: homomorphic and heteromorphic. In the great majority of self-incompatible species, no morphological distinctions exist between plants of different mating strains. Such plants are known as homomorphic. Some species are, however, heteromorphic, where the mating strain can be identified by morphological differences. Most such plants are dimorphic, with two sorts of flowers borne on separate plants, but a few are trimorphic, with three sorts. These 'sorts', or morphs, are sometimes confused by beginners with different sexes, but in fact both or all three morphs are hermaphrodite.

Homomorphic self-incompatibility

Since no morphological evidence is available, a clue to the existence of homomorphic self-incompatibility can be the lack of seed-set in a colony of plants, especially when it is likely to comprise a single clone, or in a single plant isolated from others of the same species, as in a garden. Perhaps the best-known example of this among our aliens is Slender Speedwell (Fig. 254), a species introduced into Britain from the Caucasus and Anatolia around 1800 as an attractive rock-garden plant. In the 1950s and 1960s, Ted Bangerter and Douglas Kent carefully documented the spread of Slender Speedwell here after it was first noticed in the wild in 1927 (there is actually one isolated record from much earlier). By 1950, it had reached most parts of the British Isles, and since then it has consolidated its coverage so that it is now common in many areas. Despite its attractive blue flowers, Slender Speedwell has lost favour as a garden plant because of the aggressive spread by its creeping stems, which root readily and invade lawns and other grassy places. So far as we know, all of this spread is vegetative, and the species' dispersal to new areas is in contaminated soil, turf and garden plants, and on lawnmowers. Very few records of seed formation have been made; Bangerter & Kent (1965) had seen only a dozen or so specimens. Lehmann (1944) showed by experiment that the species is self-incompatible, and the lack of seed production is surely due to this. Whether more than one mating strain is present in this country is unknown, and offers an interesting line of research for a keen naturalist. The production of a small amount of seed on a few specimens is not evidence of the existence of different mating strains, because few self-incompatibility systems are absolute.

Another example of homomorphic self-incompatibility is provided by the genus *Calystegia*, whose two common species, the native Hedge Bindweed and the alien Large Bindweed, frequently produce no fruits. Both are highly self-incompatible and show prodigious vegetative growth above and below ground, quickly covering many metres of hedgerow so that a large area can be occupied by one clone (Fig. 143). In many urban areas the alien Large Bindweed is the commoner species, yet its colonies are often widely separated and many occur too far from sources of bee-dispersed compatible pollen. This species is also mostly transported in contaminated soil and horticultural material, often entwined in rootballs. It is notable that three of the alien species most detested by gardeners (Large Bindweed, Slender Speedwell and Japanese Knotweed) should be mainly or entirely vegetatively propagated because of the lack or paucity of seed-set.

Heteromorphic self-incompatibility

Heteromorphism occurs in only a tiny fraction of self-incompatible species. The 11 alien genera of plants in our flora that contain heteromorphic species

FIG 143. Large Bindweed (*Calystegia silvatica*) is a twining plant with pernicious white rhizomes that allow the plant to smother surrounding vegetation rapidly: (a) an all-enveloping colony, outcompeting even Japanese Knotweed (*Fallopia japonica*) for light; (b) close-up of flower showing the overlapping bracteoles that hide the sepals. (CAS)

are listed in Table 9, which is based on the important paper by Ganders (1979). As seen clearly from its taxonomic distribution, heteromorphism has evolved separately many times.

TABLE 9. Genera in the flora of the British Isles containing alien heteromorphic species (dimorphic unless marked trimorphic).

Family	Genus
Amaryllidaceae	*Narcissus* (*N. triandrus* trimorphic)
Boraginaceae	*Anchusa* *Pulmonaria*
Lythraceae	*Lythrum* (trimorphic)
Oleaceae	*Forsythia*
Oxalidaceae	*Oxalis* (trimorphic)
Plumbaginaceae	*Limonium*
Polygonaceae	*Fagopyrum* *Persicaria*
Pontederiaceae	*Pontederia* (trimorphic)
Primulaceae	*Primula*

Dimorphic plants

By far the best-known dimorphic plants belong to the genus *Primula*, involving four of our five native species and six of our seven aliens. Although Primrose (*P. vulgaris*) is a native species, it is the one primarily described here as it is the most carefully researched species and is common both in the wild and in gardens, so can very easily be observed. The essential details, however, apply equally to our alien species.

Primrose plants are all hermaphrodite, but they exist as two sorts: with either all 'pin-eyed' flowers or all 'thrum-eyed' flowers (Fig. 144). In the former, the style is long and the stigma is therefore high up (usually at the top of) the corolla-tube, and the anthers are borne well below the stigma inside the corolla tube. In thrum-eyed flowers, the position is reversed, with the anthers borne near the top of the corolla tube and the shorter style bearing the stigma deeper inside the corolla tube. This can easily be seen by looking at the flower from the top, when either a stigma occupies the corolla-tube opening or there are five anthers in a spoke-like pattern.

The designation 'pin' is easy to understand; 'thrum' apparently alludes to the unwoven loose end of a thread, as would be seen in a loom, a term that would have been more familiar to the general public a couple of centuries ago. If a population of Primroses (or of any other dimorphic species) is scored, normally approximately half of the plants will be found to be of each sort. Various fanciful explanations were once advanced to show how the pin/thrum mechanism would physically prevent pin–pin or thrum–thrum pollinations, but we now know that these are scarcely effective and, if anything, only serve to reduce wastage of pollen.

FIG 144. Many species of *Primula*, including the Primrose (*P. vulgaris*), are dimorphic, the plants bearing either 'pin' or 'thrum' flowers: (a) pin-eyed flower; (b) thrum-eyed flower. (CAS)

This topic was addressed in detail by Charles Darwin (1877), but we now understand much about the genetic and biochemical mechanisms regulating dimorphism, which are very thoroughly and well discussed by John Richards in his textbook *Plant Breeding Systems* (1997). The genetic background of dimorphism regulates the biochemical pollen–stigma reaction, which ensures that pollen from a pin-eyed plant can fertilise the ovules of only a thrum-eyed plant, and vice versa. As with most self-incompatibility mechanisms, the Primrose system is not perfect, and a small proportion of the seed produced is actually of pin–pin or (less often) thrum–thrum origin. This 'imperfection', even more strongly encountered in many homomorphic systems, can be seen as a sort of safety valve, enabling a plant produce *some* seed if compatible mating strains are not present.

Although the relative positions of the anthers and stigma are the most obvious features of Primrose dimorphism, these two characters are part of a small suite that include pollen size, form of the stigmatic papillae and anatomy of the stylar pollen-tube-conducting tissue, as well as the nature of the proteins that govern the incompatibility reaction. These characteristics are obviously governed by different regions of the plant's DNA and would normally exist as separate genes. In *Primula*, however, these pieces of DNA are positioned very close together, forming a so-called supergene, such that genetic crossing over between the individual genes normally does not occur, and the whole suite of characters is inherited as a single package.

Since the most conspicuous feature of the pin–thrum distinction is the difference in style length and anther position, heteromorphism of this kind is often called heterostyly, and Primrose is known as distylous (i.e. having two style lengths, *not* two styles). Occasionally, however, Primroses (Fig. 145) are found that are not heterostylous, but are homostylous, with the stigma and anthers both borne at the top of the corolla tube (long-homostylous). Effectively, they have the pistil of a pin-eyed plant and the stamens of a thrum-eyed plant, and because of this they are fully self-fertile. These plants are thought to have arisen by a rare crossing-over event *within* the supergene. As might be expected, they produce

FIG 145. Primrose (*Primula vulgaris*) flowers cut in half longitudinally. Left to right: thrum-eyed; long-homostylous; pin-eyed. (CAS)

FIG 146. Japanese Cowslip (*Primula japonica*), one of the species whose flowers are wholly homostylous. (CAS)

more seed than heterostylous plants, and where they occur they tend to increase in relative abundance. Homostylous Primroses are particularly common in an area of southwestern England centred on Somerset. Homostyly has arisen in many species of *Primula*, either as rare isolated events as in the Primrose, or to such an extent that the species is wholly homostylous.

Our seven alien species of *Primula* occur in damp places in marshes, by streams or on mountains. Of these seven species, six are heterostylous and one, Japanese Cowslip (*P. japonica*), from Japan (Fig. 146), is homostylous. Six of our seven alien primulas do not closely resemble the Primrose in habit, but are closer to the garden polyanthus in form, with an umbel of flowers on a common stalk, or a series of tiered whorls of flowers borne on the stem and terminating in an umbel at the top. The flowers vary between red, pink, orange, yellow and white, and are very popular with keen gardeners, mainly for waterside positions. The flowering stems can be more than 1 m tall. Four of the species are natives of China, which boasts about 300 species of *Primula*.

The lungworts (*Pulmonaria*) are spring-flowering herbaceous plants that also possess distyly of the pin-eyed/thrum-eyed type. Our two native and two naturalised species all exhibit this feature; there are apparently no homostylous lungworts. The morphological characters of the two sorts of flower broadly resemble those of *Primula*, but there are differences in detail. For example, in *Pulmonaria* the anthers of thrum-eyed flowers are larger than those of pin-eyed flowers and produce more pollen, whereas in *Primula* the anthers of both sorts of flower are the same size and, because pin-pollen is smaller, pin-eyed flowers produce more pollen grains.

Our two alien *Pulmonaria* species are Lungwort (*P. officinalis*; Fig. 147), which has white-spotted leaves (hence the name lungwort), is widespread in Europe, and is by far the commonest species both in gardens and in the wild, and Red Lungwort (*P. rubra*) from southeastern Europe, a much rarer garden escape with red flowers and usually unspotted leaves. In addition, naturalised in scattered

FIG 147. Lungwort (*Pulmonaria officinalis*), a dimorphic species in which both morphs are common in gardens and in the wild, so that there is good seed-set. (CAS)

places across Britain is a lungwort with narrow unspotted leaves and attractive blue flowers, which is referable to the cultivar 'Munstead Blue'. The origin of this is uncertain; it might be a variety of the garden species *P. angustifolia*, or a hybrid. A true cultivar of a herbaceous perennial will be a single clone, so in a distylous species only one morph will exist, in this case the thrum-eyed. Any seed set by this cultivar must have resulted from a mating with a pin-eyed plant of another cultivar, so will not breed true. There are many *Pulmonaria* cultivars available, including at least four of *P. officinalis*, and the same consideration applies to them all, as it does to true cultivars of *Primula*. In gardens, *Pulmonaria officinalis* is represented by both morphs and regeneration from seed is prolific; it would be worth surveying wild populations.

The favourite garden early-flowering shrub *Forsythia* provides another example of a genus with pin-eyed and thrum-eyed plants. The commonest garden representative is a rather stiff, upright, tall shrub, sometimes pruned into a hedge, bearing deep yellow flowers in dense clusters. This is Forsythia (*F. × intermedia*; Fig. 148), reckoned to be a horticultural hybrid of German origin

FIG 148. Forsythia (*Forsythia × intermedia*) is a dimorphic shrub in which each cultivar (by definition) is represented by only one morph; thrum-eyed plants appear to be much the commoner in our gardens. (CAS)

between *F. viridissima* and *F. suspensa*, both Chinese species, although recent molecular work failed to find evidence for such a parentage, suggesting instead that it is a separate species, named as *F. intermedia*. *Forsythia suspensa* is a weaker rambling shrub with paler, less gaudy flowers borne less densely and usually pendent. Both it and its supposed hybrid root very readily where stems meet the ground or a mossy tree trunk, and that is their method of naturalisation, because seed-set is unusual (perhaps due to their self-incompatibility) and the stems are very brittle and pieces easily take root. Both pin-eyed and thrum-eyed plants occur in our gardens, but which of these occur in the wild in the British Isles is unknown. In my experience, thrum-eyed plants are much the commoner in gardens, but in addition to pin-eyed plants there are male-sterile ones and some that could be interpreted as homostyles.

Buckwheat (Fig. 68), in the family Polygonaceae, is doubly unusual among heterostylous species because it is both an annual and an important crop plant. It also differs from the above three genera in that thrum-eyed flowers are noticeably larger than pin-eyed flowers. It is a common constituent of various bird foods and a fairly common relic where such material has been used. Plant breeders have managed to produce some self-compatible strains, which have obvious advantages. Our other alien *Fagopyrum*, the Asian Tall Buckwheat (*F. dibotrys*), is a perennial, and is known as a naturalised plant in only two localities, both in Wales. There appear to have been no studies of self-incompatibility in this species, but as it is closely related to Buckwheat it is generally supposed to be distylous like the latter. The two colonies both comprise long-stamened plants.

One other genus in the Polygonaceae contains distylous species. Fifty years ago, *Polygonum* was a very large genus with a wide range of growth forms, but it has been gradually split into a number of smaller genera since then, some of which, like *Fallopia*, are now thought not to be particularly closely related to others. Two groups, often now both included in the genus *Persicaria*, contain distylous species. One, sometimes segregated into the genus *Aconogonon*, includes the distylous Lesser Knotweed (*Persicaria campanulata*); the other, sometimes separated into the genus *Rubrivena*, includes the distylous Himalayan Knotweed (*Persicaria wallichii*, formerly known as *Polygonum polystachyum*). Both are tall, branching, leafy perennials with dense masses of white flowers, and are Himalayan in origin. They make handsome garden plants and are naturalised in usually dampish, shady spots. Despite their similar appearance they are well separated taxonomically and, together with *Fagopyrum*, demonstrate that distyly has originated several times in the Polygonaceae. Also like *Fagopyrum*, these two species of *Persicaria* differ from *Primula* in that the corolla does not have a tube, but forms an open bell shape. Hence it is not the position of the stamens on the inside of the corolla tube that determines anther height, but the length of the filaments that arise from around the ovary.

Two genera in the Plumbaginaceae, the thrifts (*Armeria*) and sea-lavenders (*Limonium*), contain dimorphic species whose breeding systems have been well studied. *Armeria* differs from *Primula* in that there is no dimorphism based on style length, or stamen position or length, so the term distyly is not appropriate. Distyly is, however, present in at least some *Limonium* species. The dimorphic species of these genera are strictly self-sterile, but the structural differences between the morphs are much more subtle than in distylous species. They concern the morphology of the pollen grains, termed types A (coarsely reticulate) and B (finely reticulate), and the prominence of the surface cells on the stigmas, termed cob (appearing like a maize cob) and papillate (with papillae that protrude more). Type A pollen can penetrate papillate stigmas, and type B pollen can penetrate cob stigmas, to effect fertilisation. Hence the dimorphic species consist of A/cob and B/papillate plants, each self-incompatible but cross-compatible. In both genera a very few homomorphic species also occur, these having either A/papillate or B/cob combinations. The alien Florist's Sea-lavender or Statice (*L. platyphyllum*, formerly *L. latifolium*) from southeastern Europe, which is naturalised in rough ground in a few places in southeastern England, is dimorphic.

Like the lungworts, discussed above, the alkanets (*Anchusa*) belong to the Boraginaceae family and are distylous. In Alkanet (*A. officinalis*), the lengths of the styles and positions of the stamens are somewhat variable (there are also some homostylous plants), and there is a marked predominance of pin-eyed plants (up to 90 per cent and usually at least 70 per cent). Even more remarkable are the facts that (a) the genes governing the morphology and the self-incompatibility system are completely unlinked, i.e. they do not exist in a supergene complex but instead are positioned on different chromosomes, and (b) although each plant is self-incompatible, it is cross-compatible not only with plants of the other morph but also with plants of the same morph. Another unusual feature of Alkanet incompatibility is that the position of the biochemical barrier to incompatible fertilisations is not in the stigma or style, but is relatively deep-seated, somewhere in the ovary, perhaps in the ovule. Our two other perennial species of *Anchusa* have apparently not been investigated regarding these topics, and nor in fact has Alkanet in the British Isles, but both Yellow Alkanet (*A. ochroleuca*) and Garden Anchusa (*A. azurea*) have corollas of a very similar shape and size to those of Alkanet, and could reasonably be expected to operate a similar genetic system. Our final species of *Anchusa*, the archaeophyte Bugloss (*A. arvensis*), is an annual and not distylous.

The nature of heteromorphy in the daffodils (*Narcissus*) has been much disputed for many decades. Heteromorphism in stigma height (due to variation in style length) has been reported from many species, for some reason mostly those with more than one flower per stem. In some cases, when it is accompanied by reciprocal differences in anther position, this is true heterostyly, but in others it comes under the heading of 'stigma-height polymorphism', where there is no variation in anther height between flowers exhibiting different style lengths. According to Alan Baker *et al.* (2000), the latter is much commoner in the genus than is heterostyly, which they claim was reliably reported in only Angel's-tears (*N. triandrus*) and the rare Moroccan *N. albimarginatus*. However, there is often some minor variation in stamen height in flowers with stigma-height polymorphism, and some workers prefer not to distinguish the two by different terms.

Bunch-flowered Daffodil (*N. tazetta*) possesses stigma-height dimorphism; the flowers, which number two to 15 on each stem and are usually white with a short yellow corona ('trumpet'), have stigmas at one of two heights, with little accompanying anther height variation. There are similarities to the unusual situation in *Anchusa*, for in *Narcissus* the genes governing dimorphy and self-incompatibility are not linked, individuals are self-incompatible yet both intra- and inter-morph cross-compatible, and pollen-tube inhibition takes place in the ovary. Since in Bunch-flowered Daffodil the upper stamen position is close to the stigma position in long-styled flowers, this morph resembles a long-homostylous plant, but it is self-incompatible. As in *Anchusa*, this long-styled morph is much commoner than the short-styled morph, and is actually the only morph present in some populations of this species.

The beautiful little Angel's-tears, with pendent, usually very pale yellow flowers, has several features in common with Bunch-flowered Daffodil, but also shows important differences. The specific epithet *triandrus* is rather misleading because this species, like all others in the genus, has six stamens. However, they are placed well apart in proximal and distal whorls of three each, so that only three distal ones are visible if the flower is viewed from the end of the corona. In many wild populations the stigma is at one of three heights, the two positions not occupied by it being taken by the two whorls of anthers. Hence these populations are tristylous, with short-, mid- and long-styled morphs. Other populations are distylous, the mid-styled morph being absent.

The details of heteromorphy in *Narcissus* outlined above relate mainly to wild populations, and the situation is somewhat different in cultivars and in naturalised wild populations derived from them. The cultivars are clonal and therefore all individuals are of the same morph. In the case of Angel's-tears, for example, cultivar 'Concolor' is mid-styled and cultivar 'April Tears' is long-styled. It is also possible that some cultivars are homostylous and/or self-compatible. *Narcissus* is a remarkable genus that includes homostylous, distylous and tristylous species, and others with stigma-height polymorphism (Barrett & Harder, 2005). It appears from phylogenetic (DNA) studies that the homostylous condition is the ancestral state, and that the various sorts of heteromorphism have each evolved separately on several occasions. *Narcissus* heteromorphy continues to offer much of interest to the enquiring mind, because we still do not know the extent of heteromorphism in even our own naturalised species, and simple, albeit careful and accurate, observations could add significantly to our knowledge.

Trimorphic plants

Tristyly has been mentioned above under Angel's-tears, but in that species it takes a rather unusual form. It is normally associated with three other families – Lythraceae, Pontederiaceae and Oxalidaceae – all of which contain alien species in our flora. It is remarkable that these three cases were known to, and investigated by, Charles Darwin (1877) nearly 140 years ago, and that, until recently, when studies of some tropical and southern hemisphere plants were at last made, no other examples were known. The widely disparate nature of these three families shows that tristyly must have evolved separately in each case, but, despite this, there is striking morphological and genetic uniformity between them.

In tristylous plants the stigmas are found in one of three positions (short, mid and long) due to their different style lengths. In each flower there are filaments of two lengths, bearing anthers that occupy the two levels where there is no stigma. Hence the stigma can be at a lower level than all the anthers (short-styled), at a higher level than all the anthers (long-styled), or at a level between the two levels of anthers (mid-styled). All three morphs are self-incompatible. Compatible matings are between pollen and stigmas of the same height. Hence, on a long-styled plant the pollen from a mid-level anther can fertilise a mid-styled plant and the pollen from a short-level anther can fertilise a short-styled plant, and so on; each morph can thus fertilise the other two.

The best-known and best-understood tristylous genus is *Lythrum*, containing the native Purple Loosestrife (*L. salicaria*) and the alien False Grass-poly (*L. junceum*). The former is a familiar tall waterside or marsh plant with spires of purple flowers, which interestingly has become a serious pest in some parts of North America where it has been introduced. The latter (Fig. 149) is an alien from various sources, mostly birdseed. It is very common in damp places in southern Europe, but is a much less attractive plant than Purple Loosestrife, with flowers that are sparser and scarcely half the size, and creeping stems. Careful studies by Dulberger (1970) in Israel showed False Grass-poly to be tristylous, with approximately equal numbers of each morph in wild populations. The morphs differ not only in anther and stigma height, but also in pollen-grain size and

FIG 149. False Grass-poly (*Lythrum junceum*) is a trimorphic species that today is most commonly found in the British Isles as a birdseed or grain alien. (Peter Greenwood)

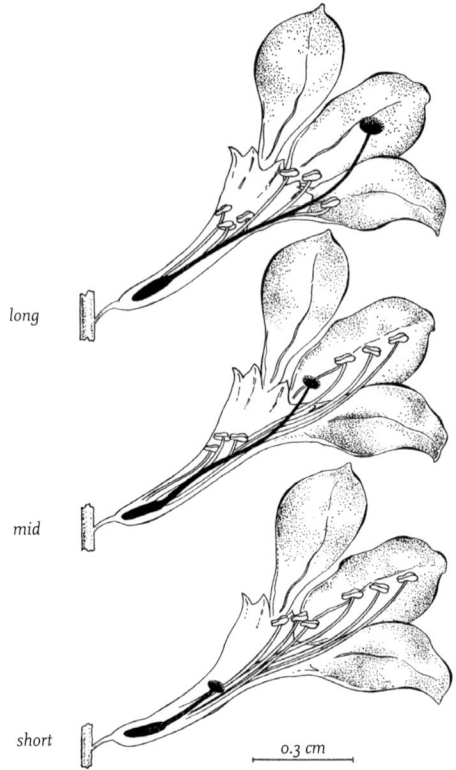

RIGHT: FIG 150. False Grass-poly (*Lythrum junceum*), showing the three flower morphs with differing relative lengths of stamens and styles. (Reproduced by permission of John Wiley & Sons, from Dulberger, (1970))

long

mid

short

0.3 cm

stigma papillae prominence, there being three grades of each (Fig. 150). By artificial crossing experiments, Dulberger showed that each plant is self-sterile, and that successful fertilisation of flowers takes place only by pollen transfer to a stigma at the same level as the anther supplying the pollen. It seems that no systematic observations on morphs have been made in the British Isles, except that a plant found on a Newport rubbish tip in 1977 was long-styled. Our other two species of *Lythrum*, the native Water-purslane (*L. portula*) and the archaeophyte Grass-poly, have less conspicuous flowers and are homomorphic, self-fertile annuals.

Pickerelweed (*Pontederia cordata*) is a North American emergent aquatic plant in the family Pontederiaceae and a favourite with gardeners. It has elongated heart-shaped leaves on long petioles, and rather late in summer produces flowering stems bearing dense spikes of blue flowers. These are slightly

zygomorphic (bilaterally rather than radially symmetrical), which is unusual
for a heteromorphic plant, as is the plant being a monocotyledon. It becomes
naturalised in ponds where neglected or discarded. Pickerelweed is tristylous,
with the same pattern and presumably the same genetic system as *Lythrum*. Other
species from Brazil were examined in great detail by Darwin (1877). Once again, I
am not aware of observations on this aspect of the plant in the British Isles.

The pink-sorrels and yellow-sorrels of the genus *Oxalis* present a more varied
manifestation of tristyly. The plants are short and often somewhat fleshy, some
without aerial stems apart from the inflorescence stalks, and most are instantly
recognisable by their trefoil-like leaves, though a few have four or more leaflets.
Tristyly exists in many species of *Oxalis*, but the genus contains about 500 species,
mainly in South America and southern Africa, and only a tiny proportion have
been critically examined. Most have either yellow or pink flowers, a few white.
Some are popular garden plants, but others are pernicious weeds, especially of
bulbfields in southwestern England, having very effective means of subterranean
propagation such as rhizomes and bulbils.

Fourteen *Oxalis* species are reckoned to be naturalised in the British Isles,
seven of them with yellow flowers and seven with pink flowers. Among the
yellow-flowered species, the most attractive garden plant is Chilean Yellow-sorrel
(*O. valdiviensis*; Fig. 151), with yellow petals finely veined with purple and much
visited by bees. It is a typically tristylous species, each plant having styles at one
of three lengths and a whorl of stamens at each of the other two levels; the three
levels are clearly separated. The breeding relationships of the three morphs
were worked out by careful crossing experiments as long ago as 1871 by the

FIG 151. Chilean
Yellow-sorrel (*Oxalis
valdiviensis*), an
attractive trimorphic
species investigated
by Charles Darwin. All
three morphs occur
in the British Isles
and seed is set here in
abundance. (CAS)

German Friedrich Hildebrand, who demonstrated the now familiar *Lythrum*-type situation. This was well before genetics was a science, and the genetic background was much later elucidated by Viviane Fyfe (1950).

Few native populations of Chilean Yellow-sorrel have been studied, and the degree of variation in tristyly in them is unknown. The species seeds very freely in my garden, but I found the unexpected morph ratio of 40 short-styled plants to 47 mid-styled and none that were long-styled. This can be explained by the fact that my garden plants were derived from a few seeds from one plant gathered 30 years previously, which by chance lacked the genetic determinant for long styles. Although this is an entirely artificial 'population', the conditions leading to a reduced range of genetic variation in my garden must have been replicated in the wild by chance dispersal to a new locality. Many populations of usually tristylous species have been found to lack one or other of the morphs, and it clearly might not always be due to mutations. This is particularly relevant to alien species, which often arrive here as an extremely small sample of the variation in their natural areas.

The large-flowered yellow Bermuda-buttercup (*Oxalis pes-caprae*; Fig. 238) – neither Bermudan nor a Buttercup – is a major nuisance in many agricultural systems throughout much of the world, including the Isles of Scilly. It is also tristylous, but it illustrates the last point above in that as far as we know only the short-styled morph occurs with us and no fruits have been found here. This does not seem to trouble it, as it enjoys extremely effective vegetative reproduction.

The four yellow-flowered members of the section *Corniculatae* in the British Isles have small flowers, are all weedy and are all self-compatible. The last is apparently also largely true of the larger-flowered species of the section, such as *Oxalis suksdorfii* in North America (not one of our species), which are morphologically distinctly tristylous. In these plants the correlation between heteromorphism and self-incompatibility has begun to break down, and this is a feature of the whole of section *Corniculatae* (Ornduff, 1972). In *O. suksdorfii*, long-styled plants are completely self-compatible, short-styled plants are fairly self-incompatible and mid-styled plants produce very little seed from any pollinations. Our weedy species of this section, Procumbent Yellow-sorrel (*O. corniculata*), Upright Yellow-sorrel (*O. stricta*; Fig. 152) and Sussex Yellow-sorrel (*O. dillenii*), are completely self-fertile and apparently homomorphic in the British Isles; they have anthers at two levels and the stigmas roughly at the level of the longer stamens. In Canada, Upright Yellow-sorrel alone of these three is represented by some other morphs as well, but it is never tristylous. In Japan, Procumbent Yellow-sorrel exists in two forms in different populations, one as

FIG 152. Upright Yellow-sorrel (*Oxalis stricta*) is self-fertile and has rather small, unattractive flowers with stamens and styles at the same level; it can be a troublesome garden weed. (CAS)

described above and one with styles longer than both tiers of anthers (probably one of the Canadian morphs), increasing the chances of outcrossing. Our fourth species, Least Yellow-sorrel (*O. exilis*), has gone a stage further in that the lower tier of stamens lacks anthers altogether.

The above examples of *Oxalis* are all yellow-flowered, but a similar situation exists among the pink-flowered species. *Oxalis* tells a complicated and fascinating story that has by no means been fully unravelled. Many species await investigation and could modify our views. Relatively little contribution could be made from British studies, because our representatives are tiny samples of each species, but it would be useful to know which morphs of each species occur here.

Apomixis: seeds without sex

Apomixis implies the production of viable seed without fertilisation, i.e. seeds containing embryos that are entirely maternal in origin. Actually, apomixis is often construed more widely, to include a whole range of vegetative propagation as well, and the more restricted definition used here is synonymous with agamospermy, which is perhaps the more obviously descriptive term. This remarkable process circumvents the usual cell reduction division, when chromosomes pair and recombine before separating (eventually leading to the production of gametes, which are haploid, i.e. have one set of chromosomes, rather than being diploid, with two sets of chromosomes), and therefore also omits the reverse process of gamete fusion (fertilisation) to reconstitute the normal diploid chromosome number.

In angiosperms, the male gametes are nuclei in the pollen grains, and the female gamete is the egg cell in the embryo sac deep in the ovary. Since the embryos and therefore the next generation are entirely maternal in origin in apomixis, it amounts to clonal reproduction, so that all the offspring are genetically identical to one another and to their seed parent (apart from any mutations that might occur). Hence large numbers of individuals essentially genetically identical can soon build up. Differences between such families of individuals might be small, but they are often morphologically recognisable and have spread over large areas, and keen-eyed taxonomists have named them as separate species. They are known as microspecies or agamospecies.

Table 10 lists the genera and indicates the species in the flora of the British Isles containing alien agamospermous species. Apomixis is also widespread in ferns (notably *Dryopteris*), but the table covers only flowering plants.

It is a moot point to what extent this concept of each agamospecies being effectively a single genetically uniform clone is realistic, but one that need not unduly concern us here. Almost certainly it is not true for any agamospecies that has existed for some time and has a wide distribution. Two factors suggest this departure from the traditional view: mutations will occur and continue to build up over time; and it is likely that any old and widespread species will have arisen more than once from the same or similar parents to produce morphologically very similar results. We need not discuss, either, the different mechanisms of agamospermy that are known, except to say that examples of all the major categories are to be found in our alien flora.

In the British Isles we have four genera in which large numbers of such agamospecies have been named. These are the hawkweeds (*Hieracium*, with 419 species currently recognised in the *H. murorum* group, only one of which is sexual), Blackberries or Brambles (*Rubus*, 334 species in the *Rubus fruticosus*

TABLE 10. Genera in the flora of the British Isles containing alien agamospermous species.

Family and genus	Explanation
Amaryllidaceae	
Nothoscordum	*N. borbonicum, fide* Nygren (1967)
Asteraceae	
Centaurea	*C. cyanus, fide* Nygren (1967)
Cichorium	*C. intybus, fide* Nygren (1967)
Erigeron	*E. annuus, E. karvinskianus*
Hieracium	About 54 agamospecies
Pilosella	All pentaploids and hexaploids, probably including some or all representatives of all four alien species
Rudbeckia	*R. laciniata, fide* Nygren (1967)
Taraxacum	About 111 agamospecies
Celastraceae	
Euonymus	*E. latifolius, fide* Nygren (1967)
Euphorbiaceae	
Euphorbia	*E. dulcis, fide* Nygren (1967)
Liliaceae	
Tulipa	*T. gesneriana, fide* Nygren (1967)
Plumbaginaceae	
Limonium	*L. hyblaeum*
Poaceae	
Echinochloa	*E. frumentacea, fide* Nygren (1967)
Eragrostis	*E. curvula, fide* Nygren (1967)
Paspalum	*P. dilatatum, fide* Nygren (1967)
Poa	*P. palustris, fide* Nygren (1967)
Rosaceae	
Alchemilla	All four alien species
Amelanchier	*A. lamarckii*
Cotoneaster	Probably about 73 of the 82 alien species
Crataegus	≥ Four North American polyploid species
Malus	*M. hupehensis, fide* Nygren (1967)
Potentilla	*P. intermedia, P. recta, fide* Nygren (1967), *P. inclinata*
Rosa	Dog-roses produce some sexual and some apomictic seed; possibly true of *R. ferruginea*
Rubus	Five agamospecies in the subgenus *Rubus: R. allegheniensis, R. armeniacus, R. canadensis, R. elegantispinosus, R. laciniatus*
Sorbus	All seven alien species
Saururaceae	
Houttuynia	*H. cordata, fide* Nygren (1967)

group, only one of which is sexual), dandelions (*Taraxacum*, with 237 species, none normally sexual) and buttercups (*Ranunculus*, probably with a few hundred species, none sexual, not yet fully worked out, in the *Ranunculus auricomus* group). There are no alien species of the *Ranunculus auricomus* group as far as we know, and only a small number (perhaps five) in the *Rubus fruticosus* group. It is impossible to know exactly how many of the species in the other two groups are aliens, but according to the specialists in each genus it is probably around 111 (47 per cent) in *Taraxacum* and 54 (13 per cent) in *Hieracium* (although perhaps the latter is an underestimate).

Two of the alien *Rubus* species are popular cultivated brambles or blackberries. One is the distinctive *R. laciniatus*, of unknown (perhaps garden) origin, which is much less grown today than formerly, but still found in many places in rough ground, where it is bird-sown or persists as a relic of cultivation. The other is the very robust, deliciously fruited, wickedly spiny *R. armeniacus* (Fig. 153), from eastern Europe, known by its misleading cultivar name of 'Himalayan Giant'. Today, more and more sorts of blackberries are being grown for fruit, especially spineless cultivars, and we should expect some of these to become naturalised in due course.

Just over half (121) of our dandelions belong to the section *Ruderalia*, and these are the commonest species of roadsides, lawns and lowland fields (Fig. 154). Nearly 100 of these agamospecies are thought to be aliens, so, when one next marvels at hectares of field or kilometres of roadside turned yellow in spring by swathes of dandelions (occasionally as many as 20 species in a large field, according to John Richards, pers. comm., 2012), followed by clouds of dandelion 'clocks', it is worth

FIG 153. The apomictic garden bramble known as Giant Blackberry 'Himalayan Giant' (*Rubus armeniacus*) has wicked spines but large, succulent fruits. (CAS)

FIG 154. The most abundant dandelions (*Taraxacum*) belong to section *Ruderalia*; all are apomictic and most are aliens, being invasive in open ground and grassland alike. (CAS)

remembering that all, or nearly all, are apomictic, and that the bulk of them is likely to be alien species. The proportion of *Hieracium* species considered as aliens is much smaller, but in this genus, too, the alien species are commoner on roadsides, railway banks and other marginal ground in the lowlands, so are more frequently encountered than might be guessed from the bare figures.

Nygren (1967) usefully provided a list of apomictic flowering plants. The list is not exhaustive, and is inconsistent in its treatment of the large apomictic genera such as *Taraxacum* and *Hieracium*, but it is an excellent starting point for a list of alien apomicts in the British Isles. An attempt at this is made in Table 10, from which it can be seen that, apart from the four genera discussed above, 22 other genera are involved. It is important to note that this list is almost certainly incomplete, but that, conversely, a number of species on it require careful confirmation (many of Nygren's supporting references date from the first half of the twentieth century). It is also the case that several species exhibit variation in their apomictic/sexual expressions, and that apomixis might not be their invariable condition. Such facultative apomixis is probably commoner than we realise.

Europe has 300 or so species of lady's-mantle (*Alchemilla*), and there are many more beyond. In Britain, however, we have only 12 native species, only five of these in Ireland, and all are apomictic. Moreover, they are all northern rather than southern in their distribution in Britain, and none is frequently encountered by many southern botanists. The most likely lady's-mantle the latter are liable to find is one of the naturalised species, Soft Lady's-mantle (*A. mollis*), by far the largest of our species and very popular in garden borders, where it self-sows prodigiously. Another popular garden plant, chiefly on rockeries, is Silver Lady's-mantle, often confused with our native Alpine Lady's-mantle (*A. alpina*) and at one time considered possibly native here (*see* Chapter 5). One feature in which *Alchemilla* contrasts strongly with *Rubus*, *Hieracium* and *Taraxacum* (but shares with *Ranunculus*) is a lack of any long-distance dispersal mechanism, suggesting both a potentially much slower spread and the likelihood that most of the many localities of Soft Lady's-mantle represent separate escape events.

In the rosaceous woody plants *Sorbus* and *Cotoneaster*, there is a higher proportion of sexual diploids (four out of 50, and *c.* nine out of 86, respectively, including both natives and aliens) in our flora than in the four genera discussed above, and some of these sexual relatives have given rise through past hybridisation to many of the apomictic agamospecies. One major difference between *Sorbus* and *Cotoneaster* is that in the former the sexual diploids are natives that have presumably given rise to agamospecies within our territory, whereas all of the sexual *Cotoneaster* species and virtually all the agamospecies are aliens introduced from China and other parts of Asia. Nevertheless the observed pattern of variation in our flora is similar: a few morphologically variable sexual diploids, and many almost invariable apomictic polyploids. Polyploids are plants or animals with more than the normal two sets of chromosomes; alien *Sorbus* polyploids are triploids or tetraploids, with three or four sets of chromosomes.

In *Sorbus* we have four native sexual diploids and 37 native polyploid agamospecies, several described very recently and many of them very restricted in distribution. In addition, there are seven alien agamospecies, all of them horticultural subjects introduced for ornament and with varying frequency becoming established from bird-sown seed in the wild. There is also one hybrid between the native sexual Rowan (*S. aucuparia*) and the introduced Swedish Whitebeam (*S. intermedia*), which like the latter is apomictic. A full treatment of all our native and alien taxa is provided in *Whitebeams, Rowans and Service Trees of Britain and Ireland* (Rich *et al.*, 2010). There are more than 200 *Sorbus* species worldwide, with about half of them in Europe.

We have only one native species of *Cotoneaster*, the tetraploid apomictic Wild Cotoneaster (*C. cambricus*), confined (in the world) to a few bushes on the Great Orme in North Wales, but we have 85 aliens, comprising 82 species (Fig. 155) and three hybrids that are, like *Sorbus*, found in the wild as bird-sown naturalisations. More than 400 species have been described, mainly from central and eastern Asia,

FIG 155.
Cotoneasters are much grown in gardens, mainly for their varied architectural habits and conspicuous fruits; about 90 per cent of the 84 alien species and hybrids in the British Isles are apomictic, of which Showy Cotoneaster (*Cotoneaster splendens*) is one example: (a) flowers; (b) fruits. (CAS)

of which over one-quarter are now cultivated in the British Isles. Many marginal habitats have been colonised, such as waste ground, spoil heaps, roadsides, railway banks and sidings, wall tops, canal sides, churchyards and cemeteries, neglected shrubberies and woodlands, rough grassland, hedgerows and chalk pits. In addition, several species have invaded natural habitats, such as chalk and limestone grassland (Fig. 248), sand dunes and scrub. On the North Downs in Kent and Surrey, huge areas that were species-rich chalk grassland half a century ago have become overgrown with scrub and woodland comprised of wind- and bird-dispersed species, in which cotoneasters figure prominently along with the likes of Hawthorn, Ash and Sycamore. In view of this success, it is perhaps surprising that *Sorbus* species have not been more successful in the wild.

As implied in Table 10, about nine of the 82 species of *Cotoneaster* are sexual diploids, and the rest are polyploid agamospecies. Waterer's Cotoneaster

(*C.* × *watereri*), sexual and fertile like its two parents, is our most commonly grown and naturalised large (i.e. more than *c.* 4 m high) *Cotoneaster*. Most species in the genus have an unknown chromosome number, so we cannot be certain of the total number of sexual species, but it is likely to be less than 15 per cent. In practice, without a chromosome count, those species that produce uniform progeny in a seed collection from one plant, as opposed to those species that show variation and segregation in their offspring, are likely to be apomictic and sexual, respectively. A thorough account of the cotoneasters in cultivation is available in *Cotoneasters. A Comprehensive Guide to Shrubs for Flowers, Fruit and Foliage* (Fryer & Hylmö, 2009).

The mouse-ear hawkweeds (*Pilosella*) are sometimes included in the same genus as the hawkweeds (*Hieracium*), but they are morphologically distinct and both the mechanism and pattern of apomixis they display are quite different. In general, in *Pilosella* the diploids are sexual, the tetraploids can be sexual or apomictic, and the higher polyploids are apomictic. In our only common native species, Mouse-ear Hawkweed (*P. officinarum*), diploids are known only in East Anglia and sexual plants are common only in the south; most of the populations that have been investigated have been found to be apomictic. Our four alien species also show a range of chromosome numbers, mostly varying from tetraploid to hexaploid (i.e. from four to six sets of chromosomes), and again from sexual to apomictic. The apomictic plants can produce good pollen and hybridise with sexual plants, causing major taxonomic problems on the Continent. In the British Isles, however, hybridisation is only a minor complication, and in *Pilosella* the apomictic species have not differentiated into the huge range of agamospecies characteristic of *Hieracium*.

The last statement holds true for the other genera in Table 10; in those cases the field botanist might well remain blissfully unaware whether the plant is sexual or apomictic, because for some reason the species concerned have not complicated their taxonomy by differentiating into numerous agamospecies, at least in our flora. In North America, the hawthorns (*Crataegus*) have produced hundreds of agamospecies, of which we have no more than four as aliens, and in Europe the same is true of the sea-lavenders (*Limonium*), of which our only apomictic alien representative is Rottingdean Sea-lavender (*L. hyblaeum*). Finally, the North American Tall Fleabane (*Erigeron annuus*) is usually divided into the sexual diploid subsp. *strigosus* and the apomictic triploid subsp. *annuus*.

Inbreeders

The degree of inbreeding (production of seed by self-fertilisation) versus outbreeding (production of seed by cross-pollination between two plants of the same species) helps to determine the pattern of variation shown by a species. Inbreeding species tend to exist as relatively uniform populations, which nonetheless often differ considerably from one another, even over quite short distances, since gene exchange between them is absent or low. In outbreeding species each population is variable but similar to other nearby populations, due to the gene exchange occurring between them. The relatively uniform populations found in strictly inbreeding species can in extreme cases come to approximate to pure genetic lines, which is indeed the principle behind standard plant-breeding technique aimed at achieving that end. These approximations to pure lines in some species become morphologically distinguishable and then, like agamospecies, can be recognised as microspecies.

If microspecies *can* be recognised, some taxonomists *will* have done so! Species that are consistently inbreeders are generally weedy annuals. Genera in our flora in which inbreeding microspecies have been recognised include the whitlow-grasses (*Erophila*), Groundsel and shepherd's-purses (*Capsella*; Fig. 156). The first two of these are natives but the third is considered to be an archaeophyte. In shepherd's-purse, about 70 microspecies have been described in northwestern Europe; in the second edition of Druce's *British Plant List* (1928), 28 microspecies are listed, with two more added in the 'Additions and Corrections' (1929).

Although these microspecies are often as distinct as the agamospecies of *Rubus*, *Hieracium* or *Taraxacum*, they do not remain as genetically distinct because some cross-pollination between them can occur, and when it does the distinctions break down. For this reason, inbreeding microspecies are today not normally recognised as distinct species by taxonomists. A knowledge of them is, however, still of great value in our efforts to understand the taxonomy of some aliens in the British Isles. The pepperworts (*Lepidium*) and fleabanes (*Conyza*) might be relevant examples. In these genera, which have not been fully researched in their native South America, we have a number of separate introductions that represent only a few facets of the native variation. They have been given species names (some coined in Europe), but if we were to see the full range of variation across the genera it is likely that we would find that it is more continuous than now evident, and we would recognise fewer species. The goosefoots (*Chenopodium*) and pigweeds (*Amaranthus*) possibly provide further examples.

The flowers of annual inbreeding weedy species are usually very small and inconspicuous, and do not attract many visits by potential pollinators. In some genotypes (e.g. of some *Lepidium* and *Capsella* microspecies) petals are lacking altogether, and sometimes the flowers never open (examples are seen in the same

FIG 156. Shepherd's-purse (*Capsella bursa-pastoris*) is a highly inbreeding species that has evolved many different variants existing in uniform populations; two are shown here. (CAS)

two genera), self-pollination occurring at the bud stage. This phenomenon is known as cleistogamy, and such flowers as cleistogamous.

Founder effects

Many botanists have studied natural migration in plants and considered the genetic, as opposed to the ecological, factors determining their subsequent success. It was pointed out long ago that a founding population of a species in a new locality will possess a much smaller range of genetic variation than that present in its parent population (the founder effect), at least initially, and that what variation that small founding sample does possess might well be a matter of chance determined by the colonising plants. A commonly encountered situation in the field is that species near the edge of their range are less variable than those in the centre of their range. British botanists, for example, often notice that a species with a known range of morphological variation and ecological tolerance here shows a much wider expression of these attributes in, say, the Mediterranean region. Usually we do not know why this is so. It could be that only part of the variation has reached our shores in the post-glacial period, or that our climate and soils allow only a part of the variation to thrive here. In general, it is rarely possible to decide whether a restricted range of variation of a species in an area is due to chance or to selection, but both must play important roles in different or even in the same migratory events.

A smaller range of genetic, hence of morphological, physiological and ecological, variation in founding populations is as true of newly arrived aliens as it is of colonising natives. In the former case, humans might in some cases have deliberately selected the narrow range (e.g. a particularly attractive flower, or a dwarf growth form), but even then the numerous other attributes with which they were not concerned (e.g. one morph in a distylous species, or one sex in a dioecious species) will also exhibit only part of the total range that could have been selected. The *Crocus*-like Oniongrass (*Romulea rosea*), which has been naturalised in Guernsey for several decades, is normally pink-flowered (note the specific epithet) in its native South Africa, but all the Guernsey plants are white-flowered. Several other examples are described more fully elsewhere: Japanese Knotweed is represented by only the female morph in the British Isles, and moreover all the plants are of the same clone; all our *Elodea* is female; we have only one morph of several *Oxalis* species; and Slender Speedwell seems to be represented by only one pair of incompatibility alleles. These facts are all due to chance sampling of the whole population in the native area.

The dilemma that a newly arrived species faces when represented by a narrow genetic base in a new territory is known as a genetic bottleneck. In order to be successful, the species must usually overcome this bottleneck, perhaps by increasing its genetic variation: by mutation, by the arrival of new immigrants or by genetic interaction (hybridisation) with other species already present in the new area. For example, a self-compatible allele might evolve from one of the self-incompatible alleles (a frequent mutation in some species). Or the other sex or heterostyly morph might be introduced, either deliberately or by chance when increasing the flower colours or range of growth-forms of a cultivated plant. In Japanese Knotweed we still have only one clone, which is female, but a male parent for it effectively exists in the guise of Giant Knotweed, which forms hybrids with Japanese Knotweed that are partially fertile and exist as both females and hermaphrodites (*see* Chapter 9). Success could also come if the newcomer managed by chance to become dispersed to a new habitat in which it was particularly favoured – perhaps open ground with few competitors. Climate change might also favour an alien species some time after it has arrived, especially an alien from warmer climates.

Fennell *et al.* (2010) used molecular (DNA) techniques to compare populations of Giant-rhubarb in native localities in Chile with those naturalised in various sites in western Ireland (and also in New Zealand and the Azores). As expected, they demonstrated that the alien populations contained a smaller proportion of the total recorded genetic variation than did the native populations, and they attributed this result to the founder effect.

Theoretically, a species with a very narrow genetic base is at a disadvantage in changing situations, e.g. climate warming, or competition from other encroaching species. It is possible that the local decrease in abundance of such plants as Canadian Waterweed or *Spartina* species, over the past 130 and 90 years, respectively (*see* Chapter 9), might be due to their inability to adapt, or perhaps to compete with newcomers or to combat infections. But this is simply conjecture. Japanese Knotweed, despite being genetically uniform and able to reproduce its own kind only by vegetative means, has shown, and continues to show, a rapid increase all over the British Isles during the same time span. Other aliens reproducing here only vegetatively similarly show no signs of stress. Furthermore, the supposed disadvantage of the extremely narrow range of variation possessed by apomicts and their inability to cross-breed is not supported by the rampant success of dandelions and brambles.

It is commonly observed that there is a considerable lag phase between the arrival of a new alien here and its becoming common (*see* Chapter 7). This is observed in many aliens, yet we know that these all have their own particular circumstances, making it very difficult to make generalisations. Oxford Ragwort, for example, was present in the Oxford Botanic Garden at least 75 years before it managed to escape outside in 1794, and it was more than a century more before it could be described as common. It did not reach Lancashire and Yorkshire until the 1930s. The spread of this species was undoubtedly greatly aided by the development of the railways, which provided both ideal habitats and means of dispersal. We also believe that it did not arrive in a genetically highly impoverished state, as it is of hybrid origin (see Chapter 9). In contrast, Japanese Knotweed arrived here as a single clone, yet emerged from its initial lag phase just as successfully. These two species show the same pattern of increase but could hardly have started from more different bases, nor spread by more different methods. The availability in recent years of molecular methods that can define genetic variation in populations will surely lead us to a better understanding of the spread of aliens.

CHAPTER 9

Alien Hybridisation

With all the gardener Fancy e'er could feign,
Who breeding flowers, will never breed the same.
John Keats, 'Ode to Psyche' (1819)

E ven in the twenty-first century, many people still believe that hybrids between different plant species are rare and invariably sterile. This is, of course, to some extent true of higher animals, as the mule and the liger illustrate, but it is much less true of some other animal groups and it is far from the case in plants. Vascular plants frequently form interspecific hybrids, and these vary from highly sterile to highly fertile. Douglas Kent's *List of Vascular Plants of the British Isles* (2002) includes 2,529 species (natives plus aliens, excluding agamospecies) and 728 interspecific hybrids. In a new book, *Hybrid Flora of the British Isles* (Stace *et al.*, 2015), 908 hybrids are covered. Several important and common evolutionary mechanisms involve hybridisation, and it is likely that about half the world's plant species have evolved via this process. Hybridisation in vascular plants cannot be viewed as rare or unusual, but is actually an important feature of normal plant behaviour. The frequency and significance of hybridisation in plants can scarcely be overemphasised.

In the evolutionary mechanisms mentioned above, totally sterile hybrids can be of more significance than highly fertile hybrids, since they can generate new polyploid species. We can genetically designate two different species as AA and BB, where A and B are the genomes (sets of genetic material) of those two species, each represented twice over because of their inheritance from each of their two parents. A and B can each comprise any number of chromosomes from two to many. During the formation of gametes (indirectly via the cell process of meiosis), the two genomes (both A or both B in these cases) recombine (swap genetic material) and separate, producing gametes that we can designate as A and B, respectively. The two species, each with two similar genomes, are diploids, and their derived gametes, each with one genome, are haploids. The number of chromosomes in the normal (sporophytic) tissues of a plant is designated by '2n =', and the number in the (gametophytic) tissues produced by meiosis (including the gametes) by 'n ='. The diploid chromosome number is reconstituted at fertilisation when the male and female gametes unite.

Meiosis, involving the halving of the chromosome number, can take place only if the chromosomes of the two genomes are homologous, i.e. sufficiently similar to allow them to associate in pairs, which allows for recombination and regular separation. If hybridisation has occurred, where species AA has hybridised with species BB, the reconstituted hybrid diploid will have the designation AB, and therefore in that hybrid the two genomes will be non-homologous. When that is the case, the chromosomes will not be able to associate in pairs and viable gametes will not be formed; the hybrid is sterile. Frequently the two genomes in a hybrid are not strictly homologous, but are not completely non-homologous; they are said to be homoeologous. Homoeology will allow some A–B chromosome pairing to take place, which will impart some degree of fertility on the hybrid. The more similar are A and B, the more pairing will occur, leading to more gamete formation and more fertility. In most cases the degree of homology of A and B is directly proportional to the degree of fertility of the hybrid.

Let us consider the possible fate of a completely sterile diploid hybrid AB, where A and B are fully non-homologous. Many hybrids approximate to this state. In some cases, however, full fertility can be restored by a doubling of the chromosome number, so that the diploid hybrid AB is converted to a tetraploid plant (with four genomes) with the designation AABB. Several mechanisms achieving this are known and are not rare. The new hybrid AABB can now achieve chromosome pairing, because both A and B genomes have another homologous genome in the same cell. Effectively, it behaves just like a normal diploid with twice the number of chromosomes. In fact, it is not possible just by counting the chromosome number of a plant to know whether it is a diploid like AA or BB or a tetraploid like AABB. For example, the common grasses Sweet Vernal-grass (*Anthoxanthum odoratum*) and Small Sweet-grass (*Glyceria declinata*) both have the chromosome number 2n = 20, but the former is a tetraploid with five chromosomes in each genome, while the latter is a diploid with ten chromosomes in each genome. We can distinguish such diploids and tetraploids by genetic experiments and by studying their closest relations.

Tetraploids that behave like diploids are termed amphidiploids. It is possible for an amphidiploid to hybridise with a third diploid species to form a sterile triploid hybrid, and then by doubling of the chromosome number to create a hexaploid amphidiploid. These events are summarised in Fig. 157. We do not know the limit on these processes, but it is way beyond the hexaploid level. For example, an adder's-tongue fern *Ophioglossum reticulatum* has the chromosome number 2n = 1,260 and is thought to be 84-ploid. Since something like half of all vascular plants are polyploids (tetraploids or above), we believe that the formation of amphidiploids via hybridisation is one of the standard routes by which new species evolve in plants.

A complete list of the 145 angiosperm hybrids in the flora of the British Isles involving at least one alien parent (including archaeophytes) is presented in Table 11. This list includes those hybrids not recorded since 1986 (and which are therefore excluded from Appendix 1), but is confined to those hybrids that have arisen in the wild in the British Isles, i.e. it excludes all those that have been introduced already in the hybrid state. There are 272 hybrids named in Appendix 1; most of the additional ones are garden plants that have become part of our alien flora, either introduced into gardens as hybrids or arising as such in gardens, but in either case escaping into the wild as hybrids. For example, there is no evidence that any hybridisation has taken place in the wild with us in the genera *Cotoneaster*, *Dianthus*, *Geranium*, *Spiraea* or *Narcissus*, none of which, therefore, is in Table 11. These genera alone account for 25 hybrids in Appendix 1.

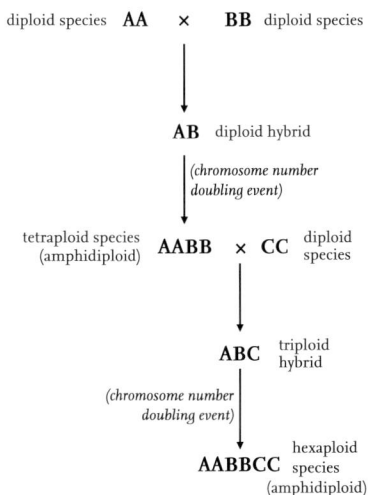

diploid species **AA** × **BB** diploid species

AB diploid hybrid

(chromosome number doubling event)

tetraploid species (amphidiploid) **AABB** × **CC** diploid species

ABC triploid hybrid

(chromosome number doubling event)

AABBCC hexaploid species (amphidiploid)

FIG 157. Diagram showing the origin of a new tetraploid amphidiploid (AABB) and a new hexaploid amphidiploid (AABBCC) from three diploid species (AA, BB, CC), following two hybridisations and two chromosome number doubling events.

TABLE 11. List of angiosperm hybrids that have arisen in the British Isles involving at least one alien parent species. Hybrids that have been introduced as such are omitted, but are listed in Appendix 1. Hybrids not recorded since 1986 are included, but are not listed in Appendix 1. Alien and neonative taxa are in italics.

Hybrid combination	Hybrid binomial
Agrostis capillaris × *castellana*	*A.* × *fouillardiana*
Agrostis capillaris × *gigantea*	*A.* × *bjoerkmanii*
Agrostis *gigantea* × stolonifera	*A.* × *murbeckii*
Agrostis stolonifera × *Polypogon viridis*	× *Agropogon robinsonii*
Alnus cordata × glutinosa	*A.* × *elliptica*
Alnus cordata × *incana*	
Alnus glutinosa × *incana*	*A.* × *hybrida*
Anagallis arvensis subsp. arvensis × *foemina*	
Anchusa ochroleuca × *officinalis*	*A.* × *baumgartenii*
Anthemis cotula × *Tripleurospermum inodorum*	× *Tripleurothemis maleolens*
Arctium *lappa* × minus	*A.* × *nothum*
Artemisia *verlotiorum* × vulgaris	*A.* × *wurzellii*
Avena fatua × *sativa*	*A.* × *hybrida*
Brassica *napus* × oleracea	
Brassica napus × rapa	*B.* × *harmsiana*
Bromus hordeaceus × *lepidus*	*B.* × *pseudothominii*
Calystegia *pulchra* × sepium	*C.* × *scanica*
Calystegia pulchra × silvatica	*C.* × *howittiorum*
Calystegia sepium × *silvatica*	*C.* × *lucana*
Capsella bursa-pastoris × *rubella*	*C.* × *gracilis*
Centaurea *jacea* × nigra/debeauxii	*C.* × *gerstlaueri*
Cerastium arvense × *tomentosum*	
Conyza bonariensis × *canadensis*	
Conyza canadensis × Erigeron acris	× *Conyzigeron huelsenii*
Conyza floribunda × Erigeron acris	
Conyza sumatrensis × Erigeron acris	
Corylus avellina × *maxima*	
Crataegus *heterophylla* × monogyna	

Hybrid combination	Hybrid binomial
Crataegus monogyna × *rhipidophylla*	C. × *subsphaerica*
Crocus *biflorus* × *chrysanthus*	
Crocus *tommasinianus* × *vernus*	
Dipsacus fullonum × *laciniatus*	D. × *pseudosilvester*
Dipsacus fullonum × *sativus*	
Epilobium *brunnescens* × *ciliatum*	E. × *brunnatum*
Epilobium *brunnescens* × lanceolatum	E. × *cornubiense*
Epilobium *brunnescens* × montanum	E. × *confusilobum*
Epilobium *brunnescens* × obscurum	E. × *obscurescens*
Epilobium *brunnescens* × palustre	E. × *chateri*
Epilobium *brunnescens* × parviflorum	E. × *argillaceum*
Epilobium *ciliatum* × hirsutum	E. × *novae-civitatis*
Epilobium *ciliatum* × lanceolatum	
Epilobium *ciliatum* × montanum	E. × *interjectum*
Epilobium *ciliatum* × obscurum	E. × *vicinum*
Epilobium *ciliatum* × palustre	E. × *fossicola*
Epilobium *ciliatum* × parviflorum	E. × *floridulum*
Epilobium *ciliatum* × roseum	E. × *nutantiflorum*
Epilobium *ciliatum* × tetragonum	E. × *mentiens*
Epilobium montanum × *pedunculare*	E. × *kitcheneri*
Fallopia *baldschuanica* × *japonica*	F. × *conollyana*
Fallopia *japonica* × *sachalinensis*	F. × *bohemica*
Festuca rubra × *Vulpia myuros*	
Fumaria muralis × *officinalis*	F. × *painteri*
Galanthus *elwesii* × *nivalis*	
Galanthus *elwesii* × *plicatus*	
Galanthus *nivalis* × *plicatus*	

continued overleaf

TABLE 11. *continued*

Hybrid combination	Hybrid binomial
Gaultheria mucronata × *shallon*	*G.* × *wisleyensis*
Geum macrophyllum × urbanum	*G.* × *convallis*
Heracleum *mantegazzianum* × sphondylium	
Hyacinthoides *hispanica* × non-scripta	*H.* × *massartiana*
Linaria purpurea × *repens*	*L.* × *dominii*
Linaria repens × *supina*	*L.* × *cornubiensis*
Linaria *repens* × vulgaris	*L.* × *sepium*
Lolium *multiflorum* × perenne	*L.* × *boucheanum*
Lolium *multiflorum* × Schedonorus arundinaceus	× *Schedolium krasanii*
Lolium *multiflorum* × Schedonorus pratensis	× *Schedolium braunii*
Lupinus arboreus × *nootkatensis* × *polyphyllus*	
Lupinus arboreus × *polyphyllus*	*L.* × *regalis*
Lupinus nootkatensis × *polyphyllus*	*L.* × *pseudopolyphyllus*
Malva *neglecta* × sylvestris	*M.* × *decipiens*
Mentha aquatica × arvensis × *spicata*	*M.* × *smithiana*
Mentha aquatica × *spicata*	*M.* × *piperita*
Mentha arvensis × *spicata*	*M.* × *gracilis*
Mimulus cupreus × *guttatus* × *luteus*	
Mimulus guttatus × *luteus*	*M.* × *robertsii*
Nothofagus alpina × *obliqua*	*N.* × *dodecaphleps*
Oenothera biennis × *glazoviana*	*O.* × *fallax*
Papaver dubium × *rhoeas*	*P.* × *hungaricum*
Pilosella *aurantiaca* × officinarum	*P.* × *stoloniflora*
Populus *balsamifera* × *deltoides* × nigra	
Prunella *laciniata* × vulgaris	*P.* × *intermedia*
Prunus *cerasifera* × spinosa	*P.* × *simmleri*
Prunus *domestica* × spinosa	*P.* × *fruticans*
Raphanus raphanistrum subsp. maritimus × *raphanistrum*	
Raphanus raphanistrum × *sativus*	*R.* × *micranthus*
Rorippa amphibia × *austriaca*	*R.* × *hungarica*
Rorippa *austriaca* × sylvestris	*R.* × *armoracioides*

Hybrid combination	Hybrid binomial
Rosa canina × *rugosa*	R. × *praegeri*
Rosa *rugosa* × spinosissima	
Rubus *cockburnianus* × idaeus	R. × *knappianus*
Rumex *confertus* × crispus	R. × *skofitzii*
Rumex *confertus* × obtusifolius	R. × *borbasii*
Rumex conglomeratus × *frutescens*	R. × *wrightii*
Rumex crispus × *cristatus*	R. × *dimidiatus*
Rumex crispus × *frutescens*	R. × *mirabilis*
Rumex crispus × obtusifolius × *patientia*	
Rumex crispus × *patientia*	R. × *confusus*
Rumex *cristatus* × obtusifolius	R. × *lousleyi*
Rumex *cristatus* × palustris	R. × *akeroydii*
Rumex cristatus × patientia	R. × *xenogenus*
Rumex *frutescens* × obtusifolius	R. × *cornubiensis*
Rumex obtusifolius × *patientia*	R. × *erubescens*
Salix *alba* × *euxina* × pentandra	S. × *meyeriana*
Salix aurita × caprea × *viminalis*	S. × *stipularis*
Salix aurita × *viminalis*	S. × *fruticosa*
Salix caprea × cinerea × *viminalis*	S. × *calodendron*
Salix caprea × *viminalis*	S. × *smithiana*
Salix cinerea × repens × *viminalis*	S. × *angusensis*
Salix cinerea × *viminalis*	S. × *holosericea*
Salix myrsinifolia × *viminalis*	S. × *seminigricans*
Salix purpurea × repens × *viminalis*	
Salix purpurea × *viminalis*	S. × *rubra*
Salix repens × *viminalis*	S. × *friesiana*
Salix triandra × viminalis	S. × *mollissima*
Senecio *cineraria* × erucifolius	S. × *thuretii*
Senecio *cineraria* × jacobaea	S. × *albescens*
Senecio squalidus × viscosus	S. × *londinensis*

continued overleaf

TABLE 11. *continued*

Hybrid combination	Hybrid binomial
Senecio *squalidus* × vulgaris	*S.* × *baxteri*
Senecio sylvaticus × *viscosus*	*S.* × *viscidulus*
Senecio *vernalis* × vulgaris	*S.* × *helwingii*
Silene dioica × *latifolia*	*S.* × *hampeana*
Solanum nigrum × *physalifolium*	*S.* × *procurrens*
Solidago *canadensis* × virgaurea	*S.* × *niederederi*
Sorbus aucuparia × *intermedia*	*S.* × *liljeforsii*
Spartina *alterniflora* × maritima	*S.* × *townsendii*
Symphytum *asperum* × officinale × *tuberosum*	
Tragopogon *porrifolius* × pratensis	*T.* × *mirabilis*
Tripleurospermum *inodorum* × maritimum	
Verbascum *blattaria* × nigrum	*V.* × *intermedium*
Verbascum *bombyciferum* × nigrum	
Verbascum *bombyciferum* × phlomoides	
Verbascum lychnitis × *pulverulentum*	*V.* × *regelianum*
Verbascum nigrum × *phoeniceum*	*V.* × *ustulatum*
Verbascum nigrum × *pulverulentum*	*V.* × *mixtum*
Verbascum nigrum × *pyramidatum*	
Verbascum nigrum × *speciosum*	
Verbascum nigrum × *virgatum*	
Verbascum phlomoides × pulverulentum	*V.* × *murbeckii*
Verbascum phlomoides × thapsus	*V.* × *kerneri*
Verbascum *pulverulentum* × thapsus	*V.* × *godronii*
Verbascum pulverulentum × virgatum	
Verbascum *pyramidatum* × thapsus	
Verbascum *speciosum* × thapsus	*V.* × *duernsteinense*
Verbascum thapsus × *virgatum*	*V.* × *lemaitrei*
Viola *arvensis* × lutea	
Viola *arvensis* × tricolor	*V.* × *contempta*

It is well documented that hybridisation is very much more common in some genera and families than others. For example, our 15 alien species in the three genera willowherbs (*Epilobium*), docks (*Rumex*) and mulleins (*Verbascum*) have given rise to 43 hybrids, whereas the 44 aliens in the genera dead-nettles (*Lamium*), peas (*Lathyrus*), yellow- and pink-sorrels (*Oxalis*) and clovers (*Trifolium*) have produced none. This simply reflects the situation in our native flora, where hybridisation occurs in those genera with weak isolating mechanisms (i.e. where mechanisms preventing crossing between species are weak). External isolating mechanisms (those imposed by the environment), in particular geographical separation, are frequently and easily overcome when species are introduced to foreign regions by people, and many of our hybrids are the result of such actions. Hybridisation is more likely to involve relatively rare species, because of the lower likelihood of their finding others of their own species, and the same is true of self-incompatible and dioecious species.

As previously implied, the impact on evolution caused by hybridisation varies enormously. Many hybrids are highly (but not necessarily completely) sterile and their existence merely adds interest to wild populations, or sometimes increases difficulties in identification. Fertile hybrids may backcross to their parents, forming a range of variation between the parents and creating a situation known as introgression, whereby characters of one species appear to have 'leaked' into another. Fertile hybrids can also cross with a third species to form a 'triple hybrid'. As described above, sterile hybrids sometimes undergo chromosome number doubling to form a new amphidiploid species.

All these sorts of hybrids and their derivatives – sterile and fertile primary hybrids (with all intermediate states), and derived polyploids – are represented in the alien flora of the British Isles; some are classical cases taught to generations of biology students, while others are new cases still being unravelled. Hybridisation in the 'significant eight' cases described below has noticeably altered the pattern of variation of the species involved, and in the first three cases has produced a really profound effect.

THE SIGNIFICANT EIGHT

Japanese Knotweed *Fallopia japonica*

The intriguing history of Japanese Knotweed as a European introduction, its very rapid spread as a wild plant here, and its more recent genetic interactions with other introduced species constitute a fascinating story from which we can learn much of more general application. The details have been unravelled only in the past 25 years by the research activities of Ann Conolly, John Bailey and their associates at the University of Leicester (Bailey & Conolly, 2000). They have consulted archive material, scientific and horticultural literature, garden catalogues, reports of collecting expeditions and herbarium specimens, following this up with original taxonomic, cytological and genetic research using the most modern molecular techniques, and with extensive fieldwork. Our account is possible only due to their efforts and publications. Here we describe the hybridisation behaviour of *Fallopia* species; other aspects are dealt with in other chapters.

Japanese Knotweed is gynodioecious, and the hermaphrodite plants are actually self-incompatible, so both hermaphrodite and female morphs, or more than one genotype of the hermaphrodite morph, are required for sexual reproduction to occur. All plants in the British Isles (and, in fact, across Europe) are effectively female, the flowers bearing stamens but lacking viable pollen (Figs 201 and 273). Their genetic base is even more restricted than that, because, as explained in Chapter 11, we believe that all the European Japanese Knotweed belongs to a single, i.e. genetically invariable, clone. Hence, no viable seed of pure *Fallopia japonica* is produced. There is, however, a dwarf variety – var. *compacta*, usually less than 1 m tall, much less vigorously rhizomatous and with a distinct leaf shape – that is grown in gardens and of which both morphs occur here. This variety, unlike the normal var. *japonica*, regularly sets seed where both morphs occur, the fruiting perianth segments usually being strongly red-tinged and very pretty against the red stems and green leaves. It is, however, quite uncommon and even rarer as an escape in the wild. We can call it Compact Knotweed.

The next nearest relative of Japanese Knotweed is Giant Knotweed (*Fallopia sachalinensis*; Fig. 158), also gynodioecious, from Sakhalin Island (Russia) and northern Japan, where it is perhaps sympatric with Japanese Knotweed. This is truly a giant; at up to 5 m in height it is one of the tallest herbaceous perennials in our flora. Both morphs of the species occur in the British Isles, and seed is often set. Again, the hermaphrodite morphs are self-incompatible, so that good seed cannot be produced by any isolated plant. In the 1970s, when Ann Conolly was first investigating the spread of these two species, she found some herbarium specimens that she found hard to identify. They appeared to be somewhat intermediate, and she wondered whether hybrids might occur. Although she never published that early idea, we now know that such hybrids are in fact common.

As in most taxonomically complex groups of plants, the chromosome numbers of the various species and varieties are relevant. In this group of plants the basic chromosome number (the number of chromosomes in one genome)

FIG 158. Giant Knotweed (*Fallopia sachalinensis*): a male-sterile flowering shoot. (CAS)

is 11. All three members of the group carry more than two genomes (sets of chromosomes), i.e. they are polyploids: Japanese Knotweed is an octoploid, with eight sets of chromosomes (2n = 88), while both Compact and Giant knotweeds are tetraploids, with four sets (2n = 44) (Bailey & Stace, 1992). In addition, John Bailey found tall (not compact) tetraploid plants of Japanese Knotweed to be common in Japan, but they do not occur here.

Anyone in the British Isles who cares to examine a number of patches of Japanese Knotweed on roadsides, riverbanks or waste ground near their home in October will sooner or later come across plants that are apparently producing fruits (Fig. 159). The fruits are small three-sided nutlets enclosed in the much-enlarged perianth. Some may be soft and empty, and falsely suggest fertility, but many are hard and contain an embryo, which can be germinated to produce a young plant. So how can we account for these viable fruits if no male-fertile

FIG 159. Fruiting shoot of Japanese Knotweed (*Fallopia japonica*), the result of pollination by Russian-vine (*F. baldschuanica*). (CAS)

pollen-producing plants exist in this country? The answer is that the male parent of these fruits was a pollen-producing plant of a different variety or species, such as Compact Knotweed (rarely) or Giant Knotweed (more frequently). We can deduce this because first, the chromosome number of such embryos and the plants grown from them is intermediate between that of its two parents, i.e. hexaploid with 2n = 66, and second, the morphology of the offspring, notably the leaf shape, is also intermediate between that of its parents. Moreover, the presence of one or other of these pollen-producers can often be tracked down in the vicinity. The pollen is transported by insects – mainly flies and bees – which travel some distance to the rich nectar sources and so effect cross-pollination.

The hybrid between the Japanese and Giant knotweeds is known as *Fallopia* × *bohemica*; it was named in 1983 from its area of discovery in the Czech Republic, and is commonly known as Bohemian Knotweed. It was recognised in Britain about the same time, but subsequently herbarium specimens from the wild dating back to 1954 have been discovered, and there are cultivated specimens going back as far as 1872. Today, it is found sparsely almost throughout the British Isles. The fact that the hybrid is known from areas where no Giant Knotweed is found (notably the Isle of Man and Guernsey, but more locally in numerous places) shows that by no means all of its colonies were formed *in situ*, but (probably most) must have got there by the same human-assisted routes as have dispersed its two parents. Hybrids between the two varieties of Japanese Knotweed are much rarer, but a few have been discovered in recent years. A third hybrid combination is that between Compact Knotweed and Giant Knotweed. This, as expected, is (like its parents) a tetraploid (2n = 44), and, like the last hybrid, it is only rarely encountered. All three hybrids have also been synthesised artificially.

There is also another route by which hybrids between Giant and Japanese knotweeds have formed. At Dolgellau, Merionethshire, some obvious hybrids were found to be octoploid. These are almost certainly hybrids between Japanese Knotweed, contributing female gametes with 44 chromosomes, as normal, and Giant Knotweed, which unusually contributed male gametes also with 44 chromosomes. This is possible in plants if the normal cell mechanism of meiosis (which halves the number of chromosomes prior to gamete formation during pollen formation on the male side) fails for some reason, so that the gametes (in this case called 'unreduced gametes') have the same number of chromosomes as their parent plant.

Earlier in this chapter we explained that when a diploid hybrid becomes polyploid by doubling of its chromosome number there might be a greater chance of fertility because there are more genomes offering possibilities of chromosome pairing. In the cases of Japanese and Giant knotweeds, we have two species that were already polyploid before hybridisation occurred. We can designate Japanese Knotweed as JJJJJJJJ, Giant Knotweed as GGGG, Bohemian Knotweed as JJJJGG, and the hybrid between Compact and Giant knotweeds as JJGG. Each letter represents a genome consisting of 11 chromosomes. Clearly, pairing could take place in both hybrids between chromosomes of two genomes derived from the same parent. Studies of chromosome pairing in the tetraploid hybrid between Compact and Giant knotweeds show the regular formation of 22 pairs of chromosomes, and this hybrid is highly fertile. In the hexaploid hybrids involving Japanese Knotweed and either of its two close relatives, there is also an even number of genomes, but two are derived from one parent and four from the other. Here there seems to be competition between chromosomes of different genomes to form pairs, and, instead of pairs, one sees several associations of three or four chromosomes, which leads to a big (though not complete) reduction in fertility.

Investigations with DNA have found that Bohemian Knotweed, unlike Japanese Knotweed, is genetically variable. In one very limited study, five different genotypes were detected in a single Scottish locality, in which the single genotype of Japanese Knotweed and two different genotypes of Giant Knotweed also occurred (Hollingsworth *et al.*, 1998). Bohemian Knotweed exists as both female and hermaphrodite types. It is likely that the range of genetic variation in this hexaploid hybrid is derived from (a) separate hybridisation events between Japanese and Giant knotweeds, (b) crosses between different hybrid clones, and (c) backcrosses between hybrids and either parent species. However, most offspring from hexaploid–hexaploid crosses or from backcrosses between hexaploids and parental taxa are not exact hexaploids, due to the latter not being able to produce regular gametes. New colonies of the hybrid have been found to arise by each of these methods, but more commonly by man-mediated vegetative dispersal from existing colonies. It is also likely that in the past hybrids were spread by the collection and sale or donation of seed from Japanese Knotweed plants, in the mistaken belief that such seed represented pure *Fallopia japonica*.

The most significant and perhaps most sinister consequence of hybridisation between Japanese and Giant knotweeds lies not in the formation of hybrid plants themselves, nor in the spread of yet another sort of invasive Asian knotweed, but in the genetic behaviour of the hybrids. Japanese Knotweed, which is by far the commonest and most invasive of these plants, cannot reproduce sexually in Europe (apart from forming hybrids) because of the absence of hermaphrodite plants. Even though it appears to have enormous competitive ability, the single genotype present is limited in its potential. We would do well, therefore, to prevent the introduction of pollen-bearing plants into Europe. Hermaphrodite plants of the hexaploid Bohemian Knotweed hybrid, however, can effectively act in place of the missing male Japanese Knotweed, so allowing it to reproduce sexually and to produce new recombined genotypes. These backcrossed plants might well display new physiological characteristics, enabling them to colonise new habitats, e.g. more mountainous or wetter or colder conditions, or soils with a different chemistry – even saline ones. Conceivably, Japanese Knotweed could in this way become an even more invasive and pernicious weed than it already is. The main limiting factor at present is probably the relative sterility of the hybrid and its backcrosses, but there are many cases known of hybrids becoming more fertile over time by well-understood genetic processes.

The story does not, however, end there. From about 1980, the Leicester group were gathering seed from wild Japanese Knotweed colonies and testing its viability. It soon became clear that not all the seed found on female Japanese Knotweed plants, particularly in urban areas, could be accounted for by pollen donation from Giant Knotweed. The morphology of quite young seedlings is sufficient to reveal their likely male parent, and in 1983 some seedlings appeared with an unusual leaf shape that did not suggest that either Giant or Compact knotweeds were likely male parents. The leaves were narrower and with a more tapering apex, and had a somewhat different texture. Chromosome counts revealed a new number, 2n = 54, which suggested that the male parent should have 2n = 20, a diploid based on a different base number (ten instead of 11), which would provide ten chromosomes to combine with the 44 in the female gametes of Japanese Knotweed. Consideration of the species thought to be closely related to Japanese Knotweed soon revealed the likely candidate, which was confirmed by artificial synthesis of this new hybrid.

Russian-vine (*Fallopia baldschuanica*; Fig. 160) is sometimes known as Mile-a-minute because of its prodigious growth rate; it can ascend a good-sized tree or a telegraph pole in one season. It is not a subterranean-spreading plant with strong rhizomes, but a climbing woody plant without rhizomes, and is always hermaphrodite. Its very frequent cultivation in gardens (often those too small to support it comfortably) means that most seed-set on Japanese Knotweed plants is, in fact, of this origin, rather than resulting from pollination by Compact or Giant knotweeds. Nevertheless, mature hybrids resulting from pollination by Russian-vine are much rarer than plants of Bohemian Knotweed, not for genetic reasons (for the seed is easy to germinate in a greenhouse), but probably because the seeds or young seedlings find it difficult to negotiate their first winter and become established in the wild – perhaps as a result of the conflict as a food-storing strategy between aerial stems and subterranean rhizomes. Three localities of this new hybrid are now known in Britain, the first found in London in 1987,

FIG 160. Russian-vine (*Fallopia baldschuanica*): (a) colony by a canal, with some plants of the alien Greek Dock (*Rumex cristatus*) in the background; (b) flowering shoot. (CAS)

and there are a few localities in northern and central Europe. It was described as *Fallopia × conollyana* (Conolly's Knotweed) by Bailey (2001), to acknowledge the pioneering studies of Ann Conolly in this genus. The ten chromosomes contributed to the hybrid by Russian-vine are larger than the 44 of Japanese Knotweed, and can be distinguished morphologically in good chromosome preparations (Fig. 161).

FIG 161. Chromosome preparation from a root tip of the hybrid Conolly's Knotweed (*Fallopia × conollyana*), with a chromosome number of 2n = 54. The ten larger chromosomes derived from the Russian-vine (*F. baldschuanica*) parent are arrowed; the other 44 chromosomes are derived from the Japanese Knotweed (*F. japonica*) parent. (Preparation by John Bailey, reproduced with permission)

The same female clone of Japanese Knotweed has been introduced to several other parts of the world. Interestingly, in New Zealand seed has been produced by Japanese Knotweed after pollination by a species of a different but related genus, Pohuehue (*Muehlenbeckia australis*), a rampant dioecious woody climber. This New Zealand native, like Russian-vine, is a diploid with 2n = 20, so the hybrid seedlings have 2n = 54, as in Conolly's Knotweed. However, no mature plants of this hybrid have yet been found. The discovery of further species of pollen-donor is to be expected, perhaps in eastern Europe, North America or Australia.

Oxford Ragwort *Senecio squalidus*

Just as the Japanese Knotweed story has been uncovered largely by one research group, the same can be said of Oxford Ragwort. In this case we owe our knowledge mainly to Ruth Ingram, Richard Abbott and their colleagues at the University of St Andrews, who built on the earlier work of Effie Rosser investigating the amphidiploid Welsh Groundsel (*Senecio cambrensis*) and of Douglas Kent concerning the spread of Oxford Ragwort.

Oxford Ragwort is a familiar, usually short-lived perennial of walls, rubble and rough ground, now found throughout most of the British Isles. Its yellow daisy-like flowers can be found virtually throughout the year, and it enlivens many a bleak spot even in cold, wet weather (Fig. 162). The first record of it in the wild here is of plants that had escaped from the Oxford Botanic Garden onto adjacent walls in the city in 1794 (Kent, 1956). There is dispute concerning the detailed origin of the botanic garden material (Harris, 2002). The long-accepted

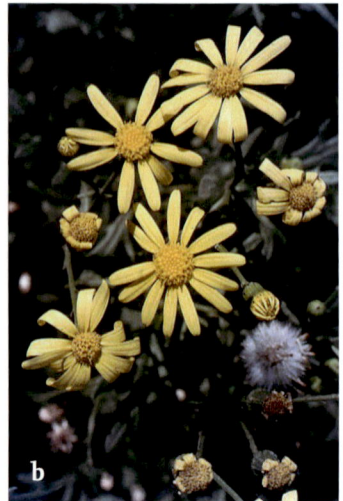

FIG 162. Oxford Ragwort (*Senecio squalidus*), one of the very few neonative species in the British Isles: (a) plant growing on waste ground in Manchester; (b) close-up showing leaves and flowers. (CAS)

story is that it was introduced to the Oxford Botanic Garden from Sicily in the latter part of the seventeenth century (Druce, 1927). However, according to Harris (2002), plant material morphologically similar to Oxford Ragwort was first grown in the British Isles in the Duchess of Beaufort's Garden at Badminton, Gloucestershire, in the early part of the eighteenth century, from seed supplied from Sicily by Francesco Cupani. This material was most likely the source of Oxford Ragwort later grown in Oxford. Whatever its route to the city, there can be no doubt that it originated from the slopes of Mt Etna in Sicily. We shall return to the situation in Sicily later.

Although the conspicuous yellow flower heads of Oxford Ragwort could not be confused with the small, dull 'petal-less' ones of the common weed Groundsel (*Senecio vulgaris*), the latter does seem to be our closest native relative of Oxford Ragwort. It should also be noted that the difference between the two species is greatly exaggerated by the presence of the ray florets (the 'petals' or ligules) in Oxford Ragwort (such plants are said to be 'ligulate'), and that there are (rare) variants of the latter lacking ray florets and (frequent) variants of Groundsel possessing them. Moreover, hybrids between the two species, known as *S. × baxteri*, occur rarely in mixed populations (Fig. 163). They are ligulate but the ray florets

FIG 163. Mixed population of groundsels (*Senecio*) growing on waste ground in Manchester: centre left, Oxford Ragwort (*S. squalidus*), with large flower heads; centre, *S. × baxteri* with deeper-yellow flower heads; and surrounding them, rayless, rayed and intermediate Groundsel (*S. vulgaris*). (CAS)

FIG 164. Flowering shoots of groundsels (*Senecio*) from the Manchester population in Figure 163 (left to right): rayless Groundsel (*S. vulgaris*), rayed Groundsel, *S.* × *baxteri*, Oxford Ragwort (*S. squalidus*). (CAS)

are usually smaller (down to 4 mm long) than in Oxford Ragwort (where they are 8–10 mm long) (Fig. 164). Hybrids were first recorded by Druce in Oxford in 1886, and later named by him as *S.* × *baxteri*, but they probably occurred from much earlier in the nineteenth century in sites supporting Oxford Ragwort. They are, however, very uncommon, and many records of them are errors of determination.

Oxford Ragwort is a highly (but not completely) self-incompatible diploid (2n = 20), whereas Groundsel is a self-fertile tetraploid (2n = 40). As expected, the F1 hybrid is a triploid (2n = 30) and highly sterile (Fig. 164). In 1948, H. E. Green found a different sort of Groundsel near Ffrith, Flintshire, which was investigated by Rosser (1955) at Manchester University. She found that this new groundsel was hexaploid (2n = 60) and that it produced a full head of viable fruits following self-pollination. The new taxon, which she named *Senecio cambrensis* (Fig. 165), is apparently a classic example of a new amphidiploid being produced by the doubling of the chromosome number of a sterile F1 hybrid, in this case *S.* × *baxteri*. That there can be little doubt of this origin is shown by the fact that colleagues of Effie Rosser produced artificial triploid hybrids between the two parents and resynthesised the hexaploid Welsh Groundsel by treatment with the drug colchicine, which can double the chromosome number by allowing chromosome duplication but preventing cell division. Later, the St Andrews group repeated this synthesis.

In the subsequent 60 years, Welsh Groundsel has not spread very far from its core localities in Flintshire and Denbighshire in northeastern Wales, and in

FIG 165. Welsh Groundsel (*Senecio cambrensis*): (a) plant showing habit; (b) close-up showing flower heads. (CAS)

fact there is evidence that it is now decreasing. A survey in 2010 found only 12 localities, compared with 35 known in the early 1980s. The latter 35 localities were in two separate areas of North Wales, but the 2010 survey found it in only one of these. Perhaps it is doomed to extinction.

In 1982, more than 100 plants of Welsh Groundsel were found growing on waste ground in Leith, Edinburgh, and a herbarium specimen collected from there in 1974 was later detected. This Scottish population has, however, since died out; the last record was apparently in 1993. Studies of the Scottish and Welsh populations (in the latter case from both main areas) have shown that they differ in several molecular characters (isozymes). This indicates that the Welsh and Scottish plants, although derived from the same two parent species, originated from separate hybridisation events, i.e. the Scottish plants did not arise by dispersal from Wales.

The multiple (polytopic) origin of Welsh Groundsel comes as no surprise. The raw materials (parental species) for its synthesis exist together in many sites, and its immediate progenitor (*S. × baxteri*) arises occasionally in different places. Polytopic origins of several other amphidiploids, e.g. among North American goat's-beards (*Tragopogon*), have been demonstrated, and this point will be raised again under *Spartina*. Perhaps we should ask why amphidiploids do not arise more often. Moreover, Oxford Ragwort also hybridises with another alien tetraploid *Senecio*, Sticky Groundsel (*S. viscosus*), and Groundsel hybridises with the rare alien diploid Eastern Groundsel (*S. vernalis*). In neither case have these sterile triploid hybrids been found to produce amphidiploid derivatives.

Returning to our common tetraploid Groundsel, one might wonder why this is not considered an archaeophyte, like Shepherd's-purse or Red Dead-nettle, which so often grow together with it. However, in the case of Groundsel there is an undoubtedly native variant, nowadays usually known as *Senecio vulgaris* subsp. *denticulatus*, which occurs in sandy places (often sand dunes) on the western coasts of Britain from Lancashire to Devon, and in the Channel Islands and the Isle of Man. If it were not for this subspecies, which differs obviously from the common Groundsel (subsp. *vulgaris*) in having ray florets, the evidence would probably suggest that the latter is indeed an alien of long standing. There are also, however, as mentioned above, other ligulate variants of Groundsel that are not coastal in distribution, but occur usually in mixed populations with the common Groundsel, mostly inland on waste ground. These plants are considered to be part of subsp. *vulgaris*, but are distinguished as var. *hibernicus*. The two ligulate variants differ in a range of minor characters concerning habit, leaf dissection, pubescence and the size of their ray florets: 2–3 mm long in subsp. *denticulatus*, and 3–5 mm long in var. *hibernicus*.

An intriguing question about var. *hibernicus* is its origin. The first known specimen in Britain is from Oxford in 1832, and in Ireland from Cork in 1853, but it did not become frequent until the 1930s and became common only in the second half of the twentieth century. From early days it was noted that ligulate variants of Groundsel generally appeared in areas that had already been populated with Oxford Ragwort (as in Oxford and Cork, above), and the suggestion was made that the origin of the ligules in Groundsel was from Oxford Ragwort. Despite the obvious logic in this, there was no hard evidence, and for a long time the alternative hypothesis – that the ligules were the result of mutation in Groundsel – seemed equally feasible. Mutation from ligulate to non-ligulate, and vice versa, is a common occurrence in the ragwort family (Asteraceae), and Oxford Ragwort and Welsh Groundsel, as well as Groundsel, are all represented by both morphs.

If the ligules of var. *hibernicus* were passed on from Oxford Ragwort, then surely some other characters also must have been inherited. The St Andrews group set out to demonstrate this and, after much frustration, their persistence paid off. Since var. *hibernicus* so closely resembles common Groundsel (var. *vulgaris*), it was a case of finding evidence of Oxford Ragwort genes in var. *hibernicus* and, as there were no morphological characters evident, this would have to be in the form of cryptic characters – most likely features of DNA or proteins. Evidence was primarily sought by investigation of enzymes, and eventually (Abbott *et al.*, 1992) it was found in a variant of the enzyme aspartate aminotransferase known as *Aat-3c*. This variant, known as an isozyme, was

present in 74 per cent of the plants of Oxford Ragwort tested, and in 47 per cent of
S. vulgaris var. *hibernicus*. Significantly, it was not found in any plants of *S. vulgaris*
var. *vulgaris* in populations where there was no Oxford Ragwort, but it was present
in 3 per cent of plants of var. *vulgaris* growing mixed with Oxford Ragwort. These
results, based on the analysis of 1,850 plants, clearly indicate that the presence of
the isozyme *Aat-3c* in some Groundsel plants is due to past hybridisation with
Oxford Ragwort, and the same origin is therefore implicated for the ligulate
character. Proof that the ligules of var. *hibernicus* came from Oxford Ragwort was
obtained by sequencing the genes responsible (Kim *et al.*, 2008). The sequences
from the two taxa are identical, but differ from that of a homologous gene in
rayless Groundsel, showing that the ligules in var. *hibernicus* cannot have arisen by
mutation. This is an extraordinarily good example of evolution in action; anyone
over about 60 years old could have witnessed it at first hand, and some of us did!

The transfer of genes must have taken place by backcrossing of the F1 hybrid
Senecio × *baxteri* to Groundsel. We do not know how many times this 'leakage'
or introgression of genes from Oxford Ragwort to Groundsel has occurred.
Extreme possibilities are that it occurred only once, and that the increase of
var. *hibernicus* has been by dispersal from that point of origin (Oxford?), or that
it has happened many times and is still occurring in new sites. The rarity of
S. × *baxteri*, the essential first step, and its high degree of sterility, as well as the
absence of var. *hibernicus* from many areas supporting both Oxford Ragwort and
Groundsel, strongly suggests that the latter alternative is most unlikely. Although
Groundsel is very common in the city of Leicester, and I found a plant of *S.* ×
baxteri there in the mid-1970s, I have never been able to find ligulate varieties of
Groundsel there. Probably the answer is somewhere between the two alternatives:
introgression has occurred in a few places (e.g. Oxford, Cork, Manchester, and in
the Welsh and Scottish sites for Welsh Groundsel, where *S. vulgaris* var. *hibernicus*
also occurs) and dispersal has occurred from those. We are not sure, either, how
much backcrossing is required to accomplish introgression. In most of the
many known examples it is achieved only by repeated backcrossing over several
generations, but in this case it is possible that one generation alone might be
sufficient. Despite the sterility of *S.* × *baxteri*, which usually results in a lack of any
offspring from self-pollination, some progeny can be obtained by backcrossing to
Groundsel. These offspring exhibit a higher level of fertility and, when some are
selfed, plants closely resembling *S. vulgaris* var. *hibernicus* are obtained.

The different sorts of ligulate plants mentioned above, often known as
'radiate Groundsels', are rather difficult to distinguish, and their characters are
difficult to describe. Nevertheless, each has a distinct appearance once one has
gained familiarity with them by detailed study. Apart from *Senecio vulgaris* subsp.

denticulatus, they can all be considered as intermediate between Groundsel and Oxford Ragwort, although they are nearly all – even their F1 hybrid *S.* × *baxteri* – closer to the former in appearance.

Occasionally, ligulate plants are encountered that are intermediate between Oxford Ragwort and Groundsel, but which differ from *Senecio vulgaris* var. *hibernicus* in having slightly larger ligules (5–6 mm long) and in being more robust plants, and hence in being closer to what one might imagine was halfway between the two parents (as is the hexaploid Welsh Groundsel). They differ from *S.* × *baxteri* and Welsh Groundsel in being tetraploid, and from *S.* × *baxteri* in being fertile. Often they have defied determination; sometimes they are described as 'fertile hybrids' between the two species. A briefly descriptive list of 31 examples was given by Lowe & Abbott (2003). One such variant, first found in 1979, gained particular notice due to its frequency (about 250 plants) on waste ground in the city of York, and it was described as York Groundsel (*S. eboracensis*) by Lowe and Abbott. Apart from its tetraploid chromosome number, it differs from Welsh Groundsel in having fewer and shorter phyllaries (small bracts surrounding each flower head), slightly fewer ray florets and a greater number of leaf lobes. York Groundsel is thought to have arisen from a backcross between Groundsel and *S.* × *baxteri*; *S. vulgaris* var. *hibernicus* resulted from a similar backcross but the two have different morphologies. It seems, however, that York Groundsel has gone the same way as most of the other 'fertile hybrids', since it has not been found in York since 2003. The observations of Praeger (1934) on Irish ligulate groundsels are interesting, albeit unproven: '*Senecio squalidus*, naturalised and abundant about Cork, where it appeared about a century ago, hybridises freely at Cork and sparingly at Dublin with *S. vulgaris*, giving rise by continued crossing to a progeny stretching from one species to the other, complicated at Cork by the frequent occurrence of *S. vulgaris* var. *radiatus* [= *hibernicus*].'

Despite the ephemeral prominence of York Groundsel and perhaps of Welsh Groundsel, introgression from Oxford Ragwort has undoubtedly altered the appearance and genetics of our common Groundsel for ever. Rayed Groundsel is now a conspicuous feature of waste ground in many parts of the British Isles where it was unknown 50 years ago.

Finally, let us return to the slopes of Mt Etna, whence our Oxford Ragwort arrived, perhaps by a single introduction of several plants (or seeds) over 300 years ago. It has long been known that the taxonomy of the southern European ragworts related to Oxford Ragwort is complex, with several taxa having been described and their classification varying from one authority to another. *Senecio squalidus* itself was described in 1753 by Linnaeus from plants growing in Sweden, the seed having been obtained from Oxford, so we are certain that it originated in Sicily.

On Mt Etna there are two related taxa, both diploids (2n = 20) like Oxford Ragwort, nowadays considered as separate species: *Senecio chrysanthemifolius*, found up to about 1,000 m altitude, and *S. aethnensis*, occurring at 1,600 m and above. At altitudes of 1,000–1,600 m there is a hybrid zone between these two species, where the plants show a big range in morphology in different combinations (Fig. 166). Morphological characters suggest that Oxford Ragwort came from this hybrid zone rather than from lower or higher on the mountain. When randomly selected regions of the DNA were analysed at St Andrews, each of the two Etnean species was found to possess segments (by chance 13 in each case) not present in the other. Analysis of British Oxford Ragwort found that it carries

FIG 166. Plant in the *Senecio* population on Mt Etna from which ancestors of our Oxford Ragwort (*S. squalidus*) were derived, with the volcano in the background. (CAS)

11 and 10 of these 13 unique segments from *S. chrysanthemifolius* and *S. aethnensis*, respectively. The existence in our Oxford Ragwort of DNA markers unique to each of the two Etnean species clearly shows that our species is derived from the hybrid zone.

There is also one other clue indicating the origin of Oxford Ragwort. It is a very variable species, as shown particularly by its great range in leaf shape. This indicates that the original material included a wider genetic range than might be expected from a single introduction of one species, but one more likely to have developed from a hybrid origin.

The evolution of Oxford Ragwort differs from that of Welsh Groundsel and many other known examples of amphidiploids in that it all occurred at one chromosome number level, rather than involving polyploidy. Such a phenomenon is known as homoploid hybrid speciation, which can occur when hybrid offspring become genetically isolated from their parents, in this case spatially, mutate along their independent pathways, and become selected under different conditions from those experienced by their progenitors. Few examples of this evolutionary route are known; Oxford Ragwort is the only one to have been confirmed in the British Isles (but see the discussion on *Oenothera* hybridisation later in this chapter), making it a very special neonative member of our flora (Abbott *et al.*, 2009).

Common Cord-grass *Spartina anglica*

Only one native species of the saltmarsh genus *Spartina* occurs in the British Isles, namely Small Cord-grass (*S. maritima*), an uncommon plant found in southeastern England from Lincolnshire round to Dorset. An alien species from northeastern North America, Smooth Cord-grass (*S. alterniflora*), was first recorded from the wild in Britain in 1816, by the River Itchen, Hampshire, although it might have arrived earlier, as it had in southwestern France. Both of these strongly rhizomatous species are pioneers of bare saline mud, and can be valuable as mud-binders around ports and estuaries, and in the first stages of land reclamation for agriculture. However, Smooth Cord-grass probably arrived in Europe by accident, as a contaminant of ship's ballast. From its original site it spread over a larger area of the Southampton region, although it has since receded to become a very local species. The hybrid between Smooth Cord-grass and Small Cord-grass was described as Townsend's Cord-grass (*S. × townsendii*) in 1881. At the time and until recently, the first collections of the hybrid (in the Kew Herbarium) were thought to date from 1870, but in 2011 specimens were discovered by Tom Cope and Mark Spencer in the herbarium of the Natural History Museum dating back to 1846.

Species of *Spartina* in general have high numbers of chromosomes, which are difficult to count exactly, and early reports were not accurate. The situation was clarified by Chris Marchant (1963), who found that the basic number of chromosomes in a genome is ten, and that the above three taxa are hexaploids or near hexaploids (2n = ± 60). At these higher ploidy levels, the loss or gain of a few chromosomes often has little impact, and not all species, or plants within a species, necessarily exist at an exact ploidy level. In fact, Small Cord-grass has 2n = 60 and Smooth Cord-grass 2n = 62; Townsend's Cord-grass mostly has 2n = 62, but wider samples have since found numbers varying from 49 to 66. This variation in chromosome number arises because meiosis is irregular and divides the number of chromosomes only *approximately* in half, and perhaps there has been a small amount of backcrossing. Hence, the F1 hybrid Townsend's Cord-grass is highly but not completely sterile – the anthers are not or only tardily dehiscent, and the pollen grains within them are mostly empty and not dispersed.

Townsend's Cord-grass spread very little until the late 1880s, when it was still confined to the Southampton area, but after that time it started to spread rapidly and by the turn of the century it had reached Sussex, Dorset and the Isle of Wight. Otto Stapf, writing in a local journal in 1913, perceptively stated that 'towards the end of the [18]80s something occurred that favoured the spreading of the grass'. It was not known what this event entailed until 1956, when Charles Hubbard, the grass specialist at Kew, realised that there were two plants included under the name *Spartina townsendii*, the sterile F1 mentioned above and a fertile relative of it. The latter proved to be the 12-ploid amphidiploid derivative of Townsend's Cord-grass, mostly with 2n = 124 but sometimes with 2n = 120 or 122. It seems highly likely that the spread of the cord-grasses, within South Hampshire at first but soon very widely in the British Isles, was accomplished by the amphidiploid, which we now call Common Cord-grass (*S. anglica*; Fig. 167), a name not validated until 1978. The earliest specimen known dates from 1887. Although the latter is fertile, having dehiscent anthers releasing pollen grains full of cytoplasmic content, and some of the plant's spread was probably from seed distribution, much of the longer-range dispersal was carried out by deliberate transplants of vegetative material, in order to achieve mud stabilisation all around the British Isles and beyond. Today, Common Cord-grass occurs over most temperate saline areas of the northern and southern hemispheres. Where it has been introduced, it has the potential of spreading locally both vegetatively and sexually, but the little evidence we have suggests that it cannot cross with either of its two progenitor species.

The presence of cord-grasses is not, however, always considered a bonus. Although they colonise bare saline mud, once established they readily invade

FIG 167. Common Cord-grass (*Spartina anglica*), which arose in England towards the end of the nineteenth century: (a) plant portrait; (b) total dominance of estuarine saltmarshes by the River Stour, Suffolk. (a, CAS; b, Richard Stace)

saltmarshes at their earlier stages of succession, ousting the native species – not only annuals like Annual Sea-blite (*Suaeda maritima*) and species of glasswort (*Salicornia*), but also perennials such as Sea Aster (*Aster tripolium*) and Common Saltmarsh-grass (*Puccinellia maritima*). The encroachment of cord-grasses is setting off many conservation alarms around our coasts. The total area occupied by the genus was estimated at 30,000 acres (>12,000 ha) in 1959, and the same figure was given for Britain alone in 1967, when 200–400 ha was estimated for Ireland. Compared with 50 years ago, the range on the east coast of Britain has not changed much, but *Spartina* has increased its area considerably on the west coast and in Ireland. It is still rare in Scotland apart from the extreme south. The total coverage, however, might not have changed greatly in that period, because stands on the south and east coasts of Britain have suffered die-back. It is thought that this is not due to pollution or to a pathogen, but is a situation where 'a newly evolved species has dramatically altered the sedimentary and drainage characteristics of the marshes, and created the anaerobic, waterlogged conditions which led to its own destruction' (Gray *et al.*, 1991). Since this process depends much on local soil conditions, die-back might not become a universal factor. As the dodecaploid Common Cord-grass has increased, so the three hexaploid taxa have decreased, and there can be little doubt that they have been ousted (outcompeted) by the amphidiploid. Where Common and Townsend's cord-grasses occur on the same saltmarsh, however, the latter tends to occupy the higher, drier parts and the former the lower, wetter ones.

Common Cord-grass is able to undergo a normal meiosis, good pollen is produced, and any seeds formed are fertile and can be germinated easily enough. The plants are to some degree self-compatible, but a higher level of seed-set is achieved by crossing. Many colonies actually produce little seed, perhaps because of weather conditions in November and December when the seeds ripen, and increase is mainly vegetative by the rhizomes. Protein studies, involving seed proteins and isozymes, have detected virtually no genetic variation in British Common Cord-grass. It is tempting to deduce from this that Common Cord-grass has been formed only once in the British Isles, unlike the amphidiploids in American salsifies (*Tragopogon*) and British ragworts (*Senecio*), but, since both its parents are equally invariable, the *Spartina* could have arisen more than once from identical parents.

Modern molecular techniques now allow further investigation of the origin of the *Spartina* hybrid and its derived amphidiploid. In offspring of flowering plants, chloroplasts are derived (in almost all cases) from the maternal parent alone. Therefore, if differences can be found between the DNA of the chloroplasts of the male and female parents, the origin of the chloroplasts in the hybrid can be determined and the female parent identified. It was found by this method that the female parent of both the hybrid Townsend's Cord-grass and the amphidiploid Common Cord-grass was Smooth Cord-grass (Ferris *et al.*, 1997).

As mentioned above, Smooth Cord-grass was naturalised in southwestern France earlier than in England – in 1803, in fact. In the Bay of Biscay, hybrids between Smooth Cord-grass and the native Small Cord-grass were discovered in 1892 and, as these were slightly different in morphology from Townsend's Cord-grass, they were described under the name *Spartina × neyrautii*. This practice is nomenclaturally illegitimate; all hybrids derived from the same two parents have to go under the same (earlier) name, so *S. × neyrautii* is actually a synonym of *S. × townsendii*. The French and English hybrids have the same chromosome number and both are sterile, with empty pollen and indehiscent anthers, but the chromosomal meiotic behaviour differs between the two and there are minor morphological differences: on the whole, the English plant is taller and more robust, with slightly larger floral parts but shorter leaves. It was suggested that the differences might be due to reciprocal hybridisations (i.e. Small Cord-grass was the male parent in one country and the female in the other). However, this is a naive – albeit frequently employed – explanation, often in the absence of any evidence. One has only to look at siblings of the same sex to see that the same mating can give different products without reversing the parentage. In fact, chloroplast DNA analysis has shown that the French hybrid also has Smooth

Cord-grass as its female parent (Baumel *et al.*, 2003), and nuclear DNA analysis proved that the parents were genetically distinct in the two countries.

Although Small Cord-grass and Smooth Cord-grass have independently hybridised on both sides of the English Channel, it is only in Britain that the hybrid evolved further to produce an amphidiploid. This has been introduced and has spread naturally to alter radically the saltmarsh vegetation in several parts of the world.

It is interesting to note that Smooth Cord-grass was introduced into San Francisco Bay, California, in 1975, and has since hybridised there with the native California Cord-grass (*Spartina foliosa*). In contrast to the situation with Townsend's Cord-grass, the American hybrid is fertile, and backcrossing is occurring to both parents, producing a broad spectrum of intermediates. Indications are, also, that crossing between the two species has taken place in both directions. The greater ease of crossing is illustrated by the fact that artificial hybrids have been made between the American species (California Cord-grass as female), but all attempts to resynthesise Townsend's Cord-grass have so far failed.

Monkeyflowers *Mimulus* species and hybrids

The predominantly yellow-flowered monkeyflowers (*Mimulus* species and hybrids) constitute a complex and taxonomically difficult group, all of which are aliens from America and are found in the British Isles by running water, both slow lowland rivers and streams, and rushing mountain becks (Fig. 168). The first

FIG 168. Hybrid Monkeyflower (*Mimulus × robertsii*), the commonest member of the genus in upland Britain, is typically found on river gravel or shingle. (CAS)

alien here was the North American Monkeyflower (*M. guttatus*) in the 1820s, soon
to be followed by the South American Blood-drop-emlets (*M. luteus*). Hybrids
between these two arise readily in gardens and in the wild, but are highly sterile
(Fig. 169). In 1990, they were described as *M. × robertsii* to commemorate the Welsh
botanist R. H. (Dick) Roberts, who first studied them in the wild and pointed
out that this hybrid is the commonest *Mimulus* in upland Britain. The genus as
a whole thrives in a damp climate, and is much commoner in the north of the
British Isles than in the south, notably Shetland, where the varied colours form
spectacular displays. Later, another South American species, the copper-coloured
M. cupreus, and various hybrid combinations were grown in gardens and escaped,
but none except perhaps the hybrid involving all three species has arisen here
naturally. The immediate parentage of the latter is *M. guttatus* × (*M. luteus* × *M.
cupreus*), since hybrids between the South American taxa are highly fertile yet all
the North × South American combinations are sterile.

FIG 169. Monkeyflowers (*Mimulus*):
(a) Monkeyflower (*M. guttatus*), the most
widespread member of the genus in the British
Isles (note the prominent bosses on the lower
lip, which almost conceal the opening to the
corolla tube); (b) the variant of Blood-drop-
emlets (*M. luteus*) known as *M. smithii* (note the
scarcely concealing bosses); (c) *M. × robertsii*, the
hybrid between these two species. (a and c, CAS;
b, Mario Vallejo-Marín)

FIG 170. The fertile amphidiploid *Mimulus peregrinus* growing in its type locality in Scotland. (Mario Vallejo-Marín)

In 2011, a much more exciting situation in our alien *Mimulus* taxa was discovered by Mario Vallejo-Marín, a Mexican botanist working at the University of Stirling (Vallejo-Marín, 2012). He found that in a large population of Hybrid Monkeyflower (*M.* × *robertsii*) growing on the banks of Shortcleugh Water in the Lowther Hills, Lanarkshire, there were some highly fertile individuals. They have mostly well-formed pollen grains (more than 80 per cent, compared with less than 5 per cent in Hybrid Monkeyflower) and produce a good crop of viable seed. The chromosome number of the fertile plant is 2n = 92, a hexaploid, which is twice the number of many Hybrid Monkeyflower plants (2n = 46, triploids), the latter being derived from the diploid Monkeyflower and the tetraploid Blood-drop-emlets. Clearly, the fertile plant is a new amphidiploid, now named *M. peregrinus* (Fig. 170), to rank alongside those in the genera *Spartina* and *Senecio* discussed above. Whether *M. peregrinus* becomes as successful in the wild as the two older amphidiploids remains to be seen; it will be fascinating to study its progress. A second colony was found a few kilometres away in 2012.

Large Bindweed *Calystegia silvatica*

There are two white-flowered species of *Calystegia* in Europe, of which one, Hedge Bindweed (*C. sepium*), is native in the British Isles. Its distinctive trumpet-shaped flowers, among the largest in our flora, are very familiar and scarcely need description. Its natural habitat is probably reedbeds and fens, but it has become characteristic of ditches and damp hedgerows throughout most of the region. In 1948, Ted Lousley noticed that another European species, Large Bindweed (*C. silvatica*; Fig. 143), was also present in the wild here, having been confused with our native plant. It was introduced to this country as a garden plant about 200 years ago, and specimens from the wild have now been found from as long ago as 1863. Both species are extremely vigorously rhizomatous, and the long, contorted white underground stems ('devil's-guts') are notorious with gardeners. Small fragments are easily transported as contaminants of soil or among the roots of garden plants, and are very effective in dispersal. A small patch rapidly becomes extensive, and many hedgerow metres of bindweed can consist of a single clone. In many patches seed-set is very sparse or even non-existent. The reason for this is that both species are highly self-incompatible (Stace, 1961); only where several clones exist close together is seed-set good, pollination being effected mainly by bees (especially bumblebees) (Stace, 1965).

The differences between Hedge Bindweed and Large Bindweed are almost entirely quantitative, and they share the same chromosome number (2n = 22). The flowers of Large Bindweed are somewhat larger, and at the base of each long-stalked flower the two bracteoles (sepal-like outgrowths immediately below the true sepals) are very broad and wrap round the flower base, overlapping at their edges and completely obscuring the sepals (Fig. 143b). In Hedge Bindweed, the bracteoles are smaller and much flatter, revealing the sepals at each side of the flower where the bracteoles do not quite meet. In 1956, it was noted that intermediates were encountered 'not infrequently', and subsequent analysis has shown them to be widespread, the earliest specimens dating from the beginning of the twentieth century. They have also been known from Italy, where both species are native, for almost 200 years, and are correctly named *Calystegia × lucana*.

To measure 'sepium-ness' and 'silvatica-ness', a simple numerical index (the 'hybrid index') was constructed on a scale of 4 to 40 (Stace, 1961), and when this was applied to colonies of bindweed in one part of southeastern England it showed that not only were there colonies midway between the values of the two species, but there were also colonies closer to each of the parents (Fig. 171). The hybrids are highly, possibly fully, fertile, and clearly can produce F2 generations and backcrosses to both parents. It is true that some hybrid colonies produce little or no seed, but this is equally true of the parents and is due to the self-incompatibility of the hybrids, as with the parents. A wider study using the same method was carried out in the Greater London area by the London Natural History Society (Bangerter, 1967), at the time one of the largest and most successful surveys of its kind, involving more than 60 recorders and over 1,200 bindweed colonies. The results mirrored those of the more local survey. Using very strict limits for hybrid plants, about 3 per cent of the total were designated as hybrids, but this figure excludes backcrosses, which merge into the parental values. It was found that in Inner London, Large Bindweed outnumbered Hedge Bindweed, whereas in Outer London the reverse prevailed; this is true of the

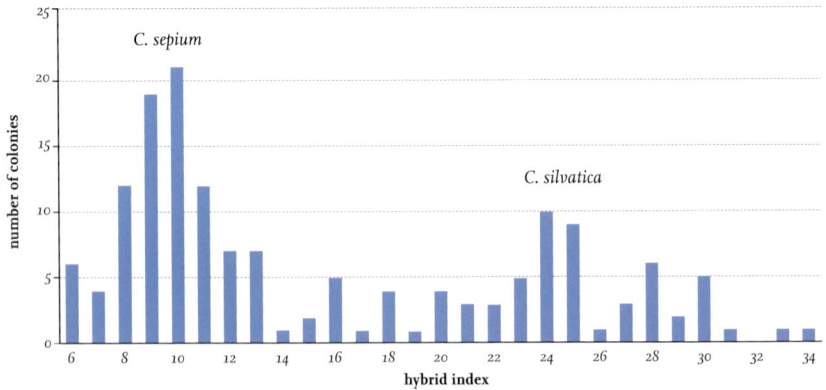

FIG 171. Histogram of hybrid indices of bindweed (*Calystegia*) colonies, arranged on a potential scale of 4–40, showing relative degrees of similarity to Hedge Bindweed (C. *sepium*; towards the left) and Large Bindweed (C. *silvatica*; towards the right). (Taken from Stace, 1961)

situation in the British Isles, where the alien is generally commoner in urban situations.

The wide range of variation of bracteole shape in Large Bindweed, both here and in southern Europe where it is native, led to the recognition of two subspecies: *Calystegia silvatica* subsp. *silvatica*, which represents the more extreme morphology; and subsp. *disjuncta*, which is morphologically less distant from Hedge Bindweed and on the above numerical scale would fall into the hybrid range or that occupied by backcrosses to Large Bindweed.

Molecular studies carried out by Brown *et al.* (2009), using parts of the nuclear DNA, identified different segments of DNA that are unique to each of the *Calystegia* taxa studied. Unsurprisingly, they found that the data were 'consistent with an origin for [C. *silvatica*] subsp. *disjuncta* from a cross between' subsp. *silvatica* and Hedge Bindweed. The production of highly fertile hybrids has enabled each species of *Calystegia* to introgress into the other, widening their range of variation and obscuring their true limits, and in the case of subsp. *disjuncta* prompting the description of a new subspecies to accommodate the extra range of variation. This is comparable to the case of Groundsel, where hybridisation with Oxford Ragwort has produced rayed Groundsels (*Senecio vulgaris* var. *hibernicus*), so enlarging the morphological range of the former.

Spanish bluebell *Hyacinthoides hispanica*
The wild Bluebell (*Hyacinthoides non-scripta*; Fig. 172a) is one of our most well-known, beautiful, delightfully scented and abundant native wildflowers, absent

only from Orkney, Shetland and the northern parts of the Outer Hebrides. On a world scale, however, it is a very local species, confined to a very narrow Atlantic slice of Europe from the Netherlands to northern Portugal. In Spain, Portugal and North Africa, it is replaced by the Spanish Bluebell (*H. hispanica*; Fig. 172c), which differs in having a more erect flower spike, less nodding flowers with more open petals, a usually paler blue colour and fainter scent, and blue rather than cream-coloured unopened anthers.

FIG 172. The two species of bluebell (*Hyacinthoides*) and their hybrid: (a) Bluebell (*H. non-scripta*); (b) Hybrid Bluebell (*H. × massartiana*); (c) Spanish Bluebell (*H. hispanica*). (CAS)

Both species are grown in gardens, for some reason Spanish Bluebell more commonly than the native. Both are partially self-fertile, but can hybridise as easily as reproducing with their own kind, and the resultant hybrids are fully fertile. By selfing, crossing and backcrossing haphazardly, the full range of morphology from one species' extreme to the other can be built up, so that the precise limits of the two species become impossible to discern. This process is not very rapid, because it takes at least five years for a bluebell seed to produce flowers, but over the past several decades the complete spectrum of bluebells has been established in a great many places. Such a situation is known as a hybrid swarm. In urban and suburban areas, the Hybrid Bluebell (*H.* × *massartiana*; Fig. 172b) has become far commoner than either parent, so that now most commercially available seed and bulbs are hybrid in nature. Hybrids prevail in all urban and rural gardens, and along grass verges and hedgerows in villages and beyond where Bluebells have been dumped or planted to 'beautify' the countryside.

Many stories about the threats to our native Bluebells, from both humans and Spanish Bluebells, have appeared in the press in recent years, and successful prosecutions have been made. Native Bluebells are specifically protected by Schedule 8 of the UK's Wildlife and Countryside Act 1981, whereby it is illegal to offer them for sale, and they are, of course, protected from being dug up from the wild by the general provisions of the Act. Apart from this, the main danger comes from introducing Spanish Bluebells into wild woods or to areas adjacent to them, when pollinators (mainly bees and butterflies) might well transfer foreign pollen into the native populations. Contamination of our wild Bluebells is a major concern for conservationists, and because of this bluebells should not be planted outside gardens. Where alien bluebells have been introduced to roadsides and similar habitats in the wild, they have proved to be long-persistent, but it has not been found that they can spread into the interior of the wood, either through dispersal of their own seed or through pollen flow to native Bluebells.

Rhododendron *Rhododendron ponticum*
The beautiful mauvish-purple flowers of the evergreen shrub Rhododendron (*Rhododendron ponticum*; Fig. 271) are too well known to need further introduction. The severely damaging effects on natural vegetation of the plant's rampant spread both vegetatively (layering) and by seeds is discussed in Chapter 15. Rhododendron has a disjunct distribution, occurring in the extreme southwestern part of the Iberian Peninsula and in Turkey and adjacent countries; the two separate populations are recognised as different subspecies. The species was introduced into cultivation in the British Isles in 1763, as far as we can tell

from the western population (subsp. *baeticum*) alone. It began to be recorded in the wild from 1894. During its cultivation and use as a rootstock for other rhododendrons, *R. ponticum* was crossed with several other *Rhododendron* species to produce a range of rootstocks called the 'iron-clads' or Hardy Hybrids (Milne & Abbott, 2000). These hybrid plants were further used in cultivation and plantings. The severely cold winter of 1879/80 killed many evergreen plants, including *Rhododendron* species, being planted at that time, but some of the hybrids appear to have survived particularly well. These surviving hybrid bushes produced viable seed and enterprising gardeners collected their seedlings for distribution around the country.

During the latter part of the twentieth century, recorders began to remark that not only was wild Rhododendron often much more vigorous and aggressive here than in the Iberian Peninsula, but also that it seemed to be more variable, particularly in floral features. This greater range of variation apparent in some of our Rhododendron includes characters that are diagnostic of three other species, all from North America and cultivated here alongside Rhododendron for the past 150–250 years: Great Rhododendron (*Rhododendron maximum*), Catawba Rhododendron (*R. catawbiense*) and Pacific Rhododendron (*R. macrophyllum*). All the species are closely related and readily cross to produce fertile hybrids. The main characters that occur in some of our Rhododendron but not in that species in its native areas are: leaves that are broader than usual and more rounded at the base and apex; presence of hairs on the lower leaf surface; flower stalks with very few glandular hairs or with branched hairs; calyx with unequal lobes; corolla with orange spots inside; and ovary rounded at the apex and/or densely pubescent or glandular-pubescent. It seems reasonable to assume that these extra characters of Rhododendron have been acquired by hybridisation with the other three species, both artificially by gardeners and naturally by bees, and a preliminary molecular study seems broadly to confirm this.

Hybridised individuals can show the features of one, two or all three of the American species. Out of 138 populations of our 'Rhododendron ponticum' examined, 33 (24 per cent) exhibited these morphological characters (Cullen, 2011), whereas 27 out of 260 (10.4 per cent) populations molecularly surveyed showed evidence of genetic material from Catawba Rhododendron (Milne & Abbott, 2000). It therefore might be the case that three-quarters of our plants have *not* been introgressed by the American species. In order to recognise the novel nature of our 'R. ponticum', James Cullen (2011) described these introgressed plants as a new hybrid, *R. × superponticum*, representing a diverse amalgam of hybridised individuals. It is tempting to attribute the extreme aggressiveness of our Rhododendron to hybrid vigour, but it has not been shown that plants referable to *R.* × *superponticum* behave differently in the British Isles in this respect from pure *R. ponticum*. In its native Iberian habitats, *R. ponticum* is not aggressive, but there are many other factors possibly underlying this difference. It is not known to what extent, or whether at all, hybridisation between the four above species continues today.

Brown Knapweed *Centaurea jacea*

The group of species centred on *Centaurea nigra*, commonly known as hardheads or knapweeds, is taxonomically very complex, and treatment of our native representatives has varied over the years. One or two species – Common Knapweed (*C. nigra*) and Chalk Knapweed (*C. debeauxii*) – may be recognised, but both have distinctive, very dark brown phyllaries (small bracts surrounding each flower head) that are deeply and finely divided in a very regular comb-like pattern. A related alien species, Brown Knapweed (*C. jacea*), which is very common on the Continent, has mid-brown phyllaries that are irregularly and coarsely divided. This alien species has been quite common in rough grassland and road verges in Britain at various times during the twentieth century, probably introduced as a contaminant of grass seed, but today we have no known localities. All three taxa are highly interfertile and produce fertile hybrids.

There are plants still to be found, however, that are close in characters to our native plants but possess some features reminiscent of Brown Knapweed. Most often the phyllaries are not as dark brown in colour, and their margins are less regularly dissected, than in Common or Chalk knapweeds. These plants are examples of the latter two species introgressed by Brown Knapweed, this process perhaps dating from a time when this species occurred in the area and hybridised with native plants, followed by backcrossing to them. Introgression of this sort, rather than the development of hybrid swarms, takes place when one of the participating species is much rarer than the other, so that most of its sexual reproduction is with other species rather than with its own species. On the Continent, even in northern France, Brown Knapweed hybridises with Common Knapweed and Chalk Knapweed, and there hybrid swarms are quite common, as a picnic on a laneside verge will often reveal. Because of this, it is likely to be the case that some of the alien plants found in the British Isles became introgressed abroad and were imported as such. A detailed monograph of these three species, with special reference to hybridisation, was published by Marsden-Jones & Turrill (1954). At the present time other related species of *Centaurea* are being imported as grass-seed contaminants, and we must expect to be confronted by further taxonomic headaches involving their hybrids with native species.

OTHER HYBRIDISING ALIENS

White Campion *Silene latifolia*

White Campion (*Silene latifolia*; Fig. 173a) is an archaeophyte closely related to the native Red Campion (*S. dioica*; Fig. 173b); both are dioecious. Although they are very similar in habit and general appearance, they differ markedly in one major character: the capsule teeth of White Campion are erect while those of Red Campion are strongly curved back. This is, of course, of no use for distinguishing male plants, when petal colour becomes the sole criterion. The petals of White Campion are always white, but albino plants of Red Campion are not rare, complicating the issue. Both species, however, are usually suffused strongly with purple coloration (anthocyanin) on the stems and sometimes leaves. This purple colour is absent from albino Red Campions, producing a somewhat anaemic appearance and affording a distinction between male plants of these and male White Campions. It is a moot point as to whether albino White Campions exist!

FIG 173. Two dioecious species of campion (*Silene*): (a) White Campion (*S. latifolia*); (b) Red Campion (*S. dioica*); (c) their pink-flowered hybrid *S. × hampeana*. (CAS)

The natural habitat of Red Campion is varied but well defined, in woodland and hedgerows, and open grassland on coasts and in the north, but the species is easily dispersed by seed and often appears in marginal ground, disturbed soil and new roadsides, especially on light soils where White Campion is characteristic. In those places the two readily form highly fertile hybrids that produce later generations and backcrosses, potentially giving a full spectrum of variation (a hybrid swarm). These situations are easily spotted because this cross is a classic example of incomplete dominance of the red colour, F_1 plants having pink flowers and others filling in the spectrum from red to white (Fig. 173c). Sometimes one encounters colonies of White Campion in which a few plants have very pale pink flowers; this is presumably where Red Campion pollen was carried to the site a few years previously and has since been progressively diluted, resulting in introgression.

American Fleabanes *Conyza* species

The four species of *Conyza* that have become aliens in the British Isles have arrived successively: Canadian Fleabane (*C. canadensis*) in 1690; Argentine Fleabane (*C. bonariensis*) in 1909; Guernsey Fleabane *C. sumatrensis* in 1961; and Bilbao's Fleabane (*C. floribunda*) in 1992. All are rather unattractive, dull-flowered, weedy plants of waste ground (Fig. 122), and all but Argentine Fleabane have hybridised with our native Blue Fleabane (*Erigeron acris*), which often grows in similar places. The hybrids are found as isolated individuals among mixed populations of the two parents, and all are evidently rare. In addition, a single plant of the hybrid *C. canadensis* × *C. bonariensis* was found in Middlesex in 1993. The fact that rather unexpected hybrids occur in *Conyza* shows that isolating mechanisms must be weak, and probably the only factor preventing frequent hybridisation is the highly inbreeding nature of these species.

Willows *Salix* species

Willows are very well known for their ability to hybridise promiscuously. It is possible that all of the world's 450 or so species could cross in virtually any combination. Moreover, except in cases where the chromosome numbers dictate otherwise, the hybrids are generally highly fertile. This means that the hybrids themselves can hybridise further; the Swede Nils Nilsson (1954) presented the parentage of a hybrid that he had derived by controlled pollinations from 13 different species. In the flora of the British Isles, there are 24 native and alien species and 68 hybrids, of which 20 are hybrids involving three species.

Osier (*Salix viminalis*) is considered to be an archaeophyte, and it is involved in seven hybrids, five formed here but two introduced as such. Of more interest, however, are the hybrids involving our two main tree species, White Willow

(*S. alba*) and *S. euxina*, and the hybrid between them, Crack Willow (*S.* × *fragilis*), which we now consider to be aliens. The uncertainty about their taxonomy and origin in the British Isles is discussed in Chapter 2.

Garden plants hybridising with wild species

There are several cases of ornamental garden introductions escaping into the wild and then hybridising with native species. Large Bindweed and Spanish Bluebell are examples already discussed; some others will be mentioned here.

The popular rock-garden plant Snow-in-summer (*Cerastium tomentosum*), from Italian mountains, has become thoroughly naturalised in many places in the British Isles on light soils. Where it meets the much less conspicuous native Field Mouse-ear (*C. arvense*), it readily forms a partially fertile hybrid. On the coastal dunes in northern Norfolk, many hundreds of square metres are turned white in May by these three taxa; all have the same chromosome number.

In most cases when found in the wild, Fox-and-cubs or Orange Hawkweed (*Pilosella aurantiaca*) is obviously a garden escape, but sometimes its 'parachuted' achenes are carried far by the wind to more remote places. When this happens it can hybridise with our native Mouse-ear Hawkweed (*P. officinarum*) to form often fertile hybrids (Fig. 174), which in suitable conditions – such as the open ground provided on the Cheshire lime-waste dumps – can form a complete range of intermediates. Occasionally, these intermediates are seen in gardens. In other cases, the hybrids are sterile; their sterility or fertility depends on the chromosome number and reproductive behaviour (sexual or apomictic) of the very variable Mouse-ear Hawkweed.

FIG 174. *Pilosella* × *stoloniflora*, the hybrid between Mouse-ear Hawkweed (*P. officinarum*) and Fox-and-cubs (*P. aurantiaca*), with some Mouse-ear Hawkweed plants in the background. (CAS)

The much-maligned Giant Hogweed (*Heracleum mantegazzianum*; Fig. 97), from southwestern Asia, which can reach over 5 m tall, is not much grown in gardens these days owing to its deserved reputation as a harmful plant, and it is more often seen by rivers and roads where it has escaped or been dumped. The juice sensitises the skin to sunlight, and brings up huge, painful blisters when exposed. Its attraction to youngsters for making pea-shooters adds to its danger. Its hybrid with our native Hogweed (*H. sphondylium*), which it often meets in marginal habitats, has been recorded mainly in Scotland and the London area. Although both species have the same chromosome number, the hybrid has very low fertility.

The very popular, but not quite fully hardy, garden shrub from southern Europe, Silver Ragwort (*Senecio cineraria*), is naturalised on cliffs and other semi-open ground mostly by the sea. Here, where it meets our native, fully herbaceous Common Ragwort (*S. jacobaea*), hybrids are often formed. They are fertile and, where they persist, backcrossing takes place and a full range of intermediates can develop. The hybrid was originally described (as *S. × albescens*) by Burbidge & Colgan (1902), from sea cliffs in Killiney Bay, Dublin. At that time the authors concluded that the hybrid existed in two forms, each nearer to one or the other of the two parents, but a resurvey 78 years later (Murphy, 1981) found that the complete range of intermediates existed. The hybrid still thrives there. Murphy observed that the conspicuous black and yellow larvae of the Cinnabar Moth (*Tyria jacobaeae*) fed solely on Common Ragwort and those hybrids that had only a sparse hair-covering approaching the state in the latter species, but in my garden I have found them devouring Silver Ragwort. The latter is also naturalised on waste ground in more urban areas, and hybrids can develop there too. In such a place its hybrid with a related native, Hoary Ragwort (*S. erucifolius*), was found in Kent in 1978.

The little-grown vegetable Salsify (*Tragopogon porrifolius*), much commoner on the Continent than with us, has long been known to hybridise with our native Goat's-beard (*T. pratensis*). The products, known as *T. × mirabilis*, are easily recognised because of their mixture of flower colours in each flower head, combining the yellow of Goat's-beard and the purple of Salsify; usually, the flower head is yellow in the centre and purple towards the circumference. There is a small degree of fertility, allowing the possibility of backcrossing, but most populations of Salsify and of the hybrid are very short-lived and populations do not build up. In 1758, Linnaeus published a thesis on this hybrid, describing some hybridisation experiments that he had carried out – some of the first such experiments to have been undertaken. In North America, both Goat's-beard and Salsify are naturalised aliens, and both have hybridised with the native *T. dubius* there. Each of these hybrids has undergone a doubling of chromosome number to produce a fertile amphidiploid, known as *T. miscellus* and *T. mirus*, respectively. It would be most interesting to know why this process has not occurred in *T. × mirabilis*.

Of the eight species of toadflax (*Linaria*) on our standard list, probably only the yellow-flowered Common Toadflax (*L. vulgaris*) is native to the British Isles. Three hybrids have arisen, all involving the archaeophyte Pale Toadflax (*L. repens*), which is widely distributed in Britain and has often been considered a native. Its hybrid with the native species is common where the two species meet; the hybrid is highly fertile and produces further generations as well as backcrosses to both parents.

At any one site a range of situations might develop along with both parents, as can be seen from the colour of the flowers ranging between the yellow of Common Toadflax and the pale mauve with violet veins of Pale Toadflax. Sometimes the hybrids appear to be F1s, being approximately halfway between the parents in appearance, whereas in other places a hybrid swarm has developed, with the whole range of intermediates between the two parents. In other cases, however, introgression has clearly occurred to one parent or the other; introgressed Common Toadflax has yellow flowers with violet veins, while introgressed Pale Toadflax has pale mauve flowers tinged with yellow. It is presumably the conditions of the habitat that determine the pattern of variation at each site, favouring only part of the spectrum of variation, but we can only guess at these (climate, soil, pollinators, etc.) in the absence of any detailed investigation.

The term introgressive hybridisation, later shortened to introgression, was proposed in 1938 by Edgar Anderson and Leslie Hubricht in relation to their work on *Tradescantia*, and these two authors are often credited with being the first to detect this process. They were, in fact, beaten to it by 40 years. In 1892, G. C. Druce encountered a hybrid swarm of *Linaria* hybrids near Oxford, which he described as 'a most complete chain of intermediates' following introduction of Pale Toadflax to the area in 1890 (Druce, 1893). Later, in 1897 in his *Flora of Berkshire*, he wrote: 'after hybridisation had gone on for some time', Common Toadflax exhibited a greater degree of variation than had previously been the case, as if 'the hybrid influence had been gradually neutralised by successive pollinations by a similar species'. These details given by Druce, well before the rediscovery of Mendelism, are a perfect description of introgression.

The other two *Linaria* hybrids represent crosses between Pale Toadflax and the neophytes Prostrate Toadflax (*L. supina*) and Purple Toadflax (*L. purpurea*; Fig. 175).

FIG 175. Hybridising toadflaxes (*Linaria*): (a) Purple Toadflax (*L. purpurea*); (b) Pale Toadflax (*L. repens*); (c) segregating F2 family sown from one plant of their F1 hybrid, *L. × dominii*. (CAS)

Our Hazel (*Corylus avellana*) is a very well-known plant intimately woven into country lore and crafts. Its nuts are edible and can be dried for eating in the winter, although almost all of those sold in shops at Christmas are imported from southern Europe, where the trees fruit more dependably and the nuts can be dried more easily. A related species, Filbert or Cobnut (*C. maxima*), from southeastern Europe and southwestern Asia, differs almost exclusively in having longer nuts that are more fully enclosed in a longer, frillier husk. It is grown commercially in orchards mainly in Kent, where the nuts are picked in late August to early September when still green, and they are preferably eaten in this state rather than dried. The nut orchards are called plats – hence the village named Platt, which has pictures of hops and cobnuts on its village sign and was the home of David McClintock, one of our most renowned plant alien specialists.

FIG 176. Fruits of left, Filbert (*Corylus maxima*); right, Hazel (*C. avellana*); and centre, their hybrid, 'Kentish Cob'. (CAS)

Both Hazel and Filbert are usually self-incompatible (see below), but can successfully pollinate each other with ease to enable fruit formation. The nuts of both are extremely attractive to Grey Squirrels, which plant millions every year, and many of them grow to maturity. Hence both Filberts and the hybrid are found wild in copses and hedges near where the former are grown (Fig. 176). There are hundreds of cultivars of Filbert, some of which are Hazel–Filbert hybrids, but only one of which is correctly known as 'Kentish Cob', the name usually used to cover them all. Exceptionally, 'Kentish Cob' itself is widely claimed as being self-fertile.

Hybrids between alien plant species
Poplars *Populus*

Poplars (*Populus* species and hybrids) are discussed in Chapter 8. All are dioecious and most cultivars are represented by one clone of a single sex. Therefore, any seed set by one of these cultivars will be the result of its hybridisation with a different cultivar. Brian Wurzell found thousands of seedlings of *P. × canadensis* on waste ground at Hackney in the hot summers of 1975 and 1976, these evidently having been produced by the pollination of the female clone 'Marilandica' by the male clone 'Serotina'. The trispecific hybrid *P. nigra × P. deltoides × P. balsamifera* has been discovered in two separate localities, having arisen intriguingly in two different ways. Brian Wurzell found it in 1984, again at Hackney, as a result of hybridisation between Hybrid Black Poplar (*P. × canadensis = P. nigra × P. deltoides*) 'Serotina' males and Balm-of-Gilead (*P. × jackii = P. deltoides × P. balsamifera*) females, the hybrids therefore involving two 'doses' of Eastern Cottonwood (*P. deltoides*). One year earlier, Oliver Gilbert (1993) had found hybrids between male Black Poplar (*P. nigra*) and female Balm-of-Gilead on waste ground in Leeds, in this case involving only one 'dose' of each species. In both the localities many seedlings developed over more than a decade into tall, vigorous saplings.

Mugworts Artemisia

The Mugworts (*Artemisia*) provide a good example of the need for accurate chromosome counts in order to understand fully hybridising populations. Mugwort (A. *vulgaris*; Fig. 177) is an extremely common archaeophyte of rough and waste ground almost throughout the British Isles, whereas Chinese Mugwort (A. *verlotiorum*) is a much rarer neophyte scattered in Britain in similar places. They differ in rather minor features of leaves, inflorescence and stem anatomy. Although Mugwort flowers rather late in the season (usually July to September),

FIG 177. The common archaeophyte Mugwort (*Artemisia vulgaris*), formerly put to many minor medicinal and culinary uses. (CAS)

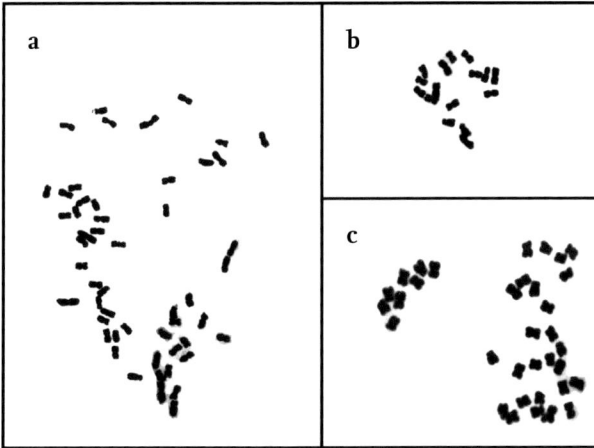

FIG 178. Chromosomes of *Artemisia*: (a) Chinese Mugwort (*A. verlotiorum*), 2n = 52; (b) Mugwort (*A. vulgaris*), 2n = 16; (c) hybrid A. × *wurzellii*, 2n = 34. (Preparations by Celia James, taken from James *et al.*, 2000)

Chinese Mugwort is later still, in October and November, and often does not manage to produce seeds before being killed by the frosts. In 1987, Brian Wurzell discovered the sterile hybrid between them, *A.* × *wurzellii*, new to science, in Tottenham, London. Subsequently, 15 other colonies were found in southeastern England, and it has since been detected in one or two other English sites and in one in the Netherlands.

With easily transported (rhizomatous) perennials such as mugworts, it is always difficult to decide whether hybrids arose where they are found or were introduced there. Mugwort is a diploid with 2n = 16, whereas its Chinese relative has known chromosome numbers of 2n = 48, 50, 52 and 54, i.e. it is hexaploid or hyperhexaploid (with a chromosome number just above the hexaploid level). The Tottenham hybrid has the chromosome count 2n = 34 (hypertetraploid), which would be expected only from a parent Chinese Mugwort with 2n = 52 (Fig. 178). Although it is far from proof, the evidence does support the case for considering the Tottenham hybrid to have arisen *in situ*, because the local Tottenham Chinese Mugwort does indeed have a 2n = 52 chromosome count (James *et al.*, 2000).

Shallons Gaultheria

Our modern concept of the genus *Gaultheria*, a shrubby member of the heather family Ericaceae, includes both of the former genera *Gaultheria* and *Pernettya*. The two commonest members are Shallon (*G. shallon*), from western North America, and Prickly Heath (*G. mucronata*), from Chile. They are both common garden shrubs and are well naturalised in the wild on acid sandy soils throughout the British Isles. They are distinguished by many characters, not least the small prickly, holly-like leaves of the latter and the almost laurel-like leaves of the former. The hybrid between the two, which formerly went under the hybrid genus name of × *Gaulnettya*, is grown in gardens, where it originated at Wisley in the 1920s, more as a curiosity than a thing of beauty – a natural product of two such different plants. It has also arisen once in the wild, in the New Forest, where both parents are naturalised and where it was discovered in 1981. The wild plant is closer in morphology to Shallon than is the garden plant. It has a low level of fertility.

Lupins Lupinus

Lupins (*Lupinus*) are justifiably among our most popular garden flowers, with their beautiful colours and scent. By far the commonest garden lupin is the Russell Lupin (*L.* × *regalis*; Fig. 179a), which is a hybrid between Garden Lupin (*L. polyphyllus*) and Tree Lupin (*L. arboreus*; Fig. 179b), both (like all the lupin species mentioned here) from western North America. Many cultivars of this hybrid were raised during a 25-year breeding programme by George Russell, and introduced into horticulture as recently as 1937. Before then, the common horticultural lupin was Garden Lupin, a less vigorous and usually unbranched species, almost always a rather dull blue in colour. The wide range in colours of the Russell Lupin is due to segregation and selection of crosses between the blue Garden Lupin and the yellow Tree Lupin. Garden Lupin is hardly ever seen in English gardens any more, but it is still grown in Scotland and is common in

FIG 179. Lupins (*Lupinus*): (a) Russell Lupin (*L.* × *regalis*), by far the commonest garden and naturalised lupin; (b) Tree Lupin (*L. arboreus*), one of its parents. (a, CAS; b, Richard Stace)

FIG 180. A mixed population of lupins (*Lupinus*) on shingle by the River Tay, Perthshire, with a predominance of Nootka Lupin (*L. nootkatensis*) and its hybrid with Russell Lupin (*L. × regalis*). (Alistair Godfrey)

Scandinavia, perhaps because it is more hardy. Russell Lupins have been widely planted on roadsides, and brighten many a tedious motorway verge, but they usually do not last many years as they get crowded out by coarse grasses, brambles and shrubs.

Lupins appeared in the wild on shingle by certain Scottish rivers, notably the Tay, Spey and Dee, from the early nineteenth century, and within a hundred years they had formed extensive populations. They had found a habitat resembling that of their native areas, and they still thrive there (Fig. 180). The first species to arrive there was the Nootka Lupin (*Lupinus nootkatensis*), which resembles Garden Lupin but has dense, shaggy hairs; it apparently originated from the gardens of Balmoral Castle. Garden and Tree lupins appeared later, and hybrids followed: *L. polyphyllus × L. nootkatensis*, spontaneous *L. × regalis*, and a hybrid with all three parents that arose from Russell Lupin crossing with Nootka Lupin. All these hybrids are highly fertile. Tree Lupin is characteristic of maritime shingle areas, especially in East Anglia.

Evening-primroses Oenothera

The evening-primroses (*Oenothera*), so called because of their evening-opening primrose-yellow petals, are also taxonomically complex and they possess an almost unique cytogenetic condition equally studied by researchers and struggled with by students. They are diploids with a chromosome number of 2n =14, but the seven chromosomes of each genome have undergone extensive reciprocal exchanges so that, when homologous chromosomes from the two genomes pair at meiosis, any one chromosome needs to pair with separate parts of more than one chromosome in the other genome. This results in four or more chromosomes being held together during pairing. The 14 chromosomes form varying sizes of pairing groups according to the nature and extent of the reciprocal exchanges that have occurred. One group of six and one of eight is not uncommon, and in extreme cases all 14 can be linked together. There is also a series of lethal genes present on the chromosomes that ensures that any recombined genomes are non-viable, so that only parental genomes are passed through to the next generation.

Each different pair of genomes in a plant produces a distinct morphology, and all these different sorts of evening-primrose have been recognised as distinct species by some botanists. However, their distinctions break down following hybridisation, which seems to be rife in naturalised populations in the British Isles; all combinations can arise and all are fertile. Another complication arises from the fact that, of the two genomes in each plant, one can be transmitted only through the male gametes and the other only through the female gametes. This means that reciprocal crosses produce distinctly different offspring. The upshot is that today it is much more common for the different genotypes *not* to be given species status.

The most variable species and the one most split up in the past is Common Evening-primrose (*Oenothera biennis*). This was cultivated in the eighteenth century and escaped into the wild early in the following century. Its anthers and stigmas are held close together and self-pollination is the norm. A related species, Large-flowered Evening-primrose (*O. glazioviana*), bears the stigmas above the anthers and is more likely to outbreed, and hence to form hybrids (Fig. 181). It was cultivated by the middle of the nineteenth century and found in the wild soon after; today, it is the commonest species in gardens and in the wild. It and Common Evening-primrose are particularly abundant on coastal sand dunes, where huge populations can develop and hybrids occur between them very readily (Fig. 252). Large-flowered Evening-primrose is not exactly matched by any North American populations, and there is the intriguing possibility that this species was not introduced from across the Atlantic but evolved in Europe as a

FIG 181. Evening-primroses (*Oenothera*): (a) Large-flowered Evening-primrose (*O. glazioviana*); (b) Common Evening-primrose (*O. biennis*); (c) their hybrid (*O. × fallax*). (CAS)

neonative by mutation and hybridisation. If so, this might represent a second British case of homoploid hybrid speciation, in addition to that of Oxford Ragwort (*see* above). The same may be true of some of the distinctive segregates of Common Evening-primrose once recognised as separate species, notably Welsh Evening-primrose (*O. cambrica*), which differs from Common Evening-primrose in having slightly longer petals and capsules, and in the lower capsules lacking glandular hairs.

Goldenrods Solidago *and Michaelmas-daisies* Aster

The possibility that an alien species might follow a different course of evolution from that in its native area, as exemplified by *Oenothera*, might also apply to two other extremely common and popular garden plants. In the British Isles there are two very common species of goldenrod – *Solidago canadensis* and *S. gigantea* – and there are three Michaelmas-daisies – *Aster novi-belgii*, *A. laevis* and *A. lanceolatus* –

plus two hybrids between them. In both genera large populations form on waste ground, building sites, roadsides and railway yards, and a wide range of variation is exhibited that does not always match that described in natural North American populations. It is most unlikely that more hybrids are not being formed, and some of these could well be in combinations that are producing new taxa. This topic has been almost totally neglected by researchers in Europe; the advent of DNA technology might change that, especially in the light of the now fashionable study of transgressive segregation discussed later in this chapter.

HYBRIDISATION WITH CROPS

Over the last 50 years, Oilseed Rape (*Brassica napus* subsp. *oleifera*) has become one of the commonest crops in the lowlands (Fig. 135a). As a result, many miles of roadsides and central reservations are now covered in its yellow flowers in spring and summer, growing from seed spilled from farm trailers the previous autumn (Fig. 135b). The Swede is *B. napus* subsp. *rapifera*. A related species is *B. rapa*, which includes the Turnip (subsp. *rapa*) and Bargeman's Cabbage or Wild Turnip (subsp. *campestris*), which grows wild, not as a crop, on river and stream banks in England. *Brassica napus* is the tetraploid amphidiploid derived from the diploids *B. oleracea* (Cabbage) and *B. rapa*, the primary F1 hybrid of which is unknown in the wild. The amphidiploid probably arose naturally by chance in cultivation many centuries ago. The above three species – *B. napus*, *B. rapa* and *B. oleracea* – are actually part of what is known as the Triangle of U (after Dr Nagaharu U, who worked out these relationships in the 1930s and established the genus *Brassica* as one of the classical examples of cytogenetic analysis). The triangle is a diagram positioning three diploids (each with different chromosome numbers) and the three amphidiploids derived from them (Fig. 182). All six species are on our list of aliens, except *B. oleracea*, which is probably a native, and the figure also shows two additional hybrids that occur with us.

Farmers have long known that crops of *Brassica napus* (Swedes or Oilseed Rape), when growing close to a source of *B. rapa* pollen (Turnips or Bargeman's Cabbage), occasionally produce some hybrid plants, formed by natural bee pollination. The hybrids, known as *B. × harmsiana*, closely resemble *B. napus* and are told from it only by their triploid chromosome number and sterility. The proportion of the crop so contaminated is minuscule and hardly affects its value. In recent years, however, hybridisation between these two species has taken on a much more significant role with the advent of genetically modified crops and the

hybrid
2n = 28

B. oleracea
2n = 18

B. napus
2n = 38

B. carinata
2n = 34

B. × harmsiana
2n = 29

B. rapa
2n = 20

B. nigra
2n = 16

B. juncea
2n = 36

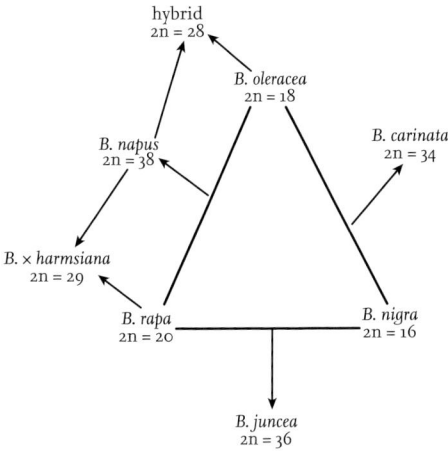

FIG 182. The Triangle of U, showing how three diploid species of *Brassica* have given rise to three tetraploid amphidiploids, one of which (*B. napus*) has in turn hybridised with each of its two parents to produce two sterile triploid hybrids.

fear that the alien genes in the crop plant could escape into wild populations by natural pollinations, i.e. in the reverse direction of the gene flow just mentioned, with *B. napus* being the pollen donor.

The extent of gene leakage from *Brassica napus* has been studied particularly by Mike Wilkinson and colleagues at the universities of Reading and Aberystwyth. They have found that hybridisation takes place at a very limited level between *B. napus* on the one hand and adjacent stands of both *B. rapa* and *B. oleracea*. In these three very similar species, accurate detection of hybrids is best achieved using molecular methods, which can detect fragments of DNA specific to one species that have been transferred to another. These have been found among Bargeman's Cabbage populations in southern and central England, mostly at the interface between the riverside habit of the latter and the crop fields, by the rivers Thames, Avon and Nene. To give an idea of the degree of hybridisation, Wilkinson *et al.* (2003) estimated, by extrapolation from their results, that there are probably *c.* 49,000 natural hybridisation events annually in the United Kingdom.

With regard to Cabbage (*Brassica oleracea*), one hybrid plant was found in a wild (probably native) population of Cabbage on the White Cliffs of Dover, among 842 plants screened. Oilseed Rape was being grown nearby. Clearly, hybridisation and hence gene leakage from the crop into wild populations is not a very common event, but in view of the large areas of crops being grown, and the fact that leakage will accumulate over the years, the escape of genes from genetically modified crops is not negligible, and the ridiculous notion once forwarded by some, albeit equally derided by others, that a few metres will be sufficient to isolate a crop field genetically is clearly untenable.

Apart from Oilseed Rape, the only other common crop that is sufficiently closely related to a truly wild (in this case native) species to hybridise with it is *Beta vulgaris*, which covers Sugar Beet, Beetroot, Perpetual Spinach, Chard and Mangel-wurzel. Their wild relative is Sea Beet (*Beta vulgaris* subsp. *maritima*), which comes into close contact with fields of Sugar Beet, especially in East Anglia. In this case we are dealing with hybrids between subspecies of a single species, but the chances of gene transfer are much higher than with *Brassica* because *Beta* is wind-pollinated, and pollen will surely be blown between the coastal Sea Beet and the inland crop fields, in either direction according to their relative positions, especially in the flatlands of East Anglia. In the case of both *Brassica* and *Beta*, seeds of crop plants are easily transported on footwear and tyres into wild populations, so that in the following year hybridisation can readily occur.

The other crop plant found to hybridise with a wild plant is Oat (*Avena sativa*), discussed in Chapter 2. Wild-oat (*A. fatua*) is very similar to Oat in growth-form and often grows as a weed with it. It is therefore very difficult to eradicate from the crop. It has the same chromosome number (both are hexaploids) and the two can produce at least partially fertile hybrids. The hybrids resemble Oat more closely than Wild-oat in the several minute characters of the spikelets, but they are rarely common enough to be detrimental to the crop. There is a mystery concerning Oat that has exercised workers for many decades, and is still not fully resolved. From time to time, there appear in Oat crops plants that have some of the characteristics of Wild-oat. Such plants are known as fatuoids, and there has been, and still is, argument concerning their origin. It is variously claimed that they are crop plants that have produced some fatuoid characters by mutation, or are hybrid backcrosses to *A. sativa*. There seems to be strong evidence in favour of both theories, which suggests that fatuoids might arise by either route.

THE MOST PROMISCUOUS ALIENS

Apart from willows and poplars, already discussed, there are three genera of alien plants that are truly promiscuous in their reproductive behaviour: willowherbs (*Epilobium*), docks (*Rumex*) and mulleins (*Verbascum*), all of which are well known to form hybrids frequently and in many combinations.

There are 15 species of willowherb on our standard list, producing 46 confirmed hybrid combinations. There are probably no combinations that cannot succeed, and the number of hybrids is simply a measure of which species coexist. Five of the 15 species are aliens; American Willowherb (*Epilobium ciliatum*; Fig. 129) and Panicled Willowherb (*E. brachycarpum*) are erect lowland species from North America, while the other three are dwarf creeping plants from New Zealand.

American Willowherb was first found in Britain in 1891, but did not become common until the 1930s. From then onwards, hybrids with other species were discovered: today, we know hybrids between it and all the other eight lowland species, as well as with the creeping New Zealand Willowherb (*Epilobium brunnescens*). Panicled Willowherb was not discovered in the British Isles until 2004, and so far no hybrids have been detected. The only common creeping species from New Zealand, New Zealand Willowherb, has been found to hybridise with six erect species, and one of the other two creeping species, Rockery Willowherb (*E. pedunculare*), has crossed with Broad-leaved Willowherb (*E. montanum*). Hybrids between the erect and creeping species were not detected here until four years after Peter Raven, the world *Epilobium* expert at the Missouri Botanical Garden, recommended in 1976 that we look out for them because such hybrids had been found in New Zealand, where American Willowherb is an established alien, and he had produced several hybrids involving European species with ease. Most of the willowherb hybrids have very low fertility, but some seed is formed in a few cases and the existence of hybrid plants involving three species is suspected.

The true docks belong to *Rumex* subgenus *Rumex*, in which we have 17 native and naturalised species, which have produced 41 hybrids. As with the willowherbs, hybridisation can be expected wherever two species occur together. Another parallel with willowherbs is that the hybrids are largely sterile, but some exhibit a degree of fertility and so offer the possibility of hybrids involving three species; a few such examples have been detected. Six of the 17 species are aliens: Willow-leaved Dock (*R. salicifolius*), Monk's-rhubarb (*R. alpinus*), Argentine Dock (*R. frutescens*), Russian Dock (*R. confertus*), Greek Dock (*R. cristatus*) and Patience Dock (*R. patientia*). The first two have produced no hybrids, but the other four species are involved in 13 hybrid combinations (Fig. 183). Due to the comparative rarity of the alien species, *R. cristatus* × *R. patientia* is the only hybrid to involve two alien species.

The mulleins (*Verbascum*) are tall biennials or perennials with spires of mostly yellow flowers, and are familiar on light, often calcareous soils only lightly covered with vegetation (Fig. 63). There are 13 native and naturalised species in the British Isles, producing 19 hybrids, which are highly but perhaps not completely sterile. Contrary to the situation with the willowherbs and docks, most of our mullein species (ten) are aliens and 16 of the hybrids involve at least one alien parent. Analysis of the variation of the hybrids shows that they inherit the characters in unpredictable ways. For instance, hybrids between species

FIG 183. Species of dock (*Rumex*) commonly hybridise in many combinations; this is the hybrid between the native Broad-leaved Dock (*R. obtusifolius*) and the alien Greek Dock (*R. cristatus*), named *R. × lousleyi* after J. E. (Ted) Lousley, arguably the greatest British alien hunter. (Geoffrey Kitchener)

with violet-coloured stamen hairs (e.g. Caucasian Mullein *V. pyramidatum*) and those with white ones (e.g. Great Mullein *V. thapsus*) can have violet hairs, white hairs, pale violet hairs, a mixture of the two on all stamens, or violet hairs on the lower two stamens and white ones on the upper three. Hybrids involving the native white-flowered White Mullein (*V. lychnitis*) have yellow flowers, like the other parent. In gardens where several species of mullein are grown, hybrids arise in bewildering arrays, aided by the fact that there is usually plenty of bare ground to enable the seedlings to develop, and several hybrids between our species not so far found in the wild have been noted.

TRANSGRESSIVE SEGREGATION

One theory proposed to explain the possibility of increased genetic variation in a segregating hybrid population is termed transgressive segregation (Rieseberg *et al.*, 1999), which describes the appearance of extreme ranges of variation beyond those normally encountered in the two parental species or their F1 hybrid. Rieseberg *et al.* listed 70 examples from the literature, most of them crop plants, which in general have been studied more deeply than wild plants. These authors saw the main advantage of the emergence of more extreme (i.e. new) characters as enabling the hybrid to colonise new habitats ('niche divergence'). They suggested that such characters might include morphology, fecundity, biochemistry, physiology, life history, tolerance of external (especially ecological) factors, and development of breeding barriers between the hybrid and its progenitor species.

There are many known examples of new characters appearing in hybrid segregants, e.g. strange growth forms in poppy hybrids (*Papaver dubium × P.*

rhoeas), yellow petals in white-flowered water buttercup hybrids, and new chemical substances in hybrids between Red Currant (*Ribes rubrum*) and Black Currant (*R. nigrum*). If this phenomenon is indeed widespread, it bestows a significantly greater evolutionary importance on hybridisation than simply the production of amphidiploids. On the face of it, Oxford Ragwort should be an ideal example to test the theory. Abbott *et al.* (2009) concluded that there is no evidence of transgressive segregation involving morphological characters in this species, but did not doubt that there might well be metabolic and physiological characters displaying it, and found some evidence of the latter from molecular studies. The significance of this might be greatest in promoting the success of invading species, particularly aliens, as pointed out by Ellstrand & Schierenbeck (2000), but it must be remembered that the great majority of aliens do not undergo hybridisation.

A NOMENCLATURAL CONUNDRUM

There are many examples of aliens that often go under the name of particular species but are actually hybrid derivatives that have arisen in cultivation, either by deliberate breeding or by chance crossing. Since much plant breeding has been carried out by individuals or organisations keen to preserve the production secrets of their new varieties, it is not surprising that we do not know the precise origins of many of these hybrids. In a number of cases more than two species have been involved in their ancestry. Quite often, hybrid names have been provided for the hybrid complex, without specifying their parentage. This is a perfectly valid procedure. The *International Code of Nomenclature for algae, fungi, and plants* (formerly the *International Code of Botanical Nomenclature*) treats hybrid binomials exactly like species binomials, except that a multiplication sign is placed between the generic name and specific epithet. Hence, the hybrid binomial relates to its type specimen, not to its parentage, and any changes in our knowledge of parentage, or indeed of whether or not it is a hybrid at all, do not affect the hybrid binomial, into which a multiplication sign can be inserted or removed according to interpretation.

There is, however, a downside to this state of affairs. Once a binomial is published, based on a type specimen, that name can apply only to the precise parentage of the type specimen, and cannot legitimately be used to cover related hybrids with a different parentage. For example, the garden hellebore known as Lenten-rose has been named as *Helleborus × hybridus* (Fig. 184) to cover a range of hybrids between *H. orientalis* and a number of other species, particularly *H. purpurascens* and *H. multifidus*. If the exact parentage of the type specimen of

FIG 184. Lenten-rose (*Helleborus × hybridus*) is a name loosely used to cover *H. orientalis* and its hybrids with several other species. The exact parentage of garden and naturalised plants has not been worked out. (CAS)

H. × hybridus could be determined (probably more likely today with the advent of molecular techniques), *H. × hybridus* could thereafter be used only for that combination and not to the others to which is now applied. The use of such broadly defined hybrids must therefore in theory be viewed as a temporary measure until the precise hybrid origins can be elucidated. The most obvious examples of such problematic species/hybrids that have escaped from cultivation into the wild are shown in Table 12.

TABLE 12. Examples of species represented in gardens in the British Isles mainly by hybrid complexes of varied uncertain parentage.

Species	Name of hybrid complex
Chrysanthemum indicum	*C. × morifolium*
Dahlia pinnata	*D. × hortensis*
Freesia refracta	*F. × hybrida*
Helenium autumnale	*H. × clementii*
Helleborus orientalis	*H. × hybridus*
Nymphaea alba	*N. × marliacea*
Rhododendron ponticum	*R. × superponticum*
Saxifraga hypnoides	*S. × arendsii*
Sidalcea malviflora	No name available
Tradescantia virginiana	*T. × andersoniana*, invalid name
Verbena peruviana	*V. × hybrida*

CHAPTER 10

Alien Non-flowering Plants

But here there is no light, save what from heaven is with the breezes blown
Through verdurous glooms and winding mossy ways.
John Keats, 'Ode to a Nightingale' (1819)

Utilising data gained through modern techniques such as the study of the chemistry of pigments, cell walls and food-storage compounds, the ultrastructure of the cell wall and cytoplasm (especially the plastid) and, most recently, DNA sequences, there are now several ways in which one can define plants. The most robust concept to emerge is that of the Green Plant, which includes vascular plants, bryophytes and green algae (Chlorophyta and Charophyta), but excludes all the other groups of algae and the fungi. In this book, however, we are following a more traditional definition, covering organisms whose main source of energy is sunlight as captured by the pigment chlorophyll a, whether or not that is masked in the plant by other pigments. Hence our term 'non-flowering plants' covers all photosynthetic cellular organisms other than the angiosperms, or flowering plants, i.e. the pigmented algae, bryophytes (mosses, liverworts and hornworts), pteridophytes (ferns and 'fern allies') and gymnosperms (conifers).

FRESHWATER ALGAE

Only four species of alien freshwater algae have been recorded from Britain or Ireland, and as far as we know none has invaded natural waters: Braun's Stonewort (*Chara braunii*; Charophyta), *Pithophora oedogonia* and *Schizomeris leibleinii* (Chlorophyta), and *Compsopogon coeruleus* (Rhodophyta). These all inhabited a part of the Reddish Canal in Manchester that was formerly heated by the outflow of warm water from cotton mills powered by steam engines. Warmth-reliant alien plants were first found in the Reddish Canal in 1883 (Bailey, 1884), when the pondweed Egyptian Naiad (*Najas graminea*) and the green algal charophyte Braun's Stonewort were recorded. The latter was last recorded in 1955, although the supply of warm water did not end (due to the replacement of steam by electric motors) until the winter of 1959/60 (Swale, 1962), soon after which the canal was filled in. At the time of the discovery the mill in question was importing cotton from Egypt, where both the aliens are common, leading Charles Bailey to suggest that this was 'probably' their origin. The *Compsopogon* has one other record from the wild, in a highly calcareous pond in County Durham for one summer only (John *et al.*, 2002).

MARINE ALGAE *by Ian Tittley*

The larger (macroscopic) marine algae, known commonly as seaweeds, belong to three major phyla: the green algae (Chlorophyta), brown algae (Phaeophyceae, now in the Ochrophyta) and red algae (Rhodophyta), all of which have alien representatives in our flora. Mankind has utilised seaweeds for thousands of years for food, fertiliser and other purposes, but it was marine transport, mariculture (seaweed farming *in situ*) and scientific research that led to their artificial, and mostly unintentional, spread beyond their natural areas of occurrence. Man-made structures, particularly harbours, marinas, artificial tidal pools and sea walls, have, as will be seen, created habitats that have aided the spread of alien algae around our shores. In addition to the seaweeds, there are three groups of marine microscopic algae that have alien representatives here: Dinophyta (dinoflagellates), and two groups now placed in the Ochrophyta – Raphidophyceae and Bacillariophyceae (diatoms).

The artificial dispersal of seaweed species around the world is not a recent feature, but can be traced back several hundred years. Shipping has been an important vector in the spread of alien seaweeds; in the days of sail, solid ballast was taken on board for stability – often rock from the coast of departure – which was then discharged at the final port of call. On these rocks were germling and mature seaweeds that sometimes survived the voyage and, if the conditions were right at the site of discharge, would continue to grow. This led, for example, to the export of the brown seaweed Serrated Wrack (*Fucus serratus*) from Britain to North America, where it became established; investigations using molecular techniques have even pinpointed the sites of origin in Britain. Reciprocally, species were probably introduced to Britain in this way. Other means of dispersal by boat involved algae that fouled hulls and survived the physiological stresses of passing through differing temperature zones to release propagules or fragments en route or at the final destination. The later transportation and discharge of ballast water was probably also responsible for the spread of alien species. The first records of several alien seaweeds in Britain have been in or near ports, notably in the Solent.

The global trade in oysters has also encouraged the spread of alien algal species because of the lack of cleaning in quarantine; the first records in the British Isles of several seaweeds associated with the industry have been in or near oyster farms. In 1971, Europeans were warned of potential ecological problems associated with the importation of the Asian Pacific Oyster (*Crassostrea gigas*) likely to be infected with other marine organisms. It was in this way that the Japanese brown alga Wireweed (*Sargassum muticum*) was introduced to the United States, subsequently spreading along the coast south to Mexico and north to Alaska. It was predicted that the species would appear in Europe, and in 1972 it was recorded there for the first time, in Bembridge lagoon on the Isle of Wight. The warning voiced concern about Wireweed becoming established in Eelgrass (*Zostera marina*) habitats and damaging them, possibly displacing Eelgrass. Bembridge lagoon was exactly this type of habitat but fortunately Eelgrass still grows there. The sea is not a barrier to algal dispersal but connects adjacent landmasses, and alien species brought to France can easily make their own way to England; it is suspected that this happened with Wireweed.

Occasionally, species have been deliberately introduced to European waters because of their economic potential. Giant Kelp (*Macrocystis pyrifera*), of the southern oceans and California, a species of commercial importance for alginate extraction, was grown on the Atlantic coast of France in the 1980s and 1990s. This caused considerable concern and consequently in 1992 the cessation of its cultivation in the sea was demanded. Fortunately, plants were removed from the sea before the onset of fertility and the species did not become established.

Seaweeds have long been used as packing for the transportation of shellfish and to decorate oyster stalls, a practice that can be traced back to medieval times. Today, seaweeds are still used to decorate fish counters in markets and supermarkets and are used as packing material. The Knotted Wrack (*Ascophyllum nodosum*) has an amphi-Atlantic distribution and occurs commonly in North America. I have seen this species used as decoration on a supermarket fish counter in Toronto, 800 km from the Atlantic coast. More significantly, it has been used to pack shellfish moved across America from the Atlantic to the Pacific coast; after unpacking, the Knotted Wrack was disposed into the sea, where it started to grow. Fortunately,

TABLE 13. List of alien marine algae in the British Isles.

Phylum	Species	First record	Native area
Chlorophyta (green algae)	*Bryopsis muscosa*	1926 (now extinct in British Isles)	Mediterranean
	Codium fragile subsp. *fragile* (subsp. *tomentosoides*)	1939	Pacific
	Umbraulva olivascens	1960	Pacific
	Ulva californica	Early 2000s	Pacific
Rhodophyta (red algae)	*Pyropia leucosticta*	1813	Pacific
	Bonnemaisonia hamifera	1890	Pacific (Japan)
	Antithamnionella ternifolia	1906	Southern hemisphere
	Neosiphonia harveyi (formerly *Polysiphonia harveyi*)	1908	Atlantic (eastern North America)
	Antithamnionella spirographidis	1934	North Pacific
	Asparagopsis armata	1939	Australia
	Grateloupia subpectinata (formerly *G. filicina* var. *luxurians*)	1947	Pacific
	Pikea californica	1967	Pacific (western North America)
	Grateloupia turuturu	1969	Pacific
	Sarcodiotheca gaudichaudii	1970s	Pacific
	Cryptonemia hibernica	1971	Pacific
	Solieria chordalis	1976	Pacific
	Anotrichium furcellatum	1976	Mediterranean (originally Pacific?)
	Agardhiella subulata	1977	Atlantic (eastern North America)
	Antithamnion densum	1990	South Atlantic
	Pyropia yezoensis	1990	Pacific
	Pyropia drachii	1994	Pacific
	Heterosiphonia japonica	1998–9	Pacific
	Polysiphonia subtilissima	Early 2000s	Tropics
	Laurencia brongniartii	Early 2000s	Tropics
	Caulacanthus okamurae	2004	Pacific
	Gracilaria vermiculophylla	2012	Pacific

Phylum	Species	First record	Native area
Ochrophyta: Phaeophyceae (brown algae)	*Colpomenia peregrina*	1907	Pacific
	Sargassum muticum	1973	Pacific (Japan)
	Undaria pinnatifida	1994	Pacific (Japan)
Ochrophyta: Bacillariophyceae (diatoms)	*Odontella sinensis*	1906	Red Sea to China
	Thalassiosira tealata	1950	Unknown
	Pleurosigma planctonicum	1966	Indian Ocean
	Hydrosera triquetra	1971	Atlantic (Venezuela)
	Coscinodiscus wailesii	1977	Pacific, Indian Ocean
	Thalassiosira punctigera	1978	Atlantic and Pacific
Ochrophyta: Raphidophyceae	*Heterosigma akashiwo*	Early 2000s	Japan?
Dinophyta (dinoflagellates)	*Karenia mikimotoi*	1968	North America
	Karenia aureola	1976	Eastern North America?

marine biologists there recognised the species as alien and apparently successfully eradicated it, unlike an earlier attempt at eradicating Wireweed in England.

Unfortunately, but occasionally, scientific research has been responsible for the spread of alien species. The red alga Grape-pip Weed (*Mastocarpus stellatus*, once known as Carragheen Moss) is of commercial importance for extraction of carrageenan (polysaccharides used in cooking and baking) and occurs widely in the North Atlantic, often as a dense zone at low shore level in wave-washed situations. Its life history differs according to where it occurs in the North Atlantic Ocean. The German island of Heligoland is the only rocky shore on the Continental coast in the southern North Sea, and for reasons unclear, Grape-pip Weed did not occur there before 1980. Research at the island's marine laboratory in the 1980s to unravel the alga's life history was in part undertaken by growing it in the sea. Not surprisingly, Grape-pip Weed became established and within 20 years was present as a dense and permanent zone at low shore levels, most noticeably on sea and harbour walls.

The marine algal flora of the British Isles contains about 600 species of green, brown and red algae, of which 29 (approximately 4.8 per cent) are alien. There are also nine alien species in the three groups of microscopic marine algae mentioned above. All are listed in Table 13 (note that in 2011, a 30th British Isles species of alien seaweed, the red alga *Polyopes lancifolius*, was found at La Rocque, Jersey). The stories of their arrival, spread and impact are told in the following sections.

Chlorophyta

Our alien green algae include two species of sea lettuce. *Ulva californica* is very similar to the native *Ulva linza*, which grows preferentially in brackish habitats in Ireland and Scotland and probably more widely. The former was only very recently discovered in Britain by using molecular techniques to compare with foreign species specimens that were suspected of having been wrongly named, in this case with *Ulva californica* from the Pacific coast of America, from where it is believed to have originated. The second alien sea lettuce, *Umbraulva olivascens*, consists of a large olive-green blade perforated with elliptical holes, and differs from *Ulva* in the presence of the photosynthetic pigment siphonaxanthin. *Umbraulva olivascens* was first found in Europe in 1930 in a tank at the Roscoff marine laboratory, and later more widely in France, the Netherlands, Ireland and Scotland. As congeners of *Umbraulva olivascens* occur in Japan, it is suspected that the species was probably introduced to Europe from the Pacific Ocean, perhaps as a consequence of research on other imported marine organisms.

The genus *Codium* is characterised by a spongy thallus composed of a mass of non-cellular siphonous (tubular, multinucleate and with few, if any, cross walls) filaments. The outer pigmented layer of the thallus, the cortex, has externally swollen filaments called utricles, the shape of which is an important character for species identification. *Codium* thalli may take the form of a mat, a ball or cylindrical branched erect axes (called green fingers). Five species of *Codium* occur in the British Isles, of which three are green fingers (Fig. 185a) and include one alien subspecies (*C. fragile* subsp. *fragile*). Seen under the microscope, *C. fragile* has utricles with pointed tips that are longer in *C. fragile* subsp. *fragile* (>30 μm) than in our native *C. fragile* subsp. *atlanticum* (<15 μm). Subsp. *fragile* probably originated in Japan and was first recorded in European waters in the Netherlands in 1900, since when it has spread widely in Europe as well as other parts of the world. It grows on rocks and in pools from mid-shore to shallow subtidal levels (Fig. 185b). Subsp. *atlanticum* was also considered to be alien, but as recent research has found no evidence of it elsewhere in the world it is now considered to be native.

The feather-shaped Mediterranean species *Bryopsis muscosa* was discovered on the monitor HMS *Glatton*, which was deliberately sunk in Dover Harbour in 1918, but only after the hull was raised and examined in 1926. As there are no supporting specimens, there is doubt as to the authenticity of the record; it has not been found since.

Phaeophyceae

The Oyster Thief (*Colpomenia peregrina*) has a spherical, inflated thallus that resembles a small balloon, and grows on oysters, sometimes floating off with

FIG 185. The Greenfingers *Codium fragile* subsp. *fragile* is a green alga of Pacific origin first found in the British Isles in 1939 and now occurring widely here: (a) plant at low tide; (b) habitat in Walpole Bay, Thanet, North Kent, with plant in foreground. (Tony Child, Thanet District Council)

them. The species was first found in France in 1906 and in Britain soon after; it probably came from the Pacific Ocean. By 1960, it had spread to eastern North America, where it is now considered a potential threat to the oyster industry. A congener, *C. sinuosa*, native to the North Atlantic Ocean, has a warm-temperate distribution to the Bay of Biscay. The two species can be distinguished microscopically only by the shape of the sporangial sori (irregular and extensive in Oyster Thief, punctate in *C. sinuosa*). The Oyster Thief is also an epiphyte on larger algae in low shore pools and at shallow subtidal levels.

The large kelp-like brown alga Wakame (*Undaria pinnatifida*; Fig. 186) was introduced to Brittany in the 1980s for commercial cultivation. This species is imported and sold, expensively, in most health-food shops in Britain and is used in a variety of oriental recipes. It was first discovered in southern England in 1994 growing on floating pontoons in a marina in the Hamble Estuary, Hampshire. Since then it has spread west to Devon (where it grows on rocky shores at subtidal fringe levels, as well as on floating pontoons), north to the Liverpool Docks (where it grows mostly on floating pontoons in harbours) and east to Essex, as far north as the Stour Estuary. It has also spread widely in Continental Europe. The species probably came from Japan as a result of oyster importation. Wakame superficially looks like the native kelp Dabberlocks (*Alaria esculenta*), as it has a midrib along the frond, but is distinguished by the marginal finger-like outgrowths.

FIG 186. Wakame (*Undaria pinnatifida*), a Japanese kelp-like brown alga introduced to Europe and cultivated for culinary use, occurs sporadically in southern Britain. (Bryony Chapman, Kent Wildlife Trust)

FIG 187. Wireweed (*Sargassum muticum*) is a Japanese brown alga that was predicted to spread to European waters and was duly first found on the Isle of Wight in 1973; it occurs commonly in the south and west of Britain. (Ian Tittley)

Probably the most discussed alien marine alga species, apart from the green 'killer alga' *Caulerpa taxifolia* of the Mediterranean, is Wireweed (*Sargassum muticum*; Fig. 187), sometimes known colloquially as 'smut' or 'japweed'. As mentioned above, its spread to east Atlantic waters was confirmed in 1973 at Bembridge. This bushy plant grows to several metres long and may be confused with the native genus *Cystoseira*, which also forms large, bushy growths. Wireweed is distinguished by its receptacles (specialised swollen branches that bear reproductive structures in pits called conceptacles) and bladders being formed on branches (in *Cystoseira* they occur along a main axis). It is a temperate species that occurs in deep, sheltered lagoons, at shallow subtidal levels and on floating pontoons. An initial attempt to eradicate the species at Bembridge by uprooting it proved unsuccessful, as small amounts of holdfast were left, allowing new plants to develop vegetatively. Its successful and rapid dispersal is aided by natural transport of fertile unattached, floating plants. It is also a successful rafting species, with the frond acting as a sail and moved by currents along the coast together with its stone or shell substratum, to be deposited at another location. Wireweed is actively spreading around Britain and Ireland, and has reached as far north as Norway and south into the Mediterranean Sea. It has no commercial value but does provide habitat for marine fauna; there was concern that it might outcompete the native bushy Sea Oak (*Halidrys siliquosa*), which also occurs in lagoon habitats, but this has not happened.

Microscopic marine algal groups

The diatoms (Ochrophyta: Bacillariophyceae) contain an internal skeleton, a frustule, made of silicon. Five microscopic alien species have been recorded from the North Sea, plus *Hydrosera triquetra*, a colonial species comprising a chain of cells and superficially resembling a filamentous brown seaweed. It was first found in Europe in 1971 among green algae on river walls in the tidal Thames at Custom House. In 2012, I recorded it growing abundantly as a golden-brown zone at low levels on the river wall at Westminster. *Hydrosera triquetra* is of tropical origin and possibly arrived on logs imported from Venezuela. Its occurrence in the central reaches of the tidal Thames coincides with water temperature elevated by the city environment. The status of some of the five microscopic species as recent introductions is questionable; they might be natives that were overlooked in the past.

The planktonic alga *Heterosigma akashiwo* (Ochrophyta: Raphidophyceae), poisonous to fish, has been recorded recently in brackish waters in Ireland and the Netherlands.

The Dinophyta (dinoflagellates) are now classified as Protozoa and thus strictly speaking are not algae, despite containing chloroplasts. *Karenia mikimotoi* was described from Japan, and first found on the eastern coast of North America in 1957 and in Norway in 1960; it now occurs widely in northern European waters, probably spread in ballast water. This species forms red tides that have caused major kills of marine organisms. A second species, *K. aureola*, has been known from North Wales and Ireland since 1976.

Rhodophyta

More species of red algae than of the other groups of algae combined are aliens in the flora of the British Isles. The first recognised was a Japanese species, Bonnemaison's Hookweed (*Bonnemaisonia hamifera*), found in 189 on the Isle of Wight. It is a tufted plant, with branches oppositely and spirally arranged, some of which are hook-shaped. Two congeners native to the North Atlantic,

B. asparagoides and *B. clavata*, lack the hook-shaped branches. Bonnemaison's Hookweed has a heteromorphic life cycle; the tufted thallus is the gametophyte and alternates with a filamentous stage that bears tetraspores. This sporangial stage was considered to be a species in a distinct genus, *Trailliella intricata*, until studies in culture revealed it to be a stage of Bonnemaison's Hookweed and therefore a synonym of *B. hamifera*. The gametophyte is a perennial plant that grows in low-shore pools and subtidal levels on rocks and stones or entwining other algae, while the tetrasporangial stage is an epiphyte at low-shore and subtidal levels. It was the tetrasporangial stage that was discovered in 1890 on the Isle of Wight; the gametophyte was found later, in 1893, at Falmouth, Cornwall. Bonnemaison's Hookweed has subsequently spread throughout the British Isles, and from Norway to the Canary Islands. A feature of the life-history stages is that they have different temperature tolerances, with the tetrasporangial stage more acclimatised to colder waters and occurring further north in our region (to Shetland and the Faroes) than the gametophyte, which has not been found north of Sutherland.

The related Harpoon Weed (*Asparagopsis armata*) is a tufted, much-branched plant but with distinct harpoon-like branches. It, too, has a heteromorphic life cycle, with a filamentous tetrasporangial stage (formerly known as *Falkenbergia rufolanosa*) that forms small pompoms. The gametophyte grows on rocks and stones at subtidal levels, while the tetrasporophyte is an epiphyte on larger algae at low-shore and subtidal levels. Harpoon Weed probably originated in Australia and was first detected in Europe at Cherbourg in France in 1925, and subsequently at Galway in Ireland in 1939. It now occurs on the southern and western coasts of the British Isles and from France to the Canary Islands. The European native *A. taxiformis* is a warm-temperate species that occurs to the south of the Bay of Biscay but lacks the harpoon branches. As with Bonnemaison's Hookweed, the Harpoon Weed tetrasporophyte occurs further north (Shetland) than the gametophyte (Donegal).

The following two aliens exemplify taxonomic and nomenclatural confusion not unusual in unravelling species relationships. The genus *Antithamnionella* is represented in the British Isles by three species, two of which (*A. spirographidis* and *A. ternifolia*) are aliens. Both are delicate, feathery plants, monosiphonous (one cell wide) and ecorticate (without a cortex) throughout, rarely more than 40 mm tall, and distinguished by microscopic characters such as the shape of the tips of apical cells. *Antithamnionella spirographidis* was first confirmed in Britain in 1934, but the pattern of its subsequent spread has been obscured by its misidentification as *A. sarniensis*, which is now considered synonymous with *A. ternifolia*. *Antithamnionella spirographidis* occurs mainly on the southern and western coasts of Britain and Ireland, but I have recently found it in Kent. Its origin is uncertain but may be the North Pacific Ocean. *Antithamnionella ternifolia* was first discovered in European waters at Plymouth in 1906. It has been found subsequently as far north as Scotland, in Ireland and on the Isle of Man, and on the Continent it occurs from the Netherlands to Portugal; it probably originated in the Southern Ocean. The species was found on the sunken HMS *Glatton* in Dover Harbour in 1926 along with *Bryopsis muscosa* (*see* above) but was at first identified as *A. sarniensis*. *Antithamnionella ternifolia* grows on a variety of substrata, in pools and on floating pontoons in harbours at the subtidal fringe level and below. The related *A. densum*, first found in northern France in 1968 and subsequently in County Mayo, Ireland, in 1990, probably came from the South Atlantic Ocean.

Polysiphonia and related genera are characterised by being filiform or filamentous and polysiphonous (the filaments having a central cell surrounded by a ring of periaxial cells). An alien species with four periaxial cells is *Neosiphonia* (formerly *Polysiphonia*) *harveyi*. It grows as an epiphyte on larger algae in rock pools and also on man-made substrata, and was first collected at Weymouth in 1908. It has been recorded only sporadically in Britain and Ireland, having been confused with related species of *Polysiphonia*. *Neosiphonia harveyi* is thought to have originated on the east coast of the United States. *Polysiphonia subtilissima*, recently recorded in Ireland, is believed to have been accidentally introduced through aquaculture and probably had a tropical origin.

Another filamentous species that has arrived recently is *Heterosiphonia japonica*, which was first recorded in Europe in 1984 in Brittany. It spread rapidly in northern Spanish waters and was found later in the Netherlands, in areas where there are oyster farms. Subsequently, it has been found as far north as Norway and as far south as the Mediterranean coast of France, as well as in Wales (the first British record, in 1998–9), Scotland, England and Ireland. *Heterosiphonia japonica* forms soft, bushy, dark red tufts and can grow to 300 mm long. The main axes are polysiphonous with four periaxial cells (distinguishing it from the native *H. plumosa*, which has 9–12 periaxial cells) and bear monosiphonous lateral branchlets, the tips of which are dichotomously branched. The species rarely reproduces sexually or asexually in northern waters and its rapid spread has probably been by vegetative means through shedding its lateral branchlets. *Heterosiphonia japonica* grows on rocks at subtidal levels to 40 m depth.

Anotrichium furcellatum is a delicate, soft rose-pink monosiphonous filamentous alga that forms cushions to 70 mm tall. It is an epiphyte on other algae at subtidal fringe levels to 2 m depth and occurs in sheltered inlets.

Gametophytes are unknown in the British Isles, while the sporophyte bears single tetrasporangia on pedicels. *Anotrichium furcellatum* is now the sole representative of the genus in the British Isles, as the native *A. barbatum* (distinguished by its wider pyriform cells and tetrasporangia on whorls of elongate cells) has not been found since the nineteenth century. *Anotrichium furcellatum* was introduced to northern France from its Mediterranean coast in the early twentieth century, and now occurs in southwestern Britain and the Channel Islands, and in oyster farms in the Netherlands.

The more sturdy and fleshy *Laurencia brongniartii*, which resembles the native Royal Fern-weed (*Osmundea osmunda*), is included in a list of invasive species of Ireland recently recorded there. Species of these closely related genera are notoriously difficult to distinguish.

The red algal genus *Grateloupia* also contains more sturdy plants and is represented by four species in the British Isles, two of which are alien; the two natives are less than 40 mm tall, while the aliens are much larger. Grateloup's Fringe Weed (*G. subpectinata*, formerly *G. filicina* subsp. *luxurians*) is a cartilaginous and mucilaginous species that grows up to 700 mm long and has simple or branched, compressed fronds with proliferations along the margins. It is a perennial species, the fronds of which die back in winter and regenerate in spring; plants grow at low-intertidal and subtidal levels in lagoons, harbours and estuaries. Grateloup's Fringe Weed was first found on the Isle of Wight in 1947; it came from the Pacific Ocean, where it is used as a food and a source of carrageenan. It also occurs in the Channel Islands and in Brittany.

Devil's-tongue Weed (*Grateloupia turuturu*; Fig. 188) is one of the largest red algae in our flora. It has linear-lanceolate fronds to 1 m long and 200 mm wide, with occasional proliferations along the margins; plants are soapy, slippery and mucilaginous to the touch. It was first found in the Solent in 1969 and now occurs along the south coast of England as far east as Kent. It is a perennial species occurring commonly on floating pontoons in harbours and marinas, and also on the chalk shores of Sussex. Devil's-tongue Weed has also been recorded sporadically on the Continental coast and in the Mediterranean Sea. It came from the Pacific Ocean, where it is a commercial source of carrageenan.

Cryptonemia hibernica is another large plant. It grows to 600 mm long and has oval or linear-lanceolate, sometimes lobed, blades to 100 mm wide with pointed tips; the blades arise from a short stipe and discoid holdfast. The species was originally described from Ireland at two sites in County Cork (Kinsale and Cork Harbour) in 1974; in 2004, a population was found at Plymouth. *Cryptonemia hibernica* is closely related to northeastern Pacific species and may have come from that ocean. It grows on rock and stones at subtidal levels to 12 m depth.

There are several native and alien species of red algae that are cylindrical and stringy. Solier's Red String Weed (*Solieria chordalis*) is bright red with stringy fronds that arise from a branched holdfast. It is distinguished from species of similar form (e.g. *Cystoclonium purpureum, Gracilaria* spp., *Gracilariopsis longissima*) by its bright red colour. Solier's Red String Weed was first found in Falmouth Harbour and Weymouth Bay in 1976, and can occur abundantly at low-shore levels. As reproductive plants are unknown in Britain, the species has probably spread by vegetative means, but this has been relatively slow and limited to southwestern England and south Wales. Solier's Red String Weed arrived in Britain from the Atlantic coast of France, where it has been known since the 1960s. It was probably introduced to Europe through the cultivation of the Pacific Oyster.

Agardhiella subulata is a stringy alien alga with a very restricted distribution. It was first found in the British Isles in Langstone Harbour, Hampshire, in 1977 and soon after in nearby Chichester Harbour, Sussex. It is thought to have originated from the east coast of America. *Agardhiella subulata* has been confused with another alien species, *Sarcodiotheca gaudichaudii*, also a bushy, tufted, bright red plant and coincidentally known in California as Red String Seaweed; unlike *Agardhiella* it has a distinct main axis and bears longer branches towards the base and shorter branches distally. It differs from Solier's Red String Weed in having a discoid holdfast and a lumpy warty texture to the frond. *Sarcodiotheca gaudichaudii* is native to the Pacific coast of America.

FIG 188. Devil's Tongue Weed (*Grateloupia turuturu*) is one of the largest red algae in the British Isles, a Pacific species first found here in 1969 and now occurring sporadically on the south coast of Britain. (Ian Tittley)

The stringy *Gracilaria vermiculophylla* was first found in Sweden in 2003, and subsequently recorded in Denmark, Germany, the Netherlands, France, Spain, Portugal and, recently, Northern Ireland. It was probably secondarily introduced to northern Europe from southern Europe. *Gracilaria vermiculophylla* is native to Japan and widespread in the Pacific Ocean, but probably introduced to the American coast and the eastern and western Atlantic Ocean. It is of commercial importance for agar extraction. When present in abundance, *Gracilaria vermiculophylla* may have a damaging effect on the communities of sandy seabeds. It is very similar to our native *Gracilaria gracilis* and to *Gracilariopsis longissima*. *Gracilaria gracilis* has a more distinct red colour, and is smaller and more sparingly branched, with the male reproductive structures in conceptacles less than 50 μm deep (compared with 75 μm deep in *Gracilaria vermiculophylla*); male organs in *Gracilariopsis longissima* are formed on the surface of the thallus. The mode of spread of *Gracilaria vermiculophylla* to Europe is uncertain; oyster transplantation has been suggested, as has ballast water and hull fouling.

Another alien species that has not spread widely is *Pikea californica*, which occurs in Europe only on the Isles of Scilly. It is a tufted or small bushy plant that grows to 150 mm long, and has a flattened main axis that bears irregularly arranged pinnate branching and irregularly arranged spine-like branchlets near the tips. Herbarium specimens reveal the species to have been present since 1967. It occurs throughout the archipelago, growing on wave-washed rocks at the subtidal fringe level to 14 m depth, and may locally be the dominant alga. It is suggested that *P. californica* could have been brought to Britain from California by Catalina flying boats during the Second World War, as seaplane hulls were occasionally fouled by algae. The narrow sea temperature range of moderate summer and high winter temperatures in the Isles of Scilly may have favoured the establishment of a population there and not in other parts of the British Isles, where sea temperatures fluctuate more.

A small wiry, turfy or mossy pinkish-red plant with spiny branches was discovered on the southern and western coasts of France in 1986, in Britain in 2004, and the Netherlands. It was first found near oyster farms. Initial investigation suggested it to be *Caulacanthus ustulatus*, a warm-temperate species with a northern distributional limit in the Bay of Biscay. Molecular studies on plants from Brittany, however, revealed it to be another species from the Pacific Ocean, *C. okamurae* (Fig. 189), and that all *Caulacanthus* in northern European waters is this species. *Caulacanthus okamurae* superficially resembles forms of the native genus *Gelidium*, but is cylindrical while *Gelidium* is flattened. Under the microscope, a transverse section of the thallus reveals a distinct central filament, which is absent in *Gelidium*, and zonately divided tetrasporangia (cleavages parallel

FIG 189. *Caulacanthus okamurae*, a red alga of Pacific origin first found in Devon in 1994, is now common along the south coast of England. (Ian Tittley)

to one another and at right angles to the long axis of the sporangium) compared with cruciate division (first and second cleavages perpendicular to each other) in *Gelidium. Caulacanthus okamurae* has spread rapidly and grows among a turf community characterised by Pepper Dulse (*Osmundea pinnatifida*) and *Gelidium* spp. on moderately wave-washed intertidal shores; on the chalk shores of Sussex and Kent, it is locally dominant. Temperature tolerance experiments in culture suggested that *C. okamurae* was unlikely to spread further north as it died below 6° C, but in Kent it has survived winter periods with lower sea temperatures.

Finally, among the commercially important membranous laverbread or nori seaweeds are three alien species: *Pyropia* (formerly *Porphyra*) *drachii*, *P. leucosticta* and *P. yezoensis*. While microscopic characters are more reliable in distinguishing the species, they do show differences in colour, *P. drachii* being purple-red, *P. leucosticta* brownish mauve and *P. yezoensis* pink. Their origins and mode of spread are unknown. *Pyropia drachii* and *P. yezoensis* are recent arrivals in the British Isles, first recorded in the 1990s, while herbarium specimens reveal *P. leucosticta* to have been present as long ago as 1813. Molecular studies show *Pyropia* to be a North Pacific genus, with *P. leucosticta* closely related to the Pacific *P. fucicola*.

Conclusions

Alien seaweeds form only a small proportion (less than 5 per cent) of our total flora. Some have spread widely and rapidly, are aggressive invasive colonisers and may locally be a dominant species (e.g. Wireweed, *Caulacanthus okamurae*), changing the local community structure. Others (e.g. Oyster Thief) may occur widely, be more benign, and be of greater or lesser ecological impact. A few species have remained restricted in distribution, some locally impactful (e.g. *Pikea californica*), others not so (e.g. *Agardhiella subulata*). They are of little value and are ecologically disruptive.

Alien seaweeds have been have been arriving continually in the British Isles since the nineteenth century, with a wave of species in the second half of the twentieth century associated with oyster farming. Their arrival has been accompanied by marine fauna such as Slipper Limpet (*Crepidula fornicata*) in the late nineteenth century, the barnacle *Elminius modestus* in the mid-twentieth century and, more recently, the Pacific Oyster and Carpet Sea-squirt (*Didemnum vexillum*). Despite national official awareness and legislation aimed at stopping the introduction of species, and risk assessments outlining the economic and ecological negative impacts caused by the more aggressive weeds, international effort and cooperation is required to prevent the further spread of unwanted alien seaweeds. Until this is achieved, species will continue to turn up, and once present, there is little hope of controlling or managing their spread and impact.

MOSSES AND LIVERWORTS

Mosses and liverworts were traditionally placed in the phylum Bryophyta, but recent molecular (DNA) studies have shown that they represent separate evolutionary lines which should be recognised by placing them in different phyla, the Bryophyta in a stricter sense (mosses) and the Marchantiophyta (liverworts). A third phylum, the Anthocerophyta (hornworts), is represented in our flora by only four native species, and is therefore not relevant to this book. Nevertheless the informal term 'bryophytes' can still be used usefully to cover all three phyla.

Bryophytes have fascinating life cycles and an intricate structure, with many technical differences between the two main groups, but none of these aspects needs concern us here. During their life cycle they alternate between a sexual gamete-producing (gametophyte) stage, which is the main moss or liverwort plant, and a spore-producing (sporophyte) stage, which is the capsule usually produced upon the gametophyte. A true vascular (conducting) system and roots are absent from both stages, tying bryophytes to habitats that are damp for at least part of the year.

The bryophytes were the subject of Ron Porley and Nick Hodgetts' *Mosses and Liverworts* (2005) in the New Naturalist series, in which most of our alien species were well surveyed, so that a full treatment is unnecessary here. The authors listed 18 species of alien bryophyte (ten mosses and eight liverworts), and six or so others have a doubtful native/alien status. Since that time one liverwort

(*Lophocolea brookwoodiana*, first record 2004) and four mosses (*Leptodontium proliferum*, 2000; *Thamnobryum maderense*, 2004; *Bryum valparaisense*, 2006; *B. apiculatum*, 2007) have been added to our alien list. In addition to these, there are several others that have been found in greenhouses, and either have not yet escaped from them, or have found it too cold for survival outside. Almost all alien bryophytes owe their existence in the British Isles to accidental import as contaminants of soil, plant specimens or timber. The earliest discovered was the moss *Atrichum crispum* in South Lancashire in 1848. There are between 1,000 and 1,100 species of bryophyte in the British Isles, and the total of 23 aliens (2.1–2.3 per cent) is paltry compared with the case in the flowering plants, where there are more aliens than natives.

Very few of our alien bryophytes have had a significant impact on natural vegetation here, but there is one major exception and another possible example. It is remarkable that, with the exception of those two, almost all our alien bryophytes remain highly localised. *Orthodontium lineare* (Fig. 190) is a southern hemisphere species first recorded in Cheshire in 1910 and now found scattered over the British Isles, although it is still rare in the north and west. The plants are erect and reach only about 1 cm tall, but grow in tufts or patches in shady, moist places on acidic rocks, banks, peat, logs and tree trunks. On acid rocks the species has come into contact with our only other member of the genus, the very similar *O. gracile*, a native that is rare and decreasing, being extinct in most of its localities. There is circumstantial evidence that *O. gracile* is at least in part being ousted by *O. lineare*. In several localities, such as the sandstone rocks of the High Weald of East Sussex, the decrease of *O. gracile* has been accompanied by the spread of *O. lineare*. It is, however, possible that some other factors such as pollution or desiccation have caused the decline of the native, whose vacated niche is being exploited by the alien (Porley & Matcham, 2003).

FIG 190. *Orthodontium lineare*, a small tussock-forming moss first recorded in 1910, is now frequent on rocks, banks and tree trunks in most of the British Isles, where it appears to be ousting the increasingly rare native *O. gracile*. (Ron Porley)

The Heath Star Moss (*Campylopus introflexus*; Fig. 247), on the other hand, has had an undoubtedly deleterious impact on several habitats. This southern hemisphere species was first found in Europe in West Sussex in 1941, and is now locally common throughout the British Isles on acid soils, particularly on peat on heathland and moorland in the open or in woods, but also on burnt ground, coal waste, sandy or clay soils, fixed sand dunes, rotten wood, tree trunks and old roofs. It is a successful pioneer species and is resistant to atmospheric pollution. It is thought to have arrived here on unspecified ships' cargo. It is dioecious, but both sexes are present here and it reproduces both sexually and by readily detached brittle stem apices or leaves, by which it can spread rapidly and become more widely dispersed. It forms pure monospecific carpets which effectively smother other bryophytes and lichens, as well as the seedlings of flowering plants, especially heather. Human or other physical disturbance can enhance its spread by fragmenting the plants, and acid rain has been implicated in the decrease of some of the species that it has replaced. It is difficult to see how the harmful effects of this moss could be reversed, as eradication is impossible, and in any case it seems that little emphasis or resources are placed on invasive bryophytes by the conservation bodies.

FERNS, CLUBMOSSES AND HORSETAILS

Ferns, clubmosses and horsetails are three diverse plant groups that were once classified together in the Pteridophyta, but which molecular (DNA) studies have shown to be less closely related than was once thought. They are still referred to informally as pteridophytes or 'ferns and fern allies'. The three phyla all share the feature that the two generations (sporophyte and gametophyte), between which all plants alternate during their life cycles, are completely physically separate and independent of each other. The sporophytes (or spore-bearing generation) are the familiar fern, clubmoss or horsetail plants. These have well-developed vascular systems throughout their roots, stems and leaves, which enable the plants to live in relatively dry atmospheres. The gametophytes (or gamete-bearing generation, known as the prothallus) are very small, delicate structures, sometimes microscopic, which are the equivalent of the main moss or liverwort gamete-bearing plant (whereas the bryophyte capsule and its stalk are the equivalent of the main pteridophyte spore-bearing plant). Like bryophytes, pteridophyte prothalli lack roots or a vascular system, and are dependent upon the absorption of water over their whole body surface. Pteridophytes are, therefore, in general also mostly found in wet or humid places.

Vegetative spread in most pteridophytes is simply by rhizome growth in the sporophyte, propagation occurring by fragmentation, which rarely effects significant dispersal except in the horsetails (*Equisetum*), Bracken and Water Fern (*see* below). Sexual reproduction, however, occurs in the gametophyte generation, eventually resulting in the production of vast numbers of spores by the sporophyte generation; these spores are easily carried great distances. The gametophyte generations (prothalli) of most pteridophytes are hermaphrodite, but some (only *Azolla* and *Selaginella* among our alien genera) are dioecious, the spores being of two sorts (a condition known as heterospory) and giving rise to the male and female prothalli, respectively.

A list of known alien pteridophytes that are established in the wild is given in Table 14, comprising 11 ferns, Branched Horsetail (*Equisetum ramosissimum*) and Krauss's Clubmoss (*Selaginella kraussiana*). Contrasting with our rich bryophyte heritage, our pteridophyte flora is poor, with only about 74 natives, although as many as 43 hybrids have been identified. The proportion of aliens (13 out of 87 = 15 per cent), is between that of the bryophytes and flowering plants.

All but one of the 13 alien pteridophytes recorded in the British Isles were originally imported as horticultural subjects and have subsequently escaped into the wild. The exception is Branched Horsetail, a Mediterranean weed that grows in long grass by a river in South Lincolnshire, where it was discovered in 1947. It is presumed that it was introduced in soil or ballast, in which its invasive rhizomes would easily survive and later spread. Its other two localities are in rough grassland in western Britain and might well have had a similar origin, but there are some who claim that this species is actually a native.

Branched Horsetail, although only a rare introduction and absent from Ireland, has for long been of particular interest to British and Irish botanists because it is a parent of a hybrid with the native Rough Horsetail (*Equisetum hyemale*), known as Moore's Horsetail (*E.* × *moorei*). The hybrid, which is sterile, occurs on dunes and banks by the sea in Ireland (County Wexford and County Wicklow), where it is an apparently native member of the flora, discovered there in 1851. It seems that it must have arrived in Ireland by independent long-range dispersal of rhizome fragments, because it also occurs in Scandinavia, where Branched Horsetail does not occur, and around the Caspian Sea, where Rough Horsetail does not reach. The alternative hypothesis, that the hybrid arose independently in Ireland, Scandinavia and the Caspian, and that subsequently one parent became extinct in each place, seems less likely.

Recently, however, the mystery has deepened. A puzzling horsetail found in Anglesey in 2000 was identified in 2006 as the hybrid *Equisetum* × *meridionale* by Marcus Lubienski in Germany. This is another hybrid of Branched Horsetail,

TABLE 14. List of alien pteridophytes in the wild in the British Isles, including total number of hectad records from January 1987 to May 2012. Taxa native anywhere in the British Isles are excluded.

Species	First record		Native area	Number of hectads
	Date	Vice-county		
Ostrich Fern *Matteuccia struthiopteris*	1834	Fife	Europe	77
Sensitive Fern *Onoclea sensibilis*	1840	South Lancashire	North America, East Asia	42
Water Fern *Azolla filiculoides*	1883	Middlesex	Tropical America	692
Krauss's Clubmoss *Selaginella kraussiana*	1908	West Cornwall	Africa	116
Little Hard-fern *Blechnum penna-marina*	1926	West Donegal	Southern hemisphere	3
Branched Horsetail *Equisetum ramosissimum*	1947	South Lincolnshire	Europe	3
Ribbon Fern *Pteris cretica*	1950	Berkshire	Southern Europe	27
Greater Hard-fern *Blechnum cordatum*	1952	Scilly	Southern South America	20
House Holly-fern *Cyrtomium falcatum*	1956	Scilly	East Asia	40
Australian Tree-fern *Dicksonia antarctica*	1960	South Kerry	Australia	10
Kangaroo Fern *Phymatosorus diversifolius*	1969	Guernsey	Australasia	3
Western Sword-fern *Polystichum munitum*	1980	Surrey	Western North America	6
Fortune's Holly-fern *Cyrtomium fortunei*	2004	West Cornwall	East Asia	7

this time with the native Variegated Horsetail (*E. variegatum*). The absence of Branched Horsetail from anywhere near Anglesey again suggests independent long-range dispersal of the hybrid, but that two different hybrids should arrive here in that way is indeed remarkable. Some botanists prefer to explain this puzzle by suggesting that Branched Horsetail is a native here, and that current sites of its hybrids are relics of a wider distribution. Discovery of historical specimens collected from Hampstead Heath, London, around 1705, and from the Liverpool area in the nineteenth century, add to the likelihood of this possibility (Rumsey & Spencer, 2012). A thriving colony of a hybrid horsetail on the Cheshire coast, previously determined as Mackay's Horsetail (*E.* × *trachyodon* = *E. hyemale* × *E. variegatum*), has also recently been redetermined as *E.* × *meridionale*.

Of the remaining 12 alien pteridophytes, only Water Fern (*Azolla filiculoides*) can be said to have any significant ecological impact on our vegetation. Christopher Page has written two excellent modern books on our ferns – one an earlier volume, *Ferns* (1988) in the New Naturalist series, and the other *The Ferns of Britain and Ireland* (Page, 1982, 1997), the standard reference work on the subject – but somewhat unexpectedly neither covers the alien species, except for a very short section on the ecology of *Azolla* in the second work. Water Fern, from tropical America, is a very distinctive plant that floats on the water surface and would never be recognised as a fern by the uninitiated or casual observer. More likely they would pass it over as an alga, bryophyte or duckweed (*Lemna*). It occurs on ponds, dykes, canals and still parts of lake margins, and can spread rapidly by vegetative means so that it forms a complete carpet entirely covering the water surface, often to varying degrees mixed with duckweed. The stems are fragile, and pieces are readily detached and transported elsewhere by humans and perhaps birds. Water Fern occurs over most of the British Isles north to central Scotland, but is much commoner in the south. Indeed, it is not fully frost-hardy, being largely killed off in hard winters, and it is persistent over many years only in sheltered southern localities. Its lush green summer colour often turns to an autumnal red, when it attracts attention and is most likely to be recorded. It was probably introduced by aquarists, for both indoor and outdoor cultivation.

Water Fern consists of branched floating stems up to about 5 cm long that are densely covered by small leaves up to only about 2.5 mm long in surface view (Fig. 272). The leaves have a very intricate three-dimensional structure, their most interesting feature being that one part has a small cavity occupied by a filamentous blue-green photosynthetic bacterium, *Anabaena azollae*. The *Anabaena* filaments 'fix' atmospheric nitrogen (forming inorganic nitrogen compounds), and constitute a symbiotic association with the *Azolla*, analogous to that between legumes and the bacterium *Rhizobium*, or between alders (*Alnus*) or bog-myrtles (*Myrica*) and the organism *Frankia* (belonging to an obscure group known as the actinomycetes). The *Anabaena–Azolla* partnership (and the others just mentioned) is termed a symbiosis, whereby each of the two partners benefits from the activities of the other. Atmospheric nitrogen cannot be utilised by plants, but when it is fixed (initially as ammonia, NH3) it is a valuable source of that essential element, and can be used by both the *Anabaena* and the *Azolla*. Much work has been carried out on this symbiotic relationship, and mutual benefits to both partners have been identified. Since no vascular plants or bryophytes can fix nitrogen, the value of this process to the fern is obvious, as is the reliance of the bacterium on the fern when we learn that it is very difficult or impossible to grow *Anabaena* in the absence of *Azolla*. In one sense the intimacy of the association is greater than that between legumes or alders and their partners, because filaments of *Anabaena* are retained throughout all the reproductive stages of the life cycle of *Azolla*, so that the association does not have to be reconstituted in each generation as it does in legumes or alders. However, it is not known how frequently, if at all, sexual reproduction in Water Fern takes place in the British Isles.

The rapid growth rate, the ability to fix nitrogen and the high nutrient status of the symbiosis are important attributes of *Azolla* ecology. Water Fern provides a nutrient-rich food for wild birds, mammals and fish, and a habitat in which they can shelter. Experiments have been undertaken to assess its potential as a farm-animal foodstuff. In warm countries it has been found that its dense carpets deny a breeding ground for disease-carrying mosquitoes. It is grown in rice paddy-fields both for this purpose and to provide an extra source of nitrogen, in the same way that clover and other legumes are used in the British Isles. However, its frost-sensitive character suggests that it will never become important in cool-temperate climates. *Azolla* is a close relative of the tropical Giant Water-fern or Kariba Weed (*Salvinia molesta*, formerly known as *S. auriculata*), whose prodigious growth rate has caused it to become a major choking pest in some tropical waters, notably Lake Kariba in Zimbabwe. One year after the completion of the Kariba Dam in 1959, the floating fern had covered 200 km² of the lake.

The remaining 11 alien pteridophytes fall into two loose categories, both introduced as garden plants. The first category comprises four hardy ferns that are cultivated in marshy ground, streamsides or humid rock gardens. These are Ostrich Fern (*Matteuccia struthiopteris*; Fig. 191), Sensitive Fern (*Onoclea sensibilis*) and two species of hard-fern, Greater Hard-fern (*Blechnum cordatum*) and Little Hard-fern (*B. penna-marina*). None has the ability to spread more than a few metres vegetatively, and none has much impact on our wild vegetation. Despite this, the ferns sometimes occur in quite wild situations, which they must have reached by dispersed spores if they were not deliberately planted there. Examples are Little Hard-fern found on moorland on the Stiperstones in Shropshire, and Greater Hard-fern by a wild stream on the Isles of Scilly (Fig. 192). By chance, all four share a structural feature that is unusual in ferns: the leaves (or fronds, as they are called in ferns for some unknown reason) are of two sorts – the normal vegetative ones and

FIG 191. Ostrich Fern (*Matteuccia struthiopteris*), which is naturalised in marshy ground, is also known as Shuttlecock Fern because of its stiffly erect clusters of leaves. (MJC)

those that bear the spore-producing bodies or sporangia. In most ferns the sporangia are developed on the normal vegetative fronds, giving them the rust-coloured undersides in late summer. The morphology of the 'fertile' fronds varies in the four species, but in all they have narrower divisions than on the 'sterile' fronds, with reduced or no green tissue and photosynthetic activity. When the sporangia are mature and the spores are being dispersed, the latter turn the 'fertile' fronds rust-coloured.

The second category consists of seven half-hardy species that were at first grown in glasshouses and conservatories, or in sheltered walled gardens in southwestern Britain and Ireland and in the Channel Islands. Without good protection, natural or artificial, they fail to survive severe winters. Two of them perhaps deserve further mention.

FIG 192. Greater Hard-fern (*Blechnum cordatum*), from southern South America, is well naturalised in a few places in the west, here on the Isles of Scilly. (CAS)

FIG 193. Krauss's Clubmoss (*Selaginella kraussiana*) is a delicate fern ally growing in damp, shady places mainly in the southwest, where it can form mats exceeding 1 m across. (CAS)

Krauss's Clubmoss (Fig. 193) is an African species of the large genus *Selaginella*, of which we have one native species, Lesser Clubmoss (*S. selaginoides*). Whereas our native is an inconspicuous plant with little-branched upright stems reaching 15 cm tall, growing in open, boggy, upland areas, the alien is a much more luxuriant, sprawling plant that is commonly grown in the southwest in shady, humid spots in gardens, often in shrubberies. The stems creep and scramble along the ground or over short vegetation, forming patches up to a metre across. The sporangia are formed in small lateral cones borne along the length of the stems; within each cone there are a few female sporangia at the base, with many male ones above. The two species are classified in very different parts of the genus.

Our most impressive fern, native or alien, is the Australian Tree-fern (*Dicksonia antarctica*; Fig. 194). Tree-ferns consist of a dense tuft of fronds, as in many ferns such as the familiar native Male-fern (*Dryopteris filix-mas*), but the rhizome apex from which they arise is carried up by a 'trunk', which is actually a vertical rhizome. In the Australian Tree-fern the trunk may be up to 15 m tall in its native habitats, although 1–2 m is nearer the norm in the British Isles, and each frond can attain 2 m long. The trunk or vertical rhizome is not woody, but consists of a relatively thin stem surrounded by the packed bases of fallen leaves; both these and the stem are densely hairy, the whole trunk providing a suitable shaggy substratum for the growth of bryophytes and smaller ferns, as well as some delicate flowering plants. Such splendid subjects proved irresistible to Victorian gardeners in the southwest, where many spectacular plantings took place. Perhaps the best are in County Kerry, especially on Valentia Island. Today, with so many people cultivating plants in conservatories, tree-ferns have entered

FIG 194. Australian Tree-fern (*Dicksonia antarctica*) is a most impressive fern cultivated in some of the larger, very mild gardens of southwest England and Ireland, from which it is now beginning to escape; here shown on Valentia Island in County Kerry. (David Holyoak)

a new era of popularity, and are sold in many garden centres. In the southern hemisphere there are nearly 700 species of tree-ferns, and some of these are now also being imported for gardens.

As with the alien bryophytes that started here in glasshouses and later escaped into sustainable habitats outside, so is the case with some of our alien ferns. Australian Tree-fern now propagates itself by spores in humid woodland outside gardens in Cornwall and County Kerry, and some of the other cultivated tree-ferns will surely follow, e.g. *Dicksonia fibrosa* and Scaly Tree-fern (*Cyathea dealbata*). Other ferns that seem to be on the brink of making it into the wild are the tropical Delta Maidenhair-fern (*Adiantum raddianum*), Hare's-foot Fern (*Davallia canariensis*) and some species of ribbon fern (*Pteris*). One habitat that nurtures these marginally naturalised ferns is the sunken forecourt of city basement flats, often separated from the footpath above by a grille. Some botanists may be found peering down beyond these grilles, risking the deep suspicion of residents and passers-by alike. Sheltered garden walls and brick-built railway arches offer similar niches. Few of these frost-sensitive ferns are found far from human influence, so the discovery of a thriving clump of House Holly-fern (*Cyrtomium falcatum*) sheltering between spray-exposed rocks on the upper seashore on St Mary's, Isles of Scilly, was a real surprise.

CONIFERS

From pine forests to Christmas trees to Leylandii hedges, conifers are familiar everyday objects that often form topical news items. But many people are surprised when they find that the vast majority of the 270 or so species grown in the British Isles are aliens, introduced here from about 1500 onwards for forestry or horticultural purposes. In fact, we possess only three native species of conifer: Scots Pine, Juniper (*Juniperus communis*) and Yew (*Taxus baccata*). All of these are very widespread, but in the case of Scots Pine there is uncertainty and controversy about its native range here. Traditionally, Scots Pine is considered to be a native only in the ancient Caledonian forest of the Scottish Highlands, and to be an introduced tree elsewhere in Scotland and in the whole of England, Wales and Ireland. Undisputed facts are that Scots Pine became common all over the British Isles soon after the last glaciation, reaching its maximum about 6,000–8,000 years ago, but was later largely replaced by broadleaved trees as the climate ameliorated, and then suffered further by man's deforestation of the land, especially in the lowlands. By 2,000 years ago, very few if any Scots Pine trees existed except in Scotland. There has been very extensive reintroduction of the species into Britain and Ireland in the past 400 or 500 years, and we shall probably never know whether or not the Caledonian forest holds the only native stock. Either way, Scots Pine is clearly not part of the subject matter of the present book.

The list of 49 alien conifers that occur in the wild in the British Isles is given in Table 15. The criteria for inclusion of tree species differ somewhat from those applied to herbaceous plants. The list includes those species that have been found to regenerate in the wild, those that are grown in plantations, and those that are frequently planted in parks, along roads and the like, often in groups or rows. The justification for including the latter two (not regenerating) groups is that trees in quantity, especially outside the garden environment, can have profound effects on wildlife; they provide new habitats and they greatly modify the local vegetation, in the case of conifers most often by suppressing it. I am constantly reminded of the remarkable effect on wildlife that even a tiny planting can have by a small area of my garden that has three conifers growing close together: an Austrian Pine (*Pinus nigra* subsp. *nigra*), a Chinese Juniper (*Juniperus chinensis*) and a Western Red-cedar (*Thuja plicata*). Coal Tits (*Periparus ater*) and Goldcrests (*Regulus regulus*) nest in these each year and are seen around them with their young families from late summer to winter; Juniper Carpet moths (*Thera juniperata*) breed on the juniper and Grey Pine Carpets (*Thera obeliscata*) on the pine. Such experiences fully justify the approach here advocated.

In fact, the great majority of species that are listed in Table 15 have been found to regenerate naturally in some places, but the frequency with which self-sowing occurs (Fig. 195) and the extent to which seedlings are successful in attaining maturity vary greatly from species to species. The most successful species are probably European Silver-fir (*Abies alba*), Western Hemlock-spruce (*Tsuga heterophylla*), Sitka and Norway spruces (*Picea sitchensis* and *Picea abies*), European and Japanese larches (*Larix decidua* and *L. kaempferi*), Lodgepole Pine (*Pinus contorta*), Austrian Pine (*Pinus nigra* subsp. *nigra*) and Corsican Pine (*Pinus nigra* subsp. *laricio*), Lawson's Cypress (*Chamaecyparis lawsoniana*) and Western Red-cedar (*Thuja plicata*). It is a moot point as to how far conifer seed can travel before it germinates. Most of the reports of regeneration are very close to the parent tree, often at the edge of, or in clearings of, plantations. Most species have well-winged seeds that have the potential to travel far in wind currents, but they seldom seem to. According to Davies (1961), in the National Pinetum at Bedgebury, Kent, the greatest distance that Western Hemlock-spruce seedlings had been found from

FIG 195. Natural regeneration of conifers: (a) two seedlings of European Larch (*Larix decidua*) and one of Scots Pine (*Pinus sylvestris*) springing up in small clearings in a larch plantation; (b) saplings of larch and spruce developing on moorland by Loch Maree, Wester Ross. (a, CAS; b, MJC)

TABLE 15. List of alien conifers in the wild in the British Isles. Total number of hectad records from January 1987 to May 2012. Taxa native anywhere in British Isles excluded. (Note that *Pinus nigra* subspecies, indicated below with an asterisk, have often not been separated by recorders.)

Species	Date introduced	Native area	Number of hectads
Norway Spruce *Picea abies*	c. 1500	Europe	1,790
Eastern White-cedar *Thuja occidentalis*	1536	Eastern North America	25
Maritime Pine *Pinus pinaster*	pre-1596	Southern and southwestern Europe	134
European Silver-fir *Abies alba*	1603	Europe	358
Weymouth Pine *Pinus strobus*	1605	Central and eastern North America	71
European Larch *Larix decidua*	1629	Europe	2,111
Cedar of Lebanon *Cedrus libani*	1638	Middle East	252
Swamp Cypress *Taxodium distichum*	1640	Southeastern North America	65
White Spruce *Picea glauca*	c. 1700	Northern North America	19
Eastern Hemlock-spruce *Tsuga canadensis*	1736	Eastern North America	14
Oriental Thuja *Platycladus orientalis*	1752	China and Korea	25
Maidenhair-tree *Ginkgo biloba*	1758	China	37
Dwarf Mountain-pine *Pinus mugo*	1774	Europe	25
Monkey-puzzle *Araucaria araucana*	1795	Western South America	292
Chinese Juniper *Juniperus chinensis*	1804	China, Mongolia and Japan	18
Corsican Pine *Pinus nigra* subsp. *laricio*	1814	Southern Europe	361*
Bhutan Pine *Pinus wallichiana*	1823	Himalayas	48
Greek Fir *Abies cephalonica*	1824	Greece	8
Douglas Fir *Pseudotsuga menziesii*	1826	Western North America	1,156
Western Yellow-pine *Pinus ponderosa*	1828	Western North America	18
Noble Fir *Abies procera*	1830	Western North America	326
Giant Fir *Abies grandis*	1831	Western North America	467
Deodar *Cedrus deodara*	1831	Western Himalayas	201
Sitka Spruce *Picea sitchensis*	1832	Western North America	1,738
Monterey Pine *Pinus radiata*	1832	California	276
Austrian Pine *Pinus nigra* subsp. *nigra*	1835	Europe	214*
Monterey Cypress *Cupressus macrocarpa*	1838	California	321
Atlas Cedar *Cedrus atlantica*	c. 1840	Morocco	237

Species	Date introduced	Native area	Number of hectads
Japanese Red-cedar *Cryptomeria japonica*	1842	China and Japan	155
Sawara Cypress *Chamaecyparis pisifera*	1843	Japan	88
Coastal Redwood *Sequoia sempervirens*	1844	California	212
Caucasian Fir *Abies nordmanniana*	1848	Caucasus	55
Lodgepole Pine *Pinus contorta*	1851	Western North America	868
Western Hemlock-spruce *Tsuga heterophylla*	1852	Western North America	856
Western Red-cedar *Thuja plicata*	1853	Western North America	773
Wellingtonia *Sequoiadendron giganteum*	1853	California	461
Incense Cedar *Calocedrus decurrens*	1853	Southwestern North America	12
Nootka Cypress *Xanthocyparis nootkatensis*	1853	Western North America	47
Lawson's Cypress *Chamaecyparis lawsoniana*	1854	Western North America	1,218
Hiba *Thujopsis dolabrata*	1859	Japan	24
Japanese Larch *Larix kaempferi*	1861	Japan	1,030
Engelmann Spruce *Picea engelmannii*	1862	Western North America	7
Macedonian Pine *Pinus peuce*	1864	Balkans	17
Colorado Spruce *Picea pungens*	1865	Western North America	11
Fraser Fir *Abies fraseri*	1871	Eastern North America	1
Leyland Cypress × *Cuprocyparis leylandii*	1888	Cultivated origin	404
Serbian Spruce *Picea omorika*	1889	Western Balkans	59
Hybrid Larch *Larix* × *marschlinsii*	1904	Cultivated origin	1,225
Dawn Redwood *Metasequoia glyptostroboides*	1948	China	59

* Subspecies often not separated by recorders; total hectads for species 1,078

the mother tree was 275 m; Lawson's Cypress self-sows there freely, but usually within 25 m of its parent. Nevertheless, in 1960 two obviously self-sown saplings of Western Hemlock-spruce were found 3 km from Bedgebury Forest in dense semi-natural Pedunculate Oak woodland on acid sandy soil. Ison & Braithwaite (2009) reported uprooting self-sown Sitka Spruce from the margin of a raised bog in Berwickshire 2 km from the nearest plantation. Chance long-range dispersal must always be a possibility.

Outside plantations and their immediate borders, conifers rarely regenerate in any great quantity. Possibly the only exceptions are Sitka Spruce, which is colonising open moorland and bog, as well as tracksides and rocky ground, and Lodgepole Pine, which is invasive on peat bogs and wet heaths in northern Scotland. These are two of our very few alien species to regenerate in upland areas; New Zealand Willowherb and a few *Cotoneaster* species (e.g. Himalayan Cotoneaster *C. simonsii*, Entire-leaved Cotoneaster *C. integrifolius*) are other notable examples.

Some genera very seldom self-sow in this country. Cedars (*Cedrus*; Fig. 196), Monkey-puzzle (*Araucaria araucana*), Coastal Redwood (*Sequoia sempervirens*) and Wellingtonia (*Sequoiadendron giganteum*) are notable examples. Monkey-puzzle is in a unique category, because it is dioecious, so trees of both sexes need to coexist, and the seeds are wingless and nut-like, dispersed by animals, probably mainly squirrels in Europe, but, Arthur Chater tells me, also by Rooks (*Corvus frugilegus*) in Cardiganshire (pers. comm., 2012). Despite these problems, regeneration has been recorded in seven southern English and Welsh vice-counties.

Cypresses and larches are among the most frequent self-sowers, and somewhat unusually among the conifers they relatively readily hybridise to produce the only two notable conifer hybrids. Both these hybrids first arose naturally, and were noticed among seedlings derived from planted seed.

The notorious Leyland Cypress, × *Cuprocyparis leylandii* (often known colloquially as Leylandii) is the hybrid (Fig. 197) between Monterey Cypress (*Cupressus macrocarpa*) and Nootka Cypress (*Xanthocyparis nootkatensis*, previously placed in *Cupressus*, later in *Chamaecyparis*, and, at the time of writing, being

FIG 196. Branches of Cedar of Lebanon (*Cedrus libani*) with abundant female cones. (CAS)

FIG 197. Twigs with female cones of Leyland Cypress (× *Cuprocyparis leylandii*) and its two parents: (a) Monterey Cypress (*Cupressus macrocarpa*); (b) Leyland Cypress; (c) Nootka Cypress (*Xanthocyparis nootkatensis*). (CAS)

considered for reassignment to *Cupressus*). For a long time the 'accepted history' was that the hybrid was first detected as a seedling in 1892 at Haggerston Castle, Northumberland, derived from seeds produced in 1888 on a Nootka Cypress at Leighton Hall, Montgomeryshire. But it has more recently come to light that the hybrid arose earlier at Rostrevor, County Down, in around 1870, plants from which are still cultivated today. New cultivars have mostly arisen via new hybridisations and from sports of earlier ones. By 2002, at least 15 cultivars were known, and by 2010 there were more than 30. They all exhibit impressive hybrid vigour, often with a growth rate of more than 1 m per year. The biggest example known is now over 40 m tall. Some hybrids have arisen from seed on Nootka Cypress trees, as did the Leighton Hall hybrid, but others have Monterey Cypress as their female parent.

Evidence available on the fertility of Leyland Cypress is fragmentary. Several sources state that it is sterile, but others suggest that it is at least partially fertile and that new cultivars have arisen from F2 or backcross seedlings. There are a few reports of seedlings being found in wild places, but how can one tell whether these are the progeny of hybrids rather than newly arisen hybrids? And even if they were Leyland Cypress progeny, what was their male parent? When a highly sterile hybrid sets some good seed, it is less likely that the pollen came from the hybrid (because of its sterility) than from one of its parents, or even from another related species. There is certainly a notable variation between Leyland Cypress cultivars in the production of cones: some cultivars rarely bear them, while others (e.g. 'Leighton Green', 'Naylor's Blue' and 'Stapehill') often produce them in abundance. These cones contain few 'full' seeds, and very few people have claimed to have germinated them; most of the credible claims come from New Zealand.

The prodigious growth rate and extreme density of the crown of Leyland Cypress, coupled with its frequent cultivation in unsuitable places, have led to much argument between neighbours when light is excluded and views obscured. It is said that 17,000 neighbours are currently in dispute over the tree in Britain alone! Many of these cases result from rows of Leyland Cypress planted as hedging quickly becoming out of hand due to a lack of hedge-trimming. Once a Leylandii tree is overgrown, the cause is lost because, like most conifers, it will not shoot from old wood, so severe cutting back leads to ugly bare wood rather than green foliage. There have been several well-publicised lawsuits in recent years that have cost home-owners fortunes and have lined the pockets of lawyers, and one even led to a murder in 2001. This has resulted in a change in the law, whereby, once private negotiation has failed (as it too often does), appeal to the local council can lead to an order under the Anti-social Behaviour Act 2003 that the owner cut down the hedge to 2 m tall; fines follow non-compliance.

There is, however, another side to the Leylandii story. Such a dense growth provides wonderful cover for wildlife, especially for birds during winter or in the nesting season. It, and other planted cypresses, are also the food-plant of a number of insects, including the moth Blair's Shoulder-knot (*Lithophane leautieri*), an alien from southern Europe, where its natural food-plants are cypresses and junipers. Another alien insect, the Cypress Aphid (*Cinara cupressi*), is said to be one of the world's 100 most invasive species. The aphid is also a native of southern Europe; it arrived here over 100 years ago, but the first attacks on Leyland Cypress were not recorded until the 1980s. It sucks the plant's sap and can cause large areas of Leylandii hedge to turn brown and even die. Its abundance and effects vary from year to year, apparently being most marked in warm summers following a mild winter.

The two parents of Leyland Cypress are both native to North America, but their natural ranges do not overlap and no wild hybrids form there. The same is true of the Dunkeld or Hybrid Larch (*Larix* × *marschlinsii*), for its two parents are the European Larch and Japanese Larch, which in this case are separated by many thousands of kilometres. The earliest hybrids arose as seed on Japanese Larches planted in 1885 at Dunkeld House, Perthshire. The first seed was planted in 1897, but the hybrid identity of some of the progeny was not noticed until 1904, by which time seed had been distributed to other estates in Scotland. From that time onwards, hybrid larches have arisen many times in plantations and parks where both species are grown together. Indeed, there seem to be no barriers to hybridisation in cultivation. Fertility of the hybrid is high and backcrossing occurs in forestry areas. In addition, the hybrid is synthesised on a regular commercial basis by reciprocal crosses. Dispersal by seed takes place locally and, although most seedlings are found in or by plantations, they also occur some distance away on moorland, heathland and rocky broken ground.

The intermediacy of Hybrid Larch is best shown by the shape of the female cones, but also by characters of the leaves. The hybrid combines the longer cones of European Larch with the spreading to recurved cone-scale tips of Japanese Larch. The backcrossing produces some trees closer in appearance to one parent or to the other, leading to difficulties in identification. Like Leyland Cypress, Hybrid Larch shows hybrid vigour, and in its early years will outstrip both its parents, with growth rates of more than 1 m per year; one 15-year-old tree was found to have attained 16.5 m in height. The largest trees today are about 40 m tall. For this reason Hybrid Larch has gained in popularity with foresters, and for several decades it has been planted for timber more than either parent. In view of the recent severe outbreaks of disease in Japanese Larch caused by *Phytophthora ramorum* (a member of the Oomycota, colourless fungus-like algae) this trend is likely to accelerate (see Chapter 14). The three larches, together with Maidenhair-tree (*Ginkgo biloba*), Dawn Redwood (*Metasequoia glyptostroboides*) and Swamp Cypress (*Taxodium distichum*), are the only six deciduous conifers of the 49 in Table 15.

Many species of conifer have been introduced very widely and the planted populations of some now far outnumber their native ones. This is especially the case with Monterey Cypress and Monterey Pine (*Pinus radiata*), two endangered and protected species that are naturally confined to the Monterey Peninsula, California, but that are common in cultivation in southern Britain. Both are particularly characteristic of maritime locations in southwestern England, where well-grown individuals can top 40 m. The pine is easily recognised by its needles (leaves) being borne in groups of three, and by its cones being very persistent on the tree, even remaining on thick branches decades after the seeds have reached maturity. These two species, along with Scots Pine and a few others, including the Evergreen or Holm Oak, do much to enhance the scenery of many seaside resorts from Dorset to Cornwall, but, where their dense foliage blocks the sea view of householders, they can provoke anger. On the Sandbanks, Dorset's millionaires' row, several nocturnal excursions by people armed with chainsaws or poisons have recently removed spectacular but offending specimens, leading to arrests in 2011.

Table 15 indicates that 13 out of our 49 alien conifers come from northwestern North America, an area of high rainfall and humidity that supports many trees of great stature. As might be expected, the area of the British Isles most closely replicating that habitat is on and near the west coast of Scotland, and indeed that is where so many conifers are grown, as specimens and in groves and plantations, and where the tallest trees in our islands are to be found, often protected from winds in steep-sided valleys. Seven of the 13 above species attain heights there of more than 40 m, and two of them may exceed 60 m. These giants belong to Giant Fir (*Abies grandis*) or Douglas Fir (*Pseudotsuga menziesii*). Alan Mitchell, the doyen of tree measuring, once told me that it was pointless trying to locate the very tallest tree, because that changes every few years as the top is blown out and a new one overtakes it, but at any one time it is likely that one or more exceed 60 m. At the time of writing (2015), the tallest tree is said to be a Douglas Fir near Inverness at just under 66.4 m (218 ft).

By far the greatest influence on the environment by alien conifers is imposed by their use in forestry (Fig. 256), although the total area under forest in the British Isles is relatively low. Most of the following figures are taken from the Forestry Commission's National Forest Inventory (2014). About 12.9 per cent of Britain and 10 per cent of Ireland is forested, compared with about 33 per cent in the European Union overall. Only Denmark and the Netherlands have percentage cover as low as that in Britain and Ireland. At the end of the First World War (1918), the coverage in Britain was only 4 per cent. Some 1.6 million ha (16,000 km², or about 51 per cent) of the forest cover in Britain is coniferous (falling from 57 per cent in 1998), the rest broadleaved. Excluding the area under Scots Pine (albeit much of it alien in provenance), and omitting Ireland, more than 13,000 km² of the British landmass is covered by alien species of conifer. Figures for only the seven most abundant species of conifer, accounting for 96.4 per cent of the total, are individually available (Table 16).

Most of the increase in forest cover in the past century has been due to the creation of conifer plantations, particularly in Scotland. This period of extensive afforestation started after the Second World War, leading to what Oliver Rackham, formerly the leading authority on British woodlands, disparagingly called 'the locust years'. During the implementation of this policy, many superb lowland woodlands, with their rich animal and plant life, were destroyed (Fig. 198). As Rackham has pointed out, the rationale for pursuing such a course seems to have been largely the production of cheap, low-grade timber, especially pit-props, yet by the time the trees were of sufficient size the mining industry was in terminal decline. Thankfully, times have changed. The Forestry Commission and the forestry industry in general are no longer solely concerned with maximum-volume timber production. Such a policy generates only poor-quality timber, chips and pulp. Today, high-quality timber is at a premium, and forests

FIG 198. Lodgepole Pine (*Pinus contorta*), from western North America, is one of our most important plantation timber trees: (a) rows of young trees planted on old peat bog cleared of developing birch woodland; (b) shoots of young trees with abundant male cones and (red, centre top) very young female cones. (CAS)

are also acknowledged as vital to the leisure industry, outdoor pursuits and nature conservation. Consequently, an increasingly high proportion of planting is with broadleaved species; as shown by the above figures, the dominance by conifers is slowly being eroded.

The importance and value of timber-producing forests are obvious when the range of uses is considered. Timber is needed for construction, joinery, utensils, toys, ornaments and so on, but matches are no longer manufactured in this country, and pit-props are no longer required. Timber side-products (woodchips, sawdust, bark) are utilised for all kinds of boards, panels and cardboard, for construction, packaging and insulation, and in the garden and for animal bedding, and pulp is used for paper, packaging, tissues and so on. There are also some traditional uses for exotic conifers, e.g. larch wood for ladders and gates.

384 · ALIEN PLANTS

TABLE 16. Area occupied in Britain by our seven most abundant conifers. Figures from Forestry Commission's National Forest Inventory (2014).

Species	Total area ('000 hectares)	% of total coniferous woodland
Sitka Spruce *Picea sitchensis*	665	50.8
Scots Pine *Pinus sylvestris*	218	16.7
Larches *Larix decidua*, *L. kaempferi* and *L.* × *marschlinsii*	126	9.6
Lodgepole Pine *Pinus contorta*	100	7.6
Norway Spruce *Picea abies*	61	4.7
Corsican Pine *Pinus nigra* subsp. *laricio*	46	3.5
Douglas Fir *Pseudotsuga menziesii*	46	3.5
	1,262	**96.4**

Of the innumerable specialist uses, Christmas trees, of which nearly 8 million are sold annually in Britain, rank highly. In 2008, just 5 per cent of Christmas trees were imported, whereas ten years previously it was 25 per cent. Marketed trees can be derived from plantation thinnings, tops of felled timber trees or specially grown specimens. Nowadays, the last category is the norm, because tops often have too long a gap between branch whorls and thinnings tend to have poor side growth low down. The increase in purpose-grown trees and the decrease in imports have gone hand in hand as more specialist growers have become established in recent years. The production of 7.6 million trees, taking into account the time needed between successive plantings on the same site (8–11 years), requires about 12,000–13,000 ha of land. According to the British Christmas Tree Growers Association, eight species constitute 'the most popular varieties': Norway Spruce, Serbian Spruce (*Picea omorika*), Blue Spruce (*P. pungens* 'Glauca'), Noble Fir (*Abies procera*), Nordmann Fir (*A. nordmanniana*), Fraser Fir (*A. fraseri*), Scots Pine and Lodgepole Pine. However, the four of these most commonly available nowadays are Norway and Blue spruces, and Nordmann and Fraser firs.

CHAPTER 11

Our Top Fifty-two Neophytes

The great source of pleasure is variety.
Samuel Johnson, *Lives of the English Poets* (1779–81)

FREQUENCY OF ALIENS

There are several ways of measuring the frequency (extent of distribution) or abundance (density of distribution; *see* Chapter 7) of a plant species in the British Isles, and often these different methods produce different results. One measure of frequency that botanists have readily available is the number of hectads (10 × 10 km Ordnance Survey grid squares) in which the taxon has been recorded. This often does not tell us the full story, because a species might be only sparsely distributed in most hectads, or show a very 'patchy' distribution in which it can be very abundant only in relatively restricted parts of our islands. Maritime species, for instance, even common ones, inevitably have a low hectad score. Frequently, the hectad total does not indicate the most important species either. In the case of aliens, 'important' often means invasive or harmful, and some of the species that would rank highly on this score are in fact (so far) quite infrequent on a national scale, e.g. New Zealand Pigmyweed. Nevertheless, we have no other objective means of expressing frequency, and so the hectad total is utilised here; in fact, in the vast majority of cases this measure does coincide with botanists' perception of high significance because on the whole they are the species most often met with in the field. Lists compiled by more subjective criteria will often differ from ours; the 'thirty interesting aliens' of Williamson (2002), for example, included only 16 of our top 52.

Data accumulated by a large number of diverse recorders are always liable to a degree of error, but again many of these errors will even out in the long term. There are, however, some more serious errors that evidently result from commonly made misidentifications, whereby one species is over-recorded at the expense of another. Sometimes this results from confusing nomenclatural changes. The following taxa are almost certainly seriously over-recorded to the extent that *most* of their records are probably errors for the taxon placed in parentheses:

Aster novi-belgii (*A. lanceolatus* × *A. novi-belgii*)
Bergenia crassifolia (*B. ciliata* × *B. crassifolia*)
Brassica rapa subsp. *oleifera* (*B. napus* subsp. *oleifera*)
Cotoneaster microphyllus (*C. integrifolius*)
Crocus vernus subsp. *albiflorus* (*C. vernus* subsp. *vernus*)
Forsythia suspensa × *F. viridissima* (*F. suspensa*)
Hyacinthoides hispanica (*H. hispanica* × *H. non-scripta*)
Lupinus polyphyllus (*L. arboreus* × *L. polyphyllus*)
Philadelphus 'Virginalis Group' (*P.* 'Lemoinei Group')
Oxalis dillenii (*O. stricta*)
Paeonia mascula (*P. officinalis*)
Spiraea salicifolia (several species and hybrids)
Symphytum asperum (*S.* × *uplandicum*)

There are approximately 3,859 hectads in the British Isles; a more precise figure would not be more useful because some maritime hectads consist of only a square metre or so of land and are best ignored. On the basis of data to hand in May 2012 and listed against each taxon in Appendix 1, 16 (0.88 per cent) flowering plant neophytes have been recorded since 1986 in at least half (1,930) of the hectads, and a further 36 (together making 2.87 per cent) in at least one-third (1,286) of the hectads. The comparable figures for the 197 archaeophytes are 25 (12.69%) in at least half of the hectads and a further 36 (together 23.86%) in at least one-third of the hectads. The vastly higher percentages for archaeophytes than for neophytes are not surprising in view of the formers' presence here for many centuries or millennia longer. The close approximation of the distribution of archaeophytes (but not of neophytes) to that of natives is shown in Fig. 130.

Outside this 'top 52' there are some interesting statistics. Only 10 per cent of our neophytes (but 53.81 per cent of archaeophytes) have been recorded from more than 10 per cent of the total number of hectads, and only 36 per cent (but 85.28 per cent of archaeophytes) in more than 1 per cent of hectads.

THE TOP FIFTY-TWO

1. Pineappleweed *Matricaria discoidea* (3,530 hectads; 91.47%)
It is something of a disillusionment that, of all the wonderful exotics that
exemplify our alien flora, the most recorded is among the least auspicious, being
one of those weedy members of the Asteraceae lacking conspicuous ray florets
(Fig. 127), hence its alternative name Rayless Mayweed. Pineappleweed is so
called because of the pleasant smell of the finely divided foliage when crushed,
vaguely resembling that of the popular fruit. It is a much-travelled, widespread
plant first recorded here in 1869. It has no obvious means of dispersal, but the
tiny fruits are readily carried around by animals, people and vehicles, embedded
in the mud so characteristic of its habitats. Edward Salisbury (1961) tells us
that the average fruit production of a plant is nearly 7,000, with a germination
success of about 93 per cent. It is thought that it arrived here from North
America, but that its original home was in eastern Asia, and now it occurs in
all corners of the British Isles. It occurs mainly on bare ground, especially that
kept open by disturbance or trampling, notably along the edges of roads and
paths and around gateways, but it is by no means confined to such habitats.
Although the 1869 record was from beside the River Thames at Kew, most of
the early occurrences were as a grain alien, especially with imported chicken
feed. Subsequent spread has been by footwear and vehicle wheels, and the rapid
increase in the first quarter of the twentieth century was attributed by Salisbury
to the development of the motor car.

Pineappleweed was first reported as *Matricaria chamomilla* var. *discoidea*, and
in fact rayless variants of that species, and indeed of most of the mayweeds
in several genera, do exist. It was first correctly identified by Druce in 1899. A
closely related taxon, *M. occidentalis*, differs only in minor characters such as the
presence of minute scales at the apex of the fruit. It is variously considered to be
a separate species, subspecies or variety, or not worthy of recognition at any rank.
It does occur here but the relative distributions of the two are unknown. The very
ready dispersal of Pineappleweed in mud has led to tentative suggestions that it
might also have arrived with us attached to the feet of migrating birds. However,
assuming it is an alien species in North America, its transport from America to
Europe even by 'natural' means would not give it native status here.

2. Sycamore *Acer pseudoplatanus* (3,461 hectads; 89.68%)
Sycamore is one of our best-known trees, and anyone not knowing its provenance
would never suspect that it is not a native with us. It is widespread on the
Continent but did not reach the English Channel naturally. None of our native

trees has a hectad total rivalling that of Sycamore, and few can match its ability to self-regenerate by seed. Of our almost 2,000 neophytes, only this tree and Pineappleweed have been recorded in more than 75 per cent of the total number of hectads.

The name 'sycamore' has had a varied application to a number of trees with somewhat similar palmately divided leaves. The biblical 'sycamore' was a fig, *Ficus sycamorus*, and the similar-shaped leaves of our plant probably gave rise to the latter's name. In North America, 'sycamore' is applied to plane trees (*Platanus*), and our Sycamore is sometimes known there as the Sycamore Maple. Conversely, in Scotland the Sycamore is often known as Plane, leading to the English book name of the Sycamore as False Plane, a translation of the Latin epithet.

Sycamore was probably originally introduced to the British Isles in the sixteenth century (some claim earlier, without firm evidence) as a timber tree and as an impressive specimen tree. The timber is pale, hard and close-grained, usually without much of the figure found in some other maples, and can be worked to a very smooth surface. It submits well to turning, and has traditionally been used for kitchenware such as mixing bowls, spoons, bread boards and rolling pins. It has also been used in joinery and for producing veneers.

The young seedlings of Sycamore, with their pair of strap-shaped cotyledons, are a very familiar sight in parks and gardens. The tree is what foresters would call a weed species, becoming invasive in both grassland and woodland. It is particularly successful at establishing itself in hedgerows, and is often a major constituent of this habitat. In many areas of choice lowland native woodland, Sycamore has become a pest species and considerable efforts to cut it out are being made. It supports a moderate invertebrate fauna (some specific to it), e.g. 58 species of moth, 16 Coleoptera and 25 Hemiptera. In comparison, the figures for the native Alder (*Alnus glutinosa*) are 99, 35 and 52, respectively, and for the native Ash (*Fraxinus excelsior*), 46, 16 and 26, respectively. Sycamore often carries a large aphid population, which in turn can attract insect and avian predators. Two fungal diseases are mentioned in Chapter 14. One of the main assets of Sycamore as a wild tree is that it is very resistant to exposure and pollution, and has been much used in the past in shelter belts around houses and farms, along routes in urban and rural areas, especially in northern Britain, and, because of its wind and salt tolerance, on the coast. For example, it is the commonest tree on the gritstone of the Peak District (Fig. 199a) and is virtually the only deciduous tree in the Shetlands. It is often seen as a large, solitary specimen in parks and on field boundaries, when it makes an extremely handsome spreading tree attaining up to 35 m. There are various cultivars with copper, pink, yellow or variegated leaves.

The flowers of Sycamore are borne in pendulous inflorescences containing hermaphrodite and male flowers, and often female and sterile ones too (Fig. 199b).

Apart from Sycamore, we have five other alien *Acer* species. Ashleaf Maple (*Acer negundo*), known as Box-elder in its native North America, is unusual in having compound pinnate leaves and in being dioecious. The species improves its chances of fertilisation by having nectar-less male flowers in dangling wind-pollinated tassels. Another North American species, now very extensively planted, Silver Maple (*A. saccharinum*) also has unisexual flowers that may be on the same or separate trees. Norway Maple (*A. platanoides*) is number 28 below. One feature of maples that does not vary is the winged fruits, or 'keys', borne usually divergently in pairs.

FIG 199. Sycamore (*Acer pseudoplatanus*), the most common neophyte tree in the British Isles: (a) over considerable areas of the Dark Peak in Derbyshire it is the only tree in evidence, by roads and around farmsteads; (b) flowering shoot. (CAS)

3. Snowberry *Symphoricarpos albus* (2,723 hectads; 70.56%)

It is somewhat surprising that a shrub that apparently very rarely reproduces sexually in the British Isles, and is not known to be accidentally dispersed by our soil-moving activities (as is the case with Japanese Knotweed), should be our third most common neophyte, distributed from the Channel Islands to the Shetlands. The widespread occurrence of this North American deciduous shrub is almost entirely due to its being planted, primarily for hedging or council landscaping in urban areas, and as game cover in rural areas. Once planted, it slowly spreads vegetatively by means of underground woody rhizomes, which throw up suckers, eventually forming dense, continuous thickets. When well managed for game, thickets are cut to the ground every three years, but most are now neglected (Fig. 263b).

The pink flowers occur in small clusters and are not very conspicuous, but they produce abundant nectar and are very attractive to bees. From these, the distinctive pure white berries, 8–15 mm across, develop in autumn, and provide the main visual attraction to the human eye (Fig. 263a). They do not seem, however, to be attractive to birds, and most of them fall off in winter, having become brown and rotten. The fruits regularly bear seeds that apparently have well-developed embryos, but British-grown seeds have resisted all attempts at germination and extremely few records of seedlings have been made. Occasionally, plants are found in such places as wall tops or atop a pollarded willow, where they were presumably self-sown, and one seedling arose in my garden about ten years ago. In contrast, Oliver Gilbert, who wrote an informative *Biological Flora* account of the species (Gilbert, 1995), reported that in Germany the berries are an important winter food for Greenfinches (*Carduelis chloris*), and that the species is spreading by seed around many towns. Why it should behave so differently in Germany is open to speculation in the absence of real evidence. Perhaps germination is facilitated by a seed passing through a bird's gut.

Three other taxa in the genus are also grown in the British Isles and have become locally naturalised. All have pink fruits and reproduce vegetatively by arching woody stolons. They are Coralberry (*Symphoricarpos orbiculatus*), its hybrid with *S. microphyllus* (Chenault's Coralberry *S. × chenaultii*), and the hybrid of the latter with Snowberry, *S. × doorenbosii*. The last, known as 'Doorenbos Hybrids', is probably the most planted taxon in urban areas now, while in the country other genera tend to be preferred as game cover.

4. Common Field-speedwell *Veronica persica* (2,707 hectads; 70.15%)

There are three species of *Veronica* known as field-speedwell, all annual weeds primarily of cultivated ground, including both gardens and arable fields. Their small blue flowers give them the country name of bird's-eye. Sometimes all three

FIG 200. Common Field-speedwell (*Veronica persica*), now the commonest speedwell in gardens and arable ground in the British Isles. (CAS)

can be found in the same small area, but the commonest of them is Common Field-speedwell (Fig. 200), a neophyte first recorded here just under 200 years ago. It differs from the other two in its larger flowers and the more divergent lobes of its fruit. Despite its relative abundance, its homeland is further away, in southwestern Asia; it now occupies all corners of our islands.

Grey Field-speedwell (*Veronica polita*) was known here about 50 years before Common Field-speedwell; it has often been considered a native, but it is treated as a neophyte in the *New Atlas* (Preston *et al.*, 2002) and we include it in our list of native/alien uncertains. Green Field-speedwell (*V. agrestis*) has also often been treated as a native, having been first recorded 200 years earlier again, in the sixteenth century, but, as with many other weeds of cultivated ground, it is probably best designated an archaeophyte. It appears to be decreasing in abundance, and in fact Salisbury (1961) suggested that it has been replaced by Grey Field-speedwell, a process that he thought might be an example of 'competitive replacement'. One might suggest that Common Field-speedwell is replacing them both. Usually, Grey Field-speedwell prefers base-rich soils and Green Field-speedwell acidic ones, whereas Common Field-speedwell can thrive in both. According to Salisbury, Green Field-speedwell has fewer (8–16, as opposed to 10–40) but larger seeds in each fruit. All three species can be found with flowers and fruits on virtually any day of the year and, unlike Slender Speedwell (number 10 below), are highly self-fertile.

5. Japanese Knotweed *Fallopia japonica* (2,659 hectads; 68.9%)

Much is said elsewhere in this book about Japanese Knotweed, particularly in Chapters 8 and 9 concerning its sexual behaviour, and in Chapter 15 on its ecological and economic impacts. When first introduced as a garden plant, it was championed as a beautiful and graceful exotic novelty (Figs 201 and 273), the foliage was said to be palatable to cattle, and the rhizomatous nature of the plant made it suitable for soil binding, e.g. in sand dunes. These attributes are no doubt all justly applied. The arching stems, rarely under 2 m tall, with long, often reddish internodes and large but neat mid-green leaves, and bearing bunches of numerous white flowers in September long after most wildflowers have finished, are certainly very attractive. The species received a horticultural gold medal in Utrecht in 1847, and briefly commanded a high price from nurseries. It continued to be sold from nurseries in Britain until at least the middle third of the twentieth century (Bailey & Conolly, 2000). According to some, it still has merits even in the wild. It can vegetate and beautify very unattractive waste places and waysides, and its dense cover, combined with its late growth in spring, creates open ground that is suitable for spring flowers. It has also been used in trials for the production of biomass. However, the emphasis today is almost solely on efforts to eradicate it. The vast spread of its rhizomes, a mere 7 g of which can regenerate a new colony, shows that mechanical disturbance will propagate rather than eliminate the plant, and it has a well-documented resistance to herbicides.

Japanese Knotweed was named as one of only two alien species to be proscribed in the UK's Wildlife and Countryside Act 1981, whereby it became an offence to introduce it into the wild. This and other publicity the plant has

FIG 201. Male-sterile flowering shoot of Japanese Knotweed (*Fallopia japonica*). (CAS)

attracted in recent years have resulted in it becoming well known to the British public. The great expense incurred in removing Japanese Knotweed from sites such as Wembley Stadium and the 2012 Olympic Village in London have caught the imagination, and the hero in one of Jeffrey Archer's novels (*A Prisoner of Birth*, 2008) gained revenge on an adversary by contaminating a once-valuable building site with its rhizomes. As a result, there has been much effort expended in recent times in searching for biological means of control (see Chapter 15).

Japanese Knotweed is native to Japan, northern China, Korea and Taiwan. It was discovered first in Japan in the 1770s by the Swede C. P. Thunberg, and described as *Reynoutria japonica* by Houttuyn in 1777. It was rediscovered there in the 1820s by the German P. F. von Siebold, and in 1846 was described as *Polygonum cuspidatum* by Siebold and Zuccarini, who did not realise that it was the same species as Houttuyn's *R. japonica*. The introduction of living plants to Britain as garden ornamentals also occurred on two well-separated occasions. In 1825, plants were imported (possibly directly) from China by the Horticultural Society of London; they were grown at Chiswick but apparently eventually died out. Contemporary illustrations of this material show that it did indeed come from China, as the Chinese and Japanese plants differ slightly morphologically. Meanwhile, plants were brought back from Japan as a result of Siebold's travels in the 1820s and were eventually grown in Leiden in the Netherlands. They were made available for sale or exchange in 1848; some arrived at Kew in 1850, and by 1854 they were offered by at least one other English nursery.

In Japan, Japanese Knotweed is a particularly characteristic pioneer species colonising volcanic lava. Clearly, this barren well-drained habitat is mimicked to a considerable degree in this country by rubbly waste ground, such as slag heaps and the sites of demolished buildings, and to some extent by a range of other sites such as neglected gardens, allotments, estates, cemeteries and parks, as well as railway property and rough banks of roads and rivers. Japanese Knotweed is characteristic of all these habitats. It causes problems by shading out other species, but more seriously its strong, fast-growing rhizomes, which can reach a depth of 3 m and a thickness of 8 cm, damage foundations, walls, pavements and river barriers, and can cause a flood hazard by damaging drainage works and clogging channels. The plant can easily grow through tarmacadam, and force up concrete blocks. It also alters soil characters and chemistry, and has been shown to release allelochemicals that inhibit the root growth of other species.

Although the initial spread in this country was by deliberate introduction into gardens and parks, for more than half a century the main introduction has been by accidental contamination during the dumping and moving of soil in such activities as road improvements, ditch digging, bank making and landscape infilling. Many sites are by rivers, and rhizome fragments and even pieces of broken stem are easily and effectively waterborne and quickly become established. As explained in Chapter 8, reproduction by seed does not occur in Europe.

The first note of Japanese Knotweed as a wild plant in the British Isles was in 1886, when it was recorded by John Storrie in his *Flora of Cardiff* as 'very abundant on the cinder tips near Maesteg'. It is intriguing to surmise that our plant might have been deliberately planted on the cinder tips due to its reputation as a soil binder. Whether or not it was originally deliberately planted in that area, South Wales was the earliest region in which the species became common, and from which its distribution radiated to other parts of the country (Conolly, 1977). Right up to 1920, in fact, South Wales still represented the greatest concentration of records, with the rest of Britain very sparsely populated. However, by 1960 the whole of the British Isles from Shetland to the Channel Islands was quite densely covered, and since then the filling in has continued.

As explained in Chapter 8, all the Japanese Knotweed in this country (and, in fact, in the whole of Europe) is female. Clearly, lack of pollen explains the usual barren state of Japanese Knotweed, but the reason for the presence here of only one sexual morph is perhaps less obvious. The apparently invariable morphology and sex expression of the plant throughout Europe led to the idea that the single original introduction of the species by Siebold might have given rise by vegetative propagation to the whole European population of the species, which is therefore a single clone. With the advent of molecular (DNA) techniques, this hypothesis could be tested. Hollingsworth & Bailey (2000) analysed 108 different fragments of DNA in 150 samples of Japanese Knotweed from the British Isles, France, Germany and the Czech Republic. They found that all 108 fragments were identical across the whole range of material, providing very strong evidence (although admittedly not absolute proof from the technique applied) that all the samples were of the same genotype, i.e. belonged to the same clone. Moreover, three other samples from the USA were also identical with the European plants, implying that the North American material came from Europe. Probably the same is true for naturalised Australian and New Zealand material.

Closely related to Japanese Knotweed is Giant Knotweed (*Fallopia sachalinensis*), a huge plant growing up to 5 m tall; it has larger leaves of a different shape from those of Japanese Knotweed and the flower clusters are more compact. It is native in extreme eastern Russia and northern Japan. It was introduced into Europe for fodder and as an ornamental, and was first recorded in the wild in the British Isles in 1896, having been first described in 1859. Today, it is spread across the British Isles (though not yet in Orkney, the Isle of Man or the Channel Islands), but it has less than a fifth of the hectad total of Japanese Knotweed. Giant Knotweed is also gynodioecious, but both morphs occur here and sexual reproduction takes place.

So what can be said about the current status and future prospects of Japanese Knotweed? First, this is an alien species with strong invasive and destructive abilities that is very difficult to eradicate. It is already found commonly throughout the British Isles and, despite efforts to curb it, the species will probably continue to increase in abundance until all available habitats have become occupied. Second, any limits on its competitive attributes that might at present be imposed by the existence of only a single female genotype might possibly be overcome in the future by the provision of male-fertile plants of the hybrid Bohemian Knotweed, as described in Chapter 9. Of relevance here is a report from central Europe that this hybrid is now spreading faster than Japanese Knotweed itself. These facts point to a more serious problem than is posed by any other alien species of plant in this country. One approach to it has been the formation of a Japanese Knotweed Alliance of bodies, including Defra, the Environment Agency, Network Rail and British Waterways, united by their desire and need to overcome this weed. Watch this space!

6. Horse-chestnut *Aesculus hippocastanum* (2,617 hectads; 67.82%)

Although the Horse-chestnut does not become naturalised very readily, and most encountered are obvious solitary plantings, this is one of the most readily recognised trees in the British Isles. The compound pinnate leaves, fabulous pyramids of white flowers and unmistakable fruits are instantly diagnostic (Fig. 202).

Only a few species of plant have gained different names for various parts of the organism (such as haws, acorns and blackberries for the fruits of hawthorns, oaks and brambles), and these are among the most important or familiar. In the case of Horse-chestnut, not only are the seeds known as conkers, but the winter twigs, often cut for indoor decoration and observed as the leaves unfold, are referred to as 'sticky buds'. The World Conker Championships are held annually on the second Sunday in October in the village of Ashton, near Oundle in Northamptonshire, and local contests still take place in many villages throughout the land.

Despite its fame, the Horse-chestnut is virtually useless except for its value as a specimen tree or in avenues. The 'chestnut' part of the name is derived from the vague resemblance of its seeds, borne one to few in a large capsule, to the fruits of the Sweet Chestnut (*Castanea sativa*), which are borne in spiny husks. However, the similarity is entirely superficial, for each sweet chestnut is a whole fruit, and

FIG 202. Horse-chestnut (*Aesculus hippocastanum*), one of the few plants sufficiently familiar to deserve a separate name for the tree and its fruits: (a) tree in flower in the familiar setting of a village green; (b) ripe fruits opening to expose the seeds (conkers). (CAS)

the husks are derived from surrounding bracts. The 'horse' epithet is said to refer to the large horseshoe-shaped leaf scars on the twigs. In the past, minor medical uses were found for concoctions from the bark, flowers and fruits, particularly on the Continent, and the timber was used for low-quality goods. The seeds are unpalatable to man and most animals, but presumably they evolved to be attractive to some agent of dispersal; squirrels sometimes take them. The species is native in a few parts of southeastern Europe and southwestern Asia. The damaging effect of the Horse-chestnut Leaf Miner moth is described in Chapter 14.

The Red Horse-chestnut (*Aesculus carnea*) is of horticultural origin from a hybrid between Horse-chestnut and the American Red Buckeye (*A. pavia*). It has twice as many chromosomes as either parent, so presumably is an amphidiploid species. Its seeds are highly inferior as 'conkers' to those of Horse-chestnut, but they are nevertheless fertile to some degree. A third, rarely naturalised species is Indian Horse-chestnut, with white flowers variably tinged with red, pink and yellow. The genus is a favoured host of Mistletoe (*Viscum album*); the crown of an Indian Horse-chestnut tree near the Leicestershire County Cricket ground in Leicester's Grace Road is almost as green in winter as in summer.

7. Ivy-leaved Toadflax *Cymbalaria muralis* (2,535 hectads; 65.69%)

The genus *Cymbalaria* contains about ten species of small-flowered, trailing, pale to dark purple snapdragons that are more or less confined to southern Europe, where several are very restricted on certain western Mediterranean islands. By far the most familiar is Ivy-leaved Toadflax, which has been cultivated in the British Isles since about 1602 and was found in the wild not long after. The long, trailing stems act as stolons or rhizomes and can reach a metre long in one season. No uses other than as an ornamental appear to have been made of it by humans. It has effective means of reproduction, both by its trailing stems, which readily root on meeting moisture, and by seed. The normal habitat is on vertical walls (Fig. 111), where it gains a hold in the crevices or mortar joints, but it is also at home in quarries, rubbly open ground and shingle beaches. The flower stalks are positively phototropic, i.e. they grow towards the light, but after fertilisation they reverse and point towards the darkness, i.e. into the nooks on the wall, where the seeds can germinate. A wall face can become covered in the plant in a very few years.

8. American Willowherb *Epilobium ciliatum* (2,501 hectads; 64.81%)

Willowherbs are often misidentified by beginners who use coloured illustrations as their guide, because the differences between the species involve details of the stigma and leaf base, and the type and distribution of the hairs, rather than obvious features of the general habit and size. American Willowherb (Fig. 129a) grows in the same sorts of habitats as several other lowland species and, frequently, together with several of them. It was first found in the British Isles in 1891 in Leicestershire but, due to its similarity to native species, was not actually identified until plants were found in Surrey by G. M. Ash in 1932. For some reason it did not thrive in Leicestershire, but its spread from southeastern England since the 1930s has been meteoric; it now occurs from the Channel Islands to Shetland, and is the commonest willowherb in much of southern England. Its increase from nothing to becoming the commonest species of the genus in less than 100 years, jostling in many places with four or five of its congeners, is remarkable. The myriads of tiny plumed seeds are very well adapted for wind dispersal, and readily colonise open ground.

Hybrids occur regularly in *Epilobium* and, aided by the fact that it often grows with one or more other members of the genus, American Willowherb forms hybrids with all the other eight lowland species as well as with the dwarf creeping New Zealand Willowherb. It hybridises with more species in our flora than any other alien plant.

9. Rhododendron *Rhododendron ponticum* (2,419 hectads; 62.68%)

The familiar Rhododendron was not recorded as a wild plant in the British Isles until 1849, but during the twentieth century it spread rapidly over the whole of our islands except the Northern Isles. Unfortunately, its early history was not well documented. It was planted both for ornament and game cover. Like most members of the Ericaceae it is a calcifuge, growing only on acid sandy and peaty soils, typical of heathland and moorland up to 600 m altitude. Its dense evergreen thickets have become a major pest in many areas (Fig. 271a) and much effort is taken in eradicating it in order to regenerate the heathland that it has usurped (see Chapter 15).

10. Slender Speedwell *Veronica filiformis* (2,359 hectads; 61.13%)

This perennial speedwell differs in many respects from the three species discussed above under 'Common Field-speedwell' (number 4). As described in Chapter 8, Slender Speedwell rarely reproduces by seed in the British Isles, almost certainly due to its self-incompatibility and the absence of compatible mating strains in any one site (if indeed more than one occurs in the whole country). It can, however, spread rapidly by its creeping and readily rooting wire-thin stems (Fig. 254), and unlike the field-speedwells it is far from confined to open ground.

Slender Speedwell was originally introduced to the British Isles as a rockery plant from the Caucasus region and Turkey around 1800, and it is certainly an extremely attractive spring-flowering subject with sky-blue flowers. However, it readily spreads to adjacent lawns and this has caused it to be considered a pest and to be banned by most gardeners – certainly all those who take a pride in their weed-free lawns. Its main method of dispersal has been as a contaminant of rockery plants and turves used for lawn-laying. Its first wild record here was at Colchester in 1838, where it was said to be well naturalised, although no voucher specimens have come to light. Remarkably, its next wild record was 89 years later, by the River Ayr near Ayr in 1927. After that, it was dispersed rapidly across the British Isles during the second quarter of the century, but in most regions it was still uncommon even in the 1950s. In 1952, Warburg was only able to describe it as 'sometimes escaping, naturalised in a number of places and increasing'. In the second edition of CTW (1962), this was emended to 'extensively naturalised throughout the British Isles'. The map produced by Bangerter & Kent (1957) shows the species' presence from the Channel Islands to northern Scotland and across Ireland, a range almost as extensive as it is today. However, in the second half of the twentieth century that range became greatly consolidated, and extended north to Shetland (see Chapter 8). Slender Speedwell is not only found in lawns, but also in other types of short grassland, especially on riverbanks. It forms wonderful blue carpets on the grassy banks of the River Dove in upper Dovedale, Derbyshire. There is some anecdotal evidence that it has become less common in the last two decades.

11. Montbretia *Crocosmia* × *crocosmiiflora* (2,320 hectads; 60.12%)
Montbretia is a justifiably favourite garden plant, occurring as clumps of iris-like leaves from which arise the attractive spikes of tubular orange flowers at a welcome time in late summer (Fig. 203b). This is a hybrid whose origin is known in some detail. It was produced in Nancy, France, by Victor Lemoine who in 1879

FIG 203. Montbretia (*Crocosmia* × *crocosmiiflora*), a hybrid created in France in 1879 from South African parents, now thrives here in hedges and rough ground, especially in the west, where its late flowering brightens the landscape: (a) growing through brambles in an Irish hedgerow; (b) plant portrait. (CAS)

deliberately crossed two South African species, *C. pottsii* (female) and *C. aurea* (male). It was introduced into British horticulture very soon after, and was first recorded in the wild here in 1907. Not only do the corms propagate themselves, forming long 'towers' with each corm on top of the last, but they also produce short, rather fleshy rhizomes, and clumps increase in size quite rapidly. A rather small proportion of flowers produces viable seed, a third means of propagation.

The dumping of excess material in hedges and so on has been a major cause of Montbretia's spread as a wild plant, together with dispersal by soil transport, and it now occurs throughout the lowland parts of the British Isles. It prefers moist conditions, even shade, and is much more abundant in the west, particularly western Wales, southwestern England and western Ireland. In many parts of Ireland, Montbretia is a characteristic plant of hedgerows, many of which it brightens significantly (Fig. 203a). It sometimes occurs there with another colourful southern hemisphere exotic, Fuchsia, but the two together form a rather clashing combination. In his *Wildflowers of Cork City and County* (2009), Tony O'Mahony refers to Montbretia as a colourful alien pest, which in County Cork is also characteristic of gravelly ground by rivers, the latter providing another means of dispersal.

12. Gooseberry *Ribes uva-crispa* (2,308 hectads; 59.81%)

If we are honest, we should admit that we are not certain of the early history of any of the three species of *Ribes* grown for fruit in the British Isles – Red Currant (*R. rubrum*), Black Currant (*R. nigrum*) and Gooseberry. All have been claimed as natives and all as aliens. In the absence of good contrary evidence, we have followed the conclusions of the *New Atlas* (Preston *et al.*, 2002), treating Red Currant as an uncertain case and the other two as neophytes. All occur in marginal woodland, especially hedgerows, where they are probably bird-sown, but Red Currant is more often encountered in more remote damp woodland in the absence of other aliens, supporting its possibly native status. It was also recorded here earlier and now occupies more hectads than the other two. The White Currant is a variety of Red Currant.

Wild Gooseberries are common in hedges and scrub, where they are presumably bird-sown. They usually produce very small, sour fruits.

Black Currant is the only one of the three to possess the very distinctive currant scent, but the rather few species of moth (e.g. Currant Clearwing *Synanthedon tipuliformis*, Currant Pug *Eupithecia assimilata*, Magpie Moth *Abraxas grossulariata* and V-moth *Macaria wauaria*) that use this genus as a food-plant do not seem to discriminate between the species. The dreaded Gooseberry Sawfly (*Nematus ribesii*), in contrast, feeds only on Gooseberry and Red and White currants.

FIG 204. Russian Comfrey (*Symphytum × uplandicum*) is highly acclaimed for its use as a manure in organic gardening. (CAS)

13. Russian Comfrey *Symphytum × uplandicum* (2,242 hectads; 58.1%)

Russian Comfrey is a hybrid between our native Common Comfrey (*Symphytum officinale*) and the Caucasian blue-flowered Rough Comfrey (*S. asperum*), which is a rare alien in the British Isles. The former is a variable plant, having chromosome races with 2n = 24, 44 and 48 (and also 40 on the Continent), and flowers that are cream or purple. Russian Comfrey has 2n = 36 or 40 in the British Isles, evidently having arisen from hybridisation between Rough Comfrey (2n = 32) and Common Comfrey with 2n = 40 or 48, respectively. The hybrid usually has a flower colour reflecting its parentage: the flowers open purplish, becoming bluer as they mature (Fig. 204).

Russian Comfrey's early history in the British Isles is not well understood, because in much of the literature of the time, the hybrid and Rough Comfrey were confused and both went under the name 'Russian Comfrey'. It is likely that Rough Comfrey was introduced first, probably in 1799, and the hybrid later, in 1827. They were not introduced as ornamentals, but as fodder. Although the flowers are not very attractive, they are a good nectar source for bees. The hybrid thrives better than Rough Comfrey over here; it soon spread into the wild, probably aided by throw-outs from gardens where it became too rampant, and it now occupies every corner of the British Isles and has become commoner than even Common Comfrey.

In his survey of the history of Russian Comfrey in Britain, Wade (1958) concluded that 'introductions from Russia and elsewhere are undoubtedly the chief, if not the sole, origin' of our plants. We have no firm evidence that the hybrid has ever arisen here. The hybrid is partially fertile, and 'triple hybrids' resulting from crosses with Tuberous Comfrey (*Symphytum tuberosum*) have been found. Hybridisation is favoured in mixed populations because these comfreys are self-incompatible; Russian Comfrey also backcrosses with Common Comfrey.

Comfrey is inextricably associated with the name of Henry Doubleday (1810–1902), a Quaker smallholder in Essex who championed Russian Comfrey as an ideal component of compost in the cause of organic gardening. He imported several clones from Russia in the 1870s. In 1954, the Henry Doubleday Research Organisation was founded by L. D. Hills, and it is still thriving today under the name Garden Organic at its garden headquarters at Ryton-on-Dunsmore, near Coventry. Comfrey is equally championed as a liquid fertiliser, which is obtained by making a cold infusion with water, and is particularly recommended as a feed for Tomato plants. Hills also developed new cultivars of Russian Comfrey, including the one now most acclaimed, 'Bocking 14'. This has the advantage of being sterile and hence does not become too much of a pest.

In the past, comfrey was used in folk medicine, notably under the name of 'knitbone' because of its reputation in healing broken bones. In fact, comfrey plants have since been shown to contain allantoin, which is an anti-irritant, anti-inflammatory and cell proliferant. The roots of the plant are used to make oils and creams for topical application to treat burns and abrasions; formerly a tea was brewed. More recently, it has been suggested that comfrey can cause liver damage and might be carcinogenic.

14. Dame's-violet *Hesperis matronalis*
(1,995 hectads; 51.7%)

Dame's-violet or Sweet Rocket is a highly valued garden ornamental with loosely tufted stems reaching 1 m or more tall. Apart from its attractive large white to purple flowers (Fig. 205), it possesses a delightful scent, strongest at dusk, when it attracts moths. It is a rather short-lived perennial but self-sows prolifically. Remarkably, although Dame's-violet is said to have been grown in gardens in the British Isles as long ago as 1350, it was not recorded in the wild here until 1805, suggesting that it might have escaped before then but gone unrecorded. It is now found throughout our islands on woodland edges, in rough ground, open grassland, ditch banks and hedges, and on riverbanks.

The capsules of Garlic Mustard (*Alliaria petiolata*) form the usual food of the caterpillars of the Orange-tip Butterfly

FIG 205. Dame's-violet or Sweet Rocket (*Hesperis matronalis*), a classic cottage garden perennial, has beautifully scented flowers, especially in the evening, and fruits that are one food-plant of larvae of the Orange-tip butterfly (*Anthocharis cardamines*). (CAS)

(*Anthocharis cardamines*), but in many places the female butterflies choose Dame's-violet on which to lay their eggs, and in terms of nutriment this is equally successful. It is not a perfect strategy, however, because the caterpillars are less well camouflaged on the fruits of the latter food-plant and consequently suffer high levels of predation.

15. Large Bindweed *Calystegia silvatica* (1,980 hectads; 51.31%)
The distinction of this southern European alien (Fig. 143) from our native Hedge Bindweed, and the consequences of its self-incompatibility and hybridisation with other species are discussed in Chapters 8 and 9. It has now reached all parts of the British Isles except extreme northern Scotland.

16. Butterfly-bush *Buddleja davidii* (1,974 hectads; 51.15%)
The Chinese Butterfly-bush was not introduced to the British Isles until 1896 and not found in the wild here until 1922, but today it is one of the more common and well known of our aliens, and one of the most attractive. It is a deciduous woody shrub, up to at least 5 m tall, but the extremities of the branches die back after leaf-fall and gardeners cut it back low down every winter. The long spires of beautifully scented flowers arise from tips of the new growth in mid- to late summer (Figs 1, 260 and 274). Typically, and in the great majority of wild plants, the flowers are a distinctive mauve in colour, but there are cultivars with flowers of all shades of violet, mauve and purple, as well as white. The flowers produce much nectar and are extremely attractive to insects – not only butterflies, but moths, bees and flies as well. For this reason the species is much planted in urban conservation schemes.

Butterfly-bush has no method of vegetative reproduction, but extremely numerous, very light seeds are produced, and these effectively colonise any suitable cranny. Rubbly ground, rocky banks and mortared walls are typical habitats, and railways provide these in abundance. The species is also common in old quarries, especially chalk, as it is very tolerant of well-drained calcareous soils. Butterfly-bush is not a good competitor, and slowly disappears as colonisation by other plants progresses. It is frequently seen high up on walls or even at the tops of buildings in neglected guttering. In the 1940s and 1950s, it became particularly common on bombed sites in southern England, particularly London. As well as providing nectar for the imagines of insects, the foliage is the food-plant of 27 species of moth. Among these is the Mullein Moth (*Cucullia verbasci*), whose distinctive larvae more often feed on mulleins (*Verbascum*), as its names imply. In the recent realignment of plant families according to DNA sequences, the Buddlejaceae have been amalgamated with the Scrophulariaceae, a relationship the Mullein Moth obviously supports.

FIG 206. Snowdrop (*Galanthus nivalis*) is one of the most recognisable garden plants in the British Isles, so abundantly naturalised in woodland that it was thought by many to be a native. (CAS)

17. Snowdrop *Galanthus nivalis* (1,911 hectads; 49.52%)

Snowdrop is one of our favourite flowers, especially because of its very early flowering, and it has been with us since the end of the sixteenth century. It is very well naturalised in many areas of woodland and old grassland, such as orchards and churchyards, and has been considered by some to be a native. The evidence for this, however, does not stand scrutiny. It is surprisingly sparse in Ireland. Grigson (1955) recorded 18 local names for the species, but it does not seem to have had any uses beyond horticulture. It has become a much-collected subject by specialist gardeners ('galanthophiles') and there are well over 500 cultivars in existence. Although many of these are scarcely distinct from others, new or rare cultivars sometimes change hands for hundreds of pounds sterling. The most distinct variants are the so-called 'double-flowered' (*flore pleno*) ones, where the inner three petals are increased in number, often by modification of the stamens. Since these produce a greater effect en masse, they have been much planted in woods, and many naturalised populations are of this sort (Fig. 206). There is evidence that Snowdrop has hybridised with two of the four other naturalised species of the genus in places where they occur together.

Four other species of snowdrop and some hybrids have been recorded as naturalised in the British Isles. Of these, Queen Olga's Snowdrop (*G. reginae-olgae*) is distinctive for flowering in autumn, from October onwards.

18. Italian Rye-grass *Lolium multiflorum* (1,892 hectads; 49.03%)

Here we have a species that was introduced to the British Isles as a fodder plant; together with its close relative Perennial Rye-grass and their hybrid *Lolium × boucheanum*, it is one of the most important commercial grasses (including the cereals) in temperate climates. The two species are often confused, especially if the long inflorescence bristles (awns) so characteristic of Italian Rye-grass are taken as diagnostic, because long-awned variants of Perennial Rye-grass also exist. The main differences are vegetative. Perennial Rye-grass is a long-lived perennial with many non-flowering shoots (tillers), on which the young leaf blades are strongly flattened, being folded along the midline. Italian Rye-grass is an annual or biennial, and by late flowering most or all of its tillers have disappeared, having developed into flowering shoots (culms), but before that the tillers can be seen to have rolled, tubular leaf blades. Although Perennial Rye-grass is a stronger species with more basal shoots, Italian Rye-grass is on the whole a more luxuriant, lusher plant with wider leaf blades, so it is not surprising that certain hybrids, which are fortunately highly fertile, combine the best of both. Many plantings of rye-grass today, and probably many records of Italian Rye-grass, are in fact the hybrid *L. × boucheanum*.

Italian Rye-grass was introduced into agriculture in the British Isles in 1831 and had been noted in the wild within ten years of this date. It has reached all parts of our region but is sparse in western Scotland and western Ireland. Being short-lived, it is mostly found in open ground around fields, gateways and buildings, and along waysides.

19. Indian Balsam *Impatiens glandulifera* (1,884 hectads; 48.82%)

This tall, handsome annual plant was introduced to the British Isles from the Himalayas as a garden ornamental in 1839, and escaped into the wild soon after. Growing to about 3 m tall it is – apart from a few casual aliens such as Sunflower – our tallest annual plant. Despite its impending demise, it manages great vigour during the summer months, sprouting new roots wherever a stem is immersed in water (Fig. 207). The large purple flowers, with their distinctive shape (Fig. 250b) giving rise to the plant's popular name of Policeman's Helmet, are certainly very attractive and Indian Balsam clearly still has a place in the flower border. Its other common name, Himalayan Balsam, is actually far more appropriate than Indian Balsam. Leo Grindon, in an 1864 letter to fellow botanist James Britten, said 'No-one would grow the plant for its beauty, for it is a cumbersome and weedy thing at the best'. Many would disagree. For instance, John Presland (2011) said, 'It is a plant I love, creating beautiful sweeps of colour… along rivers in the West Country'.

The generic name for the balsams refers to the 'impatient' explosive capsules, which throw the seeds a good distance when touched or jogged by the wind. Salisbury calculated that an average-sized Indian Balsam plant produces about 800 seeds. This mechanism ensures a rapid colonisation of open ground by the

FIG 207. Indian Balsam (*Impatiens glandulifera*) grows to 3 m tall and continues to put out new roots wherever possible. (CAS)

species, and allows it to invade suitable ground that is moist enough to sustain the rather fleshy stems and leaves. Typically, it is a riverside plant, and the distinctive seedlings can cover large areas of bare ground when they germinate in spring following winter chilling. The ground is often kept bare by the dense shading provided by the season's growth, but more open patches can permit a good spring flora to develop. Although Indian Balsam has gained a reputation as a pest, smothering native riverside vegetation, it provides welcome attractive cover along many polluted rivers that otherwise support little vegetation (Fig. 250a). Its voluminous flowers are a valuable nectar source, being very attractive to bees and wasps – particularly bumblebees, which are fully accommodated inside the flower while foraging. Arguments about how undesirable this species is and what, if anything, should be done about it are still being made. Since 2010, it has been illegal to plant or sow Indian Balsam in the wild in the UK.

Although Indian Balsam is fully self-compatible, it does not set seed without the aid of visiting insects because of a distinctive specific pollination mechanism. The stamens form a cone that completely obscures the stigma, and pollen release occurs over about two days during the 'male phase', during which time insect visitors depart with a load of pollen on their backs. At the end of this phase, the stamen cone shrivels and drops off, exposing the stigma for the 'female phase', which lasts less than a day. Pollen-bearing bees entering during this period will effect pollination. The flowers produce nectar, and a strong, rather sickly scent, for all this period, the nectar accumulating in a broad spur at the back of the flower. There is a considerable variation in flower colour, from deep purple to almost white. Some populations show this range while others are almost invariable, but overall intermediate pink or mauve colours are commoner than the extremes.

FIG 208. Orange Balsam (*Impatiens capensis*) is smaller, less common and far less invasive than Indian Balsam (*I. glandulifera*): (a) plants in reed-swamp; (b) flower seen from the front with a worker Common Carder Bumblebee (*Bombus pascuorum*) just accommodated within it. (CAS)

Two other species of *Impatiens* occur in the British Isles. Orange Balsam (*I. capensis*) is, despite its specific epithet *capensis*, a North American species. It has handsome orange flowers built on a similar pattern to those of Indian Balsam, but they and the whole plant are smaller (Fig. 208a). It is also a riverside species, often growing with Indian Balsam, but has a much more restricted distribution, mainly in the southern half of England. In America, like most plants with red or orange flowers, it is primarily pollinated by hummingbirds, but over here it is a bee-pollinated plant. The flowers are just large enough to completely engulf a worker bumblebee (Fig. 208b). The floral mechanism also resembles that of Indian Balsam. In this species, however, a varying proportion of the flowers exhibit cleistogamy, i.e. they do not open and self-pollination occurs inside the bud. The proportion of cleistogamous flowers varies, but often they are in the majority and sometimes the only sort present.

Small Balsam is another nineteenth-century introduction to Britain, this time from central Asia, possibly with that unusual alien vector, timber. It is much smaller than the other two species in stature and all its parts, and has pale yellow flowers (Fig. 78) that are pollinated predominantly by hoverflies. It grows in damp, shady places, scattered over Britain except northwestern Scotland.

20. Canadian Waterweed *Elodea canadensis* (1,868 hectads; 48.41%)

The story of the introduction and spread of Canadian Waterweed, and the consequences of the existence here of only female plants, is told in Chapter 8.

21. Cherry Laurel *Prunus laurocerasus* (1,826 hectads; 47.32%)

Cherry Laurel was introduced to the British Isles from southeastern Europe about 400 years ago, but records from the wild were not made for more than 250 years. It was prized as a garden plant mainly because evergreen shrubs are scarce in northern Europe, and it was used as game cover and for screening, either free-grown or trimmed as hedging. The beautiful erect spires of white flowers (Fig. 133) are produced early in April and are attractive to insects, but can suffer browning by the frost. The very succulent black fruits are eagerly sought by birds, especially Blackbirds (*Turdus merula*), which presumably effect dispersal. Nevertheless, Cherry Laurel does not seem to regenerate very freely from seed, and in woodland and shrubberies gradual extension by vegetative propagation (layering) is often more frequent.

Nowadays, dwarf, ground-covering cultivars of Cherry Laurel – in particular the narrow-leaved 'Otto Luyken' – are much planted, especially in the likes of council schemes and supermarket car parks. Strangely, these often produce a good flower display in the autumn.

When assessing whether or not a plant is poisonous, it is a useful rule of thumb that species of Rosaceae are not; in fact, the great majority are edible. Cherry Laurel is the exception that proves the rule. In *Poisonous Plants in Britain*, produced by the then Ministry of Agriculture and Food in 1984, this is the only species in the family that is included. Poisoning usually occurs after animals have eaten the leaves, which contain cyanogenic glycosides. These are sugar compounds that contain a cyanide group, which is released by chemical hydrolysis as the very poisonous compound hydrogen cyanide when the leaves are crushed (or chewed). It is said that 1 kg of leaves is sufficient to kill a 500 kg cow. This poisonous property is utilised by some amateur entomologists in their killing jars. Cyanide is poisonous to all animals (actually to all aerobic forms of life), and the fact that the fleshy fruits are eaten by birds shows that this part of the plant lacks the glycosides (at least in any quantity). The kernels in the fruit stones, however, are poisonous, and the same is true of the kernels of other *Prunus* species, such as cherries, Plum, Peach (*Prunus persica*) and Bitter Almond (*Prunus dulcis* var. *amara*). These, however, are unlikely to be eaten in harmful quantities.

A related species, Portugal Laurel (*Prunus lusitanica*), is recorded from only about one-third as many hectads as Cherry Laurel, as it is planted in gardens much less often. It is more of a tree than Cherry Laurel, and it takes less kindly to pruning as a hedge, but in my experience it regenerates more freely from seed.

The name 'laurel' can be confusing, as the true Laurel (*Laurus nobilis*), usually called Bay or Bay Laurel, is a very different and unrelated plant. It is an important culinary herb, and was used to crown sporting victors; its name persists in titles such as the 'poet laureate'. It also has succulent black fruits, but only on female specimens as the species is dioecious.

22. Black Currant *Ribes nigrum* (1,811 hectads; 46.93%)

Black Currant is discussed above under Gooseberry (number 12).

23. Green Alkanet *Pentaglottis sempervirens* (1,808 hectads; 46.85%)

It is a bit of a mystery why this species (also known as Evergreen Alkanet) should be grown at all, as it is not especially attractive (Fig. 126), it soon becomes a pest in gardens and it has no well-defined uses. Grigson (1955) suggested that it was grown for the red pigment that can be obtained from the roots. 'Alkanet' was used in ancient times in herbal medicine, but whether this particular species was used in the British Isles for that purpose is apparently unknown. For whatever reason, it was introduced before 1600 and has been known in the wild for nearly 300 years. Despite its Latin and English names, the plant is entirely herbaceous, not evergreen, but it shoots up very early in the new season and there are few weeks in the year when it is not showing green. The bright blue flowers each produce four nutlets that, although not dispersed far, produce offspring extremely readily and can soon colonise considerable areas. The plant's root system is deep and not easily eradicated.

24. Red Valerian *Centranthus ruber* (1,782 hectads; 46.18%)

Hailing from the Mediterranean, Red or Spur Valerian was introduced to the British Isles in the sixteenth century and has been naturalised here for 250 years. It produces seeds, and then seedlings, in prodigious amounts. The seeds have hairy 'parachutes' and are easily taken by the wind to colonise often inaccessible sites resembling those of its native rocky ground, such as cliffs, walls and rubble. According to Williamson (2002), it is 'unwanted in some places', and its estimated cost of control is said to be £17,000 per year, but one could question whether this is money well spent. It has certainly become a pest in some areas where it has escaped from the confines of walls and entered natural communities. For example, it has become 'naturalised over perhaps 1 ha or more of the Carboniferous Limestone north escarpment of the Burren' in County Clare (Jarvis, 2011). It is often abundant on maritime shingle (Fig. 209a); in parts of East

FIG 209. Red Valerian (*Centranthus ruber*) is common in urban areas on walls and rubble, but equally at home on maritime cliffs and shingle: (a) colonising shingle at Shingle Street, Suffolk; (b) closer view of flowering plant. (CAS)

Anglia it, Sea Pea (*Lathyrus japonicus*) and Sea-kale (*Crambe maritima*) are the three most conspicuous colonisers of that habitat.

Plants commonly exhibit three flower colours: red, pink and white, of which pink is commonest. They produce abundant nectar, which is secreted into a narrow spur at the base of the corolla. The flowers are very small, but they are borne in conspicuous, dense masses which are very attractive to insects (Fig. 209b), especially Lepidoptera and Hymenoptera. Red Valerian is a very good butterfly flower, especially attractive to Brimstones (*Gonepteryx rhamni*) in late summer, and it is also especially favoured by the day-flying immigrant Hummingbird Hawkmoth (*Macroglossum stellatarum*). Grigson (1955) listed 38 local names for 'this cheerful and blowsy plant', but it appears to have no other uses than as ornament. It lacks the medicinal uses of the true valerians (*Valeriana*).

FIG 210. Winter Heliotrope (*Petasites fragrans*) flowers from December to March and has beautifully scented flowers, but spreads rapidly and is a major pest in some areas, especially the west: (a) flowering shoot; (b) roadside bank covered with its leaves. (CAS)

25. Winter Heliotrope *Petasites fragrans* (1,756 hectads; 45.5%)

The origin of Winter Heliotrope in North Africa rather than Europe and the apparent absence from Europe of female plants are described in Chapter 8. It was introduced to the British Isles as an ornamental more than 200 years ago, and escaped into the wild soon after. The young leaves and beautifully scented flowers, produced from December to February (when they can be badly frosted), are very attractive, but for the rest of the year the plant forms dense patches of leaves that smother most other vegetation (Fig. 210). It is highly invasive and considered to be the biggest alien threat to native vegetation in southwestern Ireland. Its long rhizomes can reach very deep into the soil, so that physical removal is rarely possible; they easily snap off and small pieces readily regenerate.

In this list of neophytes, presented in decreasing order of frequency, Winter Heliotrope is the first to show large gaps in the north of Britain.

26. Garden Privet *Ligustrum ovalifolium* (1,738 hectads; 45.04%)

Garden Privet was introduced to the British Isles from Japan in the middle of the nineteenth century, and proved an excellent subject for hedging. It is fully evergreen and hence vastly superior to our native Wild Privet (*Ligustrum vulgare*), which is semi-deciduous and looks a sorry sight by midwinter. Otherwise, the two species are very similar, differing in minor floral dimensions, pubescence, and leaf shape and colour. When not severely cut, Garden Privet freely produces a mass of small black berries, but there are few records of it self-sowing. In most places it is merely a persister, which probably accounts for the fact that it was not recorded as a wild plant for 100 years after its introduction.

Garden Privet produces much nectar and is highly attractive to insects (Fig. 259), but for a plant that is in the same family as the lilacs (*Syringa*) and jasmines (*Jasminum*) its scent is a great disappointment. The rather nauseous scent is to me the characteristic smell of hedge-lined city streets in July, when the best of the summer has gone. Garden Privet, as well as Wild Privet, is the main food-plant of the largest common moth of the British Isles, the Privet Hawkmoth (*Sphinx ligustri*). This moth is scarce northwest of a line from the Wash to the Bristol Channel, but southeast of that its density is greatly increased by the existence of Garden Privet in areas where Wild Privet is scarce. It also provides food for several other moths and, as many children know, stick insects.

There are handsome yellow-leaved cultivars of Garden Privet that can be used in suitable places as specimen shrubs. Those with only a washed-out yellow colour might turn out to be the hybrid between the two privets, *Ligustrum × vicaryi*, because this horticultural product is said to have involved a yellow-leaved cultivar as its Garden Privet parent.

FIG 211. Honesty (*Lunaria annua*) has attractive purple or white flowers but is grown mainly for its white disc-like fruits; despite its scientific name, it is strictly biennial. (CAS)

27. Honesty *Lunaria annua* (1,665 hectads; 43.15%)

This southeastern European biennial is a very well-known garden plant, equally welcome because of its showy purple or white flowers in early spring (Fig. 211) and its fruits, used to brighten dark corners in winter or for dried-flower arrangements. The 'fruits', shiny white oval discs, are actually the sterile centre of the flat fruit, the fruit walls and contained seeds on either side having been dispersed. Because of its biennial habit, Honesty is not a strong competitor, and in many of its sites is not long-persistent. It makes up for this by being a prolific seeder; the seeds are flat and narrowly winged, and their dispersal is wind-assisted. Honesty has peculiar swollen roots resembling those of other species of its family that are afflicted by the club-root disease; they are said to be edible.

FIG 212. Norway Maple (*Acer platanoides*) is much planted, often has copper-coloured leaves and self-sows very readily. It contributes greatly to the fine colouring of the autumnal landscape in its native Scandinavia, as well as in the British Isles. (CAS)

28. Norway Maple *Acer platanoides* (1,641 hectads; 42.52%)

The Norway Maple is a spreading tree, handsome in winter and summer, beautiful in early spring with its yellowish-green flowers appearing before the leaves, and spectacular in autumn with yellow and orange leaf colours (Fig. 212). Although common in Scandinavia, and a major contributor to the splendid autumn colours there, it is widespread in Europe. It was introduced to the British Isles in 1683 but not recorded in the wild here until the early twentieth century, more than 200 years later. Norway Maple self-sows very readily, but rarely inserts itself into existing woodland. The timber is much used on the Continent, but here the tree does not attain the size of Sycamore, which has similar timber, and it is not grown as a timber tree.

29. Greater Periwinkle *Vinca major* (1,631 hectads; 42.26%)

Unlike its smaller relative Lesser Periwinkle, Greater Periwinkle has never been considered a possible native in the British Isles and usually occurs in habitats

FIG 213. Greater Periwinkle (*Vinca major*) is a familiar garden plant doubtfully worth its space, as it spreads rapidly by means of its arching stolons. (CAS)

where it is more obviously an introduction (Fig. 213), especially near gardens; moreover, its first record is 100 years later than that of the former. It is therefore somewhat surprising that Greater Periwinkle has been recorded from slightly more hectads than Lesser Periwinkle in Britain, and many more in Ireland. Greater Periwinkle has a number of closer relatives, of which Intermediate Periwinkle (*Vinca difformis*) is also naturalised in a few places. Intermediate Periwinkle and Greater Periwinkle are from southwestern Europe and the Mediterranean region, whereas Lesser Periwinkle occurs further north in France. Both Greater and Lesser periwinkles have been used for various minor medicinal purposes.

30. Purple Toadflax *Linaria purpurea* (1,626 hectads; 42.14%)

Purple Toadflax is an attractive garden plant, with its long, erect spires of small, long-spurred purple (sometimes pink in cultivars) flowers that are extremely attractive to bees (Fig. 175a). However, it seeds prolifically into paving, walls and open ground, and can soon become a pest. Tidy gardeners do not tolerate it, but it was nevertheless introduced to the British Isles more than 350 years ago from southern Italy simply for ornament. It is most at home in barish, stony ground, such as pavement cracks and walls. Like many species in the genus it is self-incompatible, so when it occurs close to Pale Toadflax, it often hybridises. The hybrids are fertile and can segregate to form hybrid swarms.

31. Sticky Groundsel *Senecio viscosus* (1,614 hectads; 41.82%)

Although Sticky Groundsel is a well-established weed of waste ground (Fig. 214) and has been considered as a native of the British Isles by many previous authors, it was not recorded here until 1660 and is probably an alien. Possibly, as with Groundsel, some maritime or river-shingle variants might be native. Its liking for well-drained soils finds it characteristic of such places as coastal shingle, walls, roadside rubble and railway cinders, in the last often defaced by such waste as pigeon feathers and discarded tissues adhering to the sticky foliage. It is rarely found as a weed of cultivated ground. Sticky Groundsel hybridises with the native Heath Groundsel (*Senecio sylvaticus*) and the alien Oxford Ragwort.

FIG 214. Sticky Groundsel (*Senecio viscosus*) is particularly characteristic of cinder beds or heaps, often near railways; it hybridises in these habitats with two other species of the genus. (CAS)

32. Lesser Swine-cress *Lepidium didymum* (1,613 Hectads; 41.8%)

Our two species of swine-cress were formerly placed in the genus *Coronopus*, but molecular studies have shown that their distinctive growth habit, with inflorescences arising at a node on the opposite side of the stem to a leaf, represents an extreme part of the genus *Lepidium*. Swine-cress (*L. coronopus*) is an archaeophyte, but Lesser Swine-cress is a neophyte from South America, first recorded in the British Isles in 1778. It occurs on cultivated and waste ground, but is not characteristic of trodden areas as is Swine-cress. When crushed, it has a strong cress-like smell that is lacking in Swine-cress, but there seems to be no evidence of its having been put to any use. Lesser Swine-cress is well distributed in the British Isles, but scarce in the north.

33. Oilseed Rape *Brassica napus* subsp. *oleifera* (1,612 hectads; 41.77%)

The position of *Brassica napus* in the cytogenetically defined Triangle of U (Fig. 182), and its natural hybridisation with *B. oleracea* and *B. rapa*, are described in Chapter 9. In the past 50 years, the sight of vast fields of yellow flowers in April and May, and roadsides lined with plants germinated from spilled seed, have become very familiar (Fig. 135). The population dynamics of Oilseed Rape on the verges of London's infamous orbital motorway, the M25, are described in Chapter 7.

34. Hybrid Black-poplar *Populus* × *canadensis* (1,567 hectads; 40.61%)

The most commonly planted poplars still belong to this 'Euroamerican' hybrid. Many clones have been bred, starting with 'Serotina' in about 1750 in France; this was imported to Britain about 50 years later. The various clones, all by definition of one sex or the other (Fig. 215), are used for timber, paper, amenity, shelter and, formerly, matches and matchboxes. Growth is very rapid and trees often quickly become too large for their surroundings, or pose dangers by losing large branches. As a result, many are cut back or pollarded, destroying their characteristic growth forms. More is said about this hybrid in Chapter 8.

35. New Zealand Willowherb *Epilobium brunnescens* (1,565 hectads; 40.55%)

This New Zealand alien of damp, stony ground is unusual in a couple of respects: it is a creeping plant, quite unlike our native willowherbs, with only the individual flowers and their long ovaries becoming erect (Fig. 224); and it is one of those relatively rare aliens in the British Isles that occur in the uplands (reaching as high as 915 m in Wales), usually mingled with native species. It also has the distinction of being one of only two alien plants that grow on the remote archipelago of St Kilda, in the Atlantic Ocean 60 km west of the Outer Hebrides (the other is Pineappleweed). Both species were probably introduced

FIG 215. Poplars and willows are among the best-known dioecious trees of the British Isles. The commonest poplars are cultivars of the 'Euro-American' Hybrid Black Poplar (*P. nigra* × *P. deltoides* = *Populus* × *canadensis*): (a) male catkins of 'Serotina'; (b) female catkins of 'Regenerata'. (CAS)

to the islands on the tyres of construction vehicles, shipped in for building work associated with the radar station in the late 1950s.

New Zealand Willowherb hybridises quite readily with most other species of *Epilobium* that it encounters. The hybrids are quite highly sterile and are intermediate in all respects, notably their habit. The story of the discovery here of hybrids between New Zealand Willowherb and six of our native species, the first in 1980, is told in Chapter 9. Interestingly, its hybrid with another alien, American Willowherb, has also been found in New Zealand, where the latter is a naturalised alien. New Zealand Willowherb was introduced to the British Isles as a rockery plant, but its rapid spread by both creeping stems and seed quickly led to its expulsion by gardeners. In upland areas, especially in the west, it is spread by vehicles and the forestry industry, and the seeds by wind.

36. Hybrid Bluebell *Hyacinthoides* × *massartiana* (1,560 hectads; 40.42%)
The hybrid between our native Bluebell and the garden Spanish Bluebell (Fig. 172), which has become by far the commonest bluebell in cultivation and as a

naturalised escape, is discussed in Chapter 9. Spanish Bluebell has been recorded from nearly three-quarters as many hectads as its hybrid, but that is probably an overestimate due to misidentification.

37. Fox-and-cubs *Pilosella aurantiaca* (1,558 hectads; 40.37%)

Fox-and-cubs is the name given to a species of mouse-ear hawkweed that has several capitula (flower heads) clustered tightly at the apex of the erect flowering stems. The flowers are foxy orange in colour, and at an early stage of flowering there is often one capitulum wide out and the others in bud, giving rise to the species' fanciful common name. This species is attractive and often grown in gardens, especially on rockeries, but it quite quickly gets out of hand due to its abundant seeding and its vigorous surface stolons and subsurface rhizomes. It often invades lawns or taller grassland. Fox-and-cubs has been in the British Isles for nearly 400 years, and naturalised here for over 200 years; in New Zealand it has become a serious pest.

The plant is native in central Europe, where it is much more variable than here, with seven chromosome races, only one of which (tetraploid) has been detected here. Hybrids (Fig. 174) with our native Mouse-ear Hawkweed have been found in several places (*see* Chapter 9) but hybridisation is far more frequent on the Continent, involving several other species.

38. Lilac *Syringa vulgaris* (1,542 hectads; 39.96%)

Botanists are often surprised to learn that the Lilac (Fig. 5a) is native not in eastern Asia but in southeastern Europe, whence it was introduced to the British Isles in the sixteenth century. Despite its flowers being extremely attractive to bees, I know of no records of it having self-sown, and its naturalisation is by means of its strong suckers, which can arise at some distance from the main trunk. This can lead to gardeners throwing out such supplementary growth, and this and plantings to beautify road verges are its main sources in the wild. It is the food-plant of several species of moth that also feed on others members of the Oleaceae (especially privets and Ash), such as the Lilac Beauty (*Apeira syringaria*).

39. Variegated Yellow Archangel *Lamiastrum galeobdolon* subsp. *argentatum* (1,533 hectads; 39.73%)

Yellow Archangel (*Lamiastrum galeobdolon*) is an attractive native spring-flowering woodland plant found fairly commonly in England and Wales, and scattered as an escape from woodland gardens elsewhere in the British Isles. Nearly all of these occurrences refer to subsp. *montanum*, a tetraploid, but in a few woods in eastern Lincolnshire the Continental diploid subsp.

FIG 216. Variegated Yellow Archangel (*Lamiastrum galeobdolon* subsp. *argentatum*) was originally described as a separate species, but is possibly only a garden cultivar. It spreads rapidly by means of long stolons, causing it to be frequently discarded from gardens. (CAS)

galeobdolon replaces it. In gardens there is a third taxon (Fig. 216), first described in 1975, which is tetraploid like subsp. *montanum* but is slightly larger in its floral parts and has a silver patch along the centre of most of its leaves. It is often treated as a third subspecies, *argentatum*, but it is quite likely that it represents a mutation of subsp. *montanum*. One piece of evidence for this is that it seems to be invariable in isozyme characters, suggesting that it might be a single clone. Variegated Yellow Archangel has been much planted in gardens since its introduction to the British Isles in the 1960s, being valuable in that difficult habitat of dry shade. Its long, arching stolons have meant that it readily escapes and is often thrown out when it becomes too rampant. It also reproduces by seed, and now has a good distribution over most of the British Isles.

40. Alsike Clover *Trifolium hybridum* (1,530 hectads; 39.65%)

The pink and white flower clusters of this clover (Fig. 217) probably gave rise to the epithet *hybridum* used by Linnaeus, implying the Red and White clovers *T. pratense* and *T. repens* as parents, but in fact any suggestion of such an origin ends with the flower colours. Like the Russian Comfrey and Italian Rye-grass above, Alsike Clover is an agricultural introduction to the British Isles, being one of a range of legumes grown for fodder and for increasing the nitrate content of soil. However, it came here before either of those species (in 1562) and has also been known in the wild for longer than either, having now reached all regions. It does not persist for long in closed communities, and to some extent it relies on repeated introductions as it lacks effective means of vegetative spread. Since Alsike Clover is now grown less than previously, it appears to be rarer than it was 50 years ago.

FIG 217. Alsike Clover (*Trifolium hybridum*) was introduced to the British Isles as a fodder plant but is now rarely grown; its scientific name suggests that the red and white flower heads were originally thought to indicate a hybrid origin. (CAS)

41. White Poplar *Populus alba* (1,518 hectads; 39.34%)

White Poplar is discussed with other members of the genus in Chapter 8. It is also known as Abele, taken from the Dutch and being derived from *albus*, meaning 'white'.

42. Beaked Hawk's-beard *Crepis vesicaria* subsp. *taraxacifolia* (1,480 hectads; 38.35%)

The stiffly upright, branched stems bearing attractive dandelion-like flower heads of this European alien are so familiar to botanists in the southern half of England that it is a surprise to find that it is rare in Ireland and much of northern Britain, where it is the sparsest of any of the top 52 neophytes. It probably arrived here as a grass-seed alien, and grassland of various sorts is its main habitat today. It is particularly characteristic of roadside verges, field margins and railway banks, but it also occurs on walls and rough ground, where it is the third (after Dandelion *Taraxacum officinale* and Colt's-foot *Tussilago farfara*) of the common yellow composites to come into flower in spring. It has shown continuing but slow spread northwestwards in the past half-century, but in northern England it is probably reaching its climatic limits.

43. Wilson's Honeysuckle *Lonicera nitida* (1,441 hectads; 37.34%)

This small-leaved (6–16 mm) evergreen Chinese shrub, so commonly used in garden hedging, would probably not be recognised as a honeysuckle by most passers-by, since it differs so starkly from the familiar tall, twining deciduous honeysuckles. It rarely flowers in such places, but the extent to which this is due to frequent clipping is uncertain, since flowers are far from common on uncut bushes. The small cream flowers are only 5–7 mm long, borne in pairs in the leaf axils. Nevertheless, reports of the plant becoming self-sown are not rare. According to Oliver Gilbert (1995), Wilson's Honeysuckle has recently been largely favoured over the more traditional Snowberry for use in newly planted game coverts, because it requires less attention. A second similar Chinese species, Box-leaved Honeysuckle (*Lonicera pileata*), has been confused with Wilson's Honeysuckle, but it is a shorter shrub with larger, more elongated leaves. It is also being planted for game cover, but unlike Wilson's Honeysuckle, it flowers and fruits very freely. Both species have attractive small violet-coloured berries.

44. Snow-in-summer *Cerastium tomentosum* (1,439 hectads; 37.29%)

There are several other descriptive vernacular names, such as Dusty Miller, for this popular rockery plant, which prefers warm open ground, on which it spreads vigorously to form extensive mats. It is a native of the southern Italian and Sicilian mountains, where it is very variable, especially in terms of the denseness of its white hair covering and in chromosome number (level of ploidy). This variation can lead to its misidentification as one of several other species of its group that occur on mountains from the Balkan Peninsula eastwards, particularly *Cerastium biebersteinii* from the Crimea.

None of these other species, however, has been confirmed from the wild in the British Isles. Snow-in-summer is nowhere better displayed here than on fixed sand dunes on parts of the Norfolk coast, where in May the sand between the Marram-grass (*Ammophila arenaria*) is a carpet of white for hundreds of square metres. Snow-in-summer is 30–100 per cent self-compatible, but the flowers are strongly protandrous (the anthers ripen well before the stigmas) so that self-pollination is very restricted, although pollination of a flower by another on the same plant is not prevented as the flowers are attractive to bees and flies. In the areas of Norfolk mentioned above, the native Field Mouse-ear is also common, and many fertile hybrids have arisen.

45. Flowering Currant *Ribes sanguineum* (1,402 hectads; 36.33%)

The name Flowering Currant clearly refers to the fact that this tall shrub was introduced to the British Isles from western North America for entirely ornamental purposes rather than for its fruit. Its purplish-black fruits are covered with a pale bloom (of yeast), unlike the shiny black fruits of Black Currant (see under Gooseberry, number 12 above), but its foliage shares a strong currant scent with the latter species. The stems 'layer' (take root) where they touch the soil, and self-sowing sometimes occurs, but in most of its sites it simply persists. The red

FIG 218. Flowering Currant (*Ribes sanguineum*) is an attractive and very popular early-flowering shrub: (a) bird-sown plant on heathy common; (b) inflorescence. (a, MJC; b, CAS)

flowers (Fig. 218) are very attractive to bees and are a familiar sight in early spring in many suburban gardens, but they tend to clash with their frequent cohabitant, the bright yellow Forsythia.

46. Oxford Ragwort *Senecio squalidus* (1,399 hectads; 36.25%)

The fascinating story of the origin and subsequent evolution and spread of Oxford Ragwort (Fig. 162) in the British Isles is told in Chapter 9. The fact that our plants are different from those in their area of origin (Sicily), having evolved into a new species since their arrival here, indicates that they must be considered a neonative rather than a neophyte.

47. Hedgerow Crane's-bill *Geranium pyrenaicum* (1,365 hectads; 35.37%)

This rather nondescript crane's-bill, appearing at a glance like a large, erect Dove's-foot Crane's-bill (*Geranium molle*), with pinkish-purple flowers about 15 mm across, has become very much a wayside species in the British Isles, on hedge banks and road and path verges. Like most of the large genus *Geranium*, the fruits have five one-seeded compartments, and when ripe they burst explosively, flinging the seeds away as much, according to Salisbury, as 3 m. It has been with us for more than 350 years and is now widely distributed, but in most areas it is not very common.

48. Soft Lady's-mantle *Alchemilla mollis* (1,345 hectads; 34.85%)

Some general points about the genus *Alchemilla* were made in Chapter 8. All our species, both native and alien, are apomictic and set plenty of seed. Soft Lady's-mantle is the largest species we have, and the commonest in southern Britain, but it also extends throughout most of the rest of the British Isles. It has been very popular as a garden subject in recent decades, and in the same period has become much commoner as an escape.

49. Yellow Corydalis *Pseudofumaria lutea* (1,333 hectads; 34.54%)

This popular garden plant with bright yellow flowers is almost always grown on walls (Fig. 219), but given the chance its profuse yield of seeds will develop in virtually any empty space. Its method of colonisation of wall crevices is quite different from those of Ivy-leaved Toadflax (Fig. 111) or Red Valerian (Fig. 209).

FIG 219.
Yellow Corydalis (*Pseudofumaria lutea*) is characteristic of cracks and mortar joints in walls, to which it gains access by its explosive fruits. (CAS)

The shiny black seeds possess an oil body, rather like a miniature Castor-oil-plant (*Ricinus communis*) seed, which is very attractive to ants. The insects carry the seeds away to their nest, where they germinate after the oil has been consumed. Yellow Corydalis is native in the Italian and Swiss Alps, but was introduced to gardens in the British Isles at least by 1596 and was recorded in the wild here by the end of the eighteenth century.

50. Wall Cotoneaster *Cotoneaster horizontalis* (1,316 hectads; 34.1%)

The apomictic genus *Cotoneaster* was discussed in Chapter 8. A minority of the species are sexual diploids, while the rest are polyploid apomicts. The western Chinese Wall Cotoneaster (Fig. 220) falls into the latter category. It is somewhat surprising that this species should be the only cotoneaster to make the top 50, but this is probably due to its being one of the easiest species to recognise in this very difficult genus. Possibly a few other less distinctive species are commoner, although Wall Cotoneaster is certainly well spread across the British Isles. Its most distinctive feature is its branchlets, which are all flattened into one plane, especially – as so often the case – when they are pressed against a wall. The leaves are very small and deciduous. The flowers are pink and rather inconspicuous, but are extremely attractive to bees, which besiege them in large numbers. The abundant orange-red fruits provide a good source of food for birds, especially Blackbirds and other thrushes. This species was introduced to Europe (France) around 1870 and was passed on to Britain soon after, but it was not recorded in the wild until more than half a century later. It can also be very invasive (Fig. 248).

Of all the naturalised cotoneaster species, the most harm is probably done by Entire-leaved Cotoneaster (*C. integrifolius*), whose dense blanket of tangled stems can smother great swathes of previously species-rich limestone cliffs and turf. It can also invade chalky grasslands once the animals are taken away, when undergrazing leads to scrubbing up. Himalayan Cotoneaster (*C. simonsii*) is invasive of heathland in Scotland (one of the few aliens to exhibit this behaviour), where it is bird-sown on rocky outcrops, often a long way from the garden whence it originated.

51. Dotted Loosestrife *Lysimachia punctata* (1,319 hectads; 33.95%)

Erect stems and spires of yellow flowers make this a popular garden plant, which spreads laterally by its scaly rhizomes into a dense clump (Fig. 253). This can become too prolific but is easily kept in check. Isolated plants seem not to set seed, strongly suggesting that the species is self-incompatible, a useful attribute in a garden plant that could become a pest. Its very ready growth from small pieces of rhizome ensures that it is quite common in places where it, or soil containing it, has been dumped. It likes moist conditions and is commoner in areas of higher rainfall.

FIG 220. Wall Cotoneaster (*Cotoneaster horizontalis*) is one of the commonest species of this large genus in the British Isles, both in gardens and in the wild. It is instantly recognisable by its flat sprays of foliage, which often grow pressed against a vertical wall or bank: (a) flowers; (b) fruits. (CAS)

The species shows variation and plants known as *Lysmachia verticillaris* have sometimes been separated from it. The latter is, in fact, by far the commoner plant in gardens and naturalised here. It is taller and more robust, with a less leafy and more branched inflorescence, and the flowers are not clear yellow but have orange suffusion at the base of each corolla lobe. Intermediates exist, however, and it seems that only one species is involved. I have grown both sorts in my garden and both set good seed, presumably pollinated by each other and probably suggesting that it is wise to grow only one clone.

52. Turkey Oak *Quercus cerris* (1,293 hectads; 33.51%)

The Turkey Oak is an extremely handsome, widely spreading tree when fully grown, reaching up to 40 m in the British Isles. It differs from our native oaks in its more deeply and narrowly lobed leaves, its whiskery bud scales and the presence of long, spreading scales on the fruiting cupules ('acorn cups'), giving them a shaggy appearance. It agrees with the natives, however, in showing a considerable range of leaf morphology, and in its native areas (from central southern Europe eastwards to Turkey) several subspecies are sometimes recognised. The range of origin of our plants, however, is uncertain, and they have almost certainly hybridised after arrival, so attempts to recognise separate subspecies here are futile. Their leaf variation has also given rise to claims that hybrids between Turkey Oak and Pedunculate Oak occur. However, the two species belong to different subgenera of *Quercus*, between which hybridisation appears to be impossible, so such claims must be ill-based and probably refer to juvenile specimens. Despite this, F. W. Simpson (1989), the author of *Simpson's Flora of Suffolk*, said that it is claimed that mature trees cut for timber in Suffolk oak woodland were recognised as this hybrid by woodmen because of their inferior timber. The acorns of Turkey Oak take two years to mature – one of the characters separating the subgenera.

Turkey Oak grows very well on warm, well-drained soils in southern England, both acid sands and chalk and limestone, where it reproduces very effectively. In favourable sites, such as parts of Epping Forest and the New Forest, it has become the most successfully regenerating tree, and is considered a weed species in the same category as sycamore. In parts of the Cotswolds, Turkey Oak is a serious threat to rich limestone grassland (Fig. 248). It was originally planted to diversify and 'ornament' the native woodland; in the New Forest this took place mainly in the 1880s (Tubbs, 1986), also involving other species such as Red Oak (*Q. rubra*), which thrives in the same conditions. In several sensitive areas, including the New Forest, there are programmes aimed at eliminating Turkey Oak and other alien tree species.

CHAPTER 12

Distribution of Alien Plants within the British Isles

The distribution of plants over the surface of the earth has been a subject of interest ever since the emergence of botany as a science, and probably since the emergence of man as a species.
David W. Goodall, 'Quantitative aspects of plant distribution' (1952)

T hanks to the dedication and application of members of the BSBI over more than half a century, we know a great deal about the spatial distribution of vascular plant species within the British Isles. The distributions of native, neophyte and archaeophyte species in Great Britain and the Isle of Man are shown in Fig. 221, expressed as the percentage of the total native, neophyte and archaeophyte floras represented in each hectad (10 × 10 km square).

For the native species, the distribution pattern reflects the interaction of climate, geology, topography and maritime influence. Native species show several hotspots in places where the geology is highly distinctive, such as the Lizard Peninsula in Cornwall, the Westmorland limestone, East Anglian Breckland, the Wye Valley, the New Forest in Hampshire and the chalk downs in southern England. Overall, there is a clear trend of declining native species richness from southeastern England to northwestern Scotland, correlated with increasing rainfall, declining mean temperatures and increasing cover of acidic bedrock.

In contrast, the alien species show where people live, and where the land is suitable for intensive agriculture. For the neophytes, the map highlights all of the largest cities: London, Birmingham, Bristol, Southampton, Manchester, Oxford and Cambridge stand out in England, and Edinburgh and Glasgow in Scotland. As with the native species, there is a clear decreasing trend in the abundance of

natives   archaeophytes   neophytes

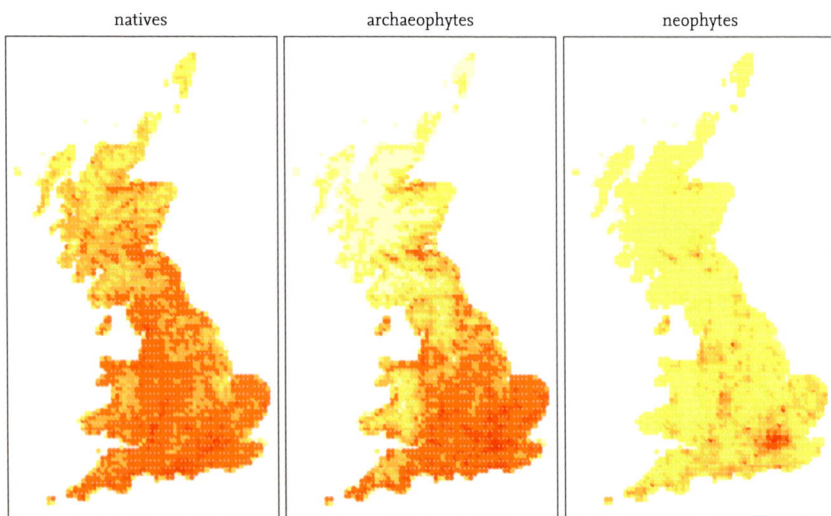

FIG 221. The distribution of native, archaeophyte and neophyte species within Great Britain and the Isle of Man. The colours indicate the percentage of species present in a given hectad on a 10-point scale from pale yellow (low) to dark red (high), with different scales for each plant status. For natives, the maximum score (the darkest red) is 65 per cent or more of total native species in that hectad; for archaeophytes, the maximum score is 90 per cent; and for neophytes it is 60 per cent. Ireland is excluded from the map because it is relatively under-recorded.

neophyte species from southeast to northwest, reflecting the same gradients of climate, altitude and geology, but exacerbated by a matching trend in human population density, which is lowest in the Highlands of Scotland and highest in southeastern England. This gradient in propagule pressure associated with human population is probably as important, if not more important, than the gradients in ecological conditions in determining the number of neophyte species found in any given place. In addition, there are hotspots for neophytes in Cornwall, the Isles of Scilly, the Isle of Man, East Anglia and on the southern shores of the Moray Firth. In part, these reflect the enthusiasm for alien species exhibited by the local vice-country recorders, but these areas also combine distinctive microclimates with above-average horticultural activity (the latter two, of course, are quite strongly correlated).

The distribution of archaeophytes closely mirrors the distribution of human impact both in terms of towns and cities (as with the neophytes), and in the pattern of intensive agriculture in the lowlands of southern and eastern England. The archaeophyte map is dominated by the effects of altitude and soil fertility, with local hotspots away from southeastern and southern England in Lancashire, North Wales and Tweedside.

CHANGES IN ALIEN PLANT DISTRIBUTION, 1962–2002

The maps of species richness in Fig. 221 are based on data gathered for the *New Atlas of the British and Irish Flora* (Preston *et al.*, 2002). The ground-breaking earlier *Atlas of the British Flora* (Perring & Walters, 1962) provided a baseline for assessing changes in the distribution over the intervening 40 years. However, because there was such a big difference in recording effort between the two surveys, and there were so many false negatives in the first *Atlas* (species were present in many squares where they were not recorded), the comparison turned out to be much less straightforward than it could have been. What might at first look like increases in geographic distribution between the two surveys often turned out to be nothing more than increases in the rate of detection as a result of greater recording effort. Declines, however, are much more likely to be genuine. To accommodate this, a rather complicated change index was adopted (*see* Preston *et al.*, 2002), the details of which need not concern us here. If there was 'no change' in distribution between 1962 and 2000, this species gets an index of zero, 'increases' are positive numbers and 'declines' are negative numbers. Analysis of these changes shows some interesting patterns when the set of explanatory variables provided by Hill *et al.* (2000) are taken into account.

The most striking changes are a large increase in the mean number of hectads occupied by neophytes (mean change score = +1.00) and a substantial decrease in the mean number of hectads occupied by archaeophytes (mean change score = -0.48). Native species showed much smaller changes on average (overall, a slight decline, with mean change score = -0.065). There was a highly significant increase in the spatial distribution of species from nitrogen-rich habitats, and an increase in the distribution of taller species. There were two smaller, but significant, responses in relation to climate: increases in the distribution of species that had higher January temperature requirements; and increases in species associated with higher precipitation totals. There were no significant interactions between native or alien status and any of these explanatory variables. If we replace plant height in the model by a factor representing life form (Chapter 7), then we see a highly significant increase in the distribution of trees and shrubs (phanerophytes) and a highly significant decrease in the distribution of annuals.

This difference between archaeophytes and neophytes is exactly what one would expect if archaeophytes have been in the British Isles long enough to reach their climatic limits, whereas some of the neophytes have not been here long, and are still actively expanding their alien range. Part of the apparent increase in neophyte species certainly reflects a wider geographic distribution of botanical recorders willing and able to survey aliens for the second atlas. The high mean rate of decline in archaeophytes reflects the fact that many of them are arable weeds, and arable weeds are one of the most rapidly declining plant groups thanks to the effectiveness of modern herbicides in eliminating weeds from arable crops. The archaeophytes that showed the largest range reductions were Corn Cleavers, Thorow-wax, Upright Goosefoot (*Chenopodium urbicum*), Darnel, Corn Buttercup (*Ranunculus arvensis*), Lamb's Succory, Shepherd's-needle and Red Hemp-nettle (*Galeopsis angustifolia*).

CLIMATE

It is important to know whether some climates within the British Isles are more conducive than others to invasion by alien plants. Obviously, if an alien species is not able to grow under the prevailing conditions of winter minimum temperature, summer maximum temperature, seasonal precipitation, snow cover and exposure, then it will simply fail to establish. Species that are unable to regenerate by vegetative means may fail to ripen any seed if the growing season is too short or summer temperatures too low, and they will be doomed to life as a casual species, reliant on repeated introduction for any semblance of permanence in the flora.

January mean temperature
The relationship between winter temperature and plant status is shown in Fig. 222. There is a substantial group of alien species with low frost-tolerance that are restricted to places with January mean temperatures in excess of 5° C. The effects of warm winter weather on alien plant distribution are well illustrated by a set of species that are confined to walls and rocks in the Channel Islands, the Isles of Scilly and southwestern Cornwall, including several succulent species from South Africa belonging to the dewplant family Aizoaceae. Frost-sensitive aliens are discussed later in this chapter.

In contrast, many of the neophytes with their distributions centred in the parts of the British Isles with the lowest winter temperatures are plantation conifers like European Larch, Hybrid Larch, Japanese Larch, Norway Spruce, Sitka Spruce and Lodgepole Pine, highlighting the fact that commercial forestry is concentrated in cool, upland habitats unsuitable for more productive agriculture. Some of the most characteristic alien species of colder places are herbaceous

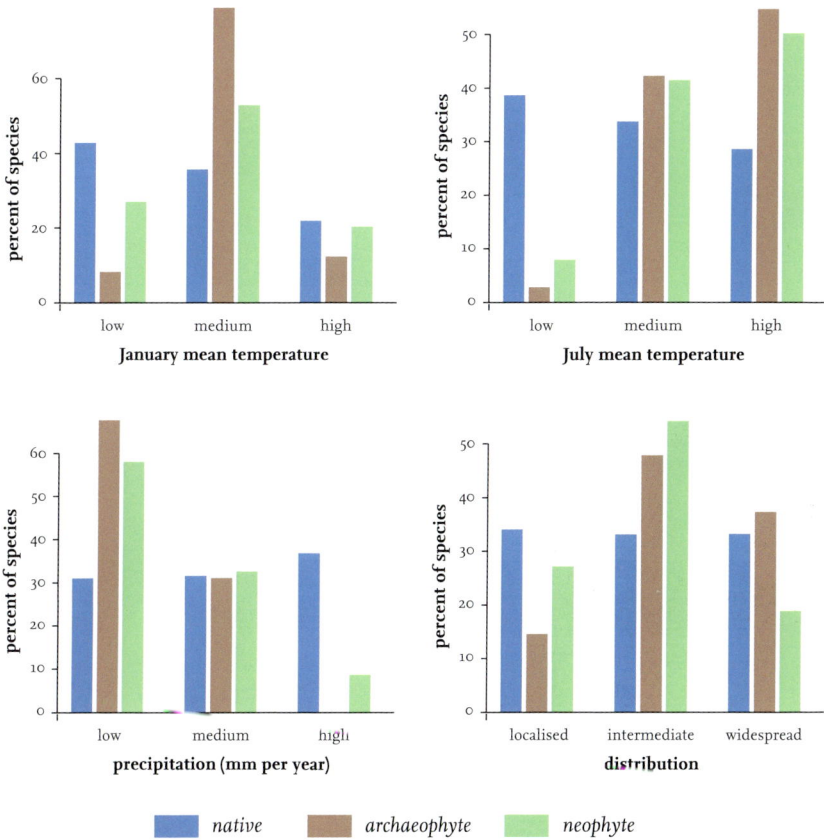

FIG 222. Plant status in relation to the climate (blue = native; brown = archaeophyte; green = neophyte). Top left – mean January temperature (<3.5° C = 'low'; 3.5–4.0°C = 'medium'; >4.0° C = 'high'); top right – mean July temperature (<14.5° C = 'low'; 14.5–15.5° C = 'medium'; >15.5° C = 'high'); bottom left – total annual precipitation (rainfall plus snowfall; annual total <900 mm = 'low'; 900–1,100 mm = 'medium'; >1,100 mm = 'high'); bottom right – geographic extent within the British Isles (<120 hectads = 'localised'; 120–1,199 hectads = 'intermediate'; >1,200 hectads = 'widespread').

perennials, which are covered in more detail under 'Northern neophytes' later in this chapter.

There are only three cold-winter archaeophytes: Large-flowered Hemp-nettle (*Galeopsis speciosa*), an annual weed of Potato, Turnip and other crops; Masterwort (Fig. 223), a wetland perennial; and Monk's-rhubarb (Fig. 32) and Good-King-Henry, relics of cultivation, typically found close to old homesteads or farm buildings.

FIG 223. Masterwort (*Imperatoria ostruthium*) is an uncommon relic of cultivation, mainly in northern Britain: (a) habitat in Sutherland; (b) plant portrait. (a, MJC; b, CAS)

July mean temperature

There is a very striking relationship between plant status and summer temperature (Fig. 222), with few alien species (either neophytes or archaeophytes) restricted to places with the lowest summer temperatures, and archaeophytes proportionately less well represented than neophytes in these cooler places. The opposite pattern is also very pronounced, with a much higher proportion of alien plant species (both neophytes and archaeophytes) compared with the native flora found in regions with the highest mean summer temperatures.

High summer temperatures might be expected to be associated with the distribution of alien late-season annual plants that need to ripen their seeds before the first frosts of autumn. The warm-summer neophytes include several geophytes like Rosy Garlic (*Allium roseum*), which require hot summers to prevent bulb-rot and associated disorders.

As with the other weather variables, this pattern is inextricably confounded by human population density, which is lowest in the Scottish Highlands (where the summers are cool) and highest in southeastern England (where the summers are warm).

Precipitation

Unsurprisingly, the pattern for alien plant distribution in relation to total precipitation is as pronounced as the pattern for summer temperatures, and consistent with it (Fig. 222). Aliens (both neophytes and archaeophytes) are strongly under-represented in parts of the British Isles with the highest rainfall and strongly over-represented in the driest parts. There are very few archaeophyte species that are confined to the wettest places although, of course, some widespread archaeophyte species like Ground-elder do grow in places with very high annual rainfall.

The aliens from the wettest parts of the British Isles are a mix of herbaceous perennial garden escapes like Pearly Everlasting (*Anaphalis margaritacea*) and Montbretia (Fig. 203); garden shrubs, including Entire-leaved Cotoneaster, Fuchsia (Fig. 243) and Rhododendron (Fig. 271); and plantation conifers like Sitka Spruce, Lodgepole Pine, Western Hemlock-spruce, Hybrid Larch and Japanese Larch. There are very few accidental introductions in this category other than Slender Rush (Fig. 64), a locally abundant weed of grass tracks in heathland and acid grassland. The neophyte associated with the highest average rainfall is New Zealand Willowherb (Fig. 224). The average annual precipitation within its alien range in the British Isles is 1,357 mm. The negative correlation between rainfall and temperature accounts for the similarity between the species lists from areas with high rainfall and low temperature.

Of the aliens that are restricted to the driest parts of the British Isles (less than 700 mm precipitation per year), the archaeophytes are all annuals (Lamb's Succory, Maple-leaved Goosefoot *Chenopodium hybridum*, Small Cudweed *Filago gallica*, Broad-leaved Cudweed and Fingered Speedwell), whereas the neophytes

FIG 224. New Zealand Willowherb (*Epilobium brunnescens*) is quite unlike any European species of the genus in that all parts of the plant are soil-hugging except for the upright flowers on their long ovaries; it hybridises with six of our erect-growing species. (Kathleen Pryce)

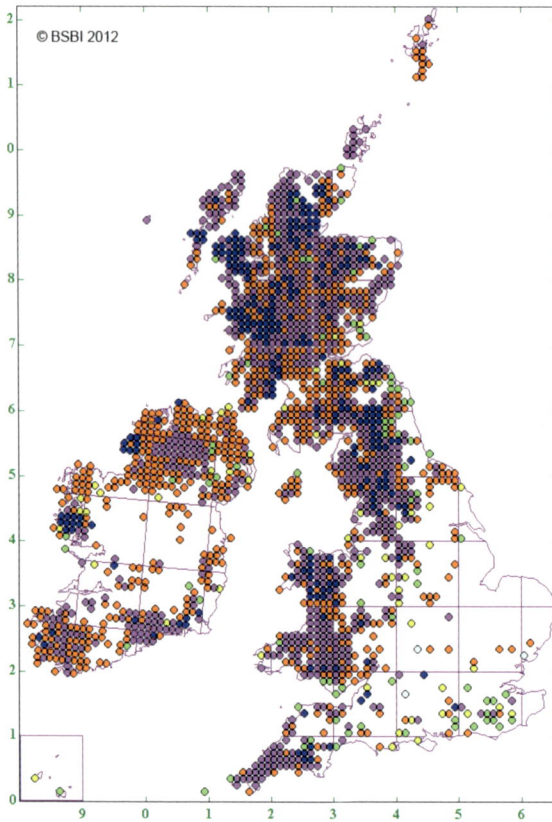

FIG 225. The distribution of New Zealand Willowherb (*Epilobium brunnescens*) is associated with habitats where the average annual precipitation exceeds 1,350 mm. The colours indicate date classes: dark blue for the most recent records and pale green and yellow for the oldest records.

are a mix of annuals (including Leafy-fruited Nightshade *Solanum sarachoides*, Dense Silky-bent *Apera interrupta*, Grey Mouse-ear *Cerastium brachypetalum* and Proliferous Pink *Petrorhagia prolifera*) and woody plants like Bladder-senna (*Colutea arborescens*). The pitfalls of this kind of correlation analysis are nicely illustrated by the fact that one of the neophytes that is statistically associated with the driest places is Floating Pennywort (Fig. 134), a rampant aquatic plant.

The neophyte that appears to be tolerant of the lowest precipitation is Breckland Speedwell (*Veronica praecox*): outside western Suffolk and western Norfolk, it is known only as a rare and declining casual. The average rainfall in its Breckland habitats is less than 620 mm per year.

Fewer neophytes are widespread within the British Isles than are either natives or archaeophytes (Fig. 222), with the result that neophytes are more likely to show correlations with local climatic conditions.

LIGHT, WATER AND SOIL CONDITIONS

It is important to know whether native, neophyte and archaeophyte species tend to be associated with different ecological conditions of water and light availability, soil nutrient status or soil acidity. The German plant ecologist Heinz Ellenberg (1913–97) proposed a simple quantitative scheme for describing the fundamental niche of each plant species in relation to these ecological factors (Ellenberg, 1953). He devised a nine-point scale for each factor (with 1 being low and 9 high), and he worked out a score for each factor for every plant species. Needless to say, the Ellenberg scores are often correlated with one another (e.g. acidic soils are often wetter, and high-nitrogen soils tend to have higher pH), but this is not a major problem for our purposes here.

Between them, these Ellenberg scores provide a simple description of the ecological niche of each species (e.g. a drought-tolerant, light-demanding species of high pH soils that are low in nitrogen, exemplified by Guernsey Fleabane). Our question here is whether natives and aliens show different representations across high, medium and low levels of each factor.

Nitrogen

For most terrestrial vegetation in the British Isles, nitrogen is the most limiting soil nutrient, with phosphorus as the next most limiting. A significantly higher proportion of both neophytes and archaeophytes is found in nitrogen-rich habitats, while a significantly higher proportion of native species comes from habitats with low soil nitrogen status (Fig. 226).

The archaeophytes requiring the highest levels of soil nitrogen include familiar farmyard or manure heap specialists like Good-King-Henry, Oak-leaved Goosefoot (*Chenopodium glaucum*), Many-seeded Goosefoot (*Chenopodium polyspermum*), Stinking Goosefoot (*Chenopodium vulvaria*) and Monk's-rhubarb (Fig. 32); common roadside plants such as Hemlock (Fig. 34) and White Dead-nettle; garden weeds like Opium Poppy (Fig. 18) and Small Nettle (*Urtica urens*); and seed-bank survivors like Henbane (*Hyoscyamus niger*), along with White Willow (Fig. 20), the most characteristic of the alien lowland riverside trees. The neophytes associated with the highest levels of soil nitrogen are a curious bunch, including species characteristic of dumped topsoil like Goat's-rue, sewage-works specialities like Tomato, horticultural relics such as Jerusalem Artichoke and seed-bank survivors like Thorn-apple (*Datura stramonium*).

Light

Very few plant species, native or alien, are adapted to deep shade (Fig. 226), and more than 60 per cent of all species require 'high' light. It is noteworthy that no

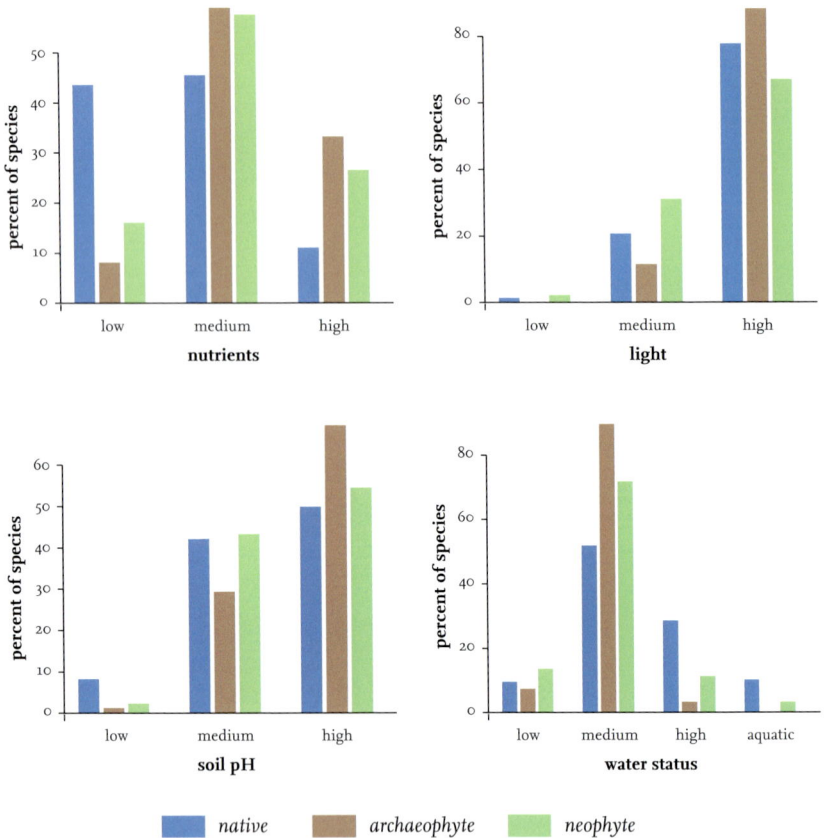

FIG 226. Plant status in relation to environmental factors. Species are categorised by status as native (blue), archaeophyte (brown) or neophyte (green), with the height of the bar indicating the percentage of species found at low, medium and high levels of each factor. Environmental factors (top left – soil nutrient content; top right – light intensity; bottom left – soil pH; bottom right – water status) are categorised by Ellenberg numbers (N, L, R and F, respectively). The Ellenberg numbers are simplified to a three-level factor: low = 1, 2, 3; medium = 4, 5, 6; high = 7, 8, 9; and, for water status, aquatic plants = 10, 11, 12 ; see text for details). In all four cases, there is a significant interaction ($p < 0.001$) between status and score, indicating that native, neophyte and archaeophyte species are distributed differently across the low, medium and high levels of each factor. The largest interaction effect involves nutrients and the smallest involves light.

archaeophytes are tolerant of deep shade, and there are relatively few archaeophytes (11 per cent of species) compared with neophytes (31 per cent) that have medium Ellenberg light scores. The relatively shade-tolerant neophytes include evergreen shrubs used for game cover, such as Shallon and Cherry Laurel (Fig. 133); herbaceous garden plants that have escaped to dominate patches of the woodland floor, including Leopard's-bane (*Doronicum pardalianches*), Winter Aconite, Small Balsam, Variegated Yellow Archangel (Fig. 216), Fringe-cups and Pick-a-back-plant; and a group of canopy tree species with shade-tolerant seedlings, both deciduous like Norway Maple and Sycamore, and evergreen like Western Red-cedar and Western Hemlock-spruce. The classic shade-tolerant urban alien is Mind-your-own-business (Fig. 110), which grows on brickwork in dank alleys and smothers shaded lawns and flower beds with its dense green blanket of tiny leaves. An invasive pest of wet woodlands and swamps is American Skunk-cabbage (Fig. 266); this can form colonies extending over many hundreds of metres and capable of excluding all native vegetation. The most shade-tolerant archaeophyte is Lesser Periwinkle.

Most alien species will grow well in high light, but relatively few absolutely demand it. Those that do require full sun are principally species with very small seeds producing tiny seedlings that are uncompetitive in closed vegetation (e.g. *Conyza* species; *see* Chapter 7). It can hardly escape one's notice how many species share groups of traits: not surprisingly, the most pronounced correlation is between high light, high temperature and low moisture.

Soil pH

There is a pronounced lack of acid-loving alien plant species. Among the archaeophytes, we find only Downy Hemp-nettle and Lamb's Succory in this category. This is presumably because early people spent very little time in acidic habitats like heaths, bogs and mountains, with the result that propagule pressure was low in these habitats. Rather more neophytes are tolerant of low soil pH, including one of our most invasive alien species, Rhododendron (*see* Chapter 15). Potentially invasive plantation conifers like Norway Spruce, Sitka Spruce, Lodgepole Pine, Douglas Fir and Western Hemlock-spruce are all grown on acidic substrates of low agricultural value.

A few garden escapes are acid-tolerant, the most successful of which are Montbretia, which dominates long stretches of roadside bank in the west and north; Garden Pink-sorrel (*Oxalis latifolia*), which can be a persistent weed; and Shallon, which can be invasive in woodland. More often, the acid-tolerant garden plants are survivors like Houseleek.

Relatively more archaeophytes are associated with soils of high pH. Early human settlements were often associated with freely draining, raised ground of relatively high pH, and so it is unsurprising that many of the archaeophytes are found in disturbed ground around buildings. Other archaeophytes, including Field Eryngo, grow in calcareous grassland, often close to the sea, and ancient mortared walls support extensive colonies of Wallflower (*Erysimum cheiri*).

Given the association between built environments, high human population density and substrates of high pH, we should not be surprised to find an abundance of neophytes in these places. Much waste ground has soil of high pH, and the group also includes species characteristic of brick and mortared stone walls, including Red Valerian (Fig. 209), Wall Cotoneaster, Ivy-leaved Toadflax (Fig. 111), Fairy Foxglove and Yellow Corydalis (Fig. 219). Arable land on chalky soils supports very few neophytes, but there are several archaeophytes found in such habitats, including Red Hemp-nettle, Night-flowering Catchfly, Venus's Looking-glass, Woad (*Isatis tinctoria*), Darnel, Rough Poppy, Cornfield Knotgrass (*Polygonum rurivagum*), Spreading Hedge-parsley (*Torilis arvensis*), Keeled-fruited Cornsalad (*Valerianella carinata*) and Broad-fruited Cornsalad (*V. rimosa*), both Sharp-leaved Fluellen (*Kickxia elatine*) and Round-leaved Fluellen (*K. spuria*), and several ramping-fumitories (*Fumaria*).

FIG 227. Greater Quaking-grass (*Briza maxima*), a handsome Mediterranean wintergreen annual, sets large amounts of seed and can become difficult to eradicate. (CAS)

Soil water

There is a distinct shortage of alien plant species (both neophytes and archaeophytes) on the wettest soils (the 'high' category in Fig. 222), and archaeophytes are even less well represented than neophytes there. Although wet soils support relatively few neophyte species, several of these are conspicuous elements of the plant communities in which they grow, including Grey Alder, Pink Purslane (*Claytonia sibirica*), Orange Balsam, Indian Balsam, Slender Rush, American Skunk-cabbage and monkeyflowers (*Mimulus*). The few wet-soil archaeophytes are mostly riverside willows like White Willow, Crack Willow, Almond Willow and Osier.

There are much smaller differences between the proportions of species found on the driest soils, but neophytes have the highest representation, with species like Pirri-pirri-bur (Fig. 73), Common Fiddleneck, Snapdragon, Greater Quaking-grass (Fig. 227), Snow-in-summer, Mexican Fleabane (*Erigeron karvinskianus*) and Hoary Mustard. Archaeophytes of dry soils include Thorow-wax (Fig. 13), Rampion Bellflower, Purple Viper's-bugloss, Field Eryngo, Small Cudweed, White Stonecrop and Rat's-tail Fescue (*Vulpia myuros*).

RAPIDLY SPREADING ALIENS

Without doubt, the aliens that most often grab the headlines are those that are spreading rapidly, and to which words such as 'invasion', 'pest', 'plague', 'peril' and 'disaster' are applied. Mere abundance alone does not appear remarkable – if a species is abundant without causing problems and is not spreading dramatically, what does it matter to the public whether it is a native like Meadow Buttercup (*Ranunculus acris*) or an alien like Beaked Hawk's-beard?

Any new alien arriving here that becomes naturalised will show a pattern of spread that is largely unpredictable and very variable in its manifestation. Often there is a lag phase with little increase (*see* Chapter 7), followed by a period of much more active colonisation, and sooner or later a levelling off as the species approaches full potential. Full potential might be reached by a very local distribution, say in the warm climate of western Cornwall or in the cold and damp of northern Scotland, or might not be realised until virtually all corners of the British Isles have been exploited. In fact, only six neophytes have been recorded in more than two-thirds of the total number of hectads (*see* Chapter 11).

A reasonable measure of whether a species has been spreading in recent years can be obtained by studying the databases of the BSBI, which give the number of hectads in which each taxon has been recorded in six date-classes. Date-class 4 covers 1987–99, and date-class 5 covers 2000–09. The latter is a slightly shorter time span, and moreover did not coincide with a period of intense recording as did the former (for the *New Atlas*). Any taxon with a higher number in date-class 5 than in date-class 4 is likely to be significantly on the increase, as otherwise one would expect the former figure to be lower.

In scrutinising the taxa that have more records in 2000–09 than in 1987–99, some discretion has to be exercised. Survivors and casuals must be excluded since any spread (or decrease) depends largely upon their continued levels of introduction. Approximately 30 naturalised species remain, but not all of them are being increasingly recorded because of their active spread. At least half are being more frequently grown in gardens, from which they are therefore escaping or being discarded in a greater number of places. These, of which the following are representative examples, are becoming more commonly cultivated by changing gardening fashions, largely driven by television gardening programmes: Adria Bellflower, Trailing Bellflower, Aunt-Eliza (*Crocosmia paniculata*) and Seaside Daisy (*Erigeron glaucus*; Fig. 228). Some of them are particularly favoured by town councils, being used in large numbers in parks and on roadsides, e.g. several *Crocus* and *Narcissus* taxa, Chenault's Coralberry, Soft Lady's-mantle, the grass Pheasant's-tail and Argentine Vervain.

Others owe their increase in hectad totals to a greater taxonomic understanding of them and a greater willingness by field botanists to record them. For example, the apparent increase in two species of *Cotoneaster*, Bullate Cotoneaster (*C. rehderi*) and Stern's Cotoneaster (*C. sternianus*), is mainly due to the realisation by recorders that these two species are commoner than Hollyberry Cotoneaster (*C. bullatus*) and Franchet's Cotoneaster (*C. franchetii*), respectively, which are closely similar to the former two and under which most records were formerly placed. Mention was made in Chapter 3 of the fact that records tend to be made only of species that are included in the current Floras. Many species illustrate this by their greatly increased records soon after they first appeared in a standard Flora.

Setting aside all the above cases, there seem to be about ten species that are genuinely showing significant independent increases at the present time. Two salt-tolerant roadside genera, *Amaranthus* and *Bassia*, are becoming more familiar by major roads, perhaps because of the increasing amounts of salt used in winter. Both Green Amaranth and Indehiscent Amaranth (*A. bouchonii*) are increasing, as is Summer-cypress. Also salt-tolerant, Foxtail Barley (*Hordeum jubatum*; Fig. 229) seems to have run out of steam and is not increasing greatly, if at all.

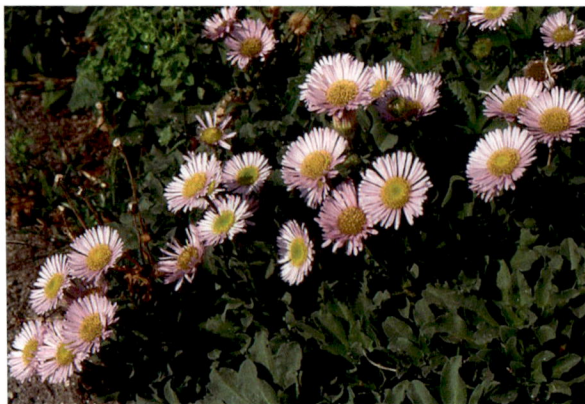

FIG 228. The North American Seaside Daisy (*Erigeron glaucus*) is a common sight on cliffs and rocky places by the sea, mainly in the south. (CAS)

FIG 229. Foxtail Barley (*Hordeum jubatum*) has become a characteristic coloniser of salted roadside verges, particularly of motorways and their interchanges: (a) roadside habitat; (b) close-up of inflorescences. (CAS)

Great Brome (*Anisantha diandra*) and Three-cornered Garlic are perhaps more surprising members of this group of ten species, but their recent increase is significant.

Much is made by conservationists of the pernicious spread by some alien aquatics, and two, Floating Pennywort (Fig. 134) and Least Duckweed (Fig. 230), support this concern with markedly increased records in the past decade. Least Duckweed was first noticed in the British Isles by chance in 1977 by the Swiss expert Elias Landolt while on a visit to Cambridge; it is now recorded in

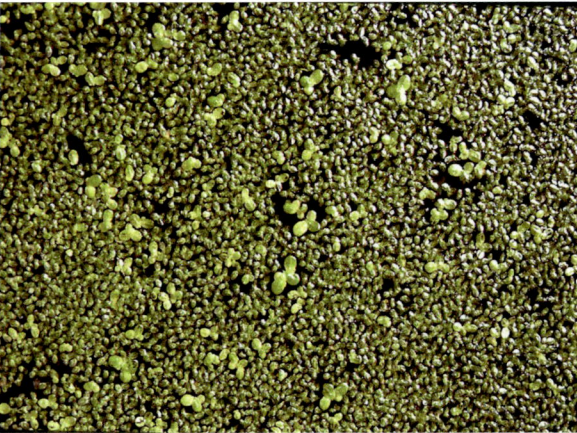

FIG 230. Least Duckweed (*Lemna minuta*) was first recorded in the British Isles as recently as 1977, but is now common in ponds and ditches, where it often grows with one or more of the native species. Here it is shown mixed with Common Duckweed (*L. minor*). (CAS)

442 · ALIEN PLANTS

921 hectads north to Banffshire in northern Scotland. It is often found mixed in with native duckweeds, and does not seem to be any more of a pest than they are. Floating Pennywort, however, not found here until 1990 and still only in 118 hectads, is causing real problems by completely clogging up canals and slow rivers, making them unusable by boats and harming wildlife by obscuring the water surface. It has so far reached north to Yorkshire and Lancashire.

Two grasses are showing remarkable increases. One of these is Cockspur (*Echinochloa crus-galli*; Fig. 231), known in North America as Barnyard Grass but found all over the warmer parts of the world. It grows up to more than a metre tall and is distinctive in having a rather one-sided inflorescence consisting of long spikes of spikelets. For some reason this species is now spreading rapidly, becoming a common sight in gardens, allotments and by roads in addition to the waste ground in which it was formerly characteristic. The *New Atlas* tells us that 'its distribution is now probably stable'; this is certainly not true any more. The other invading grass is Water Bent (*Polypogon viridis*; Fig. 232), which was formerly included in the genus *Agrostis* as *A. semiverticillata*. It was transferred to *Polypogon* because its ripe spikelets are shed whole by abscission at their base, not between the glumes and floret as in *Agrostis*. Around the turn of the nineteenth century, this grass became common in damp open ground in the Channel Islands, and this was almost the only place where one could be sure of finding it up to about 20 years ago. In the *New Atlas* it is shown as more or less confined to southern England, but it has now spread to northern Scotland and has reached Ireland, and is quite common in parts of central and northern England. As with Cockspur, the reason for its recent rapid spread is obscure.

FIG 231. Cockspur (*Echinochloa crus-galli*) is a grass of warm climates around the world that has been known as a casual in the British Isles for almost four centuries. It is now becoming increasingly common as a naturalised garden and wayside weed. (CAS)

FIG 232. Water Bent (*Polypogon viridis*, formerly known as *Agrostis semiverticillata*) was until recently naturalised only in the Channel Islands, but it has been spreading rapidly northwards in England. The bare inflorescence axes at fruiting are characteristic. (CAS)

Canadian Fleabane (Fig. 122a), a dull- and tiny-flowered relative of Michaelmas-daisies and goldenrods, is an alien from North America that has been with us since 1690. It underwent a steady and, at times, rapid increase in the second half of the twentieth century, but has shown little signs of that in this century and many judge that it is less common in some areas than it was 20 years ago, although it is still common in the Channel Islands and in Britain up to northern England. Another much rarer species, Argentine Fleabane (Fig. 122d), has been with us since 1843. Like much of this difficult genus it probably originated in South America, but is now distributed across much of the warmer parts of the Earth. This species, and more especially two other newcomers with similar world distributions, Guernsey Fleabane (Fig. 122b) and Bilbao's Fleabane (for a time known as *C. bilbaoana*; Fig. 122c), have shown remarkable increases in the past 20 years. Guernsey Fleabane was so called because it was first recorded in the wild in the British Isles on that island in 1961. Eventually it reached England, where London provided sufficiently warm conditions to support the plant, and there it joined Canadian Fleabane in its typical habitat of well-drained waste ground, notably railway cinders. It has now spread throughout England and is colonising Wales, Scotland and Ireland. In 1992, Bilbao's Fleabane was found in Southampton, and it is now spreading, so far occupying the southern halves of England and Ireland. Subsequently, a record from 1977 in Essex was uncovered. Both Guernsey Fleabane and Bilbao's Fleabane might have arrived here from Brittany, where they had been common for some time.

Finally, there is the remarkable case of Narrow-leaved Ragwort (Fig. 270) from South Africa, which has a similar growth habit and habitat to Oxford Ragwort, and, like it, is self-incompatible. From a distance they are indistinguishable, but

the leaves of Narrow-leaved Ragwort are mostly very narrow and unlobed, and the
two are actually not very closely related. The South African plant has occurred,
mainly as a wool alien, since 1836, and in the mid-twentieth century was known
from the wool alien-rich areas of Scotland, Yorkshire, Bedfordshire, inter alia, but
never commonly. In and around Bradford, however, it has shown a remarkable
increase in the past 15 years or so, and is now a common plant there. For decades
it has been very common in northern France, e.g. around Calais, but until
recently this range extended into Britain as just one patch near Dover, where it
was first observed in 1978. This remained the case for about 15 years, but since
then Narrow-leaved Ragwort has suddenly spread over many parts of southern
England. It is tempting to attribute its recent success to rail traffic through the
Channel Tunnel, which opened in 1994, and more recently due to shipping via
east coast docks. The plant is of special interest because of its origin here by two
different routes from different sources. From my own observations in East Anglia,
populations of this species are less persistent than those of Oxford Ragwort in
the same localities. The jury is still out on this one.

Scrutiny of the databases of the BSBI equally shows that a number of species
that have undoubtedly spread extensively in recent decades do *not* show a greater
number of records in 2000–09 than in 1987–99. They are showing signs that
their greatest period of increase is over and that they might be deemed as a lower
threat in years to come, although some are still increasing locally, often near the
edges of their ranges. For example, five well-publicised pestiferous aquatics do
not indicate rapid increase by the criterion that is adopted here, and possibly
are no longer markedly extending their ranges. However, the maps we have used
show distributions only at the hectad level, and it is likely that in many of the
hectads at least one of them, New Zealand Pigmyweed (Fig. 113), is 'filling in', i.e.
it is increasing in abundance rather than in range. The other four might well
have passed their period of rapid increase, and the first has certainly disappeared
from some areas: Canadian Waterweed, Nuttall's Waterweed, Curly Waterweed and
Parrot's-feather.

Examples of other invasive species that have not quite lived up to their
reputations in this century are Butterfly-bush, New Zealand Willowherb,
American Willowherb, Japanese Knotweed, Gallant-soldier, Shaggy-soldier,
Hoary Mustard, Hybrid Bluebell, Indian Balsam, Prickly Lettuce, Large-flowered
Evening-primrose, Winter Heliotrope and Oxford Ragwort. Hoary Mustard (Fig.
114), for instance, is now well naturalised in southern Britain and occurs north
to central Scotland. It showed its swiftest increase in the second half of the
twentieth century from its status in 1958, when Dandy included it in his checklist
on the basis of its being naturalised in only the Channel Islands.

NORTHERN NEOPHYTES

Plants that are commoner in the north than in the south of the British Isles are generally those requiring moister conditions, and this is no less true of aliens than of natives. For instance, I cannot maintain Sweet Cicely in my Leicestershire garden, yet it is a common wayside plant barely 80 km north in the Peak District, which has about twice the annual rainfall. A northern (or any other) regional tendency is likely to be more pronounced in aliens that have been here for some time, during which their pre-adaptation to particular environmental conditions can have taken effect. Those hunting aliens soon become aware of such geographical preferences.

Two special aliens found in the Shetland Islands are from extreme southern South America, and there is a belief (uncorroborated by hard facts) that they were transported by Shetlanders employed in South Georgia in the whaling industry or in the Falkland Islands in sheep farming. This is more plausible in the case of Tussac-grass (*Poa flabellata*; Fig. 233), known locally as Tussi-girse and often segregated as *Parodiochloa flabellata*. This is a robust, very densely tufted grass up to 2 m tall, with a dense inflorescence. It is common in the sheep lands of the Falklands, and its place in Shetland has been described by Scott & Palmer (1987). It was tested for suitability as fodder in Shetland in 1844–5, and eventually planted in crofting areas on or inside the stone walls of enclosures used for growing

FIG 233. Tussac-grass (*Poa flabellata*) is a giant tussock-forming grass from the extreme south of South America, whence it was brought to the Shetland Islands by seamen to provide green winter grazing: (a) large clumps by an old croft, with Rhubarb (*Rheum × rhabarbarum*) in the background at left; (b) close-up with inflorescences. (CAS)

FIG 234. Magellan Ragwort (*Senecio smithii*), from southern South America, is confined in the British Isles to the extreme north of Scotland and the Northern Isles, where it is well naturalised in damp grassland. (Ken Butler)

vegetables or protecting stock. It still exists there today, but there is little evidence of self-propagation; it simply seems to be a very long-lived perennial. It flowers from February to April, and 'is thus at its greenest during a very grey and harsh time in Shetland'.

Tussac-grass is confined to Shetland in the British Isles, but Magellan Ragwort (Fig. 234), a robust perennial up to 1 m tall with white-rayed, yellow-centred daisy-like flower heads about 5 cm across, occurs also in the Orkneys and some northern mainland counties, along streams or in wet pastures. Unlike Tussac-grass, it reproduces by seed and is well naturalised. There is apparently no evidence that this species, which does not occur in the Falklands or South Georgia, was introduced by southern whalers; it seems more likely that it originated here via the normal horticultural routes.

Another moisture-loving genus characteristic of Shetland is the monkeyflowers (*Mimulus*), discussed in Chapter 9. The colours of these spectacular flowers range from clear yellow to yellow with very heavy red blotching, and to coppery orange. Most of the taxa are sterile hybrids in various combinations and did not arise *in situ*, but are escaped garden plants. The yellow flowers are derived from the North American Monkeyflower (Fig. 169a), the red blotching from the South American Blood-drop-emlets and the copper colour from the South American *M. cupreus*. *Mimulus cupreus* itself does not occur in this country, and the Shetland plants with coppery flowers are the hybrid *M. × burnetii* (*M. cupreus × M. guttatus*). Most of these are what is known as a 'hose-in-hose' variety, where the calyx is petaloid, giving a flower with effectively two corollas, one inside the other. Apart from Monkeyflower, which is the commonest taxon in the genus in the British Isles, especially in the lowlands in the south, all the taxa of this group of the genus are commoner in northern Britain than in the south.

The tall, well-branched, white-flowered Aconite-leaved Buttercup (*Ranunculus aconitifolius*), from the Alps, is commonly grown in northern gardens and sometimes escapes from them. It is usually grown as a double-flowered (*flore pleno*) variety known as Fair-maids-of-France, a name that is also applied to the double-flowered variety of our native Meadow Saxifrage (*Saxifraga granulata*).

A popular genus that is represented by a different spectrum of taxa in northern than in southern gardens is the lupins (*Lupinus*). Whereas today the great majority of lupins in England and Wales, both in gardens and escaped into the wild, are Russell Lupins (Fig. 179a) in many colours, the two species Garden Lupin and Nootka Lupin, which are nearly always blue-flowered, are very frequent both in the wild and in gardens in Scotland, more so the further north you go. Here, the climate more closely approaches that of their native regions in northern North America. The only lupin recorded in the wild in Shetland is the Nootka Lupin, which is 'widely grown in gardens' there.

The extensively creeping New Zealand Willowherb (Fig. 224) is one of a group of similar New Zealand species that have caused considerable taxonomic problems both here and in their native area. It is a rarity among British and Irish aliens in that it is particularly common in upland areas, often along forest tracks and the like, but also encroaching into rocky natural habitats.

A tour of lanes in Scotland or northern England will soon reveal a list of alien plants that would be unusual in a more southern area. Three archaeophytes, all with ancient uses, have already been mentioned in Chapter 2: Masterwort (Fig. 223), Good-King-Henry and Monk's-rhubarb (Fig. 32). A neophyte in the same category is Sweet Cicely (Fig. 235), which is so common in the north that its distinction from the ubiquitous Cow Parsley (*Anthriscus sylvestris*) is an essential part of the early learning curve for beginner botanists. Its aniseed-smelling leaves and stems were once used in salads and with fruit dishes. Another is Rhubarb

FIG 235. Sweet Cicely (*Myrrhis odorata*) is commonly naturalised in the north, where it is one of the earlier umbelliferous plants to flower. Here it is seen flanking either side of a drystone wall that separates two unimproved meadows in the Peak District. (CAS)

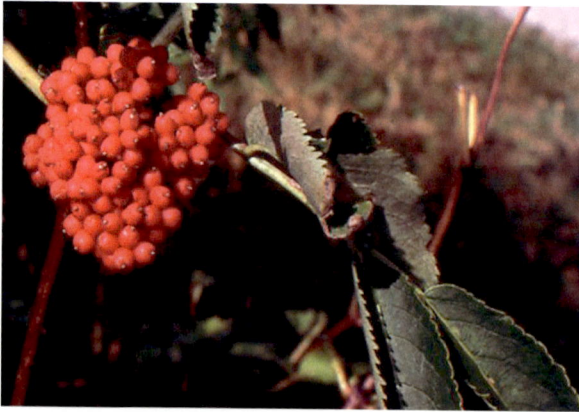

FIG 236. Red-berried Elder (*Sambucus racemosa*) is an attractive shrub of hedges and copses that is much commoner in the north than in the south, reflecting its mountainous habitats on the Continent. (CAS)

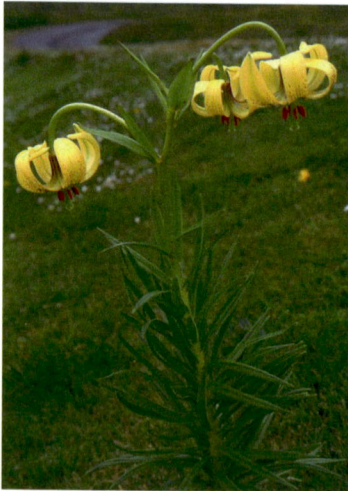

FIG 237. Pyrenean Lily (*Lilium pyrenaicum*) is a beautiful, strongly scented lily most suited to northern areas of the British Isles; it is seen here in Shetland. (CAS)

(*Rheum* × *rhabarbarum*), which is not only cultivated in the north but has more records there from the wild (most if not all are relics) too; it is frequent in old Shetland crofting areas.

Attractive garden plants that are frequently naturalised by northern roads, railways and paths include Milky Bellflower, Perennial Cornflower, Purple Crane's-bill (Fig. 96), Two-flowered Everlasting-pea (*Lathyrus grandiflorus*), Dotted Loosestrife (Fig. 253), Red-berried Elder (*Sambucus racemosa*; Fig. 236), several tall *Senecio* species such as Broad-leaved Ragwort (*Senecio sarracenicus*), and Pyrenean Valerian.

Plants that require more shaded habitats often find their ideal niche in the north, and moreover tend to be less demanding in cooler climates and so become more conspicuous there, even if not more common. Good examples are Pink Purslane, which is often white-flowered; the delicate little Fairy Foxglove; the strongly sweetly scented yellow Pyrenean Lily (Fig. 237); White Butterbur, a plant of the Alpes-Maritimes and always of the male sex here; the tall, clumped woodland grass Broad-leaved Meadow-grass; four species of yellow-flowered *Senecio*, including Broad-leaved

Ragwort; the beautiful pink-flowered, thornless Salmonberry, which is considered a problem weed in some parts of Northern Ireland; the extremely well-known London Pride (*Saxifraga × urbium*), a hybrid between *Saxifraga umbrosa* and St Patrick's Cabbage (*Saxifraga spathularis*); and those strange western North American relatives of the last species, Fringe-cups and Pick-a-back-plant. All these species, like the last group, are ornamental garden escapes or throw-outs that thrive in the local conditions.

FROST-SENSITIVE ALIENS

Probably the most significant factor contributing to the huge total of alien plants in the British Isles is that parts of the south and west are relatively frost-free, allowing species to thrive there that cannot withstand prolonged frost. Hence, we have many naturalised species that do not feature in the flora of central or even western mainland Europe. Gardens of central Europe are lacking in a huge range of our popular garden plants, the commonest source of our wild aliens.

Certain areas, notably Cornwall and County Kerry, are well known for their very mild climate, and for being able to support many 'tender' plants. Their often-used epithets 'subtropical' or 'Mediterranean' are, however, misapplied hyperboles, for nowhere in our islands can truly boast these descriptions, subtropical and Mediterranean climates differing fundamentally from anything we can offer. Nevertheless, many species that occur naturally in such conditions can withstand those in our southwestern extremities, as a slight frost for a few days can be withstood even by plants (e.g. Olive and Cork oaks) that almost never experience the likes in their native areas.

The Isles of Scilly have a special place in the annals of plant collecting. As early as 1650, the stewards of the Godolphins, governors of Scilly, were growing rare and exotic plants in their garden on St Mary's, so Augustus Smith knew exactly what he was doing in 1834 when he obtained a 99-year lease on the archipelago from the Duchy of Cornwall. A year later, he commenced building Tresco Abbey, which became his residence. What makes the Isles of Scilly so important for British botany and horticulture is that mean monthly temperature fluctuates by only 9° C between the coldest month of the year (February, at 7° C) and the warmest month (August, at 16° C). Many species grow in the open there that could only be grown under glass in other parts of the British Isles, and flowering seasons are advanced dramatically. The strength and the saltiness of the winds mean that trees will typically not grow without shelter, and those that put their heads above the shelter are ruthlessly wind-pruned. Of the trees used as shelter belts the evergreens are the most effective, including Evergreen Oak, Monterey Cypress and Monterey Pine. Shrubs used for hedging in the Isles of Scilly are mentioned in Chapter 5. Smith specialised in importing seeds and roots of plant species from South Africa, Australia and New Zealand, and many of the aliens now naturalised in the Isles of Scilly are from those countries as well as from South America, the Canary Islands and Madeira.

The Scilly bulb fields have developed a very characteristic weed flora; among the most conspicuous species are Bermuda-buttercup (Fig. 238), Garden Star-of-Bethlehem (*Ornithogalum umbellatum*), Spanish Bluebell (Fig. 172c), Eastern Gladiolus (*Gladiolus communis* subsp. *byzantinus*), Three-cornered Garlic (Fig. 238), Rosy Garlic, Tassel Hyacinth, Rough-fruited Buttercup (*Ranunculus muricatus*), fumitories (*Fumaria*), Small-flowered Catchfly (*Silene gallica*) and Lesser Quaking-grass (*Briza minor*). The dunes likewise have their collection of rare aliens, including Bugle-lily (*Watsonia borbonica*), African Lily (*Agapanthus praecox*; Fig. 106), Red-hot-poker (*Kniphofia uvaria*; Fig. 107) and Compact Brome (*Anisantha madritensis*). The granite stone walls, known locally as hedges, also display a distinctive flora, including Three-cornered Garlic, Fleshy Yellow-sorrel (*Oxalis megalorrhiza*), Wireplant (*Muehlenbeckia complexa*) and German-ivy (*Delairea odorata*).

Extreme frosts are, however, not unknown in the far southwest, and these cause immense damage to the tender alien plants and greatly affect the appearance of gardens and naturalised vegetation. One such occurrence in the Isles of Scilly was the exceptionally cold winter of 1987/8, when many plants, including trees of considerable age, were killed. However, such plants are cherished by gardeners and were sorely missed, and they were soon reintroduced, later to reappear in the wild as escapes. As Rosemary Parslow pointed out in her New Naturalist *The Isles of Scilly* (2007), some of these tender plants had developed into large stands, in the centre of which were parts

FIG 238. Aliens in damp grassland on the Isles of Scilly: white Three-cornered Garlic (*Allium triquetrum*), yellow Bermuda-buttercup (*Oxalis pes-caprae*) and, in the background, Altar-lily (*Zantedeschia aethiopica*). (CAS)

FIG 239. Cabbage-palm (*Cordyline australis*) is not a palm but a woody member of the *Asparagus* family. It grows best in the mild southwest, away from which it can be killed to the base by hard frosts. Here it is being grown as shelter for bulb fields in the Isles of Scilly. (CAS)

protected enough to survive. Others that were apparently killed eventually sprouted again from the rootstock. This proved salvation for the evergreen hedges of New Zealand, pittosporum (*Pittosporum*) and daisy-bush (*Olearia*; Fig. 93a) and even Cabbage-palm (Fig. 239), which are used to enclose the bulb fields, as many of these were killed to the base but not below soil, which allowed them to regrow within a few years without replanting.

The most 'exotic'-looking plants cultivated in the Isles of Scilly are arguably the palms and some of the other huge monocotyledons such as the centuryplants (*Agave*), Spanish-daggers (*Yucca*), New Zealand flaxes (*Phormium*) and cabbage-palm (*Cordyline*). The most cold-tolerant true palm is the Chusan Palm (*Trachycarpus fortunei*), from China, with trunks up to about 10 m high bearing a great apical tuft of palmately divided leaves, so characteristic of seafronts on the Devon and Cornish 'rivieras' and of London parks. This species regularly produces its distinctive ribbed seedlings but these do not grow to maturity, and the species is merely a survivor. It can be successfully grown much further north; the famous Royal Botanic Garden at Logan in Galloway, West Scotland, at almost 55°N, has a small avenue of the palms. The Date Palm cannot be grown in the British Isles, but it does occur here as a casual in the form of its seedlings (Fig. 240), which were formerly a characteristic sight on rubbish tips before they were routinely covered, but they never survive their first winter.

Most of the other large monocotyledons are also survivors that do not effectively reproduce here. Some of them, such as Centuryplant (*Agave americana*), are monocarpic; that is, they live for many years vegetatively, slowly building up a vast resource, and then produce a huge inflorescence. Centuryplant has no appreciable vegetative stem, but produces its huge, wickedly spiny leaves in a giant sessile rosette. After the seeds are dispersed, the plant dies, although in

FIG 240. Date Palm (*Phoenix dactylifera*) is quite frequently seen in the seedling stage on rubbish dumps in the British Isles, but it never survives its first winter with us. (CAS)

some cases small offsets are produced at the base. In the British Isles the flowers are produced only in the warmest regions.

The Centuryplant can grow to 8 m or more high, but it has an even larger Mexican relative, *Furcraea parmentieri* (formerly incorrectly known as *F. longaeva*), which is grown in the extreme southwest of the British Isles, including the famous gardens at Tresco in the Isles of Scilly. This is also monocarpic, but its inflorescence, which can allegedly exceed 12 m and is much wider than that of Centuryplant, rarely produces viable seed. However, numerous bulblets that develop into plantlets are produced among the flowers, and these drop off and can take root. Rosemary Parslow tells me that on Tresco these plantlets are being carried off by gulls, which drop them on to the dunes, where they are now growing.

The mature plant has a relatively short stem below the leaf rosette, but this is insignificant compared to the tall branched stem of the Cabbage-palm, which leads to it being wrongly referred to as a palm; its undivided leaves show that it is not.

New Zealand Flax (*Phormium tenax*) is frequent and still spreading in the Isles of Scilly (in 1898, S. W. Fitzherbert said it grew like a weed on Tresco). There had been plans to grow this species as a commercial crop (for its flax-like fibre) on the islands but they were never implemented. Lesser New Zealand Flax (*P. cookianum*) is locally well established among Bracken on St Martin's, but much less widespread.

Two other extraordinary large monocots occurring in the wild in Scilly are quite different in growth form. These are Rhodostachys (*Fascicularia bicolor*) and Tresco Rhodostachys (*Ochagavia carnea*; Fig. 117), members of the Bromeliaceae (pineapple family). The plants form giant hemispherical mounds up to 2 m across and 75 cm high, composed of branched stems bearing an impenetrable mass of very spiny leaves. Inflorescences are freely produced within these spiny mounds, although viable seed production has not been reported. The former species is well established in natural maritime, lichen-dominated dunes, where it presents an amazing spectacle.

Outside the monocotyledons there is another monocarpic giant that is naturalised in the extreme southwest, Giant Viper's-bugloss, a native of La Palma in the Canaries, where it is very rare and endangered. It thrives with us, however, and also in New Zealand, where it is well naturalised. It grows a vertical woody stem up to about 1 m tall, with leaves up to 50 cm or so long, and when mature (usually in a few years) it puts up a massive inflorescence up to 3.5 m tall with thousands of blue flowers that are very attractive to insects and produce abundant seed.

Gardens in the extreme southwest support a range of plants rarely grown in most of the British Isles. This is relevant to our present subject because it is those plants that escape and are thrown out from gardens, and in some cases become naturalised. Examples of such plants are: the shrub Marvel-of-Peru (*Mirabilis jalapa*); a number of succulents such as Aeonium (*Aeonium cuneatum*) and several species of Aizoaceae, including Hottentot-fig, Shrubby Dewplant (*Ruschia caroli*) and Pale Dewplant (*Drosanthemum floribundum*; Fig. 241); showy Iridaceae such as Chilean-iris (*Libertia formosa*), Chasmanthe (*Chasmanthe bicolor*) and Bugle-lily; Amaryllidaceae such as African Lily and Jersey Lily (*Amaryllis belladonna*, not to be confused with amaryllis grown as pot plants, which belong to the genus *Hippeastrum*); the familiar Altar-lily (Fig. 238); several attractive Asteraceae such as the treasureflowers *Gazania* and *Arctotheca*; and Giant Herb-Robert (*Geranium maderense*; Fig. 242).

FIG 241. Pale Dewplant (*Drosanthemum floribundum*) is one of the 'mesems' from South Africa that is naturalised in extreme southwestern Britain: (a) forming green 'curtains' on low cliffs in the Isles of Scilly; (b) close-up showing flowers and leaves. (CAS)

FIG 242. Giant Herb-Robert (*Geranium maderense*) is endemic to Madeira but here is seen thriving on cliffs on St Mary's, Isles of Scilly, with Three-cornered Garlic (*Allium triquetrum*), Montbretia (*Crocosmia × crocosmiiflora*) and Alexanders (*Smyrnium olusatrum*). (CAS)

These are just a few of the more conspicuous representatives of a large range of species commonly grown outdoors in the southwest in ordinary front gardens. Most of them are sometimes also grown elsewhere in the British Isles in more sheltered spots, often in pots in conservatories and on patios and forecourts. Another example that springs to mind is Cineraria, that cheerful daisy-like pot plant sold in the millions for Mothering Sunday; despite its common name, it does not belong to the genus *Cineraria*, nor is it very closely related to Silver Ragwort (*Senecio cineraria*), but is the Canary Islands plant *Pericallis hybrida*. This species abundantly reproduces itself all over the churchyard on St Mary's, Isles of Scilly.

Horticulturists have been able to select more frost-tolerant varieties of a number of the plants that were first grown in the southwest. Cape Daisy is one such, and the African lilies (*Agapanthus*) form another genus extending its garden range. But for the most part, these frost-sensitive species are seldom seen in most of our gardens, let alone naturalised in the wild.

Aside from plants largely confined to areas with frost-free conditions, many others are far more common generally in the milder western parts of Britain and in Ireland and the Channel Islands. Lanesides in much of rural Ireland are awash with hedges of Fuchsia (Fig. 243) and Montbretia, whereas further east these plants are confined to more sheltered spots. Maritime areas in general have a milder climate, and many plants are grown, and escape, far more commonly there than inland. Species such as Seaside Daisy (Fig. 228), Treasureflower and Silver Knapweed (*Centaurea cineraria*) are frequent sights by the sea in the south, but rarer elsewhere. Tender shrubs that are also evergreen and salt-tolerant are especially grown as hedging near the sea; examples are given in Chapter 5. Oleasters (*Elaeagnus*) are grown over much of the country as usually variegated, hardy garden shrubs, but they produce flowers (in the autumn) and fruits only along the milder coasts where they are used as hedging, especially in the Channel Islands.

Apart from the mild western and southwestern parts of the country, local man-made conditions elsewhere provide niches for thermophilic plants. The most obvious examples, greenhouses, are outside the scope of this book, whether artificially heated or not, despite the fact that some botanists have recorded from such places species of flowering plants, ferns and bryophytes that are not otherwise on our list of aliens. Large conurbations in general have a milder climate than surrounding rural areas, frequently escaping widespread frosts. This

FIG 243. Fuchsia (*Fuchsia magellanica*) is abundantly naturalised in the west and has become a characteristic hedgerow shrub in western Ireland. (CAS)

particularly applies to Greater London and the lower Thames Valley, which has often been the location of newly arrived warmth-seeking aliens.

Within our cities there are localised areas of even greater protection from the cold. The sunken forecourts of city basement flats have already been mentioned in Chapter 10 as a good habitat for tender ferns. Areas protected by walls, especially in narrow passages, are in a similar category. More specialised areas include rubbish tips, sewage farms and manure heaps, where rotting organic material produces heat. Nowadays such places are few and far between because perceived sanitation demands cause them to be covered rather than left open, leading to a great reduction in alien plants. Wade & Smith (1926) observed that the dumping at port areas of hot furnace slag served to provide a warmed environment for tender plants, allowing some to become naturalised.

At all times water provides a major buffer to temperature fluctuations – for example, rivers and canals in cities rarely become iced over. These waterways often receive effluent from domestic and industrial sites, further raising the temperature. The warmed water in the Greater Manchester area provided the unique habitat for at least one alien angiosperm, Egyptian Naiad, known in the Reddish Canal from 1883 to 1947, as well as for several freshwater algae and other flowering plants as described in Chapters 4 and 10. Bailey (1884) provided a long and thorough account of the naiad, which was described by the German specialist P. Magnus as a distinct variety, var. *delilei*. The species has separate male and female flowers borne on the same plant, and in the Reddish Canal mature fruits were 'produced in great abundance; scarcely a plant occurred without fruits'. There might have been one or two other alien angiosperm taxa there as well. Bailey described an unnamed horned pondweed (*Zannichellia*) with fruits bearing four or five rows of much more prominent dorsal spines than in any British material; it might have been an alien subspecies of Horned Pondweed (*Z. palustris*). In addition, Reverend Ted Shaw, that well-known alien hunter from Oldham, reported that an alien species of bladderwort (*Utricularia*) was found in the canal by the cotton mill in 1959 (Savidge *et al.*, 1963). Neither of these two additional taxa seems to have been investigated since.

The 332 casuals listed in Appendix 1 are plants that cannot survive our normal winters. The perennials among them are killed off before they can flower, but many of the annuals do flower and often produce seeds. Annuals can either germinate in the autumn and pass the winter as seedlings ('wintergreen'), or remain as seeds during winter and germinate in the spring. The seedlings of wintergreen annuals that are merely casuals with us die in the winter frosts, whereas casuals that are not wintergreen either do not produce viable seeds or the demise of the next generation occurs by seed death (often due to moist, warm conditions in our variable winters) or by seedling death for some reason after spring germination. It would be interesting to know what proportion of our casuals does produce viable seed, and which of those are wintergreen. Unfortunately, we do not have the necessary data, and it would be pointless trying to guess or generalise; for example, Greater Quaking-grass is wintergreen, while Lesser Quaking-grass mostly germinates in spring.

AN IRISH CONUNDRUM

There are three unequivocally native species that occur in the west of the British Isles and nowhere else in Europe, but are more widespread in North America. These are Pipewort (*Eriocaulon aquaticum*), American Pondweed (*Potamogeton epihydrus*) and Irish Lady's-tresses (*Spiranthes romanzoffiana*). There have been many attempts to explain such an unusual distribution pattern, none of which has been universally accepted. The explanations will not be repeated here, but in the main they involve arguments between two alternatives: that the disjunct distributions represent relics of a time when there was a connection between Europe and North America (the *land-bridge hypothesis*); and that they are the result of chance dispersal from America to Europe, probably by birds (the *long-range dispersal hypothesis*).

There are also a few other species with similar distributions, but there are varying levels of dispute as to whether they are natives, like the above three, or aliens in the British Isles. Welsh Mudwort (*Limosella australis*) occurs only in wet pond mud in a few sites in west Wales, was first found in 1897, and is mentioned in Chapter 1. One of the species most often claimed as native is Blue-eyed-grass (*Sisyrinchium bermudiana*), a beautiful member of the Iridaceae, which occurs in wet meadows and lakeside vegetation in western Ireland, where it was discovered in 1845. It is, however, also found in mainland Britain as a garden escape and, moreover, it belongs to a complex of species, the classification of which is not fully resolved. One of its closely related species, American Blue-eyed-grass (*S. montanum*), also occurs as an alien in Britain and Ireland, and the two species are often confused. In their *New Atlas*, Chris Preston *et al.* (2002) accepted Blue-eyed-grass as a native, but because of the doubts we treat it as 'uncertain'. The taxonomic problems in this genus were multiplied by the suggestion that the Irish native plants are distinct from the American aliens, and the former were accordingly described in 1961 as the endemic *S. hibernicum*, a species no longer accepted.

A second species that we treat as uncertain is the diminutive Canadian St John's-wort (*Hypericum canadense*), which was discovered in boggy ground in western Ireland as recently as 1954. A detailed study by Webb & Halliday (1973) concluded that it is probably native, but it is treated as an alien in the 2002 *New Atlas*. If it is a native, brought to Ireland by birds, its recent discovery is irrelevant, but we shall probably never know for sure.

In the classic work by J. R. Matthews (1955), classifying the flora of the British Isles into geographical elements, the above group of species was described as the 'North American Element'. Matthews included in that element Slender Rush (Fig. 64), a North American species that is now believed to be an alien in the British Isles. Although it was known earlier in Britain and mainland Europe, and there as an obvious alien, from the time of its discovery in Ireland in 1889 right up to the middle of the twentieth century, its possible native status there was much debated. The idea that it might be native was mainly based upon the fact that its early Irish localities were in wild country among native vegetation, and of course there are other American species that are native in Ireland. Native status for Slender Rush, however, can be considered no higher than feasible, and all recent authors have considered it to be an alien in Europe.

Pineappleweed (*see* Chapter 11; Fig. 127), first found in the British Isles in 1869, was at first particularly common in Ireland, prompting a suggestion that it arrived there from North America by natural means. It is, however, not a native of America, but of Asia. A variant of it, with the epithet *occidentalis* applied at various ranks, also occurs here, and this has complicated our understanding of its status (*see Sisyrinchium* above). It is likely that Pineappleweed was brought across the Atlantic by people.

FIG 244. Two species from the opposite ends of the Earth, whose discovery in boggy ground in Galway came as a total surprise, not least because their mode of origin there is a mystery: (a) Creeping Raspwort (*Haloragis micrantha*), discovered in 1988; (b) Broad-leaved Rush (*Juncus planifolius*), discovered in 1971. (Ian Denholm)

Two other Irish plants, neither found anywhere else in Europe, are not North American in distribution but are from the southern hemisphere (Southeast Asia, Australasia and South America). They are Creeping Raspwort (*Haloragis micrantha*; Fig. 244a), an inconspicuous plant allied to the water-milfoils (*Myriophyllum*) that was discovered in 1988; and Broad-leaved Rush (*Juncus planifolius*; Fig. 244b), which was first found in 1971 and when not in flower resembles a grass rather than a rush. Both occur in wet, boggy ground on peat, where alien species would not be expected; Broad-leaved Rush is spread over about 40 km^2 and occurs with Creeping Raspwort in the latter's single site. It is just possible that their distribution represents a natural bipolar disjunction, but surely it is more likely that they have been introduced (possibly together). The absence of any obvious vector, however, has persuaded Eric Clement (2010) that they are native in Ireland. Possibly the peat-cutting industry could have been a reason for human traffic between the native sites and western Ireland. We have treated them as aliens.

The presence of so many species of uncertain origin in remote, sparsely populated parts of Ireland is truly remarkable.

ALIEN ENDEMICS

Oxymorons usually demand explanation. There are some plants that as far as we know occur nowhere outside the British Isles (hence they are endemic to our region), yet they are considered by some people to be aliens here. The only way to square this circle is by supposing that the native area of these taxa is outside the British Isles, into which they have been introduced, but that they have become extinct in their original areas. Such reasoning does not follow the law of parsimony, whereby the simplest explanation (that the taxa either are not aliens or are not endemics) is considered to be the most likely one. Here we follow the law and argue that the taxa are most likely natives. Our first example, Babington's Leek (*Allium ampeloprasum* var. *babingtonii*), was considered to be an archaeophyte by Preston *et al.* (2002) and was briefly discussed in Chapter 2. There it was pointed out that this taxon is either a native here or represents a mutation that occurred in our populations of Wild Leek, which we believe have been here for more than half a millennium. In that case it would be a neonative. It could plausibly have arisen from either of the other two varieties of Wild Leek mentioned in Chapter 2, because all three differ only in the form of their inflorescence.

Plot's Elm (*Ulmus plotii*), sometimes considered as a variety or subspecies of Small-leaved Elm (*U. minor*), is a most distinctive tree with a very narrow, irregular outline. It was formerly a very characteristic feature of the landscape of parts of the East Midlands, and occurred more sparsely elsewhere in central England. Unfortunately, it proved particularly susceptible to the virulent strain of Dutch elm disease that all but wiped out field elms in the 1970s, but small numbers still exist and many hedgerows retain its 'brushwood sheaf'. It is not known on the Continent. The detailed and perceptive researches of R. H. Richens carried out in the second half of the last century on the distribution and history of elms in Britain led him to believe that all elms (termed by him 'field elms') other than Wych Elm (*U. glabra*) are aliens in the British Isles, but he failed to explain the Plot's Elm anomaly. An alternative view, followed here and in the 2002 *New Atlas*, is that Plot's Elm is a native.

Interrupted Brome (*Bromus interruptus*) is an enigmatic species, endemic to England but not seen in the wild since 1972 and presumed extinct, although it still exists in cultivation. It appears to be most closely related to the abundant Soft Brome (*B. hordeaceus*), from which it differs in its dense, unevenly lobed inflorescence and in the unique morphology of its palea, which is longitudinally split in two to the base. The species was not discovered until 1849, and most of its records were from fodder crops, mainly Sainfoin, but also clovers, cereals and grasses (especially rye-grasses *Lolium*). In the *New Atlas*, Preston *et al.* (2002) concluded that 'Its history, habitat and scattered distribution in Britain suggest that it may have been introduced as a seed contaminant from an unknown native range rather than having evolved here', and they treated it as a neophyte. On the other hand, in their more recent grass handbook, Cope & Gray (2009) decided that 'until a likely origin for it outside the U.K. can be demonstrated it must be accepted as both native and endemic', with which we agree.

Interrupted Brome is usually supposed to be a crop mimic. Such plants are dispersed and withstand eradication by mimicking in some way the crop with which they are associated. There has been, however, no convincing demonstration of mimicry of Interrupted Brome's commonest cohabitant, Sainfoin, which is now much less often grown as a crop than previously. If Interrupted Brome is a crop mimic, its most likely evolution is from Soft Brome, in which case it would be a native rather than a neonative. This might have occurred in the relatively recent past, say a few centuries ago. In Belgium, two other crop-mimicking bromes have evolved: *Bromus grossus* and *B. bromoideus* (formerly known as *B. arduennensis*). Both of these species were characteristic weeds of Spelt Wheat, a primitive cereal now rarely grown in western Europe, although it has recently experienced a slight revival in popularity (*see* Chapter 2).

Taken together, these crop mimics suggest that they did indeed arise quite recently in the areas in which they became most common, as mimics of crops that were at the time widely grown, but that changing agricultural practices have doomed them to extinction.

There is, however, another category of alien endemics for which our conclusion must be quite different. These are plants that were described as new species to the British Isles because they could not be matched with any known species, but which we know or believe to be aliens brought to this country by various means. In some cases the new species has subsequently been matched with plants growing in its native area, but sometimes this has not happened and the mystery remains. The vast number of species in the literature means that it is all too easy to miss the one that matches the newly discovered alien, especially if the native area is one for which no modern Flora exists.

In the pepperworts (*Lepidium*), the pre-eminent Swiss wool-alien expert Albert Thellung described two new species from the area around Galashiels in the Scottish Borders: *L. peregrinum* in 1913 and *L. pseudodidymum* in 1914. To judge from the species that they most closely resembled, these two were aliens from Australia and South America, respectively. When Bruno Ryves (1977) came to revise the wool-alien species of *Lepidium* in the British Isles, he continued to recognise *L. pseudodidymum*, which since its description has been found in Argentina and Chile, but considered *L. peregrinum* too poorly known for him to assess it. However, it has been found in the wild in Australia, where it is considered threatened and nearly extinct. Ryves also felt it necessary to describe as new a third alien pepperwort, from Blackmoor in Hampshire, under the name *L. fallax*; this he considered comes from South America.

FIG 245. The type specimen of *Chenopodium × haywardiae*, said (but not proven) to be a hybrid between *C. hircinum* and *C. strictum*, was discovered by Ida Hayward on the shingle at Galashiels. (From Hayward & Druce, (1919))

The goosefoot *Chenopodium × haywardiae* (Fig. 245) was described from one of the Galashiels sites by Josef Murr as the hybrid between Foetid Goosefoot (*C. hircinum*) and Striped Goosefoot (*C. strictum*), although this parentage remains untested. The nature of the flora that existed at these Scottish wool-alien sites strongly suggests that *C. × haywardiae* was an alien, imported as hybrid seed, rather than a neonative that arose here.

CHAPTER 13

Habitats of Alien Plants

Long live the weeds and the wilderness yet.
Gerard Manley Hopkins, 'Inversnaid' (1918)

I t could be said that every plant in the landscape owes its existence to people, in one way or another. Either we planted it as a crop, for shelter or for ornament, or we cherish it as a valued fragment of semi-natural vegetation. We might tolerate it because it is growing somewhere that is currently undervalued, or the plant might be growing in the wrong place but we haven't got round to destroying it yet.

Despite the ubiquitous influence of people on our landscape, there are clear patterns in the extent to which different habitats are invaded by alien plants, both in terms of the numbers of species and their abundances. It is important to know whether some habitats are inherently more invasible than others, or whether these differences are due simply to differences in propagule pressure. If there are differences in invasibility, then we need to know whether this is related to native species richness of these habitats, and if so, by what mechanisms. Equally, we need to know whether the consequences of invasion by alien plants are the same in all habitats, or whether some habitats are prone to suffering greater damage than others. The distribution of native, archaeophyte and neophyte species across habitats is shown in Figure 246. The overall pattern is crystal clear. In terms of species richness, the most invaded habitats are all man-made: waste ground, built environments and plantation woodlands. The least invaded are those habitats that are closest to pristine, with low propagule pressure and low levels of direct disturbance by people, such as heath, wetland, saltmarsh, montane, bog and marine.

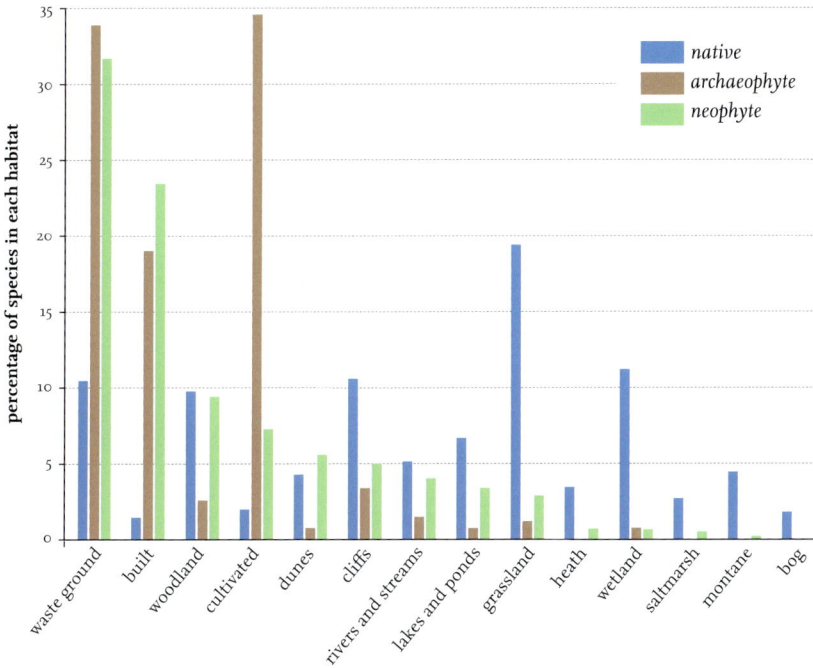

FIG 246. Habitats ranked by the fraction of their flora made up of neophytes. Many species grow in more than one habitat but, for the purpose of this illustration, each species was allocated to just one (its principal habitat), so that the total numbers of natives, neophytes and archaeophytes are preserved (Stace 2010, Appendices 1 and 2, respectively), and the percentages add to 100 within each status. Waste ground and built environments are exceptionally rich in alien plants species, while grasslands, heaths, wetlands, mountains and bogs are relatively poor.

Built environments include both hard-work and soft-landscaping in towns, villages and industrial sites. Woodlands comprise semi-natural woodlands, parkland and plantation woodlands of mostly deciduous trees (as distinct from conifer plantations). Cultivated habitats are arable fields, along with smaller-scale areas like gardens and allotments. Maritime habitats are separated into dunes, cliffs and rocky shore, and saltmarsh. Grasslands are typically unfertilised and unsown (as distinct from improved grasslands, which are fertilised, drained and sown with agricultural grasses). Heath includes lowland heath and dry heaths in the uplands. Wetlands include swamps, fens and water meadows. Montane includes all plant communities above 500 m altitude (but at lower altitudes in the far north and west). Bog comprises wet heaths, cottongrass (*Eriophorum*) and blanket bog.

Native plant species numbers peak in grasslands and wetlands, both of which are relatively poor in aliens. Archaeophytes show pronounced peaks of species richness in cultivated ground, waste ground and built environments. Neophytes peak in both species richness and abundance on waste ground and in built environments. Both neophytes and archaeophytes are rare in habitats that are characterised by high perennial plant cover, high altitude, low pH, low rates of soil disturbance and low human population density.

We describe the habitats in order of increasing alien influence (i.e. from right to left in Figure 246). Much of the information comes from recording by MJC in London, East Sutherland and Berkshire.

THE MAIN HABITATS

Heaths and mires

Numbers of alien species are uniformly low in mires and heaths. It is clear, however, that heathland communities are potentially invasible by alien species. Vast areas of heathland in central Ireland, for example, are carpeted by the alien Heath Star Moss (*Campylopus introflexus*) in quantities sufficient to interfere with regeneration of native vascular plants (Fig. 247). In Kintyre, on the southwestern coast of Scotland, Rhododendron is locally abundant in all but the most exposed heathland localities. It is capable of colonising open moorland and is fire-resistant as a seedling. The plant 'might well form a monoculture from sea level to the 200m contour' if uncontrolled (Cunningham & Kenneth, 1979). Elsewhere, shrubs like Himalayan Cotoneaster and Mountain-laurel (*Kalmia latifolia*) are locally frequent in heathland. Intentionally introduced herbs like Pitcherplant thrive over substantial areas of bog in central Ireland. In many parts of the uplands, however, one can walk all day through bog and heath without encountering a single alien vascular plant.

Saltmarshes

Saltmarsh is very poor in alien plant species, but ironically, this was the scene of one of the most impressive and best-studied hybridisation events involving native and alien plant species. The story of the origin and spread of Common Cord-grass is told in Chapter 9. The original hybridisation event appears to have taken place in a Smooth Cord-grass saltmarsh community at Marchwood in Hampshire, and led to the creation of a new, alien-dominated Common Cord-grass saltmarsh community (Fig. 167).

London's muddy seaside comprises the rather bleak, Dickensian estuaries of Essex and Kent that are the location for so much of the capital's industry: power

FIG 247. Heath Star Moss (*Campylopus introflexus*) is the only seriously invasive bryophyte in the British Isles, causing damage to native ground-level vegetation on heaths and dunes. It is seen here in central Ireland: (a) habitat; (b) dominating the ground layer; (c) close-up. (a–b, MJC; c, Ian Atherton)

stations, docks, cement works, scrapyards, landfill sites, factories, sewage works and oil refineries. Fragments of saltmarsh and reedbed sit below the sea wall, but they support few, if any, alien species. It is interesting that the list of the top ten aliens from the sea wall itself does not have a particularly maritime feel to it: Oxford Ragwort (3rd most frequent of all species), Butterfly-bush (4th), Bristly Oxtongue (6th), Mugwort (7th), Guernsey Fleabane (12th), Perennial Wall-rocket (14th), Fennel (20th), Hoary Mustard (30th) and Hoary Cress (33rd). Other aliens that are frequent enough elsewhere in London and abundant by the seaside in other parts of the British Isles, like Alexanders, are conspicuous by their absence from this list.

Grasslands

We need to distinguish between grasslands proper (habitats created and maintained by grazing domestic animals, which may or may not be cut for hay or silage) and grassy waste places (typically the early successional stages of ground destined to become dominated by scrub or woodland). Our managed grasslands are typically poor in alien species, but grassy places and lawns can be much richer (as described below under 'Waste ground' and 'Parks, gardens and lawns in towns').

Upland habitats are all exceptionally poor in aliens. The combination of remoteness and low human population density means that propagule pressure is lower here than in probably any other habitat. The harshness of the abiotic environment, the shortness of the growing season and the typically wet acidic soils further restrict the range of alien plants that could grow here.

Alien plants are very uncommon in well-managed calcareous grasslands. However, if domestic animals are taken off the grasslands or Rabbit numbers decline, then the community may be rapidly invaded by shrubs, in which case bird-dispersed alien shrubs like *Cotoneaster* sometimes become invasive (Fig. 248). Sainfoin may be native in some chalk grasslands. It was formerly grown widely as a fodder crop and in recent years has been spray-seeded in huge quantities onto chalky motorway banks, with the result that most records these days are of undoubtedly introduced plants. The only alien species recorded with any frequency from calcicolous (lime-rich) grasslands are Weld (*Reseda luteola*) and seedlings of Sycamore.

Considering that Britain's mesotrophic (neutral) grasslands were the source of so many of the most invasive alien species in other parts of the world, it is intriguing that they themselves have proved to be so resistant to invasion by alien species. Think, for example, of the havoc wreaked by Ragwort (*Senecio jacobaea*) in California, Nodding Thistle (*Carduus nutans*) in Montana, Creeping Thistle (*Cirsium arvense*) in Canada, Spear Thistle (*Cirsium vulgare*) in Japan, Perforate St John's-wort (*Hypericum perforatum*) in Australia or Mouse-ear Hawkweed in New Zealand. Of the 100 or so meadow plots on the famous Park Grass Experiment at Rothamsted in Hertfordshire, begun in 1856 but still going strong, all but two

FIG 248. Invasion of prime-quality limestone grassland in the Cotswolds, home of many rare plants and insects, by Turkey Oak (*Quercus cerris*; pale green clumps), Wall Cotoneaster (*Cotoneaster horizontalis*; large bush on the right), and Entire-leaved Cotoneaster (*C. integrifolius*; large bush on the left). (CAS)

have no alien plant species at all. Plot 17, which is closest to the manor house, has a small population of Fritillary, while Plot 10 has a single individual of Garden Star-of-Bethlehem (Crawley *et al.*, 2005).

The apparently low invasibility of our grasslands may result from the long association between British grassland plants and the domestic livestock (Cattle, Sheep, Goats and Horses) that helped create them. It is reasonable to suppose that alien plant species coming from communities with different grazing regimes would struggle to cope with the combination of selective herbivory and intense plant competition.

Many of the most intensively managed lowland grasslands are sown with alien grasses like Italian Rye-grass and alien genotypes of native grasses. Alien plants are often restricted to the most disturbed parts of these grasslands: muddy gateways are a good example, where one is likely to find species like Swine-cress, Shepherd's-purse and Pineappleweed.

Swamps

One of the 28 swamp communities in the British National Vegetation Classification (Rodwell, 2006) is named after its dominant alien, Sweet-flag (Fig. 249); the community also supports other alien species like Monkeyflower, New Zealand Pigmyweed and American Willowherb. The other most frequently recorded aliens in swamp communities include Canadian Waterweed, Dame's-violet and Orange Balsam in tall-herb fen.

FIG 249. Sweet-flag (*Acorus calamus*), a strange waterside plant formerly placed in the Araceae (arum family) but now recognised as belonging to Acoraceae, the first family to diverge from the monocot lineage. It has leaves that are sweetly scented when bruised, but it does not set seed in the British Isles. (CAS)

Shaded swamps in Alder carr are sometimes heavily invaded by aliens, with species like American Skunk-cabbage attaining dominance, along with invasive native plants such as Pendulous Sedge (*Carex pendula*). All three invasive balsams can coexist in this habitat: Orange Balsam, Small Balsam and Indian Balsam. Uncommon species of shaded swamp include Ostrich Fern and Sensitive Fern.

Still and running freshwater

Worldwide, freshwater ecosystems are among the most severely invaded by alien plant species, and several of the world's worst weeds are found in this habitat (*see* Chapter 15). For instance, Water Hyacinth (*Eichhornia crassipes*), Giant Water-fern (*Salvinia molesta*) and Parrot's-feather have all destroyed fishing-based human economies in different parts of the tropics. In the British Isles, we have experienced historical outbreaks of Canadian Waterweed and are currently suffering from invasions by New Zealand Pigmyweed, Floating Pennywort and Water Fern. In 1971, for instance, Water Fern 'formed a complete carpet over the surface of the moat [at Cardiff Castle]; the cover was so good that a child nearly drowned trying to walk on it, as a result of which the moat was drained in c. 1972 and the *Azolla* exterminated' (Wade *et al.*, 1994).

Riverbanks are a classic example of a habitat that suffers chronic disturbance without direct human intervention, as a result of storm surges, gravel deposition or bank erosion, so we might expect the habitat to be relatively rich in alien plant species (as is the case in many other parts of the world). Indeed, riverbank is the habitat for several of the British Isles' most abundant invasive alien plants, including Giant Hogweed, Indian Balsam (Fig. 250), Japanese Knotweed and Giant Knotweed.

The open gravel spits and shoals of rivers along the England–Scotland border, like the River Breamish at Ingram in Northumberland, support a colourful alien flora dominated by various monkeyflowers. Monkeyflower itself grows here in abundance but Blood-drop-emlets is uncommon. The Hybrid Monkeyflower (*M. guttatus* × *luteus* = *M.* × *robertsii*) is much more frequent than its Blood-drop-emlets parent, and Coppery Monkeyflower (*Mimulus guttatus* × *cupreus* = *M.* × *burnetii*) is found occasionally. The storm-churned gravels are also home to a distinctive community of ruderals, including aliens such as Sticky Groundsel, which is thoroughly naturalised in this disturbed open habitat.

Maritime cliffs

Sea cliffs are home to several of the most spectacular infestations of alien plants in the British Isles. Many of these are frost-sensitive species, discussed in Chapter 12. Here, we describe the more widespread alien plants of sea cliffs.

FIG 250. Indian or Himalayan Balsam (*Impatiens glandulifera*) is undeniably impressive, but loved by some for its beauty and hated by others for its smothering effect on other vegetation: (a) extensive stand on wet ground; (b) flower in side view and fruits. (CAS)

Of the herbs, the most abundant is Alexanders, forming dense colonies on the cliffs and reaching inland along roadsides for several kilometres. Colourful flowers include Red Valerian, Wallflower, Seaside Daisy, Silver Ragwort, Hoary Stock (*Matthiola incana*) and Sweet Alison. Outcasts from cliff-top gardens include various Daffodil cultivars, the most distinctive of which is the white and yellow *Narcissus* 'Princeps'. It will be interesting to follow the fate of Saltmarsh Aster (*Aster squamatus*), which is still uncommon in the British Isles but invasive on sea cliffs in Mediterranean Europe (there are records from southwestern England in 2003 and southeastern Ireland in 2006).

Numerous alien shrubs are naturalised on sea cliffs, including Escallonia, Evergreen Spindle, Tree Lupin and New Zealand Holly. Common seaside hedging plants like New Zealand Broadleaf and various hebes are also on this list, much the commonest being Hedge Hebe (the Isle of Man is a hotspot for hebes, with Koromiko *Veronica salicifolia* and Lewis's Hebe V. × *lewisii* both common).

Giant-rhubarb (Fig. 251) is an alien species that has become invasive in the west of Ireland. Although the exact date of its introduction to Ireland is unknown, Praeger first recorded it in the wild in Ireland in 1939 on Achill Island, where pollen analysis suggests that it could have been present for 70–100 years. On coastal cliffs, the main impacts of colonies of Giant-rhubarb are the threat of erosion and loss of maritime species. Apart from the ecological impacts associated with loss of biodiversity, there are also landscape impacts, including the reduction in the area of land that is suitable for agriculture and amenity

FIG 251. Giant-rhubarb (*Gunnera tinctoria*), whose huge size always attracts attention, but which is becoming a major invasive pest, especially in parts of western Ireland: (a) lakeside growth; (b) young inflorescences. (CAS)

purposes. Dense stands of Giant-rhubarb growth may also lead to the blockage of drainage channels and increased risk of flooding. Due to the size of the plant, access to sites infested with Giant-rhubarb is difficult, making control measures problematic (Armstrong *et al.*, 2009).

Sand dunes

With their combination of patchy perennial plant cover, high Rabbit density and relatively frost-free winter weather, sand dunes are a focus for alien plant establishment. The most widespread of the thicket-forming aliens on sand dunes around the coasts of the British Isles is Japanese Rose. Originating equally freely from suckers thrown out from gardens and from bird-sown seed from nearby deliberate plantings, the rose forms a chest-high phalanx that is completely impenetrable to people.

Commonest in England and Wales, members of the genus *Oenothera* can be conspicuous features of coastal sand dunes. Large-flowered Evening-primrose, Common Evening-primrose and their hybrids thrive on open but stable sand, often forming hybrid swarms of great complexity. Although the individual plants are short-lived, populations can persist for many years, spreading to occupy areas of 1 ha or more (Fig. 252). Other evening-primrose species are much more restricted in their distribution, but can be locally abundant, e.g. Fragrant Evening-primrose (*Oenothera stricta*) at Pen yr Ergyd in Cardiganshire (Chater, 2010).

FIG 252. Mixed populations of evening-primroses (*Oenothera*) colonising the extensive dune system at Ainsdale, Lancashire, in 1972. (CAS)

For Northumbrian dog owners, the greatest curse of sand dunes is Pirri-pirri-bur. Dogs are much given to rolling in patches of the plant, and when they do so the burs detach and twist the animal's fur into excruciatingly tight knots. In extreme cases, the only cure is to shave the dog from head to foot.

Because of its suckering habit and tolerance of salt-laden winds, White Poplar is often planted in coastal dunes, especially around caravan parks, where it can naturalise to form extensive stands.

Cultivated ground

The most conspicuous feature of the flora of arable land is the preponderance of annual species and the prominence of archaeophytes relative to neophytes among the alien species (Fig. 246). There have been huge changes in agricultural practice in the last 50 years and these have had enormous effects on the arable weed flora. Modern herbicides have reduced weed abundance to little more than a trace in heavily treated fields, and have brought several formerly common species to the verge of extinction (e.g. Shepherd's-needle, Corncockle and Cornflower). The rotation of cereals with Oilseed Rape has allowed the control of grass weeds using grass-specific herbicides during the occasional years of Oilseed Rape cultivation, while the broadleaved weeds are controlled by herb-specific herbicides during the years of cereal cultivation.

We discuss the archaeophytes of cultivated ground in Chapter 2. Without doubt, the most successful neophyte invader of arable land in the British Isles has been Common Field-speedwell (Fig. 200), not recorded until 1825 but now essentially ubiquitous in the lowlands. The other common arable neophytes are Winter Wild-oat, Canadian Fleabane, Lesser Swine-cress, Gallant-soldier and Pineappleweed.

Woodland and scrub

Of all the major semi-natural habitat types, it is woodland that shows the greatest range of alien plant impacts. One might walk all day through a remote birch woodland in the Scottish Highlands and never see a single alien plant. At the other extreme, a policy woodland in lowland Scotland could have a canopy made up predominantly of alien trees, a shrub layer of alien evergreens planted for game cover, and a ground layer where as many as half of all the vascular plant species are non-natives. The classic example of woodland that is highly impacted by alien plant invasion is the Atlantic oak woodland (the Celtic rainforest) that lines the western coasts of Ireland and Wales, through the Lake District and into Argyll. The problems caused by invasive Rhododendron in this habitat are described in Chapter 15.

In woodland close to human settlement, it is not just the obvious effects of planting the dominant trees, but the combination of a disturbance regime that is conducive to the regeneration of alien plants, coupled with high propagule pressure (from bird-sown garden escapes and fly-tipping of horticultural waste), that creates the ideal conditions for alien plant dominance. In a typical plantation woodland in lowland Scotland, for instance, the canopy might consist of a mix of broadleaved species like Sycamore, Horse-chestnut, Sweet Chestnut, Turkey Oak, Hybrid Black-poplar and Norway Maple, along with a set of planted ornamental conifers like Douglas Fir and Noble Fir, and commercial timber trees like Sitka Spruce, Hybrid Larch and Western Hemlock-spruce. The shrub layer is likely to consist of evergreens like Rhododendron and deciduous species like Snowberry. In the ground layer, there is a colourful patchwork of Leopard's-bane, Snowdrop, Common Blue-sow-thistle (*Cicerbita macrophylla*), Dusky Crane's-bill (*Geranium phaeum*) and Abraham-Isaac-Jacob. The steep slopes of the burn are carpeted by Few-flowered Garlic, and there are dense mats of Lesser Periwinkle and localised thickets of Salmonberry. Among the brambles and Bracken are scattered individuals of Pick-a-back-plant, Fringe-cups, Pyrenean Valerian and Buck's-beard. Tall monocots like Broad-leaved Meadow-grass and White Woodrush are the descendants of earlier intentional introductions. If left untended, self-sown thickets of Lawson's Cypress can become as dark and impenetrable as any reviled conifer planation, and even lower in associated biodiversity.

One of the main uses of woodland in lowland England these days is as cover for gamebirds. A typical rural woodland is criss-crossed by vehicle tracks lined with pheasant feeders, and the centre of the wood is often taken up by a pheasant release pen, fox-proofed with electric fences. Many evergreen species are planted in woodlands as winter cover for game. There are tall species like Cherry Laurel and Rhododendron, and lower-growing species like Wilson's Honeysuckle, Shallon, various *Cotoneaster* species and Oregon-grape (*Mahonia aquifolium*). Much the commonest of the deciduous shrubs planted as game cover in woodlands is Snowberry (Fig. 263).

Extensive areas of the woodland floor will be dominated by Small Balsam and Pink Purslane, with Variegated Yellow Archangel especially frequent near the woodland edge, where the effects of fly-tipping are most strongly felt. Of the thicket-forming woodland bamboos, the three most invasive species are the head-high Arrow Bamboo and Broad-leaved Bamboo, and the waist-high Hairy Bamboo (*Sasaella ramosa*; instantly recognisable by its softly downy leaf undersurface).

In many woodlands in southern England, the great storm of October 1987 opened up extensive canopy gaps. Poorly managed extraction of the fallen timber exacerbated the damage by creating expanses of bare soil that formed a perfect seedbed for recruitment of Butterfly-bush. More than two decades later, there are extensive thickets of *Buddleja* where once there had been a diverse woodland understorey.

The alleged threat posed to native populations of Bluebell (Pilgrim & Hutchinson, 2004) from pollination by the Spanish Bluebell or Hybrid Bluebell was greatly exaggerated by recorders failing to distinguish between Hybrid Bluebells growing on the woodland edge (which are common enough, often from fly-tipped garden waste near parking places) from those established among sheets of the native plants in the heart of the wood (where the hybrid is rare or absent in most cases).

Open vegetation

It is in open vegetation that alien plant species reach their peak frequency and their highest average abundance. The combination of open ground, low interspecific competition, high disturbance rates and plentiful propagule supply guarantees a cornucopia of exotic species. Because most open vegetation communities represent early successional stages, they are often short-lived and tend, if undisturbed, towards dominance by perennial herbs, then by scrub and woodland, with the consequent loss of many of their alien plant species.

As we saw in Figure 246, the two habitats that are richest in alien plants are waste ground and built environments. Here, we deal with built environments under several separate headings, considering first the alien flora of villages, then of transport corridors and urban streets, urban housing, urban green spaces, allotment gardens, walls and rubbish dumps. We deal finally with the habitat that represents the pinnacle of alien plant species diversity in the British Isles: waste ground in towns.

Villages

One of the earliest impacts of human settlement on alien plants involved the creation of habitats for a guild of archaeophytes that live where people live, and thrive on the kinds of trampled, nutrient-rich substrates that people create. These are some of our commonest and most familiar species: Shepherd's-purse, Field Forget-me-not (*Myosotis arvensis*), Ground-elder, Red Dead-nettle, Charlock, Cut-leaved Crane's-bill (*Geranium dissectum*), Black-bindweed, Scentless Mayweed, Hedge Mustard, Mugwort, White Campion, Ivy-leaved Speedwell (*Veronica hederifolia*), Equal-leaved Knotgrass (*Polygonum arenastrum*), Small Nettle, Petty Spurge (*Euphorbia peplus*), White Dead-nettle, Greater Celandine and Wall Barley (with the species ranked by their geographic extent within the British Isles).

The distinguishing feature of the neophyte flora of modern villages is the high frequency of Hollyhock, Rose Campion (*Silene coronaria*) and Purple Toadflax. People who live in villages clearly feel the need to cultivate 'cottage garden' plants, and these spill out in profusion onto verges and footways. After the three species already mentioned, the most frequently seen are Green Alkanet, Honesty, Dame's-violet, Montbretia, Yellow Corydalis, Red Valerian, Fox-and-cubs, Snow-in-summer, Reflexed Stonecrop, Dotted Loosestrife, Snapdragon, Shasta Daisy (*Leucanthemum* × *superbum*), Soft Lady's-mantle, Pot Marigold, Broad-leaved Everlasting-pea (*Lathyrus latifolius*), Sweet Alison, Canadian Goldenrod (*Solidago canadensis*), Trailing Bellflower, Michaelmas-daisy, Peach-leaved Bellflower, Cypress Spurge, Goat's-rue, Adria Bellflower, Mexican Fleabane, Oriental Poppy, Purple Crane's-bill and Russell Lupin. These plants are often found as fly-tipped garden waste on roadsides at the edge of the village, and as self-sown individuals on paths, banks and walls.

Transport systems

The alien floras of canal banks, railways, motorways and urban streets are compared in Table 17. There is an overall similarity, but the differences are interesting.

The canals were built in the 100 years between 1750 and 1850, and these days serve largely for recreational boating, while the tow path is popular with cyclists, joggers and dog-walkers. The influence of dog fouling is clearly seen on the tow path of the canal, with abundant nitrogen-loving plants like White Dead-nettle, Shepherd's-purse, Bristly Oxtongue and Wall Barley. As in many linear habitats in urban areas, Butterfly-bush and Guernsey Fleabane are among the most frequent species. The tow path is often bounded by a coarse hedge, and this accounts for the high frequencies of Large Bindweed, Japanese Knotweed, Garden Privet, Snowberry and Russian-vine.

The railways were built mostly during the middle third of the nineteenth century and created a massive amount of brickwork, with extensive cuttings and embankments, as well as kilometres of ballast track bed. Unsurprisingly, the most frequent alien plant on the railway, Butterfly-bush, is also the most frequent plant overall. The other distinctive feature of the railway embankment is the high frequency of garden outcasts like Green Alkanet, along with bird-sown garden escapes (most notably Firethorn *Pyracantha coccinea*). In late summer, the most conspicuous flowers of railway banks are the various kinds of Michaelmas-daisy and goldenrod. The two hybrid Michaelmas-daisies *Aster* × *salignus* and *A.* × *versicolor* are the commonest taxa, but several species are recorded, including Glaucous, Confused, Narrow-leaved and Hairy Michaelmas-daisies. Of the goldenrods, Canadian Goldenrod is much the commonest, but Early Goldenrod (*Solidago gigantea*) is indistinguishable from a passing train, and may be under-recorded.

TABLE 17. The alien floras of various linear habitats associated with transport in London: the top 20 alien species in each habitat (column 1); and the rank of the named species in the full list (native plus alien) (columns 3, 5, 7 and 9). The flora of the urban street is the most alien-dominated of the four habitats and canal the least. Native species appearing as salt adventives by roadsides are marked with an asterisk.

	Canal	Rank	Railway	Rank	Motorway	Rank	Urban street	Rank
1	Butterfly-bush *Buddleja davidii*	4	Butterfly-bush *Buddleja davidii*	1	Danish Scurvygrass *Cochlearia danica**	1	Guernsey Fleabane *Conyza sumatrensis*	2
2	White Dead-nettle *Lamium album*	5	Sycamore *Acer pseudoplatanus*	2	Hemlock *Conium maculatum*	2	London Plane *Platanus × hispanica*	5
3	Sycamore *Acer pseudoplatanus*	12	Guernsey Fleabane *Conyza sumatrensis*	4	Wild Teasel *Dipsacus fullonum*	3	Butterfly-bush *Buddleja davidii*	7
4	Guernsey Fleabane *Conyza sumatrensis*	24	Russian-vine *Fallopia baldschuanica*	11	Oilseed Rape *Brassica napus* subsp. *oleifera*	6	Bilbao's Fleabane *Conyza floribunda*	9
5	Shepherd's-purse *Capsella bursa-pastoris*	26	Oxford Ragwort *Senecio squalidus*	17	Hoary Cress *Lepidium draba*	8	Sycamore *Acer pseudoplatanus*	11
6	Large Bindweed *Calystegia silvatica*	32	American Willowherb *Epilobium ciliatum*	24	Bristly Oxtongue *Helminthotheca echioides*	9	Pellitory-of-the-wall *Parietaria judaica*	12
7	Bristly Oxtongue *Helminthotheca echioides*	40	Large Bindweed *Calystegia silvatica*	25	Common Mallow *Malva sylvestris*	13	American Willowherb *Epilobium ciliatum*	14
8	Hedge Mustard *Sisymbrium officinale*	43	Garden Privet *Ligustrum ovalifolium*	28	Oxford Ragwort *Senecio squalidus*	16	Shepherd's-purse *Capsella bursa-pastoris*	15
9	Black Horehound *Ballota nigra*	44	Bristly Oxtongue *Helminthotheca echioides*	29	Sycamore *Acer pseudoplatanus*	27	Trailing Bellflower *Campanula poscharskyana*	18

Canal	Rank	Railway	Rank	Motorway	Rank	Urban street	Rank
10 Wild Teasel *Dipsacus fullonum*	45	Firethorn *Pyracantha coccinea*	35	Guernsey Fleabane *Conyza sumatrensis*	28	Cabbage-palm *Cordyline australis*	19
11 Red Dead-nettle *Lamium purpureum*	47	Common Mallow *Malva sylvestris*	38	Beaked Hawk's-beard *Crepis vesicaria* subsp. *taraxacifolia*	31	Tree-of-heaven *Ailanthus altissima*	20
12 Crack Willow *Salix × fragilis*	48	Japanese Knotweed *Fallopia japonica*	37	White Dead-nettle *Lamium album*	33	Yellow Corydalis *Pseudofumaria lutea*	22
13 American Willowherb *Epilobium ciliatum*	56	Michaelmas-daisy *Aster × salignus*	39	Hoary Mustard *Hirschfeldia incana*	37	Oxford Ragwort *Senecio squalidus*	23
14 Japanese Knotweed *Fallopia japonica*	57	Green Alkanet *Pentaglottis sempervirens*	40	Butterfly-bush *Buddleja davidii*	40	Petty Spurge *Euphorbia peplus*	25
15 Garden Privet *Ligustrum ovalifolium*	60	White Dead-nettle *Lamium album*	43	Goat's-rue *Galega officinalis*	44	Firethorn *Pyracantha coccinea*	27
16 Wall Barley *Hordeum murinum*	65	Bilbao's Fleabane *Conyza floribunda*	41	Ribbed Melilot *Melilotus officinalis*	56	Wall Barley *Hordeum murinum*	28
17 Oxford Ragwort *Senecio squalidus*	68	Cherry Laurel *Prunus laurocerasus*	46	Prickly Lettuce *Lactuca serriola*	59	Wall Lettuce *Mycelis muralis*	29
18 Snowberry *Symphoricarpos albus*	69	Annual Mercury *Mercurialis annua*	45	Bilbao's Fleabane *Conyza floribunda*	63	Green Alkanet *Pentaglottis sempervirens*	30
19 Russian-vine *Fallopia baldschuanica*	72	Prickly Lettuce *Lactuca serriola*	48	Fennel *Foeniculum vulgare*	64	Ivy-leaved Toadflax *Cymbalaria muralis*	32
20 Bilbao's Fleabane *Conyza floribunda*	76	Evergreen Oak *Quercus ilex*	53	Reflexed Saltmarsh-grass *Puccinellia distans**	68	Annual Mercury *Mercurialis annua*	35

The trackside fence is often festooned with massive plants of Russian-vine and draped with the bright pinkish-purple flowers of Broad-leaved Everlasting-pea. On the bank beside the tracks are bulky native species belonging to the community that as children we called 'long grass', dominated by False Oat-grass (*Arrhenatherum elatius*) and Cock's-foot, with aliens including Japanese Knotweed, Hollyhock and Dotted Loosestrife (Fig. 253). Garden waste is often thrown over the fence onto the railway banks, with masses of Daffodil cultivars and Hybrid Bluebell in a wide range of colours, from white through pink to pale blue, and there are bright splashes of white from Shasta Daisy. The infrequent alien shrub Bladder-senna is most likely to be found in this habitat. The track bed itself is typically kept scrupulously clear of plant life, but where herbicide has not been applied too recently there may be Small Toadflax (*Chaenorhinum minus*), Sticky Groundsel and Guernsey Fleabane, along with Oxford Ragwort and Rat's-tail Fescue growing on the ballast.

The motorways were built in the last third of the twentieth century and their verges represent a distinctive grassland habitat with essentially no public access and no dog fouling. The motorway comprises at least three distinct sub-habitats: the immediate edge of the road surface, with its narrow strip of native salt adventives like Danish Scurvygrass (*Cochlearia danica*) and Reflexed Saltmarsh-grass (*Puccinellia distans*), and aliens like Summer-cypress, Foxtail Barley, Bugseed and Stinking Fleabane; the broader drainage zone, typically of coarse gravel and

FIG 253. Dotted Loosestrife (*Lysimachia punctata*) usually has an orange spot at the base of each corolla lobe, but this is not found in all variants of the species. (CAS)

characterised by Oilseed Rape; then the grassy bank, which is dominated by a very distinctive community of bulky dicots, including Hemlock, Wild Teasel, Hoary Cress, Bristly Oxtongue, Common Mallow (*Malva sylvestris*), Oxford Ragwort, Beaked Hawk's-beard, Hoary Mustard, Prickly Lettuce, Chicory and Fennel.

Goat's-rue is ubiquitous wherever topsoil has been imported, forming drifts of white, pale pink or bluish-mauve flowers, and Crown Vetch (*Securigera varia*) is locally dominant in similar places. White or yellow wands of White Melilot, Ribbed Melilot and Tall Melilot contrast with the deep gold of Spanish Broom (*Spartium junceum*). Note that Warty-cabbage, so abundant in this habitat in central Europe, is still an uncommon plant in the British Isles.

Urban streets

The flora of a downtown street experiences a set of conditions that are about as hostile for plant growth as one can imagine. Not only is most of the area paved with concrete, stone or tarmac, severely restricting access to rootable medium, but there is incessant pressure from hordes of people, both untargeted (the trampling of countless feet) and targeted (from tireless weeding by house-proud proprietors of shops, restaurants, theatres and apartment blocks). Samples from many of the busiest and most prosperous streets produce no vascular plant species at all, and where plants *are* found, the most frequent species are natives like Annual Meadow-grass, Common Chickweed, Smooth Sow-thistle (*Sonchus oleraceus*) and Procumbent Pearlwort. Needless to say, when growing on pavements, few of these species reach a size at which they would be likely to flower or bear seeds, and on many busy streets this plant community is probably driven entirely by immigration of seed produced elsewhere (i.e. the plant populations of pavements are better considered as casual rather than naturalised).

One of the principal sources of seed for flowering plants found on pavements is from windowboxes and hanging baskets; various colour forms of Garden Lobelia and the patterned leaves of Pink-headed Persicaria (*Persicaria capitata*), Fern-leaved Beggarticks and Petunia are common sights on city streets in late summer. These species are referred to by urban plant recorders as the 'hanging basket flora' and are particularly frequent on pavements outside public houses where money is lavished on creating particularly attractive displays.

Urban housing

The composition of the alien flora of urban housing depends upon the wealth of the house owners and on the amount of open space between the buildings. At one extreme there is essentially no soil at all, just houses and pavement. The rich live in three- or four-storey townhouses with basements, and the poor in two-storey terraced houses. The most expensive town houses in Kensington or Chelsea in central London, for example, have their basements invaded by a combination of uninvited species like Guernsey Fleabane, Ivy-leaved Toadflax and Procumbent Yellow-sorrel, along with self-sown garden escapes like Trailing Bellflower, Yellow Corydalis and Mexican Fleabane, and seedlings of street trees like London Plane, False-acacia and Tree-of-heaven. The less affluent people, living in nearby terraced houses, have many of these species, but the relative abundance is shifted in favour of weedy species like Wall Lettuce (*Mycelis muralis*), Annual Mercury and Pellitory-of-the-wall (*Parietaria judaica*), and with a different set of self-sown garden escapes, including Michaelmas-daisy, Mind-your-own-business and Japanese Anemone.

At the other ecological extreme are the properties set within substantial areas of open ground, where suburban communities are often built on higher ground, with pines and rhododendrons as the backdrop. These expensive properties receive lavish attention from teams of contract gardeners who kill most weeds on sight. The aliens that persist within these high-security environments are an altogether superior bunch, mostly planted but a few of them seeding themselves in quiet corners, overlooked by inattentive contract gardeners: Cherry Laurel, Sweet Chestnut, Horse-chestnut, all three cedars (with Atlas Cedar typically as the cultivar 'Glauca'), Chusan Palm, Foxglove-tree (Fig. 90a), Chinese Magnolia (*Magnolia × soulangiana*) and Laurustinus (*Viburnum tinus*).

High-rise council flats tend to be built on windswept sites where rows of terrace housing were demolished during slum-clearance programmes of the 1950s and 1960s. There is extensive open ground with few trees and broad swathes of mown grass. Many of the estates were victims of what is disparagingly known these days as 'prairie planning': the pointless overprovision of open space, so that everything is too far apart. Here grow the alien plants typical of soil compaction and dog fouling: Shepherd's-purse, Wall Barley, Barren Brome, Common Mallow, Bristly Oxtongue and Large Bindweed. The commonest garden escapes in such places include Snapdragon, Lavender, Red Valerian and the native Columbine (*Aquilegia vulgaris*).

Parks, gardens and lawns in towns

These green spaces are islands of biodiversity in an ocean of concrete, brick and tarmacadam. There is substantial variation from place to place in the proportion of green space in towns, but Natural England has standards for the provision of access: no person should live more than 300 m from their nearest green space; there should be at least one accessible 20 ha site within 2 km of every home; there should be one accessible 100 ha site within 5 km; and there should be one

FIG 254. The Caucasian Slender Speedwell (*Veronica filiformis*) was once a popular rockery plant but its invasive creeping stems soon threw it out of favour. It rarely sets seed in the British Isles as it is self-sterile. (CAS)

accessible 500 ha site within 10 km (English Nature, 1996). The biggest change in recent years has been the wholesale destruction of front gardens to provide off-street parking. Lawns and beds have been buried beneath concrete, asphalt or coloured brick paviours (Smith *et al.*, 2011). The paving supports a distinctive flora, with Procumbent Yellow-sorrel and Water Bent being the most frequent species, along with rarities like Jersey Cudweed (*Gnaphalium luteoalbum*), presumably imported with the builder's sand used to bed in the paving bricks.

Urban lawns are often the only green habitats in larger town centres. They are the habitat of several locally abundant alien herbs like Least Yellow-sorrel, Fox-and-cubs and Slender Speedwell (Fig. 254). Damp, shady lawns are often overrun by Mind-your-own-business. Less frequently seen are Lawn Lobelia (*Pratia angulata*), Leptinella (*Cotula squalida*) and the much rarer Hairless Leptinella (*C. dioica*), Beadplant (*Nertera granadensis*), Corsican Speedwell (*Veronica repens*) and the two lawn pennyworts, New Zealand Pennywort (*Hydrocotyle novae-zeelandiae*) and Hairy Pennywort (*H. moschata*).

Allotments

The flora of these communal gardens is rich in alien species. They are typically a mix of persistent perennial crop plants and annual garden weeds. Perennial vegetables that thrive on neglect include Beet, Globe Artichoke (*Cynara cardunculus*), Rhubarb, Fennel, Horse-radish and Jerusalem Artichoke. The identities of the annual weeds depend largely on soil type and are highly variable from place to place, but there are some constants, including archaeophytes like Shepherd's-purse, Fat-hen, Petty Spurge and Red Dead-nettle, and neophytes like Common Field-speedwell. Needless to say, Ground-elder, 'the most obstinate and detested weed in the nation's flowerbeds' (Mabey, 2010), is frequent everywhere.

Walls

Stonework and brickwork are among the most distinctive urban habitats and are often little invaded by native plant species. This is partly because urban walls are remote from natural cliff faces, but it is likely that mortared walls are simply too different from any natural habitats to have accumulated a large pool of pre-adapted native species (Woodell & Rosseter, 1959; Payne, 1978; Shimwell, 2009).

In order of frequency, London's wall aliens are Butterfly-bush, Guernsey Fleabane, Sycamore, Pellitory-of-the-wall, Oxford Ragwort, American Willowherb, Ivy-leaved Toadflax, Bilbao's Fleabane, Green Alkanet, Mugwort, Common Mallow, White Dead-nettle, Yellow Corydalis, seedlings of London Plane, Large Bindweed, Wall Lettuce, Wall Barley, Garden Privet, Shepherd's-purse and (planted) Russian-vine.

Many of the wall aliens that are so common in London are not yet found in the far north (e.g. *Conyza* spp., Pellitory-of-the-wall, Greater Celandine), with the result that Sutherland's wall flora has a more overtly horticultural feel, including such species as Honesty, Darwin's Barberry (*Berberis darwinii*), Lilac, Sweet Cicely, Feverfew, Monkeyflower, Red Dead-nettle, Caucasian-stonecrop, Japanese Rose, Garden Parsley (*Petroselinum crispum*), Mind-your-own-business, Broad-leaved Everlasting-pea, Lesser Periwinkle and Garden Arabis as the most frequent alien species, along with seedlings of several tree species, including Norway Maple, Douglas Fir and Sitka Spruce.

A number of cultivated ferns are found growing on walls in towns in the warmer parts of the British Isles. The classic habitat is on shaded walls in basements of townhouses, especially those beneath windowboxes or balcony gardens: here we may find Delta Maidenhair-fern (*Adiantum raddianum*), Sickle Fern (*Pellaea falcata*), Button Fern (*Pellaea rotundifolia*) and Ribbon Fern (*Pteris cretica*).

Rubbish dumps

Rubbish dumps have long been the haunt of botanists searching for rare alien plants, and the tips at Dagenham, Grays, Tilbury, Hackney Marsh and Yiewsley were the most celebrated sites around London. In such places, Melville and Smith found 250 adventive species in 1928, including a dense, tall forest of Giant Hogweed and Patience Dock, with Eastern Larkspur (*Consolida orientalis*), Chinese Mustard, Garden Radish, Garden Cress, Gold-of-pleasure, Beet, Cucumber, Marrow, Castor-oil-plant (*Ricinus communis*), Buckwheat, Hemp, Rough Cocklebur, Spiny Cocklebur, Niger, Pot Marigold, Lesser Star-thistle, Onion, Darnel, Rye, Bread Wheat, Six-rowed Barley, Two-rowed Barley, Oat and Common Millet. These days, Milk Thistle (*Silybum marianum*) is restricted to waste tips and soil piles, typically as a garden outcast.

Birdseed aliens are associated closely with households, and formerly a great many of them were characteristic members of council rubbish-tip floras, having come from backyard sweepings, bird-cage cleanings and kitchen waste. This habitat is no longer fruitful for the alien hunter, as rubbish, once spread over large areas, is now burnt or covered with soil or hard core very quickly. In days gone by, one could wander at will across these places, encountering many alien species and on a warm day accompanied by the shrill call of thousands of House Crickets (*Acheta domesticus*). There have been huge changes in attitudes towards waste disposal in the last two decades, with increasing emphasis on the three Rs: reduce, reuse, recycle. Much of the food and garden waste is composted, and great quantities of flammable waste are incinerated, with adverse effects on the alien flora. Access to rubbish dumps in these days of strict heath and safety laws is nigh on impossible, with high levels of security at all of the important landfill sites.

Waste ground

Waste ground in towns is typically the legacy of a land use that has come to the end of its economic life. Perhaps it was a row of terraced houses, condemned as slums beyond renovation. Or it was a factory that produced something that is now produced more cheaply overseas. Or it was railway sidings in a marshalling yard, rendered redundant by containerisation and road transport. Alternatively, it may be useful to someone as a loss-making venture to use as a write-off against tax, while the capital value of the real estate accumulates handsomely in the meantime. In due course it will be redeveloped to make someone else wealthy and the whole cycle will start all over again.

In between abandonment and redevelopment, the ground lies untended. Plants immigrate, typically on the wind or sown by birds in the vicinity of perches. Plants are imported, often with dumped topsoil and, very frequently, in fly-tipped garden waste. The combination of mechanical disturbance, rubbish dumping and topsoil movement makes for perfect conditions for establishment from seed. Here is Richard Mabey's beautifully lucid description of the atmosphere of waste ground in west London:

> There was nothing pretty or charming about this vegetation, no echo of the wild flowers of the English pastoral – or of England itself for that matter. But it pulsed with life – raw, cosmopolitan, photosynthetic life. On the tumuli of the old tips, forests of noxious hemlock shot up through the detritus. Indian balsam, smelling of lavatory cleaner but alive with insects, blanketed the thrown-out bottles. Thirty-foot high bushes of buddleja from China towered above the layered sprays of knotweed from Japan, magenta-flowered everlasting pea from the Mediterranean

and the exquisite swan-necked blooms of thornapple, a weed so spread about the
world that its original home is unknown. Beneath them a galaxy of more modest
weeds tricked out the compacted layers of plastic and glass that passed for soil.
(Mabey, 2010)

We can get an impression of the alien flora of urban waste ground in the early
years of the twentieth century from J. C. Shenstone's account of the flora of a
building site in London: he recorded Mugwort, Shepherd's-purse, Canadian
Fleabane, the uncommon native Pale Willowherb (*Epilobium roseum*), Swine-cress
and Sticky Groundsel (Shenstone, 1912). These days, there would be two extra
Conyza species (*see* below) and American Willowherb would have replaced Pale
Willowherb (but perhaps Shenstone was mistaken in 1912; could it have been an
early record of *Epilobium ciliatum*?).

Ted Lousley was one of the keenest of the London botanists after the Second
World War. Perhaps his favourite plant was the reappeared London Rocket, which
he watched for years after it was found in 1947, and which is still to be seen.
Another was the pretty, but invasive, Atlas Poppy (*Papaver atlanticum*), now to be
seen in many parts of the British Isles. This was first noted in the City of London
in the area of Gresham Street in 1946, and six years later was common. Two
other London plants are specially connected with Lousley. One is the American
Willowherb, now so abundant, which he was the first to notice there, in 1945
(Kent, 1975); the other, the 'New hybrid *Senecio* from the London area' (Lousley,
1946) *S. squalidus* × *S. viscosus*. This he first noted in 1943, on the way to tea with his
parents, just after N. Y. Sandwith, who agreed that Lousley should be the one to
name it. This he did, as *S.* × *londinensis* (McClintock, 1977).

The most distinctive of the current urban waste-ground plant communities
is the *Buddleja–Conyza* scrub community (Fig. 255). It is particularly rich in
aliens, with constant species like Butterfly-bush, Guernsey Fleabane, Mugwort,
American Willowherb, Bristly Oxtongue and Oxford Ragwort. In physiognomy
this is a heterogeneous scrub vegetation with thickets of Butterfly-bush, often
with sapling trees of Sycamore, interspersed with open clearings dominated by
Guernsey Fleabane (locally replaced by Bilbao's Fleabane) and other herbs like
Hoary Mustard or Large-flowered Evening-primrose. The perennial ground
flora of the clearings is variable, but typically forms a patchy mosaic of clonal
plants. The habitat is characteristic of sunny sites on flat, freely drained ground,
often on cinder, ballast or building rubble, but is also found in basements of
terrace houses. The community colonises bare ground within one or two years
of soil disturbance and can persist for at least 20 years. Typically, it gives way to
deciduous woodland if left undisturbed for longer periods. The dominant trees

of older communities include Sycamore and a variety of woody neophytes and natives.

In short, we can rank the habitats of the British Isles from those that are essentially uninvaded by alien vascular plants (mountain tops and mesotrophic grasslands) to those where alien plants often make up both the majority of the species and the bulk of the biomass (waste ground in towns). This continuum is correlated most closely with propagule pressure, and this in turn with human population density. The extent of bare ground and the frequency of soil disturbance are positively correlated with alien plant abundance within a given habitat, presumably because both factors facilitate seedling establishment.

FIG 255. *Buddleja–Conyza* scrub colonising abandoned railway tracks at Old Oak Common in London. The typical pattern is for Bilbao's Fleabane (*C. floribunda*) to increase and for Guernsey Fleabane (*C. sumatrensis*) to decline in relative abundance as the cover of Butterfly-bush (*Buddleja davidii*) increases over the first ten years. (MJC)

CHAPTER 14

Wildlife and Alien Plants

If you go down to the woods today you're sure of a big surprise.
Jimmy Kennedy, 'Teddy Bears' Picnic' (1932)

reen plants represent the base of the food chain upon which virtually
all fungi and animals depend for their existence, so it is inevitable
that around 2,000 alien plant species will have a significant impact
on the natural history of our wildlife, mainly in respect of the habitats, food and
diseases that they provide. We have only to consider those special aliens, our
major crop plants, and the pests that beset them, to be convinced of this.

In this chapter some of the more important and interesting aspects of this
association are discussed, first among the conifers, which are a very special group
of aliens, and then with regard to the flowering plants.

CONIFERS AND WILDLIFE

A vegetation type that covers more than 6.5 per cent of Britain – namely exotic
coniferous woodland – is bound to have an enormous influence on wildlife. The
conifers create a new sort of woodland environment (Fig. 256), which is utilised to
varying degrees by the already existent native plants, animals, fungi and micro-
organisms. In addition, the trees themselves attract their own herbivores and
diseases, both native species that can exploit the new aliens and alien species
introduced with the trees. Study of virtually any group of animals, plants or
fungi will reveal species that are associated with alien conifers. Some of them are
species- or genus-specific, while others more catholic in their tastes.

FIG 256. Interior of a conifer plantation, its dark shade permitting the growth of only 'weed species' like Bracken (*Pteridium aquilinum*) and brambles. (CAS)

As an example, we may consider the spruces (*Picea*). The two commonly grown species, Norway Spruce and Sitka Spruce, are attacked by at least 11 species of gall-forming insects, two of them flies (Diptera) and nine of them aphids (Homoptera), one of which forms the familiar pineapple gall. They are also attacked by the Green Spruce Aphid (*Elatobium abietinum*) and Spruce Spider Mite (*Oligonychus ununguis*), both of which cause leaf loss, and several fungi, including the generalist Honey Fungus (*Armillaria mellea*). Sitka Spruce has been shown to contain two chemicals (both of which are stilbene glucosides) that are highly toxic to fungi, but Honey Fungus produces enzymes that break down stilbenes and so allow it to attack the spruces (Ingram & Robertson, 1999). These trees are also infected with the hyphae of woodland basidiomycetes (toadstools), which form an ectotrophic mycorrhizal association with the roots (for a discussion of mycorrhizae, *see* Ingram & Robertson, 1999). Spooner and Roberts' New

Naturalist *Fungi* (2005) includes impressive lists of the fungal parasites of cultivated conifers.

Norway Spruce alone is used as a food-plant by 16 Coleoptera, 9 Diptera, 24 Hemiptera, 16 sawflies, 21 macro-moths and 18 micro-moths, and the larvae of about a dozen of the moths are thought to use exotic conifers exclusively. Unless they are native moths with an undetected native food-plant, these 12 or so are presumably also all aliens. The food-plants of Spruce Carpet (*Thera britannica*), Cypress Pug (*Eupithecia phoeniceata*) and Larch Pug (*E. lariciata*) speak for themselves. Perhaps the most spectacular of the moths is Blair's Shoulder-knot (*Lithophane leautieri*). This was first found, breeding on Monterey Cypress, in the British Isles in 1951 in the Isle of Wight, an area where that conifer is commonly grown. During its spread, however, it has also taken to Leyland Cypress and Lawson's Cypress, which has greatly assisted its progress northwards where Monterey Cypress is much less commonly encountered. The moth is now common in the southern half of Britain and still spreading northwards thanks to the abundance of planted cypresses. It reached Scotland in 2001 and now extends to the Firth of Forth.

According to Emmet (1988), 55 species of 'microlepidoptera' feed on exotic conifers, and 25 of them exclusively so. The genera consumed are *Abies* (17 species of micro-moth use it), *Pseudotsuga* (2), *Picea* (35), *Pinus* excluding Scots Pine (19), *Larix* (12) and *Chamaecyparis* (1). All but seven of the moths belong to the family Tortricidae. Clifton & Wheeler (2012) included 108 species of conifer-feeding moths, but this of course includes those feeding on our three native conifers, and others that are not confined to conifers.

It is clear that the above 37 (12 plus 25) or so alien moth species are able to breed in this country only because of the existence here of imported conifers. Moreover, the ranges and abundance of the non-exclusive feeders is also potentially much enhanced by their use of exotic conifers as well as their native food-plants. For example, the Juniper Carpet moth (*Thera juniperata*) feeds exclusively on junipers. Our native Juniper (*Juniperus communis*) is virtually absent from central and eastern England, and it is very rarely grown in gardens. The Juniper Carpet is, however, quite frequent in those parts of England, where it feeds on cultivated species of juniper, especially the commonly grown Chinese Juniper (*Juniperus chinensis*). To a less extreme degree, the same is true of the Grey Pine Carpet (*Thera obeliscata*), which is common in many areas where its natural food-plant, Scots Pine, is absent or rare, but where cultivated species of pine can take its place. In addition, the migratory micro-moth known as the Beautiful or Orange Pine Twist (*Lozotaeniodes formosanus*), discovered in Surrey in 1945, first appeared in my light-trap in 1976, then new to Leicestershire, and has since colonised most of England and Wales. It, too, is reliant on cultivated pines for its larval food.

Some birds have become well adapted to life in alien coniferous forest. Among the small birds, Goldcrests (*Regulus regulus*) and Firecrests (*Regulus ignicapillus*) are characteristic of spruce woodland. In Ireland, a survey found 400–600 Goldcrest territories per square kilometre, and in Buckinghamshire 65 ha of spruce plantation supported no fewer than 16 territories of the rare Firecrest (Gibbons *et al.*, 1993). On the other hand another tiny rarity, the Crested Tit (*Lophophanes cristatus*), has not strayed from its natural Scots Pine forest to any other conifers. The Common Crossbill (*Loxia curvirostra*) seems to have segregated itself from the much rarer endemic (and difficult to distinguish) Scottish Crossbill (*Loxia scotica*), since its main food in Scotland is spruce seeds, whereas the latter species relies on Scots Pine seeds (Cramp & Perrins, 1994). The opportunity afforded by the soft, spongy bark of Coastal Redwood and Wellingtonia, enabling Treecreepers (*Certhia familiaris*) to excavate small depressions for winter roosting sites, is well documented. Perhaps the most spectacular bird of spruce and other conifer forests is the elusive Goshawk (*Accipiter gentilis*), of which only 300 or so pairs exist in Britain, mainly due to continued persecution. Here and across northern Europe spruce forest is its favoured cover for roosting and nesting.

The new and developing relationships between our native wildlife and introduced conifers outlined above mainly concern birds and moths, two intensely studied groups, but similar stories could be told about many other groups of organism. Coniferous plantations are generally more inimical to flowering plants and bryophytes because of the dense shade created. Hence, on the whole, pine forest specialities do not (e.g. Twinflower *Linnaea borealis* and One-flowered Wintergreen *Moneses uniflora*) or rarely (e.g. Creeping Lady's-tresses *Goodyera repens*) stray into exotic conifer woods, even though they do so on the Continent, where woods of Norway Spruce, European Silver-fir and European Larch are native and often less dense than our plantations. Although it is certainly true that conifer plantations are vastly less rich in wildlife than deciduous or mixed woodland, they are far from sterile and devoid of interest, and in time more of our native fauna and flora might exploit them successfully. This might especially become the case as a result of a relatively new Forestry Commission policy in some areas, whereby forests are subjected to minimal management and allowed to grow to maturity, and are then felled at a time of abundant seed production, allowing natural regeneration, which removes the need for replanting in rows. The fact that abundant regeneration takes place suggests that the main barrier to its occurring more widely is the lack of open ground.

ANGIOSPERMS AND WILDLIFE

Many examples concerning angiosperms could be cited that would parallel those described above concerning conifers. There are, however, differences in emphasis. Conifers are frequently grown in large stands (forests), providing ample opportunities for wildlife, whereas most alien angiosperms similarly densely grown are crop plants that are usually treated heavily with insecticides and fungicides and do not offer the same scope for colonisation. The most important exceptions are trees, especially Sweet Chestnut, some poplars (*Populus*) and the South American southern beeches (*Nothofagus*), principally Roble and Rauli. The last are grown as timber trees, the first plantations dating from the 1930s, and they received a boost in popularity in the 1970s when they were used in many places to replace elms killed by Dutch elm disease.

Insects

The Database of Insects and their Foodplants (www.brc.ac.uk/dbif) records three beetles, six bugs, twenty-four macro-moths and four micro-moths feeding on *Nothofagus* species, but none of them is confined to that genus. All the moths are common or fairly common polyphagous species that have spread to the alien trees, often being characteristic of native Fagaceae and recorded also from Sweet Chestnut. The latter species has been here for far longer and has accrued a longer list of feeders: 8, 25, 17 and 23, respectively for the above four insect groups. Figures for Sycamore (16, 25, 33 and 25, respectively) are even higher. One other genus of trees that is grown on a small scale in forest plots, and as specimens in parks and gardens, is the gums (*Eucalyptus*). This, however, does not provide much for our wildlife; no Lepidoptera have been found feeding on gums, and the only associated gall relates to a single record. *Eucalyptus* woodland is much more of a wildlife desert than the much-derided conifer plantations, and we are fortunate that it is scarcely suited to our climate.

Although there are around 37 species of moth that feed exclusively on alien conifers (of which we list 49 species), there are only about three-quarters of that number recorded as restricted to the 2,000 or so alien angiosperms. Moreover, the majority of the alien angiosperms concerned are archaeophytes rather than neophytes; interestingly, most of the neophytes involved are woody species. The spectacular impact that can be observed when an alien insect is freed from its own natural enemies is nowhere better illustrated than by the Gracillariid moth Horse-chestnut Leaf Miner (*Cameraria ohridella*), which causes unsightly damage to Horse-chestnut leaves (Fig. 257). The moth arrived in England in 2002 after its rapid spread across Europe (Gilbert *et al.*, 2004). It was first found near Lake

FIG 257. Horse-chestnut (*Aesculus hippocastanum*) leaf in July, mined by the tiny moth Horse-chestnut Leaf Miner (*Cameraria ohridella*). (CAS)

Ohrid in Macedonia in the late 1970s and described as new to science in 1986. The trees look fine immediately after bud burst, but the first generation of moths causes a rash of blotch-like leaf mines, and by midsummer, several generations later, infected trees look as if autumn has come early, having lost most of their green leaf area. It is not yet clear what the long-term impact on seed production and tree survival will be, but effective control is available through soil application or stem injection of systemic insecticides. The search continues for parasitoids of the Horse-chestnut Leaf Miner that are sufficiently host-specific to merit consideration as biocontrol agents (Hernandez-Lopez *et al.*, 2012). The moth has now (2015) colonised most of England and is still spreading, and in many areas virtually every tree can be heavily infested. Some foresters are predicting the demise of Horse-chestnut as an ornamental tree.

It is significant that in the authoritative lepidopteran reference works that index the food-plants, no butterflies or moths are listed as associated solely with those three notorious neophytes Rhododendron, Indian Balsam and Japanese Knotweed. One moth exceptionally uses Indian Balsam and another Rhododendron among their range of food-plants, indicating that these three abundant and invasive alien plants are almost completely unexploited by our Lepidoptera. There is also a lack of predators on the three aliens in other insect groups. Rhododendron, apart from its one moth, has only five Hemiptera recorded as feeders; four of these are specific to the genus *Rhododendron* and other related alien genera. Indian Balsam supports only one fly and one aphid in addition to its moth, and Japanese Knotweed has in total only one polyphagous beetle.

FIG 258. Evergreen or Japanese Spindle (*Euonymus japonicus*) is a salt-tolerant shrub much grown in maritime areas in the south and west. Its leaves are sometimes covered with the caterpillar webs of the Spindle Ermine moth (*Yponomeuta cagnagella*). (CAS)

There are, however, many examples of Lepidoptera widening their range of food-plants to alien species they encounter, such as those mentioned above associated with *Nothofagus*. In some cases this permits them to extend their range beyond that allowed by their native foods. Perhaps one of the best-known examples is the case of the two 'cabbage-white' butterflies, Large White (*Pieris brassicae*) and Small White (*P. rapae*), whose ravages in the vegetable garden can extend to the flower garden by their use of Nasturtium instead of Cabbage. The tiny micro-moth Spindle Ermine (*Yponomeuta cagnagella*) normally feeds on Spindle (*Euonymus europaeus*), the larvae covering the bushes with their white silken protective webs. The moth is far commoner in urban areas, where Spindle is rare, than would be expected because there it utilises the common hedging plant Evergreen Spindle (Fig. 258). Four species of macro-moth that have the rather rarely grown Wild Privet as their natural food-plant are, in fact, typical garden species feeding on Garden Privet (Fig. 259) and Lilac; one of these is the Privet Hawkmoth (*Sphinx ligustri*). Mullein Moth (*Cucullia verbasci*), predictably, normally feeds on

FIG 259. Garden Privet (*Ligustrum ovalifolium*) has attractive white flowers in July, followed by black berries; it is a common food-plant of the Privet Hawkmoth (*Sphinx ligustri*). (CAS)

mulleins (*Verbascum*) and also the related figworts (*Scrophularia*), but it has also often been recorded on Butterfly-bush, potentially increasing its abundance in gardens and on waste ground.

New species of immigrant moths are recorded every year. Some of these feed on native species of plant, but others only on alien species. For example, Sombre Brocade (*Dryobotodes tenebrosa*) was first recorded in the British Isles in 2006 in the Channel Islands, and in 2008 in Dorset. It feeds on Evergreen Oak, and is now established in Dorset where that tree is commonly grown. The list of species in this category is rapidly growing as the study of moths is gaining in popularity and the global economy ensures the importation of more and more produce from an ever-widening range of sources.

New examples of other groups as well are continually coming to light, one of which will be cited here. Italian Alder (*Alnus cordata*), native to Italy and Corsica, has a specific aphid *Crypturaphis grassii*, easily recognised by its yellowish-green colour with brown markings. The aphid was first detected here in 1998 and has now spread across England and Wales, but it remains to be seen how common it will become and whether it can spread to other species of *Alnus*.

Wildlife areas are now widely established in parks, school grounds, private gardens, allotments, farm fields and the like. A major feature of these is the incorporation of nectar-providing plants in order to attract (primarily) Lepidoptera and Hymenoptera. Alien plants feature prominently in the plantings, whether they be garden annual seed mixes or floriferous shrubs like Goat Willow (*Salix caprea*) or Butterfly-bush (Fig. 260).

Gall-forming organisms usually have specific host-plants, and there are good examples among our alien angiosperms, the gall-former probably having been introduced with the host in a number of cases. There are about 11 (nine neophyte)

FIG 260. Butterfly-bush (*Buddleja davidii*) is highly nectariferous, visually attractive and well scented, making it very attractive to Hymenoptera and Lepidoptera. Here it is seen with a Red Admiral (*Vanessa atalanta*) butterfly. (David Tomlinson)

alien genera exclusively attacked by gall-forming rust fungi, ranging from grasses and herbaceous dicotyledons to woody plants such as Rhododendron. Furthermore, arthropods of all the main gall-forming groups are known to attack about 17 (13 neophyte) genera of alien flowering plants, with a similar broad taxonomic range. An example of a recent arrival is the gall midge *Obolodiplosis robiniae*, which causes marginal leaflet-rolling in False-acacia. Both host and gall-former are North American, but the latter was first found in Japan in 2002 and in Europe (Italy) in 2003. It soon spread to Britain and is now common in the London area (Brian Wurzell, pers. comm., 2012), and has extended at least as far north as the Midlands. Of the 15 species of galls known on maples (*Acer*), seven occur only on the native Field Maple (*A. campestre*), five only on alien maples (mainly Sycamore, but one on Silver Maple) and three on both.

The galls formed on our oaks (*Quercus*) are the most interesting in terms of their conspicuousness, variety and complicated life cycle. Although they are in separate subgenera, there is an intimate biological connection between the alien Turkey Oak and our two native oaks, since eight species of gall wasp in the genus *Andricus* have two generations, one sexual (on Turkey Oak) and one asexual (on Pedunculate Oak or Sessile Oak), that alternate between the two sorts of oak, forming a totally different gall on each (Redfern, 2011). In fact, all eight *Andricus* species are relative newcomers to the British Isles, arriving here from areas where both sorts of oak are native, in central and eastern southern Europe. The two best-known species are the Marble Gall Wasp (*A. kollari*), whose sexual generation develops in a small bud gall on Turkey Oak; and the Knopper Gall Wasp (*A. quercuscalicis*), whose sexual generation hatches from a tiny gall on the male catkins of Turkey Oak (Fig. 261b).

Marble galls were first detected in Britain in the 1830s, but their arrival here was largely assisted, as large quantities of the galls were imported from southeastern Europe due to their high tannin content, which was utilised in leather tanning. Knopper Gall Wasps were not seen in Britain until the 1950s, following a steady migration westwards through Europe. The asexual Knopper Gall Wasps are proficient taxonomists, because they infect only Pedunculate Oak and its hybrids with Sessile Oak, but extremely rarely Sessile Oak itself, whereas the Marble Gall Wasp attacks both species equally. A significant biological difference between them is signalled by the fact that marble galls develop in the leaf buds of native oaks, whereas knopper galls (Fig. 261a) arise from between the acorn itself and the cupule surrounding it. Affected acorns often abort early, so if infestation is heavy, which it often is, the acorn crop is badly affected. This causes problems both for animals feeding on the acorns (e.g. Jay *Garrulus glandarius* and Wood Mouse *Apodemus sylvaticus*) and for oak regeneration, and is adding to the calls to eliminate Turkey Oak.

FIG 261. Knopper galls caused by the Knopper Gall Wasp (*Andricus quercuscalicis*), whose generations alternate on two species of oak (*Quercus*): (a) autumn asexual gall on a fruit (acorn) of Pedunculate Oak (*Q. robur*); (b) spring sexual gall on a male catkin of Turkey Oak (*Q. cerris*) – two tiny galls can be seen near the top of the catkin, each with a terminal hole from which the gall wasp has escaped. (CAS)

Oomycota

The alien pestiferous fungus-like Oomycota, or Peronosporomycetes, would have been covered in this book but for a lack of space, as they are related to four algal groups, notably the brown algae or kelps (Phaeophyceae). They are all minute organisms parasitic on higher plants (and animals), and several have attracted headlines in recent years. Only six species were listed as aliens in England by Hill *et al.* (2005), but this is a gross under-representation as at least 16, and probably more than 20, are now known.

Among the most injurious Oomycota are species of the genus *Phytophthora*, notably *Phytophthora infestans*, the cause of potato blight, which changed the course of history in Ireland in the 1840s, and six or seven other species that attack a huge range of trees and shrubs. The names of the resultant diseases adequately describe the potential damage that these organisms can cause: sudden oak death, alder die-back, beech decline and bleeding canker, for example. The onset of these diseases in recent years has forced large-scale emergency felling of plantations, and has changed forestry policy. Great concern remains; in a report in the *Daily Telegraph* on 3rd February 2011, the larch disease caused by *Phytophthora ramorum* was dubbed 'Black Death'. Related genera cause downy mildews: *Pseudoperonospora humuli* on Hops, and *Plasmopara obducens* on the

FIG 262. Busy Lizzie (*Impatiens walleriana*) under attack by the downy mildew *Plasmopara obducens*: (a) healthy bedding plants; (b) plants from the same group a few weeks later. (CAS)

popular bedding plant Busy Lizzie (*Impatiens walleriana*; Fig. 262), whose future as a garden plant is seriously threatened as a result.

Fungi

As is the case with conifers (*see* above), many true fungi attack or exist in symbioses with alien angiosperms. One of the best known is the tar spot fungus *Rhytisma acerinum*, which forms conspicuous black blotches on the leaves of Sycamore. Before the Clean Air Act 1958 was passed by the UK Parliament, the fungus was much less common in urban areas because it is susceptible to sulphur pollution; the same is true of the causal organism of leaf black spot on garden roses, *Diplocarpon rosae*. Sycamore also suffers from a more serious fungal parasite, *Cryptostroma corticale*, the cause of sooty bark disease, which can kill by destruction of the bark. Brian Wurzell believes that spread of the fungus from tree to tree in London is aided by the scratchings of another alien pest, the Grey Squirrel. A further well-known fungal disease of introduced trees is peach leaf curl, caused by *Taphrina deformans*, in which the leaves become red, blistered and contorted. Often most leaves on a tree are affected, producing an unsightly appearance. Not only is the infrequently grown Peach attacked, but also the much commoner Almond, leading to the virtual abandonment of the latter as a popular roadside planting. The introduced Rauli and Roble, species of *Nothofagus* mentioned above, are hosts to a fungal rust, *Mikronegeria fagi*. The fact that its obligate alternate host is Monkey-puzzle suggests that the fungus is most unlikely to become an important pest, as it is in parts of South America.

These examples are, however, relatively unusual in that most introduced woody plants, such as Sycamore and Horse-chestnut, and indeed *Nothofagus*, harbour few specific fungi. It is more often the case that fungi found on native angiosperms have been able to extend their range to alien species. For example, the dreaded disease of Plums known as silver leaf, caused by the fungus *Chondrostereum purpureum*, also affects Rhododendrons. Common fungal diseases such as Honey Fungus (*Armillaria mellea*) attack a very wide range of alien woody plants, including *Nothofagus*. The opposite – i.e. a fungus introduced with an alien plant extending its parasitism to a native species – is less common. An example is *Puccinia malvacearum*, which causes the disfiguring leaf rust of Hollyhock, and is now also found on the native Musk-mallow (*Malva moschata*) and Tree Mallow (*M. arborea*), and on two archaeophytes in the same genus.

At the time of writing (2015), great alarm is being generated by the outbreak of deadly Ash Dieback fungus, *Chalara fraxineus* (asexual phase known as *Hymenoscyphus pseudoalbidus* or *H. fraxineus*), which was imported in February 2012 from the near Continent. By December 2014, 921 outbreaks had been detected across much of Britain, particularly in eastern England and mainly affecting young trees. The disease also kills mature trees, and those of other *Fraxinus* species. Strict monitoring of the developing situation is being maintained, with felling and burning the only feasible antidote.

Birds

Birds are probably the vertebrates most influenced by alien plants. Some cases have already been described under conifers, but, as mentioned above, alien angiosperms rarely occur in large stands providing the kind of cover found in coniferous forests. Hence it is by the provision of food that alien angiosperms offer most to birds, and succulent fruits are top of the menu. About 76 genera of plants in the British Isles have one or more alien species with succulent fruits.

Thrushes, pigeons and starlings are especially associated with alien plants because they are common birds in urban areas, where many shrubs are planted. Two Rosaceous genera, *Sorbus* and *Cotoneaster*, are particularly important food-plants as they are so commonly planted in quantity in gardens, along roads, by car parks and so on. It is very noticeable how quickly in autumn these fruits are eaten, except in areas where traffic or humans are frequently close by, as in city streets. In such places the fruits often remain well into the new year, when not only the above birds but also Waxwings (*Bombycilla garrulus*) move in.

Shrubs planted for game need to provide the two attributes of shelter and food, and, as many plantings are in woodland, ability to thrive in shade is also desirable. Typical alien plants used in this way include Snowberry (Fig. 263), Oregon-grape, Garden Privet, Himalayan Honeysuckle (Fig. 264) and Wilson's Honeysuckle.

There are no European native plants primarily adapted for bird pollination, but several of our aliens are. The commonest examples are the fuchsias (*Fuchsia*)

FIG 263. Snowberry (*Symphoricarpos albus*) is much planted as hedges, in urban shrubberies and in woods as cover for game, where its spread is evidently by rhizomes. Although it produces abundant fruits in the British Isles, it has very seldom been found self-sowing: (a) flowers and fruits; (b) woodland game cover. (CAS)

FIG 264. Himalayan Honeysuckle or Pheasant-berry (*Leycesteria formosa*), a distinctive soft-wooded shrub that often turns up in gardens and copses after birds have eaten the succulent berries and deposited the seeds: (a) habit of plant growing where bird-sown; (b) flowers. (MJC)

and montbretias (*Crocosmia*), from South America and South Africa, respectively, but these have quite specialised flower structures that our birds seem not to have been able to exploit. In the British Isles these species are pollinated by bees. However, many bird-pollinated species with large flowers are grown in Cornwall, and sparrows and starlings frequently feed on their nectar. In the Isles of Scilly, Starlings (*Sturnus vulgaris*) became so covered with pollen on the crowns of their heads that some enthusiastic birders reported them as an unknown exotic species of 'yellow-crested starling'.

Parasitic angiosperms

There are very few alien parasitic angiosperms. Bean Broomrape (*Orobanche crenata*; parasitic on leguminous plants) and Purple Toothwort (*Lathraea clandestina*, parasitic mainly on willows and poplars) are two chlorophyll-lacking root parasites, and Yellow Dodder (*Cuscuta campestris*) is a stem parasite, mainly on carrots. In the past, Flax Dodder (*C. epilinum*) was encountered sporadically in fields of Flax, from which it was actively eradicated, the last record being in 1968.

Bean Broomrape, so abundant in parts of the Mediterranean, especially parasitising Broad Beans, has been found sporadically in Essex for about 50 years, but in the past decade it has locally become much commoner, sometimes attaining pestilential status, e.g. about 750,000 plants in a Pea field. In 2013, even greater numbers were recorded across the Thames in Kent: 11.4 million plants were estimated in a Broad Bean field, almost every bean plant being parasitised.

There are signs that this species could become a major pest of the future. A second species of broomrape, *Orobanche lucorum*, which is parasitic on barberry (*Berberis*) bushes, shows signs of escaping into the wild from the gardens to which it has accidentally been introduced. Purple Toothwort (Fig. 265) always attracts attention, usually in damp places, because of its bright purple flowers arising directly from the ground in early spring.

Among the numerous semi-parasitic Orobanchaceae (formerly Scrophularicaeae) that photosynthesise as well as parasitising the roots of grasses and dicotyledons is Greater Yellow-rattle (*Rhinanthus angustifolius*), which was formerly widespread over Britain but is now extremely local, and Field Cow-wheat, with attractive pink and white flowers, which was also much commoner formerly but is now rare and always confined to southern England. The latter was reported as so common in wheat fields in the Isle of Wight around 200 years ago that it 'rendered the bread discoloured and unwholesome' (Salisbury, 1961), giving it the local name of Poverty-weed. Salisbury described the seeds as 'very like a blackish wheat-grain', and said that winnowing failed to separate the two. Grigson (1955) described the plant as standing up in the wheat 'like a purple, rose and yellow pagoda', but nowadays, and in northern France where it is locally still common, it is seen mostly on banks and verges by roads and around cornfields, from which it has presumably been successfully excluded.

FIG 265. The spectacular Purple Toothwort (*Lathraea clandestina*) grows in wet ground as a root parasite on various woody plants, especially willows and poplars. (Philip Oswald)

CHAPTER 15

Environmental and
Economic Impacts

*Even if it would be an overstatement to say that most invasions cause ecosystem
impacts, it would not be more of an overstatement than the common assertion that
very few introduced species have any significant impact.*
Daniel Simberloff, 'How common are invasion-induced ecosystem impacts?'
(2011)

E cosystems worldwide are rapidly losing taxonomic, phylogenetic,
genetic and functional diversity as a result of human appropriation of
natural resources, modification of habitats and climate, and the spread
of pathogenic, exotic and domestic plants and animals (Naeem *et al.*, 2012). In many
countries, alien plants pose serious problems for agriculture, water management,
biodiversity and ecosystem services. In agriculture, for instance, alien plants
can lead to decreased productivity of grazing lands (through the replacement of
palatable native species by unpalatable alien plants), increased costs of tillage and
weed control, damage to soil quality through depletion of limiting nutrients or
addition of plant toxins and, in the longer term, through reduced land values. In
freshwater systems, alien plants cause obstruction of boat traffic, create breeding
sites for disease vectors like mosquitoes and snails, block extraction pumps and
interfere with hydroelectricity generation. Freshwater life is threatened by reduced
light inputs and lower oxygen availability, while chemical control of waterweeds
can cause mass decomposition, leading to further oxygen depletion and tainting
of drinking water. Ecosystem services threatened by alien plants include increased
risks of flooding, landslide, fire or soil erosion, loss of water resources through
increased transpiration and lowering of the water table, reduced native plant

densities, and changes to habitat that affect native animals for nesting, feeding and dispersal. Alien plants can impact recreation and ecotourism, when spiny or thorny species create impenetrable barriers for people or animals.

The ancient biogeographic barriers that allowed the evolution of rich biological diversity have been broken down by international commerce, and this has resulted in a biological scramble of species competing for newly opened up resources on a global scale. This recent period of homogenisation of biological communities by alien species has been christened *the Homogocene* (Orians, 1986), and some pundits foresee this new epoch as a time when weedy generalist species will take over large portions of the globe, pushing out the specialist species that developed in isolation.

Scientists and policymakers need to agree on objective methods for assessing the various impacts of alien species: ecological, genetic, environmental, economic and sociological. All of these are likely to require an estimate of the geographic extent of the impact of the alien, the abundance of the alien within the impacted range and the impact per alien individual (or per unit biomass of the alien). The genetic impact caused by hybridisation needs to be quantified. With sufficient resources, it is relatively straightforward to measure the current geographic extent of the impact (it is trickier, of course, to predict its future range). Again, we know how to carry out randomised sampling to measure alien plant abundance, using either biomass (expensive) or percentage cover (cheap). We are still a long way, however, from agreeing on the best ways to measure impacts, and it is likely, for instance, that the best way for measuring ecological impact will be different from the best way of measuring economic impact. We need impact data because they would then allow us to estimate direct costs (e.g. losses of revenue from crop yield, water yield or tourism income) and indirect costs (e.g. loss of ecosystem services and biodiversity). If we had this level of quantitative information, it would be relatively straightforward to predict the practicality and likely costs of mitigation.

We also need to know whether the effect of the addition of more alien species to an ecosystem is a function of the number of alien species already present. Should we be less worried about the first few introductions than about those occurring after the pool of exotics is already large? Or should we be more worried? Much remains to be done, because most current assessments of impact are hopelessly qualitative and anecdotal. Some of the assessments amount to nothing more than proof by repeated assertion: things are bad because everyone says things are bad.

As we saw in Chapter 13, different habitats support different numbers of alien plant species, and these species are present at widely different levels of abundance, some locally devastating but most inconsequential. At Silwood Park in Berkshire, for example, alien plant biomass is greatest in wet woodland (American Skunk-cabbage, Fig. 266; Broad-leaved Bamboo, Fig. 118; Orange

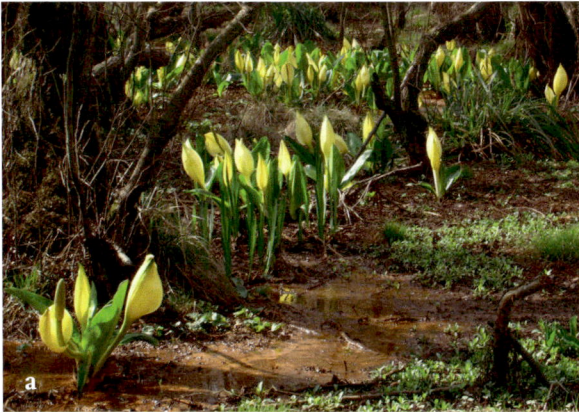

FIG 266. American Skunk-cabbage (*Lysichiton americanus*), a spectacular member of the Araceae (arum family), can dominate the ground layer in swampy woodlands of alder or willow: (a) flowers in April; (b) leaves in June. (MJC)

Balsam, Fig. 208) and in scrub (Tall Tutsan *Hypericum* × *inodorum*, Fig. 267; Green Alkanet, Fig. 126), and least in grassland (where there are usually no alien plant species at all). It is clear from this work that non-random sampling of alien abundance greatly overestimates their impact; the overestimation was greatest in built habitats where much of the area is uninhabitable by plants (buildings, concrete and tarmac, for instance; see Chapter 7).

Where objective surveys have been carried out at a national scale in the British Isles, they tend to conclude that invasive alien species like Japanese Knotweed, Indian Balsam and Rhododendron are relatively inconsequential. The Countryside Survey is an audit of rural natural resources in the UK, and has been carried out at regular intervals since 1978. Stratified random sampling within land classes allows the comparison of new results with those from previous

FIG 267. Tall Tutsan (*Hypericum androsaemum* × *hircinum* = *H.* × *inodorum*) is commonly naturalised as the cultivar 'Elstead' and can form a dense monoculture to the exclusion of native species: (a) habit and fruits; (b) flowers. (MJC)

surveys, so that gradual and subtle changes over time can be detected. From 16,851 plots surveyed in 1998 there were 123 alien plant species, and these were found mostly in habitats with anthropogenic associations, high fertility, high number of ruderal species and high diversity. Over the period 1990–8, the average total cover of alien species increased from 1.2 per cent to 1.9 per cent. The conclusion was that at the landscape scale in the British Isles, alien plant species are best considered as symptoms of disturbance and land-use change rather than as a direct threat to biodiversity (Maskell *et al.*, 2006).

This accounts for the abundance of the aliens, but what about the consequences of their continued presence? What are these, and how should we quantify them? Among many other pressing concerns, we need to understand how alien plants affect ecosystems, and how to ameliorate the worst of their negative impacts. There are many equally tricky questions to consider. How long are the effects of invasions likely to last? Will this generation of alien plants be replaced by another generation of the same species, or will succession lead to its replacement? Will the replacement be native or alien? And so on.

ECOSYSTEM SERVICES

We need to know what the alien plant does to a range of ecological processes. These functions determine the amount, forms, distribution and fluxes (import and export) of energy and various materials, including but not limited to carbon, water, macronutrients such as nitrogen and phosphorus, important trace materials and toxins. These, in turn, influence what have come to be known as ecosystem services (Dickie *et al.*, 2011). The major ecosystem services are taken to involve primary production (carbon fixation through photosynthesis), oxygen production, community stability, soil respiration, maintenance of microbial biomass and mycorrhizal infection rates, effective pollination and fruit dispersal by mutualist species, resilience and recovery rate following disturbance, functioning successional dynamics, water yield and regulation of water supply from catchments, soil retention and protection from erosion, rates of decomposition, nutrient fluxes, carbon sequestration, maintenance of trophic integrity (including functionally abundant top carnivores), sustainable yields of economic products, invasion resistance, biodiversity, aesthetics and scientific understanding. In short, everything.

It is unclear whether few or many of the species in an ecosystem are needed to sustain the provisioning of ecosystem services. In an important study, Isbell *et al.* (2011) showed that 84 per cent of the 147 grassland plant species studied in 17 biodiversity experiments Worldwide promoted ecosystem functioning at least once. But different species promoted ecosystem functioning during different years, at different places, for different functions and under different environmental change scenarios. Furthermore, the species needed to provide one function during multiple years were not the same as those needed to provide multiple functions within one year:

> Our results indicate that even more species will be needed to maintain ecosystem functioning and services than previously suggested by studies that have either considered only the number of species needed to promote one function under one set of environmental conditions, or separately considered the importance of biodiversity for providing ecosystem functioning across multiple years, places, functions or environmental change scenarios. (Isbell et al., 2011)

Although species may appear functionally redundant when one function is considered under one set of environmental conditions, many species are likely to be needed to maintain multiple functions at multiple times and places in a changing world.

We do not know the extent to which effects on ecosystem processes and services are mediated by biodiversity or by the direct and indirect effects of alien plant invasion. Nor do we understand whether novel ecosystems invaded by multiple alien species will provide all of the ecosystem services that our ancestors obtained from pristine ecosystems, and that we have come to take for granted from semi-natural ecosystems.

ECONOMIC IMPACTS

In order to calculate the full economic costs of invasive species, we need to separate the direct costs through lost production in agriculture and forestry and the management costs associated with control of the invasive species, as well as the much more tricky indirect costs arising through loss of biodiversity, loss of ecosystem services, and loss of recreational and tourism revenues. To date, however, country-wide economic evaluations of alien plant species have proceeded simply by multiplying a unit value (price) by a physical quantity (e.g. volume of product damaged) to arrive at aggregate estimates of economic impacts. This approach is inadequate for policy development because it ignores the dynamic impacts of biological invasions on the evolution of prices, quantities and market behaviour, and it fails to account for the loss in the economic value of non-market ecosystem services, such as landscape aesthetics, outdoor recreation and the knowledge that healthy ecosystems exist. Finally, policies that shift the burden of economic impacts from taxpayers and landowners onto parties responsible for introducing or spreading invasives, whether through the imposition of tariffs on products suspected of imposing unacceptable risks on native forest ecosystems or by requiring standards on the processing of trade products before they cross international boundaries, may be most effective at reducing their impacts (Holmes *et al.*, 2009).

Eschen & Williams (2010) estimated the annual costs of alien species in Britain using data that were freely available in the public domain. Many of the impacts, such as additional management costs, structural damage or crop losses have a monetary value that can be assessed, but aliens often also have indirect impacts (for example, the loss of employment or price increases) that are much harder to quantify. The total annual cost of invasive alien species to Britain was calculated to be £1.7 billion. By far the largest impacts were in the agricultural and horticultural sectors, as a combination of crop losses and control costs, which amounted to just under two-thirds of the total cost estimate. This research highlighted the lack of knowledge among conservationists of the monetary value of ecosystem services and the full economic costs of biodiversity loss.

The economic costs of the most damaging alien plants are becoming clearer. Mortgage lenders are becoming unwilling to lend money to potential house buyers when Japanese Knotweed is discovered in the garden or on the boundary of the intended purchase. House owners have been charged extortionate amounts for removal of Japanese Knotweed that has grown up through the foundations of new-build houses, in extreme cases running to half of the value of the property (£150,000 on a new £300,000 house). The costs of removing Butterfly-bush from chimneys, walls and roofs, and of repairing the damage caused to brickwork, are unknown but are likely to be massive (Fig. 274).

INTERNATIONAL CONVENTIONS ON ALIEN PLANTS

Article 8(h) of the 2000 Convention on Biological Diversity multilateral treaty states that: 'Each Contracting Party shall, as far as possible and as appropriate: Prevent the introduction of, control or eradicate those alien species which threaten ecosystems, habitats or species.' This implies (but does not define precisely) that species that cause such damage form a subset of alien species, and that methods are needed to manage such species (and certainly not *all* alien species). Executive Order 13112 on 'invasive species' issued by United States President Bill Clinton on 3 February 1999 defines invasive species as 'alien species whose introduction does or is likely to cause economic or environmental harm or harm to human health'. This represents an attempt to define the subset of alien species for which various control measures need to be implemented, and considers plants to be 'invasive' only when they cause obvious ecological or economic damage.

The mission of the Global Invasive Species Programme (GISP) was to conserve biodiversity and sustain human livelihoods by minimising the spread and impact of invasive alien species. GISP was established in 1997 as an international not-for-profit partnership dedicated to tackling the global threat of invasive species, whose impacts are thought to cost at least US$1.4 trillion annually. It was set up in response to the first international meeting on invasive species held in Trondheim in 1996, and aimed at providing policy support to international agreements of relevance to invasive species, specifically Article 8(h) of the Convention on Biological Diversity (*see* above), and to raise awareness of the threats posed by invasive species globally. The main goal of GISP was that, by 2020, a majority of countries would have the necessary policies in place to implement their national biosecurity strategies and action plans.

To the average reader in the British Isles, all this might seem like a major overreaction. We have problems with Rhododendron in our western oak woodlands, several aquatic plant species are a nuisance in freshwater ecosystems, and Japanese Knotweed is an expensive problem on urban brown-field sites required for building development, but otherwise what is all the fuss about? The truth is that the British Isles have been blessed with a very conspicuous lack of serious negative impacts from invasive alien plants. And this, of course, is despite our attempts over several centuries to import and grow every plant we could find, without a single thought to whether it might escape from cultivation and run amok.

508 · ALIEN PLANTS

To put our good fortune into perspective, it is useful to catalogue some of the world's most important invasive plants and the impacts they have had. Browsing the list in the following paragraphs, one should bear in mind that some places – Hawai'i, for instance (Fig. 268) – suffer from ten or more of these species simultaneously. The material is based on summaries from the Global Invasive Species Database (GISD), managed by the Invasive Species Specialist Group (ISSG) of the Species Survival Commission (SSC) of the International Union for Conservation of Nature (IUCN).

Trees that come to dominate the canopy of precious semi-natural woodlands are perhaps the most pernicious of all alien plant invaders. Broad-leaved Paperbark Tree (*Melaleuca quinquenervia*) can reach heights of 25 m and hold up to 9 million viable seeds per individual in a massive canopy seed bank. This fire-resistant wetland invader aggressively displaces native sawgrass and pine communities in southern Florida, alters soil chemistry and modifies Everglades ecosystem processes. The species is notoriously difficult to control, but biocontrol (integrated with herbicidal and other methods) holds a promising alternative to traditional control methods (Center *et al.*, 2000). There are several alien trees in the British Isles that have the potential to be canopy dominants (e.g. Sycamore, Turkey Oak), but they are yet to become severe problems over large areas.

Shade-tolerant trees can become severe pests in previously diverse tropical forests. Miconia (*Miconia calvescens*) is a small tree native to rainforests of tropical America, where it primarily invades tree-fall gaps and is uncommon. It is now considered one of the most destructive invaders in insular tropical rainforest habitats in its introduced range. It has invaded relatively intact vegetation and displaces native plants on various islands even without habitat disturbance.

FIG 268. These grasslands in temperate climates at high altitude on the 'Big Island' of Hawai'i have a flora that consists almost entirely of British native species; the conspicuous flowering shrub is Gorse (*Ulex europaeus*). (MJC)

Miconia has earned itself the descriptions 'green cancer of Tahiti' and 'purple plague of Hawai'i'. More than half of Tahiti is heavily invaded by this plant. Miconia has a superficial root system that may make landslides more likely, it shades out the native forest understorey and it threatens endemic species with extinction (Meyer & Florence, 1996).

An important group of alien invaders consists of fast-growing nitrogen-fixing leguminous trees that can colonise previously treeless landscapes on unproductive soils, replacing high-value, often diverse plant communities with worthless monocultures. Black Wattle (*Acacia mearnsii*), native to Australia, is often used as a commercial source of tannin or a source of firewood by local communities in semi-arid parts of Africa. It threatens species-rich native habitats by competing with indigenous vegetation, replacing grass communities, reducing native biodiversity and increasing water loss from riparian zones (Higgins *et al.*, 1999). False-acacia fills the equivalent niche in southern Europe, although it has not yet achieved pest status in the British Isles.

Evergreen shrubs are often the most damaging invasive species in woodland, as we have learned to our cost with Rhododendron. In warmer climates, Strawberry Guava (*Psidium cattleianum*), native to Brazil, has become many conservationists' candidate for the world's worst weed. It has been introduced to Florida, Hawai'i, tropical Polynesia, Norfolk Island and Mauritius (Fig. 269), where it is grown for its edible fruit. It escapes from cultivation (often dispersed by alien fruit-feeding birds like Japanese White-eye *Zosterops japonicus*) to form dense evergreen thickets, where it shades out native vegetation in tropical forests. It has had a devastating effect on native habitats in Mauritius and is considered the worst plant pest in Hawai'i, where it has invaded a variety of natural areas. It benefits from the presence of Wild Pigs (*Sus scrofa*), which by feeding on its fruit serve as a dispersal agent for its seeds. In turn, the guava provides favourable conditions for the pigs, facilitating further habitat degradation (Huenneke & Vitousek, 1990).

Thicket-forming spiny shrubs can invade grasslands, plantation understorey and early-successional stages of semi-natural vegetation. Prickly Lantana (*Lantana camara*) is a highly significant weed, occurring in more than 650 cultivars in over 60 countries. It is established and expanding in many regions of the world, often as a result of forest clearance for timber or agriculture. It impacts severely on agriculture as well as on natural ecosystems. The plants can grow individually, in clumps or as impenetrable thickets, crowding out more desirable species. In disturbed native forests, Prickly Lantana can become the dominant understorey species, disrupting succession and decreasing biodiversity. At some sites, infestations have been so persistent that they have completely stalled the regeneration of rainforest for three decades. The species' allelopathic qualities

FIG 269. Like many tropical islands, Mauritius has a vegetation that is dominated by alien plants. The evergreen understorey shrub Strawberry Guava (*Psidium cattleianum*) is considered by many conservationists to be the world's most damaging alien invasive. The emergent Traveller's Palm (*Ravenala madagascariensis*) is a more local problem. (MJC)

can reduce vigour of nearby plant species and reduce productivity in orchards, and the plant can be toxic to domestic animals. Prickly Lantana has been the focus of repeated biological control attempts for a century, yet still poses major problems in many regions, not least because of its genetic polymorphism, which means that control of one or two genotypes at a given site simply leads to expansion of other, resistant genotypes (Gentle & Duggin, 1997). We can get an impression of its effects (mercifully without the thorns) from the impact of Butterfly-bush on abandoned ground in towns in the British Isles.

Rampant vines are important plant pests in many parts of the world, where they grow up and along power lines, smother buildings and bring down canopy trees in native forests. Perhaps the most celebrated of them all is the Asian Kudzu Vine (*Pueraria montana* var. *lobata*). This plant was introduced by highway engineers for soil stabilisation on the banks of newly created freeways in southeastern USA. The roots can eventually comprise over 50 per cent of the plant's biomass, serving as an organ for carbohydrate storage for recovery after disturbance, making it difficult to control the plant with herbicides. Only in eastern USA is Kudzu considered a serious pest, although it is also established in Oregon in northwestern USA, and in Italy and Switzerland, and there is one infestation on the northern shore of Lake Erie in Canada. Kudzu is considered naturalised in the Ukraine, Caucasus, central Asia, southern Africa, Hawai'i, Hispaniola and Panama. Impacts of Kudzu in the southeastern USA include loss of productivity of forestry plantations (estimated at about US$120 per hectare per year), smothering and killing of native plants, and denying access to lands for hunting, hiking and birdwatching, as well as causing costs to infrastructure, including the downing of power lines (Pappert *et al.*, 2000; Forseth & Innis, 2004). Our alien vines are tame by comparison, but Large Bindweed fills the equivalent niche with us.

Waterweeds that choke entire lakes and rivers are among the most pernicious of the world's invasive plant species, ruining the livelihoods of local fishing-based communities and impairing navigation. Some genetically uniform taxa, like *Salvinia molesta* (Giant Water-fern, *see* Chapter 10) have proved to be relatively susceptible to effective biological control, as discussed below. Other genetically more diverse taxa have resisted repeated attempts at biological control. Water Hyacinth, originally from South America, has become one of the worst weeds in the world. It has beautiful large purple and violet flowers, which make it a popular ornamental plant for ponds. It is now found in more than 50 countries on five continents. Water Hyacinth is a very fast-growing plant, with populations known to double in biomass in as little as 12 days. Infestations of the weed block waterways, limiting boat traffic, swimming and fishing. Water Hyacinth also prevents sunlight and oxygen from reaching the water column and submerged plants. By shading and crowding native aquatic plants, it dramatically reduces biological diversity in aquatic ecosystems (Malik, 2007). Floating Pennywort fills this niche in the British Isles.

Wetlands have proved to be particularly susceptible to alien plant invasions. Purple Loosestrife is an erect perennial herb of wetlands and watersides, with a woody stem and whorled leaves, native to the British Isles. It has the ability to reproduce prolifically by both seed dispersal and vegetative propagation. Any sunny or partly shaded wetland is vulnerable to invasion, but disturbed areas with exposed soil accelerate the process by providing ideal conditions for seed germination. As an alien species, it dominates vast areas of sub-arctic Canada, where it interferes with wildfowl breeding (Malecki *et al.*, 1993). This is one of just two British native plant species that appear on the list of the world's worst weeds (the other is Gorse).

The world's grasslands have been invaded by scores of alien weeds, the most pernicious of which are unpalatable to domestic livestock and increase under heavy grazing to the point at which they make up the bulk of pasture biomass. Two life forms are particularly troublesome: herbaceous perennials that are toxic to large herbivores; and spiny shrubs that are avoided by the animals. The *Euphorbia esula* (Leafy Spurge) aggregate is native to Europe and temperate Asia.

512 · ALIEN PLANTS

These weeds are found throughout the world, with the exception of Australia. They are aggressive invaders and one of the first plants to emerge in the spring; they displace native vegetation by shading and by outcompeting natives for available water and nutrients. Leafy Spurge contains a highly irritating substance called ingenol that, when consumed by livestock, is an irritant, emetic and purgative (Dunn, 1979). Although our pastures are currently free of alien weeds, Narrow-leaved Ragwort (Fig. 270) is a serious pasture pest as an alien in South America and might occupy this niche in the British Isles in future.

There are several highly invaded habitats and ecological niches in other parts of the world that have no counterparts in the British Isles. For example, invasive shrubs have come to dominate riparian zones in arid climates in several parts of the world. Salt Cedar (*Tamarix ramosissima*) is a rampantly invasive shrub in western USA that has infested more than 4,000 km^2 of riparian land there. Typically found in conjunction with other *Tamarix* species and their resultant hybrids, Salt Cedar displaces native plants, drastically alters habitat and food webs for animals, depletes water sources, increases erosion, causes flood damage and soil salinity, and is a fire hazard (Di Tomaso, 1998).

Arid rangelands in Australia, Asia, Africa and, recently, in Mediterranean Europe have proved to be highly invasible by various cacti. Vast thickets of cactus

FIG 270. Narrow-leaved Ragwort (*Senecio inaequidens*) was a frequent South African wool alien that has become naturalised in Yorkshire, but much more recently it has invaded England via the east coast ports from its strongholds in northern France, and is now spreading rapidly northwards and westwards. It is a light-demanding species of open waste ground, here seen growing in pavement cracks at Silwood Park, Berkshire. (MJC)

effectively preclude grazing livestock from the landscape. Erect Prickly Pear (*Opuntia stricta*) is a shrubby cactus that can grow to 2 m in height and originates in Central America. It is a spiny shrub that favours habitats such as rocky slopes, riverbanks and urban areas, and was considered to be Australia's worst ever weed before being controlled successfully. The cactus is also invasive in South Africa, where biological control options are currently being explored to control the problem. The main choice is between the Cactus Moth (*Cactoblastis cactorum*) and the cochineal insect *Dactylopius opuntiae*, each of which has proved successful in one or more different countries (Foxcroft *et al.*, 2004).

Fire is a minor ecological factor in most plant communities in the British Isles (lowland heathland and heather moorland managed for grouse are the only important exceptions). In other parts of the world, however, alien plants that alter the fire regime are a major problem. In western parts of the United States, for example, the highly flammable Cheatgrass or Drooping Brome (*Anisantha tectorum*) covers more than a million hectares and has replaced virtually all of the previous rangeland species (Pellant, 1996). Cogon Grass (*Imperata cylindrica*) is native to Asia, and common in the humid tropics. This invasive grass has spread to the warmer temperate zones worldwide, where it is considered to be one of the ten worst weeds in the world (Holm *et al.*, 1977). Its extensive rhizome system, adaptation to poor soils, drought tolerance, genetic plasticity and fire adaptability make it a formidable invasive grass. Increases in Cogon Grass are of concern to ecologists and conservationists as it displaces native plant and animal species, and, because of its high flammability, it is likely to alter fire regimes (Lippincott, 2000).

ALIEN PLANT LEGISLATION IN THE BRITISH ISLES

Alien species are covered by legislation in Great Britain and Northern Ireland under the Wildlife and Countryside Act 1981. The law about alien animals is very strict: 'if any person releases or allows to escape into the wild any animal which— (a) is of a kind which is not ordinarily resident in and is not a regular visitor to Great Britain in a wild state… he shall be guilty of an offence'. The law relating to plants is much less stringent. Native plants are protected under Schedule 8 of the Act, which bans the collection or sale of a list of protected native species. Picking the flowers of all other species is legal, but uprooting native plants requires the permission of the landowner. As to alien plants, the law states under Section 14 (Introduction of new species etc.): 'if any person plants or otherwise causes to grow in the wild any plant which is included in Part II of Schedule 9, he shall be guilty of an offence' (Table 18).

OPPOSITE: **TABLE 18.** Plants carrying Department of Environment, Food and Rural Affairs (Defra) UK risk assessments (left-hand column) and plants identified as 'thugs' on the Royal Horticultural Society (RHS) website (right-hand column). Invasive Species Action Plans are used to help coordinate the response to key invasive alien species across England, Scotland and Wales. The plans provide a short and strategic overview, identifying the key aims, objectives and actions. One can find example risk assessments on the Defra website. It is not clear what action might be taken against any of these species, and the list is a curious mix of thoroughly established species like Indian Balsam (*Impatiens glandulifera*) and others like Water-lettuce (*Pistia stratiotes*) and Giant Water-fern (*Salvinia molesta*) that are currently not hardy in the British Isles. According to the RHS, the species they label 'thugs' are 'invasive plants that can quickly get out of hand in the garden, even though they are not regarded as weeds and are commonly sold in garden centres. One should think carefully about introducing these plants to the garden, and be prepared to carry out judicious pruning and digging or thinning out as required.' Notice how many of the top 52 neophytes (Chapter 11) fall into this category.

The approach to date in the UK and Ireland has been to address invasive alien species on a biogeographical basis. A Great Britain Non-native Species Programme Board (GB NNSPB) has been established, which includes representation from England, Scotland and Wales. For the island of Ireland, a joint programme of work between Northern Ireland and the Republic of Ireland, known as the Invasive Species Ireland Project, has been established since 2006 as part of this biogeographical approach. To ensure harmonisation of approach, both the GB NNSPB and the Invasive Species Ireland Project Steering Group work closely. In Northern Ireland, alien plant species are the responsibility of the Environment and Heritage Service of the Department of Environment (Northern Ireland), and in the Republic of Ireland they are the responsibility of the National Parks and Wildlife Service of the Department of Environment, Heritage and Local Government. Both have international obligations to address invasive species issues, principally the Convention on Biological Diversity, International Plant Protection Convention, Bern Convention and the Habitats Directive. Interest in the alien flora of Ireland is reflected in the publication of *A Catalogue of Alien Plants in Ireland* by Sylvia Reynolds in 2002. Legislation is in place to prevent the release of invasive species in the Republic of Ireland under Section 52(7) of The Wildlife (Amendment) Act 2000: 'any person who plants or otherwise cause to grow in a wild state in any place in the State any species of flora, or the flowers, roots, seeds or spores of flora, otherwise than under and in accordance with a licence granted in that behalf by the Minister shall be guilty of an offence'. In Scotland, the Wildlife and Natural Environment (Scotland) Act 2011 (Commencement No. 4,

Plants carrying Defra UK risk assessments	RHS 'thugs'
Acaena novae-zelandiae (Pirri-pirri-bur)	*Allium paradoxum* (Few-flowered Garlic)
Ailanthus altissima (Tree-of-heaven)	*Allium triquetrum* (Three-cornered Garlic)
Allium paradoxum (Few-flowered Garlic)	*Anemone* × *hybrida* (Japanese Anemone)
Allium triquetrum (Three-cornered Garlic)	*Cotula squalida* (Brass Buttons)
Ambrosia artemisiifolia (Ragweed)	*Crassula helmsii* (New Zealand Pigmyweed)
Azolla filiculoides (Water Fern)	*Elodea canadensis* (Canadian Waterweed)
Cabomba caroliniana (Carolina Water-shield)	*Elodea nuttallii* (Nuttall's Waterweed)
Crassula helmsii (New Zealand Pigmyweed)	*Euphorbia cyparissias* (Cypress Spurge)
Egeria densa (Large-flowered Waterweed)	*Fallopia baldschuanica* (Russian-vine)
Eichhornia crassipes (Water Hyacinth)	*Houttuynia cordata* (Fish-plant)
Elodea canadensis (Canadian Waterweed)	*Hypericum calycinum* (Rose-of-Sharon)
Elodea nuttallii (Nuttall's Waterweed)	*Kerria japonica* (Kerria)
Eucalyptus glaucescens (Tingiringi Gum)	*Lagarosiphon major* (Curly Waterweed)
Eucalyptus gunnii (Cider Gum)	*Lamiastrum galeobdolon* (Yellow Archangel)
Eucalyptus nitens (Shining Gum)	*Ludwigia hexapetala* (Water-primrose)
Fallopia japonica (Japanese Knotweed)	*Lysimachia punctata* (Dotted Loosestrife)
Fallopia sachalinensis (Giant Knotweed)	*Macleaya* × *kewensis* (Hybrid Plume-poppy)
Gaultheria shallon (Shallon)	*Myriophyllum aquaticum* (Parrot's-feather)
Heracleum mantegazzianum (Giant Hogweed)	*Passiflora caerulea* (Passion Flower)
Hydrocotyle ranunculoides (Floating Pennywort)	*Pratia pedunculata*
Impatiens glandulifera (Indian Balsam)	*Prunus laurocerasus* (Cherry Laurel)
Lagarosiphon major (Curly Waterweed)	*Rhus typhina* (Stag's-horn Sumach)
Lemna minuta (Least Duckweed)	*Robinia pseudoacacia* (False-acacia)
Ludwigia hexapetala (Water-primrose)	*Sasa palmata* (Broad-leaved Bamboo)
Lysichiton americanus (American Skunk-cabbage)	*Soleirolia soleirolii* (Mind-your-own-business)
Mahonia aquifolium (Oregon-grape)	*Solidago canadensis* (Canadian Goldenrod)
Mimulus cupreus, M. guttatus and hybrids (monkeyflowers)	*Symphoricarpos albus* (Snowberry)
Myriophyllum aquaticum (Parrot's-feather)	*Vinca major* (Greater Periwinkle)
Pistia stratiotes (Water-lettuce)	*Vinca minor* (Lesser Periwinkle)
Quercus cerris (Turkey Oak)	× *Cuprocyparis leylandii* (Leyland Cypress)
Quercus ilex (Evergreen Oak)	
Rhododendron ponticum (Rhododendron)	
Robinia pseudoacacia (False-acacia)	
Salvinia molesta (Giant Water-fern)	
Spartina anglica (Common Cord-grass)	
Undaria pinnatifida (Wakame) (seaweed)	

Savings and Transitional Provisions) Order 2012 deals with non-natives. Two further orders deal with the keeping and releasing of certain species, reporting on the presence of, and the planting of, invasive species. It also amends the Wildlife and Countryside Act 1981 with any species listed by order of Scottish Ministers prohibited from sale. A Code of Practice on Non-native Species was approved by the Scottish Parliament. The Scottish courts can use this code when considering whether or not a person is liable in criminal proceedings.

For England and Wales, Section 50 of the Natural Environment and Rural Communities Act 2006 introduced section 14ZA to the Wildlife and Countryside Act 1981, enabling the Department of Environment, Food and Rural Affairs (Defra) Secretary of State to ban the sale of invasive non-native plants. Five species of invasive alien aquatic plants were banned from sale after April 2014: Water Fern, Parrot's-feather, Floating Pennywort, New Zealand Pigmyweed and Water-primrose (*Ludwigia hexapetala*). These species had been listed in the Wildlife and Countryside Act 1981, but it was illegal only to introduce the plants into the wild. This was the first time that alien plants had been banned from sale in Britain, and is a clear sign of the tightening up of the precautionary regime against invasive species. Officials hope that banning sales will save money and help protect vulnerable habitats.

The *Invasive Non-Native Species Framework Strategy for Great Britain* (Defra, 2008) states:

> Our vision is that when this Strategy is fully implemented, biodiversity, quality of life and economic interests in Great Britain will be better protected against the adverse impacts of invasive non-native species because there will be widespread awareness and understanding of the risks and adverse impacts associated with invasive non-native species, and greater vigilance against these; a stronger sense of shared responsibility across government, key stakeholder organisations, land managers and the general public for actions and behaviours that will reduce the threats posed by invasive non-native species or the impacts they cause; and a guiding framework for national, regional and local invasive non-native species mitigation, control or eradication initiatives helping to reduce the significant detrimental impact of invasive non-native species on sensitive and vulnerable habitats and species.

As at the time of writing (2015), the European Union (EU) is expected soon to bring forward proposals for the first ever EU non-native species legislation. Species deemed to be of 'Union concern' would be placed on a list of those that should not be introduced, transported, placed on the market, offered, kept, grown or released into the environment. An invasive alien species should be considered of EU concern if the damage it is causing in the affected member states is so significant that it justifies the adoption of dedicated measures, the scope of which extends across the EU, including in the member states that are not yet affected or even unlikely to be affected.

These are worthy aims, inspired in large measure by the 2005 Millennium Ecosystem Assessment report *Ecosystems and Human Well-being: Biodiversity Synthesis*, in which one of the key messages was that: 'The most important direct drivers of biodiversity loss and ecosystem service changes are habitat change…, climate change, invasive alien species, overexploitation, and pollution.' It notes that invasive alien species continue to be major drivers of change in biodiversity, have been a major cause of extinctions, especially on islands, and that 'the introduction of non-native invasive species is one of the major causes of species extinction in freshwater systems'. This is a consciously international perspective, which might be thought to overstate the role of alien plants in the British Isles. It has, however, been the stimulus for direct action, especially in Scotland in relation to clearance of Rhododendron under The Scottish Natural Heritage (SNH) Species Action Framework (January 2007; *see* below).

The fundamental requirements of the legislation are to provide an effective decision-making framework and associated communications processes concerning control, mitigation and eradication of invasive non-native species, and to improve overall clarity and coordination of responsibilities and functions within government and its associated bodies. If and when action is taken, the requirement then is to ensure that sustainable action to control established invasive non-native species is adequately resourced and delivered. The species that have been considered for action are listed in Table 18.

MANAGEMENT OF ALIEN PLANTS

There is a growing worry that the trickle of alien invaders could become a flood as our climate changes in the coming decades. As we have seen, invasive species have been identified as one of the main causes of global biodiversity loss alongside habitat destruction and climate change, so spotting a potential invader before it is too late is seen as increasingly important. It is clear that most introduced species do *not* become invasive, and action must be targeted towards species likely to cause problems, based on thorough, transparent risk analyses. This should include impact assessment, cost estimation and cost-benefit analyses to agreed criteria (economic, biodiversity, social, animal welfare, animal and human health considerations). These analyses should provide criteria from which to prioritise actions for different species. We need to remember that only about 2.7 per cent of the 70,000 non-native plant species available to gardeners have established in the wild, and only 3.5 per cent of these are considered to be a problem, so simply banning all species but natives from gardens would be as ridiculous as it would be unenforceable.

Given the unpredictability of the impacts on biological diversity of alien species, efforts to identify and prevent unintentional introductions, as well as decisions concerning intentional introductions, are best made on the basis of a precautionary approach. Lack of scientific certainty about the environmental, social and economic risk posed by a potentially invasive alien species should not be used as a reason for not taking preventative action against the introduction of such species. Likewise, lack of certainty about the long-term implication of an invasion should not be used as a reason for postponing eradication, containment or control measures (*see* the 2000 Convention on Biological Diversity).

Risk assessment is a key tool to determine the potential impacts of alien plants on ecosystem processes, community structure, community composition, other trophic levels, genetic integrity, social nuisance and potential injuries to human health, increases in injury risks, landscape diversity and aesthetic aspects, as well as to consider detrimental impacts on agriculture, forestry or fishery productivity, and any socio-political, religious or ethical considerations (Keller *et al.*, 2011). Used properly, it can aid prioritisation, help enable effective rapid responses and underpin decision-making. Some risk assessment protocols, when tested retrospectively, have performed reasonably well. For instance, the Weed Risk Assessment in Australia, based on invasion likelihood and probable impacts, would have rejected 94 per cent of the invasive species and 50 per cent of the casual species, while 29 per cent of the casual species required further evaluation (Gasso *et al.*, 2010). Whether such systems help to evaluate the risks posed by species in advance of their introduction is a moot point. As in so many other areas, decision-makers must grapple with poor or incomplete data, inherent unpredictability and ignorance. There is frequently a high level of uncertainty surrounding the risks of entry, invasiveness and the threats of environmental, economic or social damage posed by particular species. Providing clear scientific evidence in advance of the invasive potential of a specific alien species, or of the range of species that could enter unintentionally, will often be unfeasible. By the time there is compelling information available on the invasiveness of an organism in a particular country, it will usually be too late to prevent invasion. While research has focused for some decades on seeking to understand the biology of invasiveness, this is currently not a reliably predictive science, and is unlikely to be so in the foreseeable future (Hulme, 2012).

Prevention

Prevention is generally far more cost-effective and environmentally desirable than measures taken following introduction of an alien invasive species. Clearly, assessment of aliens should include a consideration of the pests that an imported species might carry. Arguably, Dutch elm disease and ash dieback are more injurious than any alien plant is likely ever to be. Identifying future invaders and taking effective steps to prevent their dispersal and establishment constitutes an enormous challenge to both conservation and international commerce. Priority should be given to prevention of entry of known alien invasive species both between and within countries. Ideally, every country should implement border control and quarantine measures to ensure that intentional introductions are subject to appropriate authorisation, and that unintentional or unauthorised introductions of alien species are minimised. Adopting a 'guilty until proven innocent' approach would be a productive first step. That said, the high number of candidate alien species, the investment required in taxonomic support and inspection capacity and the expense of individual risk assessments may act against effective action.

Monitoring and information exchange

Monitoring is the key to early detection of new alien species, but it is expensive to carry out effectively. States should document the history of invasions (origin, pathways and time period), characteristics of the alien invasive species, ecology of the invasion, and the associated ecological and economic impacts and how they change over time. In order to develop best practice, it is essential to know which control efforts have been most successful, and what invasive plant control research best translates into successful restoration application. Successful control depends more on commitment and continuing diligence than on the efficacy of specific tools themselves, and control of biotic invasions is most effective when it employs a long-term, ecosystem-wide strategy rather than a tactical approach focused on battling individual invaders. Rapid response should be consequent on early detection but, when alien plants are rare, detection rates are compromised by low occurrence and limited power to discern significant changes in abundance.

Screening

One action is to screen plant species that are not yet part of the flora, and to prohibit importation of species that fail their screening tests. A trait that has been suggested for screening is seedling growth rate, measured under containment in the greenhouse. If the tested species' relative growth rate is higher or not significantly different from a known invasive counterpart, then it should be considered highly likely to become invasive, and excluded from further consideration as a potential import (Grotkopp *et al.*, 2010).

Eradication

Ideally one would eradicate newly established, potential future invasive species as soon as possible, and would limit the further spread of locally or regionally established aliens, whilst not ignoring the need to mitigate the impact of widespread aliens that have the highest costs. Although the cost of these control measures may appear high, it is money well spent, as without them the future costs of alien species to the British economy will be much higher. Where it is feasible and cost-effective, early eradication, when populations are small and localised, should be given priority over other measures to deal with established alien invasive species. Hence, early detection systems focused on high-risk entry points can be critically useful. Community support, built through comprehensive consultation, should be an integral part of eradication projects.

Eradication of established alien species is often regarded with scepticism by conservationists. There are at least three reasons for this: it is not seen as feasible; it is expensive; and it might have intolerable non-target impacts. Rejmánek & Pitcairn (2003) described 53 attempted plant eradications from California, in which there were *no* successes when the infested area was more than 100 ha. A success story comes from the eradication of Procumbent Pearlwort on Gough Island (part of the Dependency of Tristan da Cunha, at 40°S in the South Atlantic), which was funded by the British Foreign and Commonwealth Office in 2000–02. The flora of Gough Island runs to just 36 species of flowering plants, of which no fewer than 21 are endemic to Tristan da Cunha. The alien plants were dug up and bagged, along with the soil from the immediate vicinity. Soil was treated with boiling water to reduce the seed bank. Follow-up clearance involved hand-picking and herbicide treatment of hard-to-reach populations. The plan now is to monitor systematically to ensure that any population resurgence is eliminated before it can spread.

Containment

Management of dispersal pathways and timely detection of new foci of infestation appear to be critical (Panetta & Cacho, 2012). Theory suggests that the rate of spread is largely driven by rare long-distance dispersal, but this, by its very nature, is generally difficult to detect. Furthermore, most long-distance dispersal events do not give rise to new infestations, yet a key impediment to containment is undetected spread. Feasibility of containment should be viewed in terms of the effort required to reduce weed spread rate, as well as the effectiveness of relevant management actions. Where dispersal vectors are not readily manageable and the probability of detection is low, a much greater reliance upon fecundity control may be needed to contain a weed.

Mitigation

Where a release constitutes a criminal offence or wilful negligence, then the 'polluter pays' principle should be available, and courts should be able to impose fines bearing some relation to the cost of reparation. The statutory framework could also be revised to assist the capacity to undertake mitigation measures. Powers of compulsory access to undertake management or eradication of problem species are provided under some existing legislation, like the Destructive Imported Animals Act 1932. However, there is no general provision in respect of alien species, which would enable emergency control of a newly discovered non-native species before it becomes firmly established and much more difficult and expensive to remove (Defra, 2003).

Managed relocation (translocation of native species outside their native range) has been suggested as a more comprehensive ethical and policy response to climate change. This technique, also known as assisted colonisation or assisted migration, is one of the most controversial proposals to emerge in the ecological community in recent years, not least because of concerns about the potential invasion risk of the relocated species in their new environments, and worries about genetic pollution (Minteer & Collins, 2010).

Restoration

Management of invasive plants should focus not only on removal of aliens, but also on re-establishment of important native species (Prevey *et al.*, 2010). Removing one alien may simply allow the invasion of another alien. Management in which removal of the alien is followed by re-establishment of the intended appropriate native is much more difficult. Again, as more ecosystem components and interactions are altered, restoration of pre-invasion conditions becomes more difficult. Once formed, however, the assembled native community might be more resistant to reinvasion (MacDougall *et al.*, 2008).

BIOLOGICAL CONTROL OF ALIEN PLANTS

Classical biological control is a powerful tool for suppression of invasive plants in natural ecosystems. It will play an increasingly important part in ecological restoration because it provides a means to suppress invaders permanently over large landscapes without long-term resource commitments, and hence it is sustainable. As such, it merits use against many invasive plants that have become pests in sensitive landscapes; it promises to be cheap, effective, permanent and environment-friendly. The major problem, however, is that the success rate continues to be disappointingly low (Sheppard *et al.*, 2006). Future success rates may be even lower than historical success rates (between one in four and one in ten) if it turns out that the easy targets were dealt with first.

A recurrent problem with past biocontrol work has been the lack of funds to carry out sufficient post-release monitoring. This has meant that we have poor information on success, and on indirect unintended effects, both positive and negative. Successful biological control of alien plants is characterised by a persistent, strong reduction in the pest population following the introduction of a natural enemy. Success is typically regarded as having been achieved when the abundance of the alien plant is permanently reduced to a level at which it no longer causes economic damage. In terms of working out a consistent model for best practice, however, it is often difficult, if not impossible, to determine exactly why a release was a success or a failure (e.g. was it chance or timing), let alone to work out how species' biology and release logistics might have influenced the level of success achieved. The truth is that most practitioners would not care about the niceties of the stability of the interaction if the agent reduced the alien plant to low densities. If the weed and the insect can persist as a low-density metapopulation, then so much the better. But the need for repeated reintroduction of a successful biocontrol agent against the most recalcitrant alien weeds would be regarded as an acceptable cost in most cases.

Prospecting for potential biocontrol agents

This is where the expertise of people in museums and botanical gardens is essential. One needs to know exactly what the alien plant is, and exactly where that particular genotype came from. Only then can one tell where to go to look for potential biocontrol agents like insect herbivores or pathogenic fungi. There have been some expensive errors in the past. For example, what is now regarded as the world's most repeatable weed biocontrol success (Room *et al.*, 1981) failed miserably at first, because the weed was misidentified. The weed in question, a floating fern, was identified as *Salvinia auriculata*, and on the basis of this, the entomologists went to the wrong part of Brazil to collect potential control agents. They searched long and hard and found a likely looking weevil. The beetles were returned to base, and fed the floating fern. It turned out to be a hopeless task. They were about to give up, when a passing fern expert told them they had misidentified their plant: it was Giant

Water-fern (*S. molesta*), not *S. auriculata*. The entomologists then went back to Brazil, but to the right part of the country this time, found a likely looking biocontrol agent in the form of the Salvinia Weevil (*Cyrtobagous salviniae*), brought it back, screened it extensively, and eventually obtained permission for a field release against the floating fern in New Guinea. It was a spectacular success, and this success has been repeated in many parts of the tropical world, wherever biocontrol of Giant Water-fern has been attempted (Room *et al.*, 1981). The moral is that identification matters.

Screening for host-specificity

The screening phase is the most expensive part of classical biocontrol, involving many people and taking many years. The first phase is essentially a bureaucratic box-ticking exercise, not based on scientific understanding of herbivore host ranges, but simply a demonstration that the agent will not attack food crops or other highly valued, but arbitrarily selected, plant species. The real skill involves selecting the range of native and horticultural plants with which to confront the agent. The most extreme tests are called starvation trials: the caterpillars and adults of the agent are presented with one species, and if they won't eat it they starve to death. In more realistic trials, the herbivores are presented with a choice between the target weed and the valued plant. Any suggestion that a valued plant might be attacked (either eaten or have eggs laid on it) means that the potential agent is rejected. The screening is ruthlessly selective.

If the agent is shown to have no potentially deleterious side effects, then the decision needs to be made about its likely effectiveness. In the absence of evidence from previous releases elsewhere in the world this is exceptionally difficult to do, and often comes down to nothing more than informed guesswork. Then the government agencies need to be convinced into granting permission for the release, and there will be widespread and protracted consultation.

If permission is granted, then a breeding programme is begun in which the abundance of the agent is bulked up under quarantine until sufficient material is available for release into the field. At this stage, the key decisions are how many releases to make (lots, ideally), how many agents to release at each site (lots, ideally), and at how many sites to release them (lots, ideally). Then one waits and hopes for a benign first winter, lots of survival, effective reproduction and plenty of unassisted post-release dispersal. Needless to say, much can go wrong.

Even following successful releases, there are vocal critics. Williamson (1999) accused biocontrol practitioners of giving

> *no consideration whatever to related species of purely ecological and biodiversity interest, ignoring the potential of biological control to make such species*

endangered or extinct. With such attitudes, it is not surprising that biological control is no longer viewed universally as desirable... The only demonstrably safe biological control agents are those that are completely species-specific... Damage by biological control is almost guaranteed to be permanent... Biological control in future should be absolutely specific and guaranteed to remain specific if it is to be used at all.

This view is almost perversely risk-averse, and would put an end to weed biocontrol if it were taken literally.

Despite current concern about the safety of biological control of weeds, assessing the indirect impacts of introduced agents is not common practice. Recently, however, the use of a highly host-plant-specific, seed-feeding weed biocontrol agent introduced into Australia was shown to be associated with declines of several native insect populations in the vicinity. The agent shares predators and parasitoids with seed-feeding species from native plants, so apparent competition is the most likely cause for these losses. Both species richness and abundance in native seed herbivores and their parasitoids were negatively correlated with the abundance of the biocontrol agent, and local losses of up to 11 species took place as the abundance of the biocontrol agent increased. Ineffective biocontrol agents that remain highly abundant in the community are most likely to have persistent, indirect negative effects (Carvalheiro *et al.*, 2008).

Plant biological control in the British Isles
One might ask why Britain and Ireland have been so slow to adopt biological control of alien species, given our long history in researching and executing biological control projects overseas. There are two obvious reasons. The first is that invasive plant problems in the British Isles are so much less severe than in other parts of the world, so there has not been the pressure and therefore the financial resources to control them. The second is that we tend to be so risk averse: no one wants to be the person who gave permission for the introduction of an alien species that turned out to be a disaster.

CASE STUDIES OF ALIEN PLANT MANAGEMENT
IN THE BRITISH ISLES

We shall discuss four case studies in some detail: Rhododendron in Atlantic oak woods in western Britain as an example of expensive mechanical control; Water Fern as an example of successful natural enemy augmentation; Japanese Knotweed as Britain's first attempt at classical biological control; and Butterfly-bush on the railways as an example of the apparently endless need to employ herbicides.

Rhododendron *Rhododendron ponticum* (Fig. 271) in Atlantic oak woods

The Atlantic oak woods are identified as a habitat of high importance in the EU's Habitats Directive. These oak woods are restricted to the Atlantic coastal fringes of Britain, Ireland, France and Spain, and within the British Isles they peak in importance in southwestern Ireland and southwestern Scotland, on the rocky, exposed west coast and along the southern-facing slopes of glens. The combination of mild winters and notoriously wet weather here, with abundant rainfall throughout the year, provides ideal conditions for the development of a diverse flora of ferns, lush carpets of mosses and liverworts, and colourful assemblages of lichens. This Celtic rainforest supports many lichen and bryophyte species that are largely confined to ancient or long-established woodland. Many of these species are absent in other parts of the British Isles and Europe, and some are globally rare.

As an example of what we might lose were it not for strenuous (and in this case highly effective) efforts to eliminate Rhododendron, consider Taynish Woods Site of Special Scientific Interest in Argyll. This is part of the Taynish and Knapdale Woods Special Area of Conservation and represents an extremely diverse, multiple-interest site. It is one of the largest and most intact western deciduous oak woodlands remaining in the British Isles, where oaks and birches on shaded boulder slopes support a rich bryophyte flora and fern flora. The species list for mosses and liverworts currently stands at more than 250 (about 25 per cent of the British bryoflora). The rich lichen flora is another

FIG 271. Rhododendron (*Rhododendron ponticum*): (a) thickets spreading on acid sandy soil in Leicestershire; (b) close-up of flower head. (CAS)

internationally important feature of these western Scotland woods. There are 475 species, of which 148 are considered notable. Various lichen groups are well represented: the lobarion communities, including such species as Tree Lungwort (*Lobaria pulmonaria*); and on smooth bark, especially Hazel trees, there are abundant graphidion lichens that are hyper-oceanic (very influenced by the surrounding seas). All of these species are threatened with complete elimination by the formation of a monoculture of Rhododendron in the understorey. Regeneration of the canopy trees is also compromised.

There is no prospect of using biocontrol against Rhododendron because of the great horticultural value of this genus in the UK. The only practical options are mechanical or manual removal of the adult plants coupled with follow-up herbicide treatment of the regrowth, stumps and seedling recruitment. We know that mechanical control can lead to complete eradication of Rhododendron from the experience of the Dorset Wildlife Trust on Brownsea Island (101 ha) in Poole Harbour, where spread of the shrub had threatened the survival of the celebrated Red Squirrel (*Sciurus vulgaris*) population. After many years of clearance work by teams of volunteers starting in 1961, and inspired by the indefatigable Helen Brotherton, the last rhododendron on the island was felled in September 2011, following a raffle to determine the identity of the person to be given the privilege of dispatching the last individual. Follow-up treatment with herbicides has prevented any subsequent regeneration.

The Scottish Biodiversity Strategy sets out a medium- to long-term vision to halt and reverse loss of biodiversity. One of its aims is that 'The spread of invasive non-native species has been slowed or halted, and specific areas, regions or islands are designated as free from some invasive and non-native species' (Scottish Executive, 2004). It recognises that complete eradication of Rhododendron from the region is a long-term goal that may not be achievable. What is achievable in the short term is local eradication on smaller, isolated sites and the containment and control of the current infestations on larger sites. To this end, the Scottish government provides grants to support Rhododendron control projects through its Scottish Rural Development Programme. In the worst infestations, the government will pay £9,500 per hectare of infested land for manual clearance over five years. The cost of eradicating the current Rhododendron source populations from Argyll and Bute alone was estimated at more than £9.3 million in 2008.

The good news is that there is no persistent seed bank. Once wetted, shed Rhododendron seeds remain viable for only up to one year, but can germinate on a wide variety of substrates. They are reliant on light and constant humid conditions for germination (Cross, 1981). The principal substrate for seedling establishment appears to be thin layers of moss or litter on dead tree material such as tree stumps. Invasion progresses at about 25 m per year on average, but can jump by many kilometres in occasional cases where the seeds are moved by people (on footwear or vehicles).

The recommended initial treatment is mechanised cutting using a flail fitted to a tracked excavator to cut the bushes as close as possible to the ground to reduce the likelihood of regrowth. Alternatively, a walking excavator can be used to uproot the scrub. Felled material should be burnt, but if operations include the removal of top growth during the period of seed dispersal (late winter), caution is needed to prevent spreading seed across larger areas by inadvertently transporting seed-laden branches around the site. A high proportion of flail-cut stems will regrow and hence require herbicide treatment. Follow-up is essential to deal with regrowth from cut or flailed stumps, as well as survival of small bushes where foliar herbicide application was ineffective. Seedlings developing on the cleared site need to be treated with herbicide, as do cut or flailed branch material that has rerooted, ideally within 24 months of initial clearance. The payment above covers follow-up applications of herbicide or hand-pulling of seedlings to ensure there is no successful regeneration from seed or regrowth. By year five, there must be no Rhododendron present on the site (Edwards, 2006).

Manual control by people with chainsaws is very expensive and should be considered only for sensitive sites or for difficult areas where mechanised clearance methods are not possible. The most expensive control is required on cliffs and in ravines where the site is too steep, too rocky or too wet for mechanised clearance. In these sites, teams of qualified operators need to use ropes to abseil down the cliff, then use cordless drills to create wells in the woody base of each individual plant, into which they inject stump-killing herbicide.

Water Fern *Azolla filiculoides*

Biological control of alien plants within the British Isles has a very short history. The first target was Water Fern (Fig. 272). The issues involved in this work were much less constraining than usual, because the biocontrol agent was already on the British list, as an insect that was 'ordinarily resident' in the UK, which meant that it could be moved around legally. A single adult of the North American alien insect had been found by Oliver Janson on a visit to the Norfolk Broads to look for aquatic beetles. He reported his find of the weevil *Stenopelmus rufinasus* in 1921 as 'An Addition to the List of British Coleoptera' and, since the insect did not do any reported harm in the meantime, this saved the CABI scientists a massive amount of effort that they would normally have expected to spend in screening a potential biological control agent before release into the wild. The production

FIG 272. Water Fern (*Azolla filiculoides*), the only truly aquatic fern in the British Isles, turns red in autumn and can cover the water's surface, smothering other small aquatics. (CAS)

and distribution of *Stenopelmus rufinasus* has been ongoing for some years, and there have been many satisfied customers (McConnachie *et al.*, 2004). After release, the weevil is able to control the weed completely, often in just a few weeks. CABI are increasing their weevil-rearing facilities and hope to meet all current requests. So the score is currently 1–0 to the biocontrol practitioners for British release attempts. The next target, however, looks rather more challenging.

Japanese Knotweed *Fallopia japonica*

Japanese Knotweed (Fig. 273) was brought to England in the early nineteenth century as a prized garden plant, but soon moved 'from prize-winner to pariah' after showing its ability to spread and its resistance to control (Bailey & Conolly, 2000). The effects of the alien in breaking up road surfaces and undermining the foundations of new-build houses erected on brown-field sites and former waste ground have been widely publicised. It is easy to quantify the economic impacts of Japanese Knotweed in particular circumstances. For instance, more than £70 million was spent on its eradiation from the 2012 Olympics site in Stratford in East London. Since the start of the project in 2007, a remarkable 137,500 tonnes of soil contaminated with Japanese Knotweed was cleaned on the Olympic Park site (Olliver, 2011).

FIG 273. Colony of the notorious Japanese Knotweed (*Fallopia japonica*) in full flower in September. (CAS)

Now classed as controlled waste, any Japanese Knotweed that is dumped can result in a large fine and up to two years in prison. There are specialised Japanese Knotweed contractors who require a licence in order to dispose of the waste in a landfill site, and often the landfill site will need prior notice as the waste needs to be buried some 10 m deep. This incurs a considerable cost to the landowner wanting to get rid of the knotweed, but in many cases it is money well spent. The presence of Japanese Knotweed on a building site means that mortgage lenders are becoming increasingly reluctant to lend, making some properties practically unsellable. There is a *Japanese Knotweed Manual* (Child & Wade, 2000) to guide those intending to control the plant.

CABI's project on the biocontrol of Japanese Knotweed began in 2003 thanks to a consortium of funders including Defra, the Environment Agency, the Welsh Assembly Government, Network Rail, the South West of England Regional Development Agency, British Waterways (now the Canal and River Trust), and was coordinated by Cornwall Council. Research into potential agents for a classical biological control programme lasted many years and involved CABI scientists, in collaboration with a Japanese team at the University of Kyushu and many others. Eventually, they were able to select the psyllid *Aphalara itadori* as the most promising candidate from more than 180 insects that feed on the plant in Japan. Subsequently, six years of host range testing, following international protocols, were carried out in quarantine at Egham in Surrey, during which the sap-sucker was shown to be a true specialist with a clear preference for the target weed and an inability to develop on any of the non-target plants of concern that were presented to it.

Whilst the science was not always easy, the principal novelty of the project was in the challenge of navigating European and national legislation that had not been designed with the release of biocontrol agents in mind. For release to be authorised, two legislative pathways had to be pursued: the Wildlife and Countryside Act 1981 to legalise the release into the wild of a non-native species, and the Plant Health Act 1967 to free it from the Plant Health Quarantine Licence under which it is being held. As an organism liable to be injurious to plants in the UK, it was clear that the psyllid would be best considered under EU plant health regulations effectively as a 'beneficial pest'. Thus the European and Mediterranean Plant Protection Organization's Pest Risk Analysis template was used to provide the information necessary for consideration by various experts, independent peer reviewers and, finally, a three-month public consultation. The project is now set firmly within the policy framework of the *Invasive Non-native Species Framework Strategy for Great Britain* (Defra, 2008), and its pursuit through the relevant regulatory regimes may set a helpful precedent for other EU countries in undertaking similar projects. After eight years of research, the release of *Aphalara itadori* was authorised by UK Wildlife Minister Huw Irranca-Davies on 9 March 2010. This is the first time that an EU member state has made such a move (Shaw *et al.*, 2009).

After some successful overwintering from the original limited releases, it was decided that a larger effort was required in 2012, so around 10,000 adult psyllids were released at each of the eight approved sites in May 2012. These sites, and their local pairs, at which no psyllids have been released, continue to be monitored for any adverse effects on the receiving environment and a contingency plan remains in place. It is hoped that these larger releases, and a good summer, will result in good establishment of the psyllid in the UK and allow populations to grow large enough to damage the knotweed (Shaw *et al.*, 2011). However, as of 2015 they are failing to live up to expectations.

In parallel, a project has begun to complete the study of the leaf-spot fungus *Mycosphaerella polygoni-cuspidati*, another natural enemy of Japanese Knotweed with high potential as a biocontrol agent. In 2008, assessment of this agent was put on hold while the psyllid research was completed. Current research into the leaf-spot fungus is being carried out both in the UK and Japan and is focused on life-cycle studies and safety testing using non-target plants species that were not yet assessed (Dick Shaw, pers. comm., 2015).

Butterfly-bush *Buddleja davidii*

Butterfly-bush (Fig. 274) is much planted in parks and gardens, and spreads freely by abundant windborne seeds. It is native to China, where it is found in thickets on mountain slopes at altitudes of 800–3,000 m, from Gansu to Zhejiang.

FIG 274. Butterfly-bush (*Buddleja davidii*) is very attractive both to humans and to insects, but can be extremely destructive to brickwork; (a) young plant on a wall; (b) a wall in Reading, Berkshire, demolished by root growth. (MJC)

Seedlings, saplings and mature plants are now common all over towns and cities throughout most of the British Isles, on railways, pavements, waste ground, chalk pits, gravel workings and high up on old walls, chimney stacks and so on. The plant was first recorded as regenerating in London in 1927, but inspection of old railway photographs shows that it did not become abundant until recently (since about 1980). Its economic effects have not yet been felt in full, because the damage inflicted by the roots of Butterfly-bush growing through brickwork is not widely appreciated.

At present, the greatest expenditure on Butterfly-bush control is borne by the railways, where the plant is a pestilential invader of the ballast track bed. Network Rail spends approximately £25 million on vegetation management annually, but the costs are not broken down by target species, and this figure is purely the cost of routine vegetation management necessary to operate the railway infrastructure safely (Müller *et al.*, 2001). The problem with herbicide control is that it is a recurrent expense, and there will be serious trouble if herbicide-resistant Butterfly-bush should evolve.

It is most unlikely that Butterfly-bush would ever be selected as a target for classical biological control in the British Isles, simply because it so highly prized as a butterfly plant in gardens. But if it is ever chosen, there is an excellent prospective agent waiting in the wings. It is a defoliating weevil, *Cleopus japonicus*, from the Butterfly-bush's native home in China. In tests under quarantine containment in New Zealand, the weevil killed 30 per cent of the test plants and reduced the growth of all of them (Watson *et al.*, 2011).

CONSERVATION OF ALIEN PLANTS

A good deal of straightforward xenophobia has been involved in thinking about aliens and conservation in the past. The attitude was simply 'natives good, aliens bad'. Many Sycamores and a good few Turkey Oaks were felled on this premise. Of course, the costs or benefits of a particular plant in a particular place are a matter of opinion, and are therefore difficult or impossible to quantify in a way that stands up to critical scrutiny. For instance, Ted Green has rechristened Sycamore as 'Celtic Maple' to draw attention to its importance in Ireland, Scotland and northern England as a landscape feature and as a substrate for lichens and bryophytes (Green, 2005).

The attribution of native or alien status to a species can be consequential, because most conservation legislation is aimed specifically at natives. For example, in the third edition of the *British Red Data Books, 1. Vascular Plants*

(Wigginton, 1999), 30 species that had been included in the second edition were dropped because they were no longer considered to be native, including Small Alison (*Alyssum alyssoides*), Rampion Bellflower, Spring Crocus, Tenby Daffodil, Wild Peony and Breckland Speedwell. In so far as these species were afforded some protection by dint of being included in the *Red Data Books*, they are now stripped of that protection.

There is a small set of plants that are now regarded as aliens but are highly cherished by local people and government agencies. The county flower of Berkshire, for example, is Summer Snowflake, which the locals call Loddon Lily. That such a conspicuous plant was not recorded from the wild (in 1788) until many years after it was known in horticulture (in 1596) is taken as evidence of its neophyte status (Pearman, 2013), despite the fact that it was regarded as native in the third edition of the *British Red Data Books, 1. Vascular Plants* (Wigginton, 1999) and in the *New Atlas* (Preston *et al.*, 2002). It is arguable that the riverside habitats in which it grows are worthy of conservation in their own right, and it would be a callous manager who would argue that a stretch of riverbank is now less worthy of protection because the Loddon Lily is no longer regarded as native. On the other hand, it is doubtful that anyone would seriously argue that, because it is not native, the Loddon Lily should be removed from the banks and holts of the River Thames. The point is that we should be allowed to conserve the plants we like, irrespective of their status.

More has been written about the nativeness or otherwise of Fritillary than probably any other species. It is known to have been in cultivation at least by 1597 (and possibly by 1578), but the first record from the wild was not until 1736 from Harefield in Middlesex (Pearman, 2007). If it were native, then this would be a very late date for such a colourful and obvious flower, and it is most unlikely to have been overlooked by all of William Turner, John Gerarde, John Parkinson, Thomas Johnson, John Ray, Robert Morison and Johann Dillenius (Harvey, 1996). Despite this evidence, it had been classified as native in *Scarce Plants in Britain* (Stewart *et al.*, 1994).

Given that most alien plant species are innocuous, it is clearly not sensible to spend limited resources on their management. Equally, pouring money into the conservation of a native species that is doomed to extinction as a result of climate change is just as ridiculous as persecuting innocuous alien species. Where resources are to be deployed, they should be cost-effective.

'Sharing or sparing' is one of the big issues in conservation planning at present. The question is whether biodiversity is best served by having very intensive agriculture in the smallest possible area and minimal impacts in the remainder ('sparing' the relatively pristine landscapes), or spreading the impact more widely and sacrificing productivity in agricultural lands to make them more biodiverse ('sharing' the pain across a broader landscape). Where the idea has been tested in places with reasonably pristine habitats, the clear consensus is that land sparing is superior to land sharing in terms of biodiversity conservation (Phalan *et al.*, 2011). But in environments like the British Isles, where there simply *are no* pristine habitats, attempts to increase the biodiversity of intensive agricultural land by sharing it with conservation projects might well be the best way forward. In truth, however, most arable farmers in England make few concessions to biodiversity conservation over and above the odd token beetle bank, conservation headland, wildflower strip for pollinators or cover, and food for wintering gamebirds. As we have seen, many of the species in these conservation features are aliens.

The principal focus for the conservation of rare alien plants in the British Isles concerns arable weeds, which are the group suffering the greatest rate of decline in the vascular plant flora. These archaeophytes include several UK Biodiversity Action Plan priority species like Cornflower, Shepherd's-needle and Small-flowered Catchfly. Uncropped, cultivated field margins are compensated under the stewardship schemes and are seen as the key mechanism for the future recovery and conservation of these plants. Where there is a viable soil seed bank of these rare species, then annual cultivation, coupled with cessation of herbicide and fertiliser application, can be an effective strategy, so long as sufficient financial compensation continues to be paid to the farmers. However, in locations where there is no surviving seed bank, then cultivated headlands simply end up being dominated by pernicious native weeds, and this is clearly a waste of scarce conservation resources. Cultivating and sowing the seeds of these species each year in wildflower mixtures seems to us to be nothing more than gardening.

Perhaps we need a subtle change of focus? We know that we can't afford to preserve all of our native species in the face of environmental change or, even if we wanted it, to get rid of all of the aliens. If it can be shown that an alien plant is genuinely detrimental to native biodiversity, then by all means persecute it vigorously (as in the case of Rhododendron in Atlantic oak woodlands described earlier). But for the bulk of alien plant species that are clearly inconsequential, we should celebrate them as an addition to total biodiversity. And finally, for aliens that we really cherish like Fritillary and Loddon Lily, let's look after them and their habitats just as carefully as we would if they were natives.

List of Angiosperm (Flowering Plant) Neophytes and Neonatives Recorded in the British Isles from January 1987 to May 2012

A ppendix 1 is intended to provide a definitive working list of our angiosperm neophytes that attain certain agreed levels of occurrence (*see* below). Appendices 1 and 2 are also available in Excel on the New Naturalist website: www.newnaturalists.com/page/AdditionalResources, enabling users to manipulate, analyse and add to the lists as they wish. This preamble explains the criteria and conventions used in compiling the list.

CRITERIA, CONVENTIONS AND ABBREVIATIONS

- Alien non-flowering plants are listed in Chapter 10, and angiosperm archaeophytes in Appendix 2.
- The choice of 1987 as the starting point is explained in Chapter 1.
- Naturalised species are included in the list even if they occur in only one locality, but casual species are accepted only after their repeated recording in one or more localities. This seems justified because a naturalised species at one site can be recorded year after year, whereas casuals need to be repeatedly introduced in order to achieve a comparable impact on recording and the environment.

- Preston *et al.* (2002) considered 42 species to be of uncertain native/alien status (*see* Chapter 1). We have increased this figure to 55, as explained here. These 55 species are included in Appendix 1, indicated with an asterisk (*); they are additional to the 1,809 neophytes (*see* Chapter 1). It should be noted that a number of the doubtful cases on our list would be archaeophytes rather than neophytes if they are indeed not natives.

We agree that for most species on the 'native or alien' list produced by Preston *et al.* the evidence is equivocal. Our 55 include all but the following 12 of the 42 listed by Preston *et al.*:

Aethusa cynapium subsp. *cynapium*	*Helleborus viridis* subsp. *occidentalis*
Brassica nigra	*Phyteuma spicatum*
Crassula aquatica	*Pinguicula alpina*
Daphne mezereum	*Schoenoplectus pungens*
Festuca lemanii	*Scorzonera humilis*
Gastridium ventricosum	*Solanum nigrum*

For these 12 we believe that the evidence, as assessed on the criteria outlined in Chapter 1, is persuasively in favour of native status, or at least gives no reason to conclude otherwise. In two cases (the *Aethusa* and the *Solanum*) there is fossil evidence for their presence in the post-glacial period prior to the Neolithic, but this is not the place to argue each case separately.

Some of the 25 taxa that we have added to the remaining 30 taxa on the list of Preston *et al.*'s *New Atlas* were treated as natives by them, and others as aliens, as shown below:

Native in New Atlas	Alien in New Atlas	Not in New Atlas
Helleborus foetidus	*Hypericum canadense*	*Artemisia campestris* subsp. *maritima*
Lathyrus sylvestris	*Juncus subulatus*	
Medicago minima	*Malva pseudolavatera*	
Medicago polymorpha	*Ornithogalum umbellatum* subsp. *campestre*	
Mycelis muralis	*Petrorhagia prolifera*	
Nymphoides peltata	*Rhinanthus angustifolius*	
Parietaria judaica	*Scrophularia scorodonia*	
Reseda lutea	*Spergularia bocconei*	
Sisyrinchium bermudiana	*Valerianella eriocarpa*	

Native in New Atlas	*Alien in* New Atlas	*Not in* New Atlas
Sorbus domestica	*Veronica polita*	
Symphytum tuberosum	*Vulpia unilateralis*	
Tanacetum vulgare		
Vicia bithynica		

In addition, there is one species said in the *New Atlas* to be an alien that we feel should be designated as native, and this is not included in Appendix 1, namely Wall Germander (*Teucrium chamaedrys*). This species is convincingly native in its East Sussex sites in chalk turf on the South Downs, exactly mimicking many sites just across the English Channel (particularly in the Somme Valley); elsewhere it is an obvious escape or relic on walls and the like.

Conversely, Summer Snowflake (*Leucojum aestivum* subsp. *aestivum*), treated as a native in the *New Atlas*, is now considered more likely to be a neophyte (Pearman, 2013), and we have treated it as such here.

- In addition to the 1,809 neophytes listed in Appendix 1, there are at least a further 2,000 aliens that are considered beyond the scope of this work. These are taxa that either have not been recorded here since 1986, or have been recorded since then but only as very rare casuals or survivors. Many of the excluded survivors are only marginally 'in the wild', and none of them has an important impact on our environment and wildlife.

- The neophytes, naturalised or casual, in *Rubus* subgenus *Rubus* (brambles), *Hieracium* (hawkweeds) and *Taraxacum* (dandelions) exhibit apomictic reproduction, whereby all the embryos produced are entirely maternal in origin (*see* Chapter 10). In these three genera apomixis has resulted in the evolution of many different genotypes, which are traditionally recognised as species (agamospecies or microspecies). In our flora there are 334, 417 and 237 microspecies currently recognised in these three taxa, respectively. Experts in these genera reckon that the above totals include about 5, 54 and 111 alien species, respectively. The alien apomictic brambles and dandelions all belong to sections containing British natives, and are not included in Appendix 1. The same is true of most hawkweeds, but four species of *Hieracium* feature in Appendix 1 because they represent non-British sections: *Hieracium amplexicaule, H. pulmonarioides* and *H. speluncarum* (section *Amplexicaulia*), and *H. lanatum* (section *Andryaloidea*).

- This inventory represents an update of the taxa contained in the third edition of *New Flora of the British Isles* (Stace, 2010), itself a partial update of those in the first edition (Stace, 1991). The following new criteria were applied:

o   any taxa with no records since 1986 were deleted (57 taxa);
o   any naturalised or neonative taxa recorded since 1986 were added (32 taxa);
o   any casual or survivor taxa not previously included but with five or more
    records between 1930 and 1986 and at least one since 1986 were added (71
    taxa); and
o   any casual or survivor taxa not previously included, and with no more
    than four records between 1930 and 1986, but with at least 18 records since
    1986, are added (15 taxa). The high number of 18 records required for
    qualification is a reflection of the fact that most of these species are trees
    and other horticultural specimens that have been deliberately planted in
    parks, on roadsides, etc., a category that is increasingly being recorded
    by field botanists. That number was chosen by inspection of the actual
    species falling above and below it.

These adjustments therefore amount to a net addition of 61 taxa.
- Status (naturalised, casual, survivor) is colour-coded in the appendix:

Naturalised	Casual	Survivor

* = Native/alien status uncertain
There are 1,138 naturalised, 342 casual and 329 survivor taxa designated.
- 57 taxa that are invasive at least in some places, as defined in Chapter 1, are
  given in boldface.
- Number of hectads (of which there are 3,859 in total) recorded from January
  1987 to May 2012 is taken from the BSBI Distribution Database
  (http://bsbidb.org.uk).
- First record is taken largely from the unpublished database maintained by
  David Pearman, with his kind permission.
- Vectors listed are the known or most likely means of introduction to the
  British Isles; in many cases others have probably operated. Once arrived,
  many taxa have become dispersed by quite different vectors, not given here.
  Data mainly from Clement & Foster (1994) and Ryves et al. (1996), with many
  additions. Key to codes:

A   = as or with fodder or other foodstuffs or straw for animals
As   = with agricultural or horticultural seeds
B   = with ships' ballast
Bm   = for biomass production
Bs   = as or with birdseed

C = with raw cotton

E = with Esparto-grass

F = as human foodstuffs, including herbs and spices, as plants, seeds or fruits

G = with grain

Gs = with grass-seed and wildflower mixes

H = horticultural introductions, mainly ornamentals, but also grassland, hedging, game cover and green manure

I = contaminants of horticultural introductions

M = introduced medicinal, drug and cosmetic plants

Neo = neonatives (141 designated here)

O = oilseed plants and their contaminants

S = with imported soil, sand and rocks (not ballast)

T = with timber

Tn = with tannery raw materials

V = inadvertently with people, animals or vehicles

W = with raw wool

? = data unknown or very uncertain

- Life forms are given according to the Raunkiaer (1934) system, as defined in Table 4, with abbreviations as follows:

Ch = chamaephyte

Ge – geophyte

Hc = hemicryptophyte

Hy = hydrophyte

Ph = phanerophyte

Th = therophyte

- S/V status indicates whether the taxon reproduces in this country from seed (S) and/or from vegetative propagation (V). A dash in this column indicates that the taxon does not reproduce with us, and is therefore a casual or survivor.

Scientific name	Family	No. of hectads	Introduced from	First record	Vector	Life form	S/V
Abutilon theophrasti	Malvaceae	73	Southeast Europe	1887	Bs,O,W	Th	–
Acacia melanoxylon	Fabaceae	2	Australia	1959	H	Ph	V
Acaena anserinifolia	Rosaceae	20	New Zealand	1914	H,W	Ch	S/V
Acaena anserinifolia × inermis	Rosaceae	2	Horticulture	1969	H	Ch	V
Acaena caesiiglauca	Rosaceae	4	New Zealand	1996	H	Ch	S/V
Acaena inermis	Rosaceae	18	New Zealand	1968	H	Ch	S/V
Acaena novae-zelandiae	Rosaceae	118	Antipodes	1901	H,W	Ch	S/V
Acaena ovalifolia	Rosaceae	55	South America	1921	H	Ch	S/V
Acanthus mollis	Acanthaceae	173	Mediterranean	1820	H	Hc	S/V
Acanthus spinosus	Acanthaceae	34	Mediterranean	1942	H	Hc	S/V
Acer cappadocicum	Sapindaceae	104	Southwest Asia	1977	H	Ph	S/V
Acer negundo	Sapindaceae	83	North America	1913	H	Ph	–
Acer palmatum	Sapindaceae	33	East Asia	1950	H	Ph	–
Acer platanoides	Sapindaceae	1,641	Europe	1905	H	Ph	S
Acer pseudoplatanus	Sapindaceae	3,461	Europe	1632	H	Ph	S
Acer saccharinum	Sapindaceae	150	North America	1959	H	Ph	–
Achillea distans subsp. tanacetifolia	Asteraceae	5	Europe	1844	H	Ch	S/V
Achillea filipendulina	Asteraceae	27	West & central Asia	1909	H	Ch	S/V
Achillea ligustica	Asteraceae	1	Mediterranean	1900	H	Ch	S/V
Achillea nobilis	Asteraceae	5	Europe	1896	H	Ch	S/V
Aconitum lycoctonum subsp. vulparia	Ranunculaceae	8	Europe	1880	H	Ge	S/V
*Aconitum napellus subsp. napellus	Ranunculaceae	636	Europe	1819	H	Ge	S/V
Aconitum napellus subsp. vulgare	Ranunculaceae	6	Europe	1998	H	Ge	S/V
Aconitum napellus × variegatum = A. × stoerkianum	Ranunculaceae	165	Horticulture	1905	H	Ge	V
Acorus calamus	Acoraceae	451	North temperate zone	1668	H	Hy	V
Acorus gramineus	Acoraceae	6	East Asia	1986	H	Hy	V
Acroptilon repens	Asteraceae	1	North temperate zone	1950	W	Hc	S/V
Aeonium cuneatum	Crassulaceae	4	Canary Islands	1963	H	Ch	V
Aesculus carnea	Sapindaceae	292	Horticulture	1955	H	Ph	S
Aesculus hippocastanum	Sapindaceae	2,617	Europe	1870	H	Ph	S
Aesculus indica	Sapindaceae	25	Himalayas	1989	H	Ph	–
Agapanthus praecox subsp. orientalis	Amaryllidaceae	18	South Africa	1939	H	Ge	S/V
Agave americana	Asparagaceae	18	Mexico	1826	H	Ph	–
Ageratum houstonianum	Asteraceae	19	Mexico	1914	H	Th	–
Agrostis avenacea	Poaceae	3	Antipodes	1908	E,W	Th	–
Agrostis capillaris × castellana = A. × fouilladeana	Poaceae	2	Neonative	1975	Neo	Hc	V
Agrostis capillaris × gigantea = A. × bjoerkmanii	Poaceae	2	Neonative	1855	Neo	Hc	V
Agrostis castellana	Poaceae	92	Europe	1924	Gs,H,W	Hc	S/V
Agrostis gigantea × stolonifera	Poaceae	10	Neonative	1962?	Neo	Hc	V

Scientific name	Family	No. of hectads	Introduced from	First record	Vector	Life form	S/V
Agrostis lachnantha	Poaceae	4	Central & South Africa	1909	W	Hc	S
Agrostis scabra	Poaceae	19	North America	1896	G,T,W	Hc	S
Agrostis stolonifera × Polypogon viridis = × Agropogon robinsonii	Poaceae	2	Neonative	1924	Neo	Hc	–
Ailanthus altissima	Simaroubaceae	124	China	1935	H	Ph	S/V
Aira elegantissima	Poaceae	1	Europe	1912	Gs,W	Th	–
*Ajuga chamaepitys	Lamiaceae	23	Europe	1551	?native	Th	S
Alchemilla conjuncta	Rosaceae	61	Europe	1812	H	Hc	S
Alchemilla mollis	Rosaceae	1,345	Carpathians	1948	H	Hc	S
Alchemilla tytthantha	Rosaceae	5	Crimea	1956	H	Hc	S
Alchemilla venosa	Rosaceae	1	Caucasus	1963	H	Hc	S
Allium carinatum	Amaryllidaceae	116	Europe	1867	H	Ge	S/V
Allium moly	Amaryllidaceae	30	Western Europe	1838	H	Ge	S/V
Allium neapolitanum	Amaryllidaceae	74	Mediterranean	1864	H	Ge	S/V
Allium nigrum	Amaryllidaceae	23	Southern Europe	1866	H	Ge	S/V
Allium paniculatum subsp. fuscum	Amaryllidaceae	1	Southeast Europe	2004	H	Ge	S/V
Allium paniculatum subsp. paniculatum	Amaryllidaceae	1	Europe	2010	H	Ge	S/V
Allium paradoxum	Amaryllidaceae	396	Caucasus	1849	H	Ge	S/V
Allium pendulinum	Amaryllidaceae	2	Italy	1977	H	Ge	S/V
Allium roseum	Amaryllidaceae	189	Mediterranean	1837	H	Ge	S/V
*Allium sphaerocephalon	Amaryllidaceae	12	Europe	1836	?native	Ge	S/V
Allium subhirsutum	Amaryllidaceae	83	Mediterranean	1954	H	Ge	–
Allium triquetrum	Amaryllidaceae	633	Mediterranean	1847	H	Ge	S/V
Allium unifolium	Amaryllidaceae	2	North America	1985	H	Ge	–
Alnus cordata	Betulaceae	586	Italy	1935	H	Ph	S
Alnus cordata × A. glutinosa = A. × elliptica	Betulaceae	2	Neonative	2006	Neo	Ph	–
Alnus cordata × incana	Betulaceae	4	Neonative	2007	Neo	Ph	–
Alnus glutinosa × incana = A. × hybrida	Betulaceae	67	Neonative	1950	Neo	Ph	V
Alnus incana	Betulaceae	1,026	Europe	1920	H	Ph	S/V
Alnus rubra	Betulaceae	36	North America	1968	H	Ph	S
Alnus viridis	Betulaceae	27	Europe	1987	H	Ph	–
Aloe aristata	Xanthorrhoeaceae	1	South Africa	1995	H	Ph	–
Alstroemeria aurea	Alstroemeriaceae	43	Chile	1950	H	Ge	S/V
Alyssum alyssoides	Brassicaceae	11	Europe	1837	As,W	Th	S
Amaranthus albus	Amaranthaceae	70	North America	1872	As,Bs,C,E, G,O,Tn,W	Th	SW
Amaranthus blitoides	Amaranthaceae	19	North America	1929	As,Bs,C, O,W	Th	–
Amaranthus blitum	Amaranthaceae	19	Southern Europe	1771	Bs,W	Th	S
Amaranthus bouchonii	Amaranthaceae	69	Europe	1949	As,Bs,W	Th	S
Amaranthus bouchonii × retroflexus	Amaranthaceae	1	Neonative	2005	Neo	Th	–
Amaranthus capensis	Amaranthaceae	1	South Africa	1908	W	Th	–
Amaranthus caudatus	Amaranthaceae	25	South America	1908	Bs,H	Th	–

Scientific name	Family	No. of hectads	Introduced from	First record	Vector	Life form	S/V
Amaranthus crispus	Amaranthaceae	1	South America	1958	W	Th	–
Amaranthus cruentus	Amaranthaceae	26	Central America	1929	A,As,Bs,E, G,O,W	Th	S
Amaranthus deflexus	Amaranthaceae	10	South America	1874	S,Tn,W	Th	S
Amaranthus graecizans	Amaranthaceae	2	Mediterranean	1905	Bs,W	Th	–
Amaranthus hybridus	Amaranthaceae	200	America	1876	A,As,Bs,E, G,O,W	Th	S
Amaranthus hybridus × retroflexus = A. × ozanonii	Amaranthaceae	3	North America	1959	A,As,Bs,E, G,H,O,W	Th	–
Amaranthus hypochondriacus	Amaranthaceae	2	North America	1900	A,As,Bs,E, G,O,W	Th	–
Amaranthus palmeri	Amaranthaceae	1	North America	1959	G,O	Th	–
Amaranthus powellii	Amaranthaceae	7	America	1982	O	Th	–
Amaranthus quitensis	Amaranthaceae	1	South America	1916	Bs,O,W	Th	–
Amaranthus retroflexus	Amaranthaceae	486	North America	1853	As,Bs,C,G, O,W	Th	S
Amaranthus standleyanus	Amaranthaceae	1	South America	1932	Bs,O,W	Th	–
Amaranthus thunbergii	Amaranthaceae	3	South Africa	1909	Bs,E,W	Th	–
Amaranthus viridis	Amaranthaceae	1	South America	1923	Bs,O,W	Th	–
Amaryllis belladonna	Amaryllidaceae	14	South Africa	1957	H	Ge	–
Ambrosia artemisiifolia	Asteraceae	223	North America	1836	As,Bs,G,O	Th	–
Ambrosia psilostachya	Asteraceae	3	North America	1902	G	Hc	–
Ambrosia trifida	Asteraceae	5	North America	1893	G,O	Th	–
Amelanchier lamarckii	Rosaceae	197	North America	1887	H	Ph	S
Ammi majus	Apiaceae	152	Europe	1821	As,Bs,G,W	Th	–
Ammi visnaga	Apiaceae	12	Mediterranean	1881	Bs,G,W	Th	–
Amsinckia lycopsoides	Boraginaceae	14	North America	1877	As,G	Th	S
Amsinckia micrantha	Boraginaceae	384	North America	1910	As,Bs,G, Gs,W	Th	S
Anaphalis margaritacea	Asteraceae	153	North America	1698	H	Hc	S/V
Anchusa azurea	Boraginaceae	29	Southern Europe	1866	H	Hc	–
Anchusa ochroleuca	Boraginaceae	4	Eastern Europe	1922	H	Hc	S
Anchusa ochroleuca × officinalis = A. × baumgartenii	Boraginaceae	1	Neonative	1950	Neo	Hc	S
Anchusa officinalis	Boraginaceae	65	Europe	1799	H	Hc	S
Anemanthele lessoniana	Poaceae	36	New Zealand	1974	H	Hc	S
Anemone apennina	Ranunculaceae	170	Mediterranean	1724	H	Ge	S/V
Anemone blanda	Ranunculaceae	98	Balkans	1983	H	Ge	S/V
Anemone hupehensis × vitifolia = A. × hybrida	Ranunculaceae	180	Horticulture	c. 1900	H	Hc	–
Anemone ranunculoides	Ranunculaceae	22	Europe	1778	H	Ge	S/V
Angelica archangelica	Apiaceae	120	Europe	1597	F	Hc	S
Angelica pachycarpa	Apiaceae	1	Southwest Europe	1993	?	Hc	S
Anisantha diandra	Poaceae	499	Europe	1835	G,W	Th	S
Anisantha madritensis	Poaceae	83	Europe	1716	E,G,W	Th	S
Anisantha rigida	Poaceae	93	Europe	1834	As,B,G,W	Th	S
Anisantha rubens	Poaceae	2	Europe	1908	G,E,W	Th	–

Scientific name	Family	No. of hectads	Introduced from	First record	Vector	Life form	S/V
Anisantha tectorum	Poaceae	37	Europe	1847	G,Gs,W	Th	S
Anoda cristata	Malvaceae	3	North America	1939	As,Bs,O,W	Th	–
Anthemis cotula × Tripleurospermum inodorum = × Tripleurothemis maleolens	Asteraceae	1	Neonative	1966	Neo	Th	–
Anthemis austriaca	Asteraceae	33	Europe	1878	Gs,W	Th	–
Anthemis punctata subsp. cupaniana	Asteraceae	50	Sicily	1923	H	Hc	S
Anthemis tinctoria	Asteraceae	120	Europe	1661	Bs,H	Hc	S
Anthoxanthum aristatum subsp. puelii	Poaceae	8	Europe	1872	As,G,T	TH	–
Anthyllis vulneraria subsp. carpatica	Fabaceae	34	Europe	1895	As,Gs	Hc	S
Anthyllis vulneraria subsp. polyphylla	Fabaceae	22	Europe	1956	A,As,Gs	Hc	S
Antirrhinum majus	Veronicaceae	1,208	Europe	1698	H	Ch	S
Apera interrupta	Poaceae	71	Europe	1843	As,Gs	Th	S
Aponogeton distachyos	Aponogetonaceae	58	South Africa	1889	H	Hy	–
Aptenia cordifolia	Aizoaceae	9	South Africa	1928	H	Ch	V
Aquilegia pyrenaica	Ranunculaceae	5	Southwest Europe	1895	H	Hc	S
Arabidopsis arenosa	Brassicaceae	2	Europe	1863	G,S	Th	–
Arabis caucasica	Brassicaceae	316	Southern Europe	1855	H	Ch	S/V
Arabis collina	Brassicaceae	3	Southern Europe	1926	H	Hc	S
Arabis procurrens	Brassicaceae	1	Europe	2002	H	Hc	V
Arachis hypogaea	Fabaceae	10	Brazil	1937	F	Th	–
Aralia chinensis	Araliaceae	5	China	1990	H	Ph	V
Aralia elata	Araliaceae	13	East Asia	1980	H	Ph	V
Aralia racemosa	Araliaceae	1	North America	1926	H	Hc	S
Arctium lappa × minus = A. × nothum	Asteraceae	5	Neonative	1955	Neo	Hc	S
Arctium tomentosum	Asteraceae	4	Europe	1908	G	Hc	–
Arctotheca calendula	Asteraceae	2	South Africa	1876	H	Th	–
Aremonia agrimonioides	Rosaceae	7	Europe	1852	H	Hc	S
Arenaria balearica	Caryophyllaceae	68	Mediterranean	1805	H	Hc	S/V
Arenaria montana	Caryophyllaceae	2	Southwest Europe	1865	H	Hc	S/V
Argemone mexicana	Papaveraceae	5	Central America	1798	H	Th	–
Arisarum proboscideum	Araceae	5	Southern Europe	1872	H	Ge	V
Aristea ecklonii	Iridaceae	1	South Africa	1977	H	Ge	V
Aristolochia clematitis	Aristolochiaceae	22	Europe	1685	H,M	Ge	V
Aristolochia hirta	Aristolochiaceae	1	Turkey	1969	H	Ge	V
Aristolochia rotunda	Aristolochiaceae	3	Europe	1901	H	Ge	V
Aronia arbutifolia	Rosaceae	3	North America	1975	H	Ph	S/V
Aronia melanocarpa	Rosaceae	3	North America	1980	H	Ph	S/V
Artemisia abrotanum	Asteraceae	22	Horticulture	1872	H	Ph	–
Artemisia annua	Asteraceae	5	Eurasia	1903	Bs,W	Th	–
Artemisia biennis	Asteraceae	15	Asia, America	1903	G,O,W	Th	–
*Artemisia campestris subsp. maritima	Asteraceae	2	Western Europe	1865	?native	Ch	S

Scientific name	Family	No. of hectads	Introduced from	First record	Vector	Life form	S/V
Artemisia dracunculus	Asteraceae	15	Russia	1906	H,W	Ch	–
Artemisia stelleriana	Asteraceae	16	Northeast Asia	1888	H	Ch	V
Artemisia verlotiorum	Asteraceae	81	China	1908	?	Hc	S/V
Artemisia verlotiorum × vulgaris = A. × wurzellii	Asteraceae	7	Neonative	1987	Neo	Hc	V
Arum italicum subsp. italicum	Araceae	445	Europe	1870	H	Ge	S/V
Aruncus dioicus	Rosaceae	72	Europe	1950	H	Hc	S
Asarina procumbens	Veronicaceae	50	Europe	1938	H	Hc	S/V
Asarum europaeum	Aristolochiaceae	13	Europe	1640	H	Ge	V
Asperugo procumbens	Boraginaceae	2	Europe	1660	G,W	Th	S
Asperula arvensis	Rubiaceae	9	Europe	1841	Bs,G	Th	–
Asperula taurina	Rubiaceae	9	Southern Europe	1856	H	Hc	S/V
Asphodelus albus	Xanthorrhoeaceae	2	Europe	1970s	H	Ge	S/V
Aster concinnus	Asteraceae	3	North America	1990	H	Hc	S/V
Aster laevis	Asteraceae	12	North America	1894	H	Hc	S/V
Aster laevis × novi-belgii = A. × versicolor	Asteraceae	353	Horticulture	1800	H	Hc	S/V
Aster lanceolatus	Asteraceae	237	North America	1865	H	Hc	S/V
Aster lanceolatus × novi-belgii = A. × salignus	Asteraceae	490	Horticulture	1846	H	Hc	S/V
Aster novae-angliae	Asteraceae	59	North America	1823	H	Hc	S/V
Aster novi-belgii	Asteraceae	543	North America	1860	H	Hc	S/V
Aster pilosus	Asteraceae	1	North America	1932	H	Hc	S/V
Aster puniceus	Asteraceae	3	North America	1915	H	Hc	S/V
Aster squamatus	Asteraceae	3	Central & South America	1992	?	Th	S
Astilbe chinensis	Rosaceae	2	Central Asia	1906	H	Hc	S/V
Astilbe chinensis × japonica = A. × arendsii	Rosaceae	28	Horticulture	1966	H	Hc	S/V
Astilbe japonica	Rosaceae	15	Japan	1908	H	Hc	S/V
Astragalus cicer	Fabaceae	1	Europe	1913	G	Hc	S
Astragalus odoratus	Fabaceae	1	Mediterranean	1930	G	Hc	S
Astrantia major subsp. major	Apiaceae	87	Europe	1825	H	Hc	S
Atriplex halimus	Amaranthaceae	40	Southern Europe	1900	?	Ph	V
Atriplex sagittata	Amaranthaceae	1	Eastern Europe	1905	H,W	Th	S
Aubrieta deltoidea	Brassicaceae	609	Southeast Europe	1913	H	Ch	S/V
Aucuba japonica	Garryaceae	278	Japan	1978	H	Ph	–
Aurinia saxatilis	Brassicaceae	157	Europe	1912	H	Ch	S
Avena barbata	Poaceae	6	Mediterranean	1903	G,W	Th	S
Avena fatua × sativa = A. × hybrida	Poaceae	4	Neonative	1871	Neo	Th	–
Avena sterilis subsp. ludoviciana	Poaceae	198	Europe	1903	G,W	Th	–
Baccharis halimifolia	Asteraceae	2	North America	1924	H	Ph	–
Ballota acetabulosa	Lamiaceae	3	Southeast Europe	1930	H	Ch	–
Barbarea intermedia	Brassicaceae	832	Europe	1836	?	Hc	S
Barbarea stricta	Brassicaceae	58	Europe	1840	?	Hc	S
Barbarea verna	Brassicaceae	444	Europe	1745	F	Hc	S

Scientific name	Family	No. of hectads	Introduced from	First record	Vector	Life form	S/V
Bassia scoparia	Amaranthaceae	125	Asia	1866	H	Th	S
Beckmannia syzigachne	Poaceae	4	North America	1911	Bs,G	Th	–
Berberis aggregata	Berberidaceae	24	China	1969	H	Ph	–
Berberis aggregata × wilsoniae	Berberidaceae	3	Horticulture	2003	H	Ph	–
Berberis buxifolia	Berberidaceae	6	South America	1961	H	Ph	S
Berberis darwinii	Berberidaceae	351	South America	1915	H	Ph	–
Berberis darwinii × empetrifolia = B. × stenophylla	Berberidaceae	115	Horticulture	1935	H	Ph	–
Berberis gagnepainii	Berberidaceae	46	China	1984	H	Ph	–
Berberis glaucocarpa	Berberidaceae	16	Himalayas	1907	H	Ph	S
Berberis julianae	Berberidaceae	50	China	1984	H	Ph	–
Berberis thunbergii	Berberidaceae	151	Japan	1971	H	Ph	–
*Berberis vulgaris	Berberidaceae	608	Europe	1597	?native	Ph	S
Berberis wilsoniae	Berberidaceae	41	China	1941	H	Ph	–
Bergenia ciliata × crassifolia = B. × schmidtii	Saxifragaceae	10	Horticulture	1979	H	Ch	–
Bergenia cordifolia	Saxifragaceae	35	Siberia	1974	H	Ch	–
Bergenia crassifolia	Saxifragaceae	128	Siberia	1961	H	Ch	–
Berteroa incana	Brassicaceae	9	Europe	1798	G	Th	S
Beta trigyna	Amaranthaceae	4	Southeast Europe	1891	As	Hc	–
Beta vulgaris subsp. cicla	Amaranthaceae	64	Agriculture	1955	F	Hc	–
Beta vulgaris subsp. vulgaris	Amaranthaceae	313	Agriculture	1905	F	Hc	–
Betula papyrifera	Betulaceae	16	North America	1968	H	Ph	–
Betula populifolia	Betulaceae	1	North America	1972	H	Ph	–
Betula utilis	Betulaceae	3	Himalayas	1998	H	Ph	–
Bidens bipinnata	Asteraceae	1	South America	1899	Bs,W	Th	–
Bidens connata	Asteraceae	20	North America	1977	?	Th	S
Bidens ferulifolia	Asteraceae	43	Mexico	1915	H	Th	–
Bidens frondosa	Asteraceae	91	North & South America	1899	?	Th	S
Bidens pilosa	Asteraceae	2	South America	1911	Bs,C,W	Th	–
Bifora radians	Apiaceae	1	Europe	1905	G	Th	–
Borago pygmaea	Boraginaceae	14	Corsica	1932	H	Hc	S
Brachiaria platyphylla	Poaceae	4	North America	1970	Bs,O	Th	–
Brachyglottis compacta × laxifolia = B. × jubar	Asteraceae	128	Horticulture	1981	H	Ph	–
Brachyglottis monroi	Asteraceae	13	New Zealand	1985	H	Ph	–
Brachyglottis repanda	Asteraceae	7	New Zealand	1971	H	Ph	–
Brachypodium hybridum	Poaceae	2	Mediterranean	1903	Bs,E,G,W	Th	–
Brassica carinata	Brassicaceae	3	Abyssinia	1941	Bs,O	Th	–
Brassica elongata	Brassicaceae	2	Southeast Europe	1881	G	Hc	–
Brassica juncea	Brassicaceae	99	Asia	1876	As,Bs,G,W	Th	–
Brassica napus subsp. oleifera	Brassicaceae	845	Agriculture	1975	F	Th	S
Brassica napus subsp. rapifera	Brassicaceae	21	Agriculture	1905	F	Hc	S
Brassica napus × oleracea	Brassicaceae	1	Neonative	2004	Neo	Th	–
Brassica napus × rapa = B. × harmsiana	Brassicaceae	7	Neonative	1975	Neo	Th	–

Scientific name	Family	No. of hectads	Introduced from	First record	Vector	Life form	S/V
*Brassica oleracea	Brassicaceae	315	Europe	1548	?native	Ch	S
Brassica rapa subsp. oleifera	Brassicaceae	51	Agriculture	1905	Bs,O	Th	–
Brassica tournefortii	Brassicaceae	4	Mediterranean	1904	E,G,W	Th	–
Briza maxima	Poaceae	319	Mediterranean	1860	E,H,W	Th	S
Bromopsis inermis subsp. inermis	Poaceae	231	Europe	1890	A,Gs,W	Hc	S/V
Bromopsis inermis subsp. pumpelliana	Poaceae	1	North America	1967	Gs	Hc	S/V
Bromus arvensis	Poaceae	23	Europe	1763	As,G,W	Th	S
Bromus hordeaceus subsp. molliformis	Poaceae	15	Western Europe	1927	G,W	Th	S
Bromus hordeaceus × lepidus = B. × pseudothominei	Poaceae	416	Neonative	1881	Neo	Th	S
Bromus lanceolatus	Poaceae	4	Southern Europe	1878	Bs,E,W	Th	–
Bromus lepidus	Poaceae	187	Uncertain	1836	Gs	Th	S
Bromus pseudosecalinus	Poaceae	6	Uncertain	1890	Gs	Th	S
Bromus squarrosus	Poaceae	1	Southern Europe	1873	Gs,W	Th	–
Brunnera macrophylla	Boraginaceae	126	Caucasus	1926	H	Hc	S
Buddleja alternifolia	Scrophulariaceae	11	China	1979	H	Ph	–
Buddleja davidii	Scrophulariaceae	1,974	China	1922	H	Ph	S
Buddleja davidii × globosa = B. × weyeriana	Scrophulariaceae	42	Horticulture	1976	H	Ph	–
Buddleja globosa	Scrophulariaceae	171	South America	1964	H	Ph	–
Bunias orientalis	Brassicaceae	83	Europe	1825	B,Bs,G	Hc	S
Bupleurum falcatum	Apiaceae	6	Europe	1831	?	Hc	S
Bupleurum fruticosum	Apiaceae	7	Southern Europe	1905	H	Ph	S
Bupleurum subovatum	Apiaceae	21	Mediterranean	1839	Bs	Th	–
Cabomba caroliniana	Cabombaceae	2	North America	1969	H	Hy	V
Calceolaria chelidonioides	Calceolariaceae	5	Central & South America	1867	Bs,H,W	Th	S
Calendula arvensis	Asteraceae	26	Europe	1823	E,H	Th	S
Calla palustris	Araceae	41	Europe	1861	H	Hy	S/V
Callistephus chinensis	Asteraceae	11	China	1903	H	Th	–
Calystegia pulchra	Convolvulaceae	666	Uncertain	1850	H	Hc	S/V
Calystegia pulchra × sepium = C. × scanica	Convolvulaceae	11	Neonative	1960	Neo	Hc	S/V
Calystegia pulchra × silvatica = C. × howittiorum	Convolvulaceae	35	Neonative	1959	Neo	Hc	S/V
Calystegia sepium × silvatica = C. × lucana	Convolvulaceae	174	Neonative	1901	Neo	Hc	S/V
Calystegia silvatica	Convolvulaceae	1,980	Southern Europe	1863	H	Hc	S/V
Camelina microcarpa	Brassicaceae	3	Europe	1867	G	Th	–
Campanula alliariifolia	Campanulaceae	12	Turkey	1943	H	Hc	S
Campanula cochleariifolia	Campanulaceae	4	Europe	1980	H	Hc	S
Campanula fragilis	Campanulaceae	1	Italy	1976	H	Hc	S/V
Campanula garganica	Campanulaceae	3	Mediterranean	2002	H	Hc	S/V
Campanula lactiflora	Campanulaceae	40	Turkey	1954	H	Hc	S
Campanula medium	Campanulaceae	82	Southern Europe	1870	H	Hc	S

Scientific name	Family	No. of hectads	Introduced from	First record	Vector	Life form	S/V
Campanula persicifolia	Campanulaceae	410	Europe	1802	H	Hc	S/V
Campanula portenschlagiana	Campanulaceae	513	Yugoslavia	1922	H	Hc	S/V
Campanula poscharskyana	Campanulaceae	720	Yugoslavia	1957	H	Hc	S/V
Campanula pyramidalis	Campanulaceae	10	Mediterranean	1883	H	Hc	S
Campanula rapunculoides	Campanulaceae	272	Europe	1708	H	Hc	S
Campanula rhomboidalis	Campanulaceae	1	Europe	1978	H	Hc	S
Capsella bursa-pastoris × rubella = C. × gracilis	Brassicaceae	6	Neonative	1956	Neo	Th	–
Capsella rubella	Brassicaceae	22	Europe	1928	G	Th	–
Capsicum annuum	Solanaceae	9	Central & South America	1930	Bs,F	Th	–
Cardamine corymbosa	Brassicaceae	118	New Zealand	1975	I	Th	S
Cardamine heptaphylla	Brassicaceae	9	Europe	1907	H	Ge	S/V
Cardamine quinquefolia	Brassicaceae	1	Eastern Europe	2005	H	Ge	S/V
Cardamine raphanifolia	Brassicaceae	54	Southern Europe	1930	H	Hc	S
Cardamine trifolia	Brassicaceae	5	Europe	1903	H	Hc	S
Carduus acanthoides	Asteraceae	1	Europe	1908	?	Hc	–
Carduus pycnocephalus	Asteraceae	5	Southern Europe	1868	Bs,?I,W	Th	S
Carex buchananii	Cyperaceae	8	New Zealand	1990	H	Hc	S
Carex vulpinoidea	Cyperaceae	4	North America	1880	A,As,W	Hc	S
Carpobrotus acinaciformis	Aizoaceae	3	South Africa	1911	H	Ch	S/V
Carpobrotus edulis	Aizoaceae	93	South Africa	1886	H	Ch	S/V
Carpobrotus glaucescens	Aizoaceae	12	Australia	1947	H	Ch	S/V
Carrichtera annua	Brassicaceae	1	Mediterranean	1846	E,G,W	Th	–
Carthamus lanatus	Asteraceae	5	Europe	1876	Bs,G,W	Th	–
Carthamus tinctorius	Asteraceae	68	Southwest Asia	1899	Bs,G,O	Th	–
Catananche caerulea	Asteraceae	4	Europe	1988	H	Hc	–
Celastrus orbiculatus	Celastraceae	4	East Asia	1985	H	Ph	–
Cenchrus echinatus	Poaceae	1	America	1969	O,W	Th	–
Centaurea aspera	Asteraceae	10	Europe	1788	?	Hc	S
Centaurea cineraria	Asteraceae	1	Italy	1928	H	Hc	–
Centaurea diluta	Asteraceae	12	Mediterranean	1904	Bs,G,W	Hc	–
Centaurea hyalolepis	Asteraceae	2	Mediterranean	1908	Bs,G,W	Hc	–
Centaurea jacea	Asteraceae	14	Europe	1775	As,G,Gs	Hc	S
Centaurea jacea × debeauxii	Asteraceae		Neonative	1928	Neo	Hc	S
Centaurea jacea × nigra = C. × gerstlaueri	Asteraceae	7	Neonative	1930	Neo	Hc	S
Centaurea macrocephala	Asteraceae	5	Caucasus	1920	H	Hc	S
Centaurea melitensis	Asteraceae	2	Southern Europe	1876	As,Bs,E,G,W	Th	–
Centaurea montana	Asteraceae	914	Europe	1845	H	Hc	S/V
Centaurea solstitialis	Asteraceae	9	Europe	1778	As,Bs,G,W	Th	–
Centranthus calcitrapae	Valerianaceae	1	Southern Europe	1770	H	Th	S
Centranthus ruber	Valerianaceae	1,782	Europe	1698	H	Hc	S
Cephalaria gigantea	Dipsacaceae	71	Caucasus	1915	H	Hc	–
Cerastium arvense × tomentosum	Caryophyllaceae	40	Neonative	1968	Neo	Ch	S/V

Scientific name	Family	No. of hectads	Introduced from	First record	Vector	Life form	S/V
Cerastium brachypetalum	Caryophyllaceae	3	Europe	1947	?	Th	S
Cerastium tomentosum	Caryophyllaceae	1,439	Europe	1909	H	Ch	S/V
Ceratochloa carinata	Poaceae	225	North America	1919	A,G,H,W	Hc	S
Ceratochloa cathartica	Poaceae	131	Central & South America	1854	A,Bs,G,W	Hc	S
Ceratochloa marginata	Poaceae	2	North America	1907	G,W	Hc	S
Cercis siliquastrum	Fabaceae	19	Mediterranean	1980	H	Ph	–
Cerinthe major	Boraginaceae	33	Europe	1943	H	Th	–
Chaenomeles japonica	Rosaceae	20	Japan	1972	H	Ph	–
Chaenomeles japonica × speciosa = C. × superba	Rosaceae	1	Horticulture	1988	H	Ph	–
Chaenomeles speciosa	Rosaceae	86	China	1963	H	Ph	–
Chaenorhinum origanifolium subsp. crassifolium	Veronicaceae	20	Europe	1899	H	Hc	S
Chaerophyllum aureum	Apiaceae	7	Europe	1810	H	Hc	S
Chaerophyllum hirsutum	Apiaceae	3	Europe	1970	H	Hc	S
Chamaemelum mixtum	Asteraceae	2	Mediterranean	1832	Bs,G,S,W	Th	–
Chasmanthe bicolor	Iridaceae	4	South Africa	1960	H	Ge	S/V
Chenopodium berlandieri	Amaranthaceae	3	North America	1904	G,O	Th	–
Chenopodium bushianum	Amaranthaceae	1	North America	1938	Bs,O	Th	–
Chenopodium capitatum	Amaranthaceae	17	Widespread	1888	Bs,E,W	Th	–
Chenopodium giganteum	Amaranthaceae	14	India	1926	Bs,W	Th	–
Chenopodium hircinum	Amaranthaceae	1	South America	1898	W	TH	–
Chenopodium nitrariaceum	Amaranthaceae	1	Australia	1961	W	Ph	–
Chenopodium opulifolium	Amaranthaceae	9	Europe	1853	Bs,G,O,W	Th	S
Chenopodium pratericola	Amaranthaceae	4	North America	1901	G,W	Th	–
Chenopodium probstii	Amaranthaceae	20	North America	1930	Bs,C,E,G,O,W	Th	–
Chenopodium quinoa	Amaranthaceae	80	South America	1916	A,Bs,F,W	Th	–
Chenopodium strictum	Amaranthaceae	13	Europe	1905	Bs,G,O,W	Th	–
Chenopodium suecicum	Amaranthaceae	10	Europe	1844	Bs,G,W	Th	S
Chimonobambusa quadrangularis	Poaceae	1	China	1993	H	Ph	V
Chloris truncata	Poaceae	1	Australia	1915	W	Hc	–
Choisya ternata	Rutaceae	22	Mexico	1980	H	Ph	S/V
Chrysocoma coma-aurea	Asteraceae	2	South Africa	1914	H	Ch	–
Cicerbita bourgaei	Asteraceae	8	Turkey	1940	H	Hc	S
Cicerbita macrophylla subsp. uralensis	Asteraceae	656	Urals	1915	H	Hc	S/V
Cicerbita plumieri	Asteraceae	9	Europe	1917	H	Hc	S
Cirsium erisithales	Asteraceae	2	Europe	1980	H	Hc	S
Cirsium oleraceum	Asteraceae	9	Europe	1861	H	Hc	S
Citrullus lanatus	Cucurbitaceae	8	South Africa	1921	F,W	Th	–
Clarkia amoena	Onagraceae	40	North America	1902	H	Th	–
Clarkia unguiculata	Onagraceae	14	North America	1912	H	Th	–
Claytonia perfoliata	Montiaceae	609	North America	1849	I	Th	S
Claytonia sibirica	Montiaceae	1,100	North America	1837	H	TH	S

Scientific name	Family	No. of hectads	Introduced from	First record	Vector	Life form	S/V
Clematis cirrhosa	Ranunculaceae	3	Southern Europe	1969	H	Ph	–
Clematis flammula	Ranunculaceae	12	Mediterranean	1927	H	Ph	S/V
Clematis montana	Ranunculaceae	39	Asia	1928	H	Ph	–
Clematis tangutica	Ranunculaceae	39	China	1965	H	Ph	–
Clematis viticella	Ranunculaceae	15	Southern Europe	1880	H	Ph	–
Cleome sesquiorgyalis	Cleomaceae	2	South America	1967	H	Th	–
Cochlearia acaulis	Brassicaceae	2	Portugal	1908	H	Th	–
Cochlearia megalosperma	Brassicaceae	2	Southwest Europe	1910	H	Th	S
Coincya monensis subsp. cheiranthos	Brassicaceae	52	Europe	1852	B	Hc	S
Colutea arborescens	Fabaceae	134	Europe	1900	H	Ph	S
Colutea arborescens × orientalis = C. × media	Fabaceae	16	Horticulture	1969	H	Ph	S
Conringia orientalis	Brassicaceae	1	Mediterranean	1724	As,Bs,G	Th	–
Consolida ajacis	Ranunculaceae	256	Europe	1650	G,H	Th	S
Consolida hispanica	Ranunculaceae	3	Europe	1853	G	Th	–
Consolida regalis	Ranunculaceae	4	Europe	1870	G	Th	–
Convolvulus tricolor	Convolvulaceae	1	Europe	1798	Bs,H	Th	–
Conyza bonariensis	Asteraceae	33	South America	1843	W	Th	S
Conyza bonariensis × canadensis	Asteraceae	2	Neonative	1993	Neo	Th	–
Conyza canadensis	Asteraceae	1,129	North America	1690	?V	Th	S
Conyza canadensis × Erigeron acris = × Conyzigeron huelsenii	Asteraceae	8	Neonative	1884	Neo	Th	–
Conyza floribunda	Asteraceae	211	South America	1977	V	Th	S
Conyza floribunda × Erigeron acris	Asteraceae	1	Neonative	2010	Neo	Th	–
Conyza sumatrensis	Asteraceae	414	South America	1961	V	Th	S
Conyza sumatrensis × Erigeron acris	Asteraceae	1	Neonative	2006	Neo	Th	–
Coprosma repens	Rubiaceae	4	New Zealand	1963	H	Ph	–
Cordyline australis	Asparagaceae	119	New Zealand	1965	H	Ph	–
Coreopsis grandiflora	Asteraceae	2	North America	1912	H	Hc	S
Coreopsis tinctoria	Asteraceae	4	North America	1905	H,W	Th	–
Corispermum intermedium	Amaranthaceae	4	Europe	1962	?S	Th	S
Cornus alba	Cornaceae	204	East Asia	1875	H	Ph	V
Cornus mas	Cornaceae	68	Europe	1927	H	Ph	–
Cornus sanguinea subsp. australis	Cornaceae	2	Caucasus	1995	H	Ph	–
Cornus sericea	Cornaceae	665	North America	1838	H	Ph	V
Coronilla scorpioides	Fabaceae	1	Europe	1859	Bs,Tn,W	Th	–
Coronilla valentina subsp. glauca	Fabaceae	19	Mediterranean	1949	H	Ph	S
Correa backhouseana	Rutaceae	1	Tasmania	2005	H	Ph	–
Cortaderia richardii	Poaceae	103	New Zealand	1990	H	Hc	S
Cortaderia selloana	Poaceae	381	South America	1925	H	Hc	S
Corydalis cava	Papaveraceae	11	Europe	1656	H	Ge	S/V
Corydalis cheilanthifolia	Papaveraceae	29	China	1981	H	Hc	–
Corydalis solida	Papaveraceae	71	Europe	1796	H	Ge	S/V
Corylus avellana × maxima	Betulaceae	21	Neonative	1974	Neo	Ph	S
Corylus colurna	Betulaceae	14	Southwest Asia	1995	H	Ph	–

Scientific name	Family	No. of hectads	Introduced from	First record	Vector	Life form	S/V
Corylus maxima	Betulaceae	35	Southwest Asia	1975	F,H	Ph	S
Cosmos bipinnatus	Asteraceae	56	North & Central America	1930	Bs,C,H,W	Th	S
Cotinus coggygria	Anacardiaceae	24	Europe	1966	H	Ph	–
Cotoneaster adpressus	Rosaceae	2	China	1984	H	Ch	–
Cotoneaster affinis	Rosaceae	3	Himalayas	1971	H	Ph	–
Cotoneaster amoenus	Rosaceae	7	China	1981	H	Ph	–
Cotoneaster apiculatus	Rosaceae	1	China	1992	H	Ph	S
Cotoneaster ascendens	Rosaceae	7	China	1995	H	Ph	S
Cotoneaster astrophoros	Rosaceae	2	China	1994	H	Ch	–
Cotoneaster atropurpureus	Rosaceae	20	China	1963	H	Ph	–
Cotoneaster atrovirens	Rosaceae	1	China	1999	H	Ph	–
Cotoneaster bacillaris	Rosaceae	6	Himalayas	1923	H	Ph	–
Cotoneaster boisianus	Rosaceae	4	China	1995	H	Ph	–
Cotoneaster bradyi	Rosaceae	1	China	2003	H	Ph	–
Cotoneaster bullatus	Rosaceae	344	China	1952	H	Ph	S
Cotoneaster calocarpus	Rosaceae	2	China	1982	H	Ph	–
Cotoneaster cashmiriensis	Rosaceae	5	Kashmir	1991	H	Ch	–
Cotoneaster cochleatus	Rosaceae	3	China	1954	H	Ch	–
Cotoneaster congestus	Rosaceae	7	Himalayas	1988	H	Ph	–
Cotoneaster conspicuus	Rosaceae	35	Tibet	1984	H	Ph	S
Cotoneaster conspicuus × dammeri = C. × suecicus	Rosaceae	83	Horticulture	1984	H	Ch	S
Cotoneaster cooperi	Rosaceae	2	Himalayas	1985	H	Ph	–
Cotoneaster dammeri	Rosaceae	51	China	1976	H	Ch	–
Cotoneaster dammeri × salicifolius = C. 'Hybridus Pendulus'	Rosaceae	5	Horticulture	1984	H	Ph	–
Cotoneaster dielsianus	Rosaceae	211	China	1965	H	Ph	S
Cotoneaster divaricatus	Rosaceae	78	China	1974	H	Ph	S
Cotoneaster ellipticus	Rosaceae	3	Himalayas	1959	H	Ph	–
Cotoneaster fangianus	Rosaceae	2	China	1990	H	Ph	–
Cotoneaster franchetii	Rosaceae	177	China	1968	H	Ph	S
Cotoneaster frigidus	Rosaceae	166	Himalayas	1918	H	Ph	S
Cotoneaster frigidus × salicifolius = C. × watereri	Rosaceae	235	Horticulture	1968	H	Ph	S
Cotoneaster froebelii	Rosaceae	1	China	1984	H	Ph	–
Cotoneaster fruticosus	Rosaceae	1	China	2002	H	Ph	–
Cotoneaster glabratus	Rosaceae	1	China	2002	H	Ph	–
Cotoneaster glaucophyllus	Rosaceae	1	China	1997	H	Ph	–
Cotoneaster hedegaardii	Rosaceae	1	Nepal	1989	H	Ph	–
Cotoneaster henryanus	Rosaceae	4	China	1985	H	Ph	S
Cotoneaster hissaricus	Rosaceae	1	Central Asia	1985	H	Ph	–
Cotoneaster hjelmqvistii	Rosaceae	94	China	1981	H	Ph	S
Cotoneaster hodjingensis	Rosaceae	1	China	1997	H	Ph	–
Cotoneaster horizontalis	Rosaceae	1,316	China	1940	H	Ph	SW
Cotoneaster hsingshangensis	Rosaceae	4	China	1986	H	Ph	–

Scientific name	Family	No. of hectads	Introduced from	First record	Vector	Life form	S/V
Cotoneaster hummelii	Rosaceae	1	China	1987	H	Ph	–
Cotoneaster hurusawanus	Rosaceae	1	China	2003	H	Ph	–
Cotoneaster hylmoei	Rosaceae	4	China	1993	H	Ph	–
Cotoneaster ignescens	Rosaceae	1	China	2001	H	Ph	S
Cotoneaster ignotus	Rosaceae	6	Himalayas	1983	H	Ph	–
Cotoneaster induratus	Rosaceae	2	China	1991	H	Ph	S
Cotoneaster insculptus	Rosaceae	2	China	1984	H	Ph	–
Cotoneaster integrifolius	Rosaceae	574	Himalayas	1895	H	Ph	S
Cotoneaster lacteus	Rosaceae	91	China	1968	H	Ph	S
Cotoneaster laetevirens	Rosaceae	3	China	1983	H	Ph	–
Cotoneaster lidjiangensis	Rosaceae	2	China	2004	H	Ph	–
Cotoneaster lucidus	Rosaceae	3	Siberia	1983	H	Ph	–
Cotoneaster mairei	Rosaceae	22	China	1978	H	Ph	S
Cotoneaster marginatus	Rosaceae	8	Himalayas	1923	H	Ph	–
Cotoneaster microphyllus	Rosaceae	78	Himalayas	1905	H	Ph	S
Cotoneaster monopyrenus	Rosaceae	2	China	1981	H	Ph	–
Cotoneaster moupinensis	Rosaceae	1	China	1966	H	Ph	–
Cotoneaster mucronatus	Rosaceae	1	China	1987	H	Ph	–
Cotoneaster nanshan	Rosaceae	5	China	1985	H	Ch	–
Cotoneaster nitens	Rosaceae	7	China	1982	H	Ph	S
Cotoneaster nohelii	Rosaceae	1	China	2005	H	Ph	–
Cotoneaster obscurus	Rosaceae	2	China	1994	H	Ph	–
Cotoneaster obtusus	Rosaceae	11	Himalayas	1987	H	Ph	S
Cotoneaster pannosus	Rosaceae	9	China	1972	H	Ph	–
Cotoneaster perpusillus	Rosaceae	1	China	1988	H	Ph	–
Cotoneaster prostratus	Rosaceae	5	Himalayas	1989	H	Ch	–
Cotoneaster pseudoambiguus	Rosaceae	5	China	1990	H	Ph	–
Cotoneaster radicans	Rosaceae	1	China	1996	H	Ch	–
Cotoneaster rehderi	Rosaceae	297	China	1981	H	Ph	S
Cotoneaster rotundifolius (nitidus)	Rosaceae	3	Himalayas	1985	H	Ph	–
Cotoneaster salicifolius	Rosaceae	159	China	1966	H	Ph	S
Cotoneaster serotinus	Rosaceae	1	China	1998	H	Ph	–
Cotoneaster shannanensis	Rosaceae	2	China, Tibet	1985	H	Ph	–
Cotoneaster sherriffii	Rosaceae	3	Tibet	1985	H	Ph	–
Cotoneaster simonsii	Rosaceae	1,272	Himalayas	1910	H	Ph	S
Cotoneaster splendens	Rosaceae	9	China	1981	H	Ph	S
Cotoneaster sternianus	Rosaceae	198	China	1969	H	Ph	S
Cotoneaster tengyuehensis	Rosaceae	1	China	1990	H	Ph	–
Cotoneaster thymifolius (linearifolius)	Rosaceae	2	Himalayas	1934	H	Ch	–
Cotoneaster tomentellus	Rosaceae	1	China	1993	H	Ph	–
Cotoneaster transens	Rosaceae	4	China	1971	H	Ph	–
Cotoneaster uva-ursi (rotundifolius auct.)	Rosaceae	3	Himalayas	1957	H	Ph	–
Cotoneaster villosulus	Rosaceae	7	China	1983	H	Ph	–
Cotoneaster vilmorinianus	Rosaceae	8	China	1978	H	Ph	–

Scientific name	Family	No. of hectads	Introduced from	First record	Vector	Life form	S/V
Cotoneaster wardii	Rosaceae	3	Tibet	1994	H	Ph	–
Cotoneaster zabelii	Rosaceae	8	China	1977	H	Ph	S
Cotula alpina	Asteraceae	5	Australia	1995	H	Ch	V
Cotula australis	Asteraceae	3	Antipodes	1908	H	Th	S
Cotula coronopifolia	Asteraceae	59	Southern hemisphere	1869	H,W	Th	S
Cotula dioica	Asteraceae	2	New Zealand	1956	H	Ch	–
Cotula squalida	Asteraceae	25	New Zealand	1913	H	Ch	V
Crambe cordifolia	Brassicaceae	7	Caucasus	1966	H	Hc	–
Crambe hispanica subsp. abyssinica	Brassicaceae	6	East Africa	2004	F	Th	–
Crassula decumbens	Crassulaceae	1	Southern hemisphere	1959	I,W	Th	S
Crassula helmsii	Crassulaceae	808	Antipodes	1956	H	Hy	V
Crassula pubescens subsp. radicans	Crassulaceae	3	South Africa	1970	H	Ch	–
Crataegus coccinea	Rosaceae	13	North America	1982	H	Ph	–
Crataegus coccinioides	Rosaceae	10	North America	1970	H	Ph	–
Crataegus crus-galli	Rosaceae	32	North America	1912	H	Ph	–
Crataegus heterophylla	Rosaceae	13	Caucasus	1930	H	Ph	S
Crataegus heterophylla × monogyna	Rosaceae	4	Neonative	1990	Neo	Ph	S
Crataegus laevigata × Mespilus germanica = × Crataemespilus grandiflora	Rosaceae	6	Horticulture	1915	H	Ph	–
Crataegus monogyna × rhipidophylla = C. × subsphaerica	Rosaceae	4	Europe	1980	H	Ph	–
Crataegus orientalis	Rosaceae	19	Southwest Asia	1918	H	Ph	S
Crataegus persimilis	Rosaceae	120	North America	1934	H	Ph	–
Crataegus rhipidophylla subsp. lindmanii	Rosaceae	1	Europe	2010	H	Ph	–
Crataegus rhipidophylla subsp. rhipidophylla	Rosaceae	4	Europe	2007	H	Ph	–
Crataegus submollis	Rosaceae	7	North America	1957	H	Ph	–
Crataegus succulenta	Rosaceae	6	North America	1958	H	Ph	–
Crepis nicaeensis	Asteraceae	6	Mediterranean	1874	As,Bs	Th	–
Crepis setosa	Asteraceae	73	Southern Europe	1843	As,G,Gs	Th	–
Crepis tectorum	Asteraceae	21	Europe	1872	G,Gs	Th	–
Crepis vesicaria subsp. taraxacifolia	Asteraceae	1,480	Europe	1713	As,G,Gs	Hc	S
Crinum bulbispermum × moorei = C. × powellii	Amaryllidaceae	8	Horticulture	1975	H	Ge	V
Crocosmia aurea × masoniorum × pottsii	Iridaceae	1	Horticulture	1999	H	Ge	S/V
Crocosmia aurea × pottsii = C. × crocosmiiflora	Iridaceae	2,320	Horticulture	1907	H	Ge	S/V
Crocosmia masoniorum	Iridaceae	13	South Africa	1982	H	Ge	S/V
Crocosmia paniculata	Iridaceae	232	South Africa	1961	H	Ge	S/V
Crocosmia paniculata × pottsii	Iridaceae	9	Horticulture	1994	H	Ge	S/V
Crocosmia pottsii	Iridaceae	76	South Africa	1982	H	Ge	S/V
Crocus ancyrensis	Iridaceae	3	Turkey	1992	H	Ge	S/V

Scientific name	Family	No. of hectads	Introduced from	First record	Vector	Life form	S/V
Crocus angustifolius × flavus = C. × luteus	Iridaceae	313	Horticulture	1848	H	Ge	S/V
Crocus biflorus subsp. adamii	Iridaceae	8	Southeast Europe	1830	H	Ge	S/V
Crocus biflorus subsp. biflorus	Iridaceae	1	Southeast Europe	1992	H	Ge	S/V
Crocus biflorus × chrysanthus	Iridaceae	7	Neonative	1992	Neo	Ge	S/V
Crocus chrysanthus	Iridaceae	71	Southeast Europe	1983	H	Ge	S/V
Crocus kotschyanus	Iridaceae	4	Turkey	1929	H	Ge	S/V
Crocus longiflorus	Iridaceae	2	Italy	1992	H	Ge	–
Crocus nudiflorus	Iridaceae	93	Southwest Europe	1738	H	Ge	S/V
Crocus pulchellus	Iridaceae	1	Southeast Europe	1983	H	Ge	–
Crocus serotinus subsp. salzmannii	Iridaceae	1	Mediterranean	1992	H	Ge	S/V
Crocus sieberi	Iridaceae	10	Balkans	1984	H	Ge	–
Crocus speciosus	Iridaceae	42	Southwest Asia	1972	H	Ge	S/V
Crocus tommasinianus	Iridaceae	354	Southeast Europe	1963	H	Ge	S/V
Crocus tommasinianus × vernus	Iridaceae	26	Neonative	1991	Neo	Ge	S/V
Crocus vernus subsp. albiflorus	Iridaceae	21	Europe	1908	H	Ge	S/V
Crocus vernus subsp. vernus	Iridaceae	506	Europe	1763	H	Ge	S/V
Cucumis melo	Cucurbitaceae	21	Africa	1928	F	Th	–
Cucumis sativus	Cucurbitaceae	9	India	1909	F,H	Th	–
Cucurbita maxima	Cucurbitaceae	15	Central America	1928	F	Th	–
Cucurbita pepo	Cucurbitaceae	45	Central America	1927	F,H	Th	–
Cuscuta campestris	Convolvulaceae	17	North America	1927	As,Bs,W	Th	–
Cyclamen coum	Primulaceae	78	Mediterranean	1920	H	Ge	S
Cyclamen hederifolium	Primulaceae	481	Europe	1597	H	Ge	S
Cyclamen repandum	Primulaceae	49	Mediterranean	1925	H	Ge	S
Cydonia oblonga	Rosaceae	28	Asia	1878	H	Ph	–
Cymbalaria hepaticifolia	Veronicaceae	11	Corsica	1965	H	Hc	S/V
Cymbalaria muralis subsp. muralis	Veronicaceae	2,535	Europe	1640	H	Ch	S/V
Cymbalaria muralis subsp. visianii	Veronicaceae	2	Mediterranean	1970	H	Ch	S/V
Cymbalaria pallida	Veronicaceae	108	Italy	1924	H	Hc	S/V
Cynara cardunculus	Asteraceae	36	Mediterranean	1904	F,H	Hc	–
*Cynodon dactylon	Poaceae	51	Widespread	1688	?native	Hc	S/V
Cynodon incompletus	Poaceae	3	South Africa	1950	W	Hc	–
Cynoglossum amabile	Boraginaceae	1	China	1963	H	Hc	–
Cynoglottis barrelieri	Boraginaceae	7	Southeast Europe	1963	H	Hc	–
Cynosurus echinatus	Poaceae	53	Europe	1778	G,I,W	Th	S
Cyperus eragrostis	Cyperaceae	153	Central & South America	1909	H,W	Hc	S
Cytisus multiflorus	Fabaceae	25	Southwest Europe	1957	H	Ph	–
Cytisus nigricans	Fabaceae	1	Europe	1970	H	Ph	S
Cytisus striatus	Fabaceae	66	Southwest Europe	1963	H	Ph	S/V
Dactylis polygama	Poaceae	1	Europe	1934	H	Hc	S
Dactyloctenium radulans	Poaceae	1	Australia	1915	W	Th	–
Dahlia × hortensis	Asteraceae	15	Horticulture	1969	H	Ge	–
Darmera peltata	Saxifragaceae	78	North America	1926	H	Hc	V
Datura ferox	Solanaceae	25	Mexico	1953	W	Th	–

Scientific name	Family	No. of hectads	Introduced from	First record	Vector	Life form	S/V
Datura stramonium	Solanaceae	536	America	1777	Bs,G,H,O,W	Th	S
Daucus carota subsp. sativus	Apiaceae	63	Agriculture	1928	F	Hc	–
Delairea odorata	Asteraceae	32	South Africa	1923	H	Hc	S
Deutzia scabra	Hydrangeaceae	49	East Asia	1963	H	Ph	–
Dianthus barbatus	Caryophyllaceae	158	Southern Europe	1805	H	Ch	S
Dianthus caryophyllus	Caryophyllaceae	9	Southern Europe	1745	H	Ch	S
Dianthus caryophyllus × gratianopolitanus	Caryophyllaceae	1	Horticulture	1980	H	Ch	S
Dianthus caryophyllus × gratianopolitanus × plumarius	Caryophyllaceae	1	Horticulture	1991	H	Ch	S
Dianthus caryophyllus × plumarius	Caryophyllaceae	3	Horticulture	1937	H	Ch	S
Dianthus gallicus	Caryophyllaceae	1	Western Europe	1892	H	Hc	S
Dianthus gratianopolitanus × plumarius	Caryophyllaceae	1	Horticulture	1991	H	Ch	S
Dianthus plumarius	Caryophyllaceae	19	Southeast Europe	1724	H	Ch	S
Dicentra eximia	Papaveraceae	10	North America	1912	H	Ge	S/V
Dicentra eximia × formosa	Papaveraceae	1	North America	1979	H	Ge	S/V
Dicentra formosa	Papaveraceae	77	North America	1859	H	Ge	S/V
Digitalis lutea	Veronicaceae	26	Europe	1919	H	Hc	S
Digitaria ciliaris	Poaceae	11	Old World tropics	1859	Bs,C,O,W	Th	–
Digitaria ischaemum	Poaceae	17	Southern Europe	1805	As,Bs,I,W	Th	S
Digitaria sanguinalis	Poaceae	142	Southern Europe	1690	As,Bs,G,I,O,W	Th	S
Diplotaxis erucoides	Brassicaceae	15	Southern Europe	1859	G,W	Th	–
Diplotaxis muralis	Brassicaceae	737	Europe	1778	B,Bs	Th	S
*Dipsacus fullonum	Dipsacaceae	1,839	Europe	1538	?native	Hc	S
Dipsacus fullonum × laciniatus = D. × pseudosilvester	Dipsacaceae	3	Neonative	1992	Neo	Hc	S
Dipsacus fullonum × sativus	Dipsacaceae	1	Neonative	1945	Neo	Hc	S
Dipsacus laciniatus	Dipsacaceae	23	Europe	1905	Gs,H	Hc	S
Dipsacus sativus	Dipsacaceae	25	Uncertain	1762	H	Hc	S
Dipsacus strigosus	Dipsacaceae	5	Russia	1828	H	Hc	S
Disphyma crassifolium	Aizoaceae	18	South Africa	1936	H	Ch	S/V
Dittrichia graveolens	Asteraceae	9	Southern Europe	1913	W	Th	S
Dittrichia viscosa	Asteraceae	6	Southern Europe	1876	?	Ph	S
Doronicum columnae	Asteraceae	12	Southeast Europe	1979	H	Hc	S/V
Doronicum columnae × pardalianches × plantagineum = D. × excelsum	Asteraceae	62	Horticulture	1916	H	Hc	S/V
Doronicum pardalianches	Asteraceae	750	Europe	1633	H	Hc	S/V
Doronicum pardalianches × plantagineum = D. × willdenowii	Asteraceae	35	Horticulture	1800	H	Hc	S/V
Doronicum plantagineum	Asteraceae	52	Europe	1800	H	Hc	S/V
Downingia elegans	Campanulaceae	5	North America	1978	As,Gs	Th	–
Dracocephalum parviflorum	Lamiaceae	3	North America	1908	Bs,G,W	Hc	–
Dracunculus vulgaris	Araceae	40	Mediterranean	1918	H	Ge	S/V
Drosanthemum floribundum	Aizoaceae	6	South Africa	1921	H	Ch	S/V

Scientific name	Family	No. of hectads	Introduced from	First record	Vector	Life form	S/V
Dysphania ambrosioides	Amaranthaceae	2	Central & South America	1866	Bs,O,W	Th	S
Dysphania botrys	Amaranthaceae	5	Eurasia	1779	W	Th	–
Dysphania pumilio	Amaranthaceae	1	Australia	1911	W	Th	–
Ecballium elaterium	Cucurbitaceae	9	Southern Europe	1906	H,W	Ge	–
Echinochloa colona	Poaceae	42	Widespread	1906	Bs,C,G,O,W	Th	–
Echinochloa crus-galli	Poaceae	509	Widespread	1620	As,Bs,C,G,O,W	Th	S
Echinochloa esculenta	Poaceae	39	Japan	1971	A,Bs	Th	–
Echinochloa frumentacea	Poaceae	26	India	1918	A,Bs	Th	–
Echinops bannaticus	Asteraceae	164	Southeast Europe	1913	H	Hc	S
Echinops exaltatus	Asteraceae	72	Southeast Europe	1931	H	Hc	S
Echinops sphaerocephalus	Asteraceae	70	Southern Europe	1908	H	Hc	S
Echium pininana	Boraginaceae	57	Canary Islands	1971	H	Ch	S
Echium rosulatum	Boraginaceae	1	Southwest Europe	1927	?	Hc	S
Egeria densa	Hydrocharitaceae	22	South America	1950	H	Hy	V
Elaeagnus commutata	Elaeagnaceae	3	North America	1995	H	Ph	V
Elaeagnus macrophylla	Elaeagnaceae	13	East Asia	1984	H	Ph	–
Elaeagnus macrophylla × pungens = E. × submacrophylla	Elaeagnaceae	18	Horticulture	1988	H	Ph	–
Elaeagnus pungens	Elaeagnaceae	16	Japan	1950	H	Ph	–
Elaeagnus umbellata	Elaeagnaceae	30	East Asia	1971	H	Ph	–
Eleusine indica subsp. *africana*	Poaceae	7	Africa	1917	Bs,W	Th	–
Eleusine indica subsp. *indica*	Poaceae	3 (7)†	India	1872	B,Bs,C,G,O	Th	–
Elodea callitrichoides	Hydrocharitaceae	4	South America	1948	H	Hy	V
Elodea canadensis	Hydrocharitaceae	1,868	North America	1842	H	Hy	V
Elodea nuttallii	Hydrocharitaceae	958	North America	1966	H	Hy	V
Elymus scabrus	Poaceae	2	Australia	1957	W	Hc	–
Epilobium brachycarpum	Onagraceae	2	North America	2004	I	Th	S
Epilobium brunnescens	Onagraceae	1,565	New Zealand	1904	H,I	Hc	S/V
Epilobium brunnescens × ciliatum = E. × brunnatum	Onagraceae	10	Neonative	1981	Neo	Hc	S/V
Epilobium brunnescens × lanceolatum = E. × cornubiense	Onagraceae	3	Neonative	1995	Neo	Hc	S/V
Epilobium brunnescens × montanum = E. × confusilobum	Onagraceae	9	Neonative	1996	Neo	Hc	S/V
Epilobium brunnescens × obscurum = E. × obscurescens	Onagraceae	9	Neonative	1980	Neo	Hc	S/V
Epilobium brunnescens × palustre = E. × chateri	Onagraceae	1	Neonative	1995	Neo	Hc	S/V
Epilobium brunnescens × parviflorum = E. × argillaceum	Onagraceae	1	Neonative	2002	Neo	Hc	V
Epilobium ciliatum	Onagraceae	2,501	North America	1891	I,T	Hc	S/V
Epilobium ciliatum × hirsutum = E. novae-civitatis	Onagraceae	77	Neonative	1936	Neo	Hc	V
Epilobium ciliatum × lanceolatum	Onagraceae	15	Neonative	1949	Neo	Hc	S/V
Epilobium ciliatum × montanum = E. × interjectum	Onagraceae	344	Neonative	1930	Neo	Hc	S/V

Scientific name	Family	No. of hectads	Introduced from	First record	Vector	Life form	S/V
Epilobium ciliatum × obscurum = E. × vicinum	Onagraceae	235	Neonative	1934	Neo	Hc	S/V
Epilobium ciliatum × palustre = E. × fossicola	Onagraceae	33	Neonative	1934	Neo	Hc	S/V
Epilobium ciliatum × parviflorum = E. × floridulum	Onagraceae	188	Neonative	1934	Neo	Hc	V
Epilobium ciliatum × roseum = E. × nutantiflorum	Onagraceae	13	Neonative	1943	Neo	Hc	S/V
Epilobium ciliatum × tetragonum = E. × mentiens	Onagraceae	52	Neonative	1936	Neo	Hc	S/V
Epilobium komarovianum	Onagraceae	7	New Zealand	1961	H,I	Hc	S/V
Epilobium montanum × pedunculare = E. × kitcheneri	Onagraceae	2	Neonative	1996	Neo	Hc	S/V
Epilobium pedunculare	Onagraceae	18	New Zealand	1931	H,I	Hc	S/V
Epilobium tetragonum subsp. tournefortii	Onagraceae	1	Mediterranean	2007	I	Hc	S/V
Epimedium alpinum	Berberidaceae	8	Southern Europe	1746	H	Hc	V
Eragrostis cilianensis	Poaceae	16	Southern Europe	1904	B,Bs,G,O,W	Th	–
Eragrostis curvula	Poaceae	4	Africa	1915	B,W	Hc	S
Eragrostis minor	Poaceae	1	Southern Europe	1898	B,Bs,G,W	Th	–
Eragrostis neomexicana	Poaceae	1	America	1969	B,Bs,W	Th	–
Eragrostis pilosa	Poaceae	1	Mediterranean	1926	Bs,G,O,W	Th	S
Eragrostis tef	Poaceae	1	Africa	1961	Bs,F,H,W	Th	–
Eranthis hyemalis	Ranunculaceae	614	Europe	1838	H	Ge	S/V
Erepsia heteropetala	Aizoaceae	1	South Africa	1957	H	Ch	S/V
Erica arborea	Ericaceae	7	Mediterranean	1984	H	Ph	–
Erica carnea	Ericaceae	18	Europe	1905	H	Ch	–
Erica carnea × erigena = E. × darleyensis	Ericaceae	28	Horticulture	1930s	H	Ph	–
Erica lusitanica	Ericaceae	14	Southwest Europe	1905	H	Ph	S
Erica terminalis	Ericaceae	4	Mediterranean	1908	H	Ph	S
Erigeron annuus	Asteraceae	14	North America	1902	Gs,H	Hc	S
Erigeron glaucus	Asteraceae	246	North America	1942	H	Ch	S
Erigeron karvinskianus	Asteraceae	524	Mexico	1860	H	Hc	S
Erigeron philadelphicus	Asteraceae	17	North America	1916	H	Hc	S
Erigeron speciosus	Asteraceae	22	North America	1928	H	Hc	–
Erinus alpinus	Veronicaceae	359	Southwest Europe	1867	H	Hc	S
Erodium botrys	Geraniaceae	4	Mediterranean	1909	W	Th	–
Erodium brachycarpum	Geraniaceae	2	Mediterranean	1917	W	Th	–
Erodium chium	Geraniaceae	3	Mediterranean	1858	W	TH	–
Erodium crinitum	Geraniaceae	5	Australia	1930	W	Th	–
Erodium cygnorum subsp. cygnorum	Geraniaceae	3	Australia	1865	W	Th	–
Erodium cygnorum subsp. glandulosum	Geraniaceae	1	Australia	1958	W	Th	–
Erodium malacoides	Geraniaceae	2	Southern Europe	1888	W	Th	–
Erodium manescavii	Geraniaceae	3	Southwest Europe	1964	H	Hc	S
Erucastrum gallicum	Brassicaceae	89	Europe	1863	Bs,G	Th	–

Scientific name	Family	No. of hectads	Introduced from	First record	Vector	Life form	S/V
Eryngium amethystinum	Apiaceae	5	Mediterranean	1963	H	Hc	S
Eryngium giganteum	Apiaceae	12	Caucasus	1937	H	Hc	S
Eryngium planum	Apiaceae	36	Europe	1951	H	Hc	S
Erysimum decumbens × perofskianum = E. × marshallii	Brassicaceae	26	Horticulture	1929	H	Hc	–
Erysimum repandum	Brassicaceae	1	Southern Europe	1867	As,G,W	Th	–
Erythronium dens-canis	Liliaceae	27	Southern Europe	1965	H	Ge	S/V
Escallonia rubra var. macrantha	Escalloniaceae	334	Chile	1905	H	Ph	S
Eschscholzia californica	Papaveraceae	486	North America	1864	H	Hc	S
Eucalyptus globulus	Myrtaceae	10	Australia	1979	H	Ph	S
Eucalyptus gunnii	Myrtaceae	55	Australia	1887	H	Ph	S
Eucalyptus johnstonii	Myrtaceae	1	Australia	1996	H	Ph	–
Eucalyptus niphophila	Myrtaceae	3	Australia	1987	H	Ph	S
Eucalyptus pulchella	Myrtaceae	2	Australia	1963	H	Ph	S
Eucalyptus urnigera	Myrtaceae	5	Australia	1996	H	Ph	S
Eucalyptus viminalis	Myrtaceae	3	Australia	1996	H	Ph	–
Euonymus japonicus	Celastraceae	282	Japan	1897	H	Ph	S
Euonymus latifolius	Celastraceae	40	Europe	1903	H	Ph	–
Euphorbia amygdaloides subsp. robbiae	Euphorbiaceae	267	Turkey	1977	H	Ch	S/V
Euphorbia characias subsp. characias	Euphorbiaceae	30 (101)†	Mediterranean	1797	H	Ph	–
Euphorbia characias subsp. veneta	Euphorbiaceae	24	Mediterranean	1953	H	Ph	–
Euphorbia corallioides	Euphorbiaceae	15	Italy	1808	H	Hc	S
Euphorbia cyparissias	Euphorbiaceae	288	Europe	1799	H	Hc	S/V
Euphorbia cyparissias × esula = E. × pseudoesula	Euphorbiaceae	11	Europe	1904	G,Gs,T	Hc	S/V
Euphorbia dulcis	Euphorbiaceae	73	Europe	1849	H	Hc	S/V
Euphorbia esula	Euphorbiaceae	40	Europe	1768	G,Gs,T	Hc	S/V
Euphorbia esula × waldsteinii = E. × pseudovirgata	Euphorbiaceae	133	Europe	1887	G,Gs,T	Hc	S/V
Euphorbia maculata	Euphorbiaceae	17	North America	1917	I	Th	–
Euphorbia mellifera	Euphorbiaceae	34	Canary Islands	1971	H	Ph	–
Euphorbia oblongata	Euphorbiaceae	68	Mediterranean	1938	H	Hc	S/V
*Euphorbia stricta	Euphorbiaceae		Europe	1773	?As,?I	Th	S
Euphorbia waldsteinii	Euphorbiaceae	5	Europe	1937	G,Gs,T	Hc	S/V
Fagopyrum dibotrys	Polygonaceae	1	Asia	1961	H	Hc	S
Fagopyrum esculentum	Polygonaceae	335	Asia	1597	As,Bs,G,I	Th	S
Fagopyrum tataricum	Polygonaceae	6	Asia	1905	Bs,G,I,W	Th	–
Falcaria vulgaris	Apiaceae	35	Europe	1858	As,G	Hc	S
Fallopia baldschuanica	Polygonaceae	837	Asia	1936	H	Ph	–
Fallopia baldschuanica × japonica = F. × conollyana	Polygonaceae	4	Neonative	1987	Neo	Ph	–
Fallopia japonica	Polygonaceae	2,659	East Asia	1886	H	Hc	V
Fallopia japonica × sachalinensis = F. × bohemica	Polygonaceae	247	Neonative	1954	Neo	Hc	V
Fallopia sachalinensis	Polygonaceae	452	Japan	1896	H	Hc	V

Scientific name	Family	No. of hectads	Introduced from	First record	Vector	Life form	S/V
Fargesia spathacea	Poaceae	19	China	1964	H	Ph	V
Fascicularia bicolor	Bromeliaceae	4	Chile	1962	H	Ch	–
Fatsia japonica	Araliaceae	38	Japan	1989	H	Ph	–
Felicia bergeriana	Asteraceae	1	South Africa	2011	H	Th	–
Ferula communis	Apiaceae	7	Mediterranean	1956	H	Hc	S
Festuca brevipila	Poaceae	154	Europe	1825	H	Hc	S
Festuca gautieri subsp. scoparia	Poaceae	2	Southwest Europe	1992	H	Ch	V
Festuca glauca	Poaceae	13	France	1999	H	Hc	S
Festuca heterophylla	Poaceae	49	Europe	1874	H	Hc	S
Festuca rubra subsp. megastachys	Poaceae	152	Europe	1966	H	Hc	S/V
Festuca rubra × Vulpia myuros	Poaceae	5	Neonative	1951	Neo	Hc	V
Ficaria verna subsp. chrysocephala	Ranunculaceae	22	Mediterranean	1994	H	Ge	S/V
Ficaria verna subsp. ficariiformis	Ranunculaceae	35	Southern Europe	1913	H	Ge	S/V
Ficus carica	Moraceae	241	Southwest Asia	1918	F	Ph	–
*Filago lutescens	Asteraceae	25	Europe	1846	?native	Th	S
Filipendula camtschatica	Rosaceae	19	East Asia	1965	H	Hc	S/V
Filipendula camtschatica × ? = F. × purpurea	Rosaceae	7	Horticulture	2000	H	Hc	S/V
Forsythia suspensa	Oleaceae	27	China	1982	H	Ph	V
Forsythia suspensa × viridissima = F. × intermedia	Oleaceae	335	Horticulture	1970	H	Ph	V
Fragaria ananassa	Rosaceae	465	Agriculture	1900	F	Hc	S/V
Fragaria moschata	Rosaceae	59	Europe	1810	F	Hc	S/V
Fraxinus angustifolia subsp. angustifolia	Oleaceae	11	Mediterranean	1991	H	Ph	–
Fraxinus angustifolia subsp. oxycarpa	Oleaceae	10	Europe	1986	H	Ph	–
Fraxinus ornus	Oleaceae	75	Europe	1923	H	Ph	–
Freesia × hybrida	Iridaceae	5	South Africa	1975	H	Ge	–
*Fritillaria meleagris	Liliaceae	223	Europe	1736	?native	Ge	S/V
Fuchsia cordifolia × globosa = F. 'Corallina'	Onagraceae	8	Horticulture	1996	H	Ph	–
Fuchsia magellanica	Onagraceae	973	South America	1857	H	Ph	S
Fumaria muralis × officinalis = F. × painteri	Papaveraceae	1	Neonative	1896	Neo	TH	S
Fumaria reuteri	Papaveraceae	11	Europe	1904	?	Th	S
Furcraea parmentieri	Agavaceae	1	Mexico	1995	H	Ph	V
Gaillardia aristata × pulchella = G. × grandiflora	Asteraceae	13	Horticulture	1959	H	Hc	S
Galactites tomentosus	Asteraceae	6	Mediterranean	1798	Bs	Th	–
Galanthus elwesii	Amaryllidaceae	187	Caucasus	1957	H	Ge	S/V
Galanthus elwesii × nivalis	Amaryllidaceae	15	Neonative	1984	Neo	Ge	S/V
Galanthus elwesii × plicatus	Amaryllidaceae	3	Horticulture	1992	Neo	Ge	S/V
Galanthus nivalis	Amaryllidaceae	1,911	Europe	1776	H	Ge	S/V
Galanthus nivalis × plicatus	Amaryllidaceae	87	Horticulture	1947	Neo	Ge	S/V
Galanthus plicatus subsp. byzantinus	Amaryllidaceae	22	Turkey	1986	H	Ge	S/V

Scientific name	Family	No. of hectads	Introduced from	First record	Vector	Life form	S/V
Galanthus plicatus subsp. plicatus	Amaryllidaceae	105	Eastern Europe	1947	H	Ge	S/V
Galanthus reginae-olgae	Amaryllidaceae	2	Mediterranean	1995	H	Ge	S/V
Galanthus woronowii	Amaryllidaceae	90	Mediterranean	1984	H	Ge	S/V
Galega officinalis	Fabaceae	328	Europe	1640	H	Hc	S
Galeopsis ladanum	Lamiaceae	1	Europe	1827	?	Th	–
Galinsoga parviflora	Asteraceae	362	South America	1861	H,I,W	Th	S
Galinsoga quadriradiata	Asteraceae	512	South America	1909	Bs,I,W	Th	S
*Galium parisiense	Rubiaceae	58	Europe	1690	?native	Th	S
Galium spurium	Rubiaceae	10	Europe	1763	Bs,G,W	Th	S
*Gaudinia fragilis	Poaceae	64	Europe	1903	?native	Th	S
Gaultheria mucronata	Ericaceae	177	Chile	1903	H	Ph	S
Gaultheria mucronata × shallon = G. × wisleyensis	Ericaceae	1	Neonative	1981	Neo	Ph	–
Gaultheria procumbens	Ericaceae	6	North America	1938	H	Ch	S
Gaultheria shallon	Ericaceae	201	North America	1914	H	Ph	S
Gazania rigens	Asteraceae	12	South Africa	1866	H	Ch	S/V
Genista aetnensis	Fabaceae	6	Sicily, Sardinia	1977	H	Ph	S
Genista hispanica subsp. occidentalis	Fabaceae	61	Southwest Europe	1927	H	Ph	S
Genista monspessulana	Fabaceae	17	Mediterranean	1915	H	Ph	S
Gentiana acaulis	Gentianaceae	1	Europe	1960s	H	Hc	V
Gentiana asclepiadea	Gentianaceae	6	Europe	1867	H	Hc	S
*Gentianopsis ciliata	Gentianaceae	1	Europe	1875	?native	Hc	S
Geranium endressii	Geraniaceae	597	Southwest Europe	1906	H	Hc	S/V
Geranium endressii × versicolor = G. × oxonianum	Geraniaceae	651	Horticulture	1954	H	Hc	S/V
Geranium herrerae	Geraniaceae	4	South America	1926	?	Hc	S
Geranium himalayense	Geraniaceae	13	Himalayas	1976	H	Hc	S/V
Geranium ibericum	Geraniaceae	14	Caucasus	1922	H	Hc	S/V
Geranium ibericum × platypetalum = G. × magnificum	Geraniaceae	295	Horticulture	1932	H	Hc	V
Geranium macrorrhizum	Geraniaceae	211	Southern Europe	1888	H	Hc	S/V
Geranium macrorrhizum × dalmaticum = G. × cantabrigiense	Geraniaceae	1	Horticulture	2005	H	Hc	V
Geranium maderense	Geraniaceae	11	Madeira	1978	H	Hc	S
Geranium nodosum	Geraniaceae	38	Southern Europe	1801	H	Hc	S/V
Geranium phaeum	Geraniaceae	364	Europe	1724	H	Hc	S/V
Geranium phaeum × reflexum = G. × monacense	Geraniaceae	5	Horticulture	1899	H	Hc	S/V
Geranium platypetalum	Geraniaceae	1	Caucasus	1987	H	Hc	S/V
Geranium psilostemon	Geraniaceae	14	Caucasus	1966	H	Hc	S
Geranium pyrenaicum	Geraniaceae	1,365	Europe	1762	?	Hc	S
Geranium reuteri	Geraniaceae	2	Canary Islands	2002	H	Hc	S
Geranium versicolor	Geraniaceae	276	Mediterranean	1820	H	Hc	S/V
Geranium yeoi	Geraniaceae	10	Madeira	1930s	H	Hc	S
Geropogon glaber	Asteraceae	1	Southern Europe	1922	Bs,F,G	Th	–
Geum macrophyllum	Rosaceae	39	North America	1908	H	Hc	S/V

Scientific name	Family	No. of hectads	Introduced from	First record	Vector	Life form	S/V
Geum macrophyllum × urbanum = G. × convallis	Rosaceae	2	Neonative	2011	Neo	Hc	V
Geum quellyon	Rosaceae	5	South America	1980	H	Hc	–
Gilia capitata	Polemoniaceae	2	North America	1888	H	Th	–
Gladiolus communis subsp. byzantinus	Iridaceae	210	Mediterranean	1862	H	Ge	S/V
Glebionis coronaria	Asteraceae	23	Europe	1897	Bs,G,H,W	Th	–
Glyceria canadensis	Poaceae	1	North America	1996	H	Hy	V
Glycine max	Fabaceae	20	Uncertain	1923	Bs,F,O	Th	–
*Gnaphalium luteoalbum	Asteraceae	36	Europe	1690	?native	Th	S
Gnaphalium purpureum	Asteraceae	2	North America	1913	H,S,W	Th	S
Gnaphalium undulatum	Asteraceae	18	South Africa	1888	H,W	Th	S
Grindelia stricta	Asteraceae	3	North America	1961	H	Hc	S
Griselinia littoralis	Griseliniaceae	96	New Zealand	1957	H	Ph	–
Guizotia abyssinica	Asteraceae	128	Africa	1861	Bs,G,O,W	Th	–
Gunnera manicata	Gunneraceae	58	South America	1935	H	Ge	–
Gunnera tinctoria	Gunneraceae	201	South America	1908	H	Ge	S/V
Gypsophila elegans	Caryophyllaceae	9	Caucasus	1928	H	Th	–
Gypsophila paniculata	Caryophyllaceae	9	Europe	1884	H	Hc	S
Haloragis micrantha	Haloragaceae	1	Southeast Asia	1988	?	Ch	S
Hedera algeriensis	Araliaceae	21	North Africa	1963	H	Ph	V
Hedera colchica	Araliaceae	192	Caucasus	1959	H	Ph	V
Helenium × clementii	Asteraceae	6	Horticulture	1955	H	Hc	–
Helianthus annuus	Asteraceae	610	North America	1891	A,,Bs,F,H,O,W	Th	–
Helianthus annuus × decapetalus = H. × multiflorus	Asteraceae	15	Horticulture	1950	H	Hc	S/V
Helianthus pauciflorus × tuberosus = H. × laetiflorus	Asteraceae	131	Horticulture	1902	H	Hc	S/V
Helianthus petiolaris	Asteraceae	24	North America	1908	G,O,W	Th	–
Helianthus tuberosus	Asteraceae	125	North America	1897	Bs,F	Hc	S/V
Helichrysum bellidioides	Asteraceae	1	New Zealand	1975	H	Ch	V
Helichrysum petiolare	Asteraceae	6	South Africa	1985	H	Ch	S
Heliotropium europaeum	Boraginaceae	1	Europe	1905	O,W	Th	–
Helleborus argutifolius	Ranunculaceae	64	Corsica	1991	H	Ch	S
*Helleborus foetidus	Ranunculaceae	532	Europe	1597	?native	Ch	S
Helleborus orientalis	Ranunculaceae	79	Turkey	1925	H	Hc	S
Hemerocallis fulva	Xanthorrhoeaceae	242	Horticulture	1905	H	Ge	S/V
Hemerocallis lilioasphodelus	Xanthorrhoeaceae	33	East Asia	1894	H	Ge	S/V
Hepatica nobilis	Ranunculaceae	7	Europe	1836	H	Hc	S
Heracleum mantegazzianum	Apiaceae	1,109	Southwest Asia	1828	H	Hc	S
Heracleum mantegazzianum × sphondylium	Apiaceae	48	Neonative	1951	Neo	Hc	S
Hermodactylus tuberosus	Iridaceae	22	Mediterranean	1839	H	Ge	S/V
Herniaria hirsuta	Caryophyllaceae	5	Europe	1789	E,G,Tn,W	Th	–
Hesperis matronalis	Brassicaceae	1,995	Europe	1805	H	Hc	S
Heuchera sanguinea	Saxifragaceae	55	North America	1967	H	Hc	–

Scientific name	Family	No. of hectads	Introduced from	First record	Vector	Life form	S/V
Hibiscus trionum	Malvaceae	22	Old World	1852	Bs,H,O,W	Th	–
Hieracium amplexicaule	Asteraceae	2	Europe	1820	H	Hc	S
Hieracium lanatum	Asteraceae	1	Europe	1954	H	Hc	S
Hieracium pulmonarioides	Asteraceae	6	Europe	1885	H	Hc	S
Hieracium speluncarum	Asteraceae	9	Europe	1885	H	Hc	S
Hippocrepis emerus	Fabaceae	13	Europe	1937	H	Ph	S
Hirschfeldia incana	Brassicaceae	509	Southern Europe	1837	Bs,G,W	Hc	S
Hoheria populnea	Malvaceae	3	New Zealand	1970	H	Ph	–
Holodiscus discolor	Roscaeae	31	North America	1961	H	Ph	–
*Homogyne alpina	Asteraceae	2	Europe	1814	?native	Hc	S/V
Hordeum euclaston	Poaceae	1	South America	1961	W	Th	–
Hordeum geniculatum	Poaceae	3	Southern Europe	1838	E,G,Gs,W	Th	–
Hordeum jubatum	Poaceae	244	North America	1890	Bs,G,Gs, O,W	Th	S
Hordeum murinum subsp. glaucum	Poaceae	19	Mediterranean	1952	E,W	Th	–
Hordeum murinum subsp. leporinum	Poaceae	21	Southern Europe	1950	E,W	Th	–
Houttuynia cordata	Saururaceae	9	East Asia	1992	H	Hc	S/V
Hyacinthoides hispanica	Asparagaceae	1,152	Southwest Europe	1875	H	Ge	S/V
Hyacinthoides hispanica × non-scripta = H. × massartiana	Asparagaceae	1,560	Neonative	1953	Neo	Ge	S/V
Hyacinthoides italica	Asparagaceae	12	Southwest Europe	1896	H	Ge	S/V
Hyacinthus orientalis	Asparagaceae	162	Southwest Asia	1957	H	Ge	–
Hydrangea macrophylla	Hydrangeaceae	73	Japan	1982	H	Ph	–
Hydrocotyle moschata	Hydrocotylaceae	1	New Zealand	1858	H	Hc	S/V
Hydrocotyle ranunculoides	Hydrocotylaceae	118	North America	1990	H	Hy	S/V
Hypericum androsaemum × hircinum = H. × inodorum	Hypericaceae	205	Horticulture	1803	H	Ph	V
Hypericum calycinum	Hypericaceae	653	Southeast Europe	1809	H	Ch	S/V
*Hypericum canadense	Hypericaceae	3	North America	1906	?native	Hc	S/V
Hypericum forrestii	Hypericaceae	1	China	1987	H	Ph	V
Hypericum × hidcoteense	Hypericaceae	35	Horticulture	1990	H	Ph	–
Hypericum hircinum subsp. majus	Hypericaceae	141	Mediterranean	1856	H	Ph	S/V
Hypericum nummularium	Hypericaceae	2	Southwest Europe	1943	H	Hc	S
Hypericum olympicum	Hypericaceae	23	Balkans	1981	H	Ch	S
Hypericum pseudohenryi	Hypericaceae	3	China	1979	H	Ph	S/V
Hypericum xylosteifolium	Hypericaceae	8	Caucasus	1978	H	Ph	V
Hyssopus officinalis	Lamiaceae	7	Southern Europe	1852	H	Ph	S
Iberis sempervirens	Brassicaceae	72	Southern Europe	1928	H	Ch	–
Iberis umbellata	Brassicaceae	251	Southern Europe	1858	H	Th	–
Ilex aquifolium × perado = I. × altaclerensis	Aquifoliaceae	299	Horticulture	1968	H	Ph	–
Impatiens capensis	Balsaminaceae	313	North America	1822	H	Th	S
Impatiens glandulifera	Balsaminaceae	1,884	Himalayas	1855	H	Th	S
Impatiens parviflora	Balsaminaceae	370	Asia	1848	A,I,T	Th	S
Inula hookeri	Asteraceae	7	Himalayas	1986	H	Hc	S
Inula oculus-christi	Asteraceae	5	Eastern Europe	1967	H	Hc	–

Scientific name	Family	No. of hectads	Introduced from	First record	Vector	Life form	S/V
Inula orientalis	Asteraceae	2	Caucasus	1981	H	Hc	S
Ipomoea batatas	Convolvulaceae	2	Central & South America	1964	F	Ge	–
Ipomoea hederacea	Convolvulaceae	10	North America	1932	G,O	Th	–
Ipomoea lacunosa	Convolvulaceae	2	North America	1947	G,O	Th	–
Ipomoea purpurea	Convolvulaceae	14	North America	1921	Bs,G,O	Th	–
Iris ensata	Iridaceae	2	East Asia	1985	H	Hc	–
Iris filifolia × tingitana = I. × hollandica	Iridaceae	9	Horticulture	c. 1970	H	Ge	–
Iris germanica	Iridaceae	250	Horticulture	1849	H	Hc	V
Iris laevigata	Iridaceae	9	East Asia	1999	H	Hc	V
Iris latifolia	Iridaceae	11	Southwest Europe	1848	H	Ge	S/V
Iris orientalis	Iridaceae	68	Mediterranean	1915	H	Hc	V
Iris sibirica	Iridaceae	91	Europe	c. 1770	H	Hc	V
Iris spuria	Iridaceae	10	Europe	1836	H	Hc	V
Iris unguicularis	Iridaceae	5	Mediterranean	1989	H	Hc	–
Iris versicolor	Iridaceae	41	North America	1893	H	Hc	V
Iris versicolor × virginica = I. × robusta	Iridaceae	2	Horticulture	1965	H	Hc	V
Iris xiphium	Iridaceae	8	Europe	1849	H	Ge	–
Islandsmelia carinata	Asteraceae	3	Africa	1928	Bs,H	Th	–
Iva xanthiifolia	Asteraceae	4	North America	1901	As,Bs,G,W	Th	–
Ixia campanulata	Iridaceae	1	South Africa	1959	H	Ge	V
Ixia paniculata	Iridaceae	2	South Africa	1961	H	Ge	–
Jasminum beesianum	Oleaceae	9	China	1999	H	Ph	S
Jasminum nudiflorum	Oleaceae	37	China	1976	H	Ph	S/N
Jasminum officinale	Oleaceae	29	Asia	1967	H	Ph	V
Juglans nigra	Juglandaceae	12	North America	1994	H	Ph	S
Juncus anthelatus	Juncaceae	1	North America	1955	?	Ph	–
Juncus dudleyi	Juncaceae	1	North America	1915	?,W	Ph	S
Juncus ensifolius	Juncaceae	2	North America	1956	?	Hc	S/V
Juncus pallidus	Juncaceae	3	Antipodes	1944	Bs,W	Ph	–
Juncus planifolius	Juncaceae	4	Southern hemisphere	1971	?native	Hc	S/V
*Juncus subulatus	Juncaceae	1	Mediterranean	1954	?native	Hc	S/V
Juncus tenuis	Juncaceae	1,193	North & South America	1795	?A	Hc	S
Kalmia angustifolia	Ericaceae	6	North America	1913	H	Ph	S
Kalmia latifolia	Ericaceae	3	North America	1989	H	Ph	S
Kalmia polifolia	Ericaceae	3	North America	1910	H	Ph	S
Kerria japonica	Rosaceae	199	China, Japan	1965	H	Ph	S/V
Kniphofia uvaria	Xanthorrhoeaceae	164	South Africa	1961	H	Hc	S/V
Kniphofia × praecox	Xanthorrhoeaceae	70	Horticulture	1950	H	Hc	S/V
Koelreuteria paniculata	Sapindaceae	15	China	1908	H	Ph	–
Laburnum alpinum	Fabaceae	57	Europe	1905	H	Ph	–
Laburnum alpinum × anagyroides = L. × watereri	Fabaceae	66	Horticulture	1970	H	Ph	–

Scientific name	Family	No. of hectads	Introduced from	First record	Vector	Life form	S/V
Laburnum anagyroides	Fabaceae	1,206	Europe	1879	H	Ph	S
Lactuca tatarica	Asteraceae	5	Eastern Europe	1884	G	Hc	S/V
Lagarosiphon major	Hydrocharitaceae	517	South Africa	1944	H	Hy	V
Lagurus ovatus	Poaceae	127	Southern Europe	1791	G,H,W	Th	S
Lamiastrum galeobdolon subsp. argentatum	Lamiaceae	1,533	Uncertain	1974	H	Hc	S/V
Lamium maculatum	Lamiaceae	901	Europe	c. 1730	H	Hc	S/V
Lampranthus aureus	Aizoaceae	1	South Africa	2006	H	Ch	S/V
Lampranthus falciformis	Aizoaceae	4	South Africa	1908	H	Ch	S/V
Lampranthus roseus	Aizoaceae	19	South Africa	1924	H	Ch	S/V
Lamprocapnos spectabilis	Papaveraceae	21	East Asia	1922	H	Ge	–
Lappula squarrosa	Asteraceae	20	Europe	1866	Bs,G,Gs,O,W	Th	S
*Lapsana communis subsp. communis	Asteraceae	3,211	Europe	1597	?native	Th	S
Lapsana communis subsp. intermedia	Asteraceae	19	Southeast Europe	1945	?	Hc	S
Lathraea clandestina	Orobanchaceae	106	Western Europe	1908	H	Ge	S/V
Lathyrus annuus	Fabaceae	4	Mediterranean	1897	Bs,G,Tn	Th	–
*Lathyrus aphaca	Fabaceae	113	Europe	1632	?native	Th	S
Lathyrus grandiflorus	Fabaceae	163	Mediterranean	1903	H	Hc	S
Lathyrus heterophyllus	Fabaceae	4	Europe	1956	H	Hc	S
Lathyrus hirsutus	Fabaceae	20	Europe	1666	Bs,G	Th	S
Lathyrus inconspicuus	Fabaceae	1	Mediterranean	1908	Bs,Tn	Th	–
Lathyrus latifolius	Fabaceae	792	Europe	1670	H	Hc	S
Lathyrus niger	Fabaceae	2	Europe	1821	H	Hc	S
Lathyrus odoratus	Fabaceae	35	Italy	1887	H	Th	–
Lathyrus sativus	Fabaceae	1	Mediterranean	1883	Bs	Th	–
*Lathyrus sylvestris	Fabaceae	324	Europe	1597	?native	Hc	S
Lathyrus tuberosus	Fabaceae	67	Europe	1859	Bs,G,H	Hc	S
Lathyrus vernus	Fabaceae	1	Europe	1934	H	Hc	S
Laurus nobilis	Lauraceae	304	Mediterranean	1924	H	Ph	S
Lavandula angustifolia	Lamiaceae	182	Mediterranean	1973	H	Ph	S
Legousia speculum-veneris	Campanulaceae	2	Europe	1868	G,H	Th	S
Lemna minuta	Lemnaceae	921	North & South America	1977	H	Hy	V
Lemna turionifera	Lemnaceae	10	North temperate zone	2007	H	Hy	V
Leonurus cardiaca	Lamiaceae	42	Europe	1597	G,H	Hc	S/V
Lepidium africanum	Brassicaceae	1	South Africa	1913	W	Th	–
Lepidium bonariense	Brassicaceae	1	South America	1911	Bs,W	Th	–
Lepidium didymum	Brassicaceae	1,613	South America	1778	?,B,W	Th	S
Lepidium draba subsp. chalepense	Brassicaceae	20	Southwest Asia	1915	G	Hc	S/V
Lepidium draba subsp. draba	Brassicaceae	990	Southern Europe	1802	A,B,G,W	Hc	S/V
Lepidium graminifolium	Brassicaceae	1	Southern Europe	1857	G,W	Hc	–
Lepidium perfoliatum	Brassicaceae	2	Europe	1888	G,Gs	Th	–
Lepidium virginicum	Brassicaceae	21	North America	1881	Bs,G,T,W	Th	–

Scientific name	Family	No. of hectads	Introduced from	First record	Vector	Life form	S/V
Leucanthemella serotina	Asteraceae	55	Southeast Europe	1909	H	Hc	S/C
Leucanthemum lacustre × maximum = L. × superbum	Asteraceae	954	Horticulture	1913	H	Hc	S/V
Leucojum aestivum subsp. aestivum	Amaryllidaceae	263	Europe	1788	H	Ge	S/V
Leucojum aestivum subsp. pulchellum	Amaryllidaceae	216	Mediterranean	1848	H	Ge	S/V
Leucojum vernum	Amaryllidaceae	86	Europe	1834	H	Ge	S/V
Leucothoe fontanesiana	Ericaceae	1	North America	1960s	H	Ph	V
Levisticum officinale	Apiaceae	45	Iran	1836	F,H	Hc	S
Leycesteria formosa	Caprifoliaceae	801	Himalayas	1905	H	Ph	S
Libertia elegans	Iridaceae	2	Chile	1989	H	Hc	–
Libertia formosa	Iridaceae	36	Chile	1916	H	Hc	S/V
Ligularia dentata	Asteraceae	13	China, Japan	1925	H	Hc	S
Ligularia przewalskii	Asteraceae	1	China	1982	H	Hc	S
Ligustrum ovalifolium	Oleaceae	1,738	Japan	1939	A,H	Ph	–
Ligustrum ovalifolium × vulgare = L. × vicaryi	Oleaceae	1	Horticulture	2004	H	Ph	–
Lilium candidum	Liliaceae	12	Greece	1965	H	Ge	–
Lilium martagon	Liliaceae	212	Europe	1782	H	Ge	S/V
Lilium pyrenaicum	Liliaceae	251	Southwest Europe	1853	H	Ge	S/V
Limnanthes douglasii	Limnanthaceae	230	North America	1870	H	Th	S
Limonium hyblaeum	Plumbaginaceae	10	Sicily	1979	H	Hc	S
Limonium platyphyllum	Plumbaginaceae	2	Southeast Europe	1973	H	Hc	S
Linaria arenaria	Veronicaceae	3	Western Europe	1893	H	Th	S
Linaria dalmatica	Veronicaceae	4	Southeast Europe	1915	H	Hc	S
Linaria maroccana	Veronicaceae	68	Morocco	1927	H	Th	–
Linaria pelisseriana	Veronicaceae	1	Europe	1837	?	Th	S
Linaria purpurea	Veronicaceae	1,626	Italy	1830	H	Hc	S
Linaria purpurea × repens = L. × dominii	Veronicaceae	67	Neonative	1950	Neo	Hc	S
Linaria repens × vulgaris = L. × sepium	Veronicaceae	85	Neonative	1830	Neo	Hc	S/V
Linaria repens × supina = L. × cornubiensis	Veronicaceae	1	Neonative	1925	Neo	Th	–
Linaria supina	Veronicaceae	18	Southwest Europe	1847	B	Th	S
Liriodendron tulipifera	Magnoliaceae	74	North America	1933	H	Ph	–
Lobelia erinus	Campanulaceae	545	South Africa	1905	H,W	Th	S
Lobularia maritima	Brassicaceae	726	Southern Europe	1804	Bs,H,W	Hc	S
Lolium multiflorum	Poaceae	1,892	Southern Europe	1840	Bs,	Th	S
Lolium multiflorum × perenne = L. × boucheanum	Poaceae	277	Neonative	1903	Neo	Hc	S/V
Lolium multiflorum × Schedonorus arundinaceus = × Schedolium krasanii	Poaceae	11	Neonative	1942	Neo	Hc	V
Lolium multiflorum × Schedonorus pratensis = × Schedolium braunii	Poaceae	31	Neonative	1933	Neo	Hc	V
Lolium rigidum	Poaceae	3	Southern Europe	1864	As,Bs,E,G,O,W	Th	–

Scientific name	Family	No. of hectads	Introduced from	First record	Vector	Life form	S/V
Lonicera caprifolium	Caprifoliaceae	64	Southern Europe	1763	H	Ph	S/V
Lonicera caprifolium × etrusca = L. × italica	Caprifoliaceae	58	Horticulture	1967	H	Ph	S/V
Lonicera henryi	Caprifoliaceae	36	China	1950	H	Ph	S/V
Lonicera involucrata	Caprifoliaceae	50	North America	1952	H	Ph	–
Lonicera japonica	Caprifoliaceae	341	East Asia	1937	H	Ph	S/V
Lonicera nitida	Caprifoliaceae	1,441	China	1955	A,H	Ph	–
Lonicera pileata	Caprifoliaceae	193	China	1959	H	Ph	–
Lonicera tatarica	Caprifoliaceae	37	Asia	1867	H	Ph	–
Lonicera xylosteum	Caprifoliaceae	155	Europe	1770	H	Ph	S
Ludwigia hexapetala	Onagraceae	10	North & South America	1998	H	Hy	V
Ludwigia palustris × repens = L. × kentiana	Onagraceae	3	Horticulture	1927	H	Hy	V
Luma apiculata	Myrtaceae	17	South America	1970	H	Ph	S
Lunaria annua	Brassicaceae	1,665	Southeast Europe	1597	H	Hc	S
Lupinus albus	Fabaceae	6	Balkans	1905	A	Th	–
Lupinus angustifolius	Fabaceae	6	Mediterranean	1890	A,G,W	Th	–
Lupinus arboreus	Fabaceae	332	North America	1926	H	Ph	S
Lupinus arboreus × nootkatensis × polyphyllus	Fabaceae	2	Neonative	1991	Neo	Hc	S
Lupinus arboreus × polyphyllus = L. × regalis	Fabaceae	391	Neonative	1955	Neo	Hc	S
Lupinus luteus	Fabaceae	6	Mediterranean	1928	A,W	Th	–
Lupinus nootkatensis	Fabaceae	36	North America	1862	H	Hc	S
Lupinus nootkatensis × polyphyllus = L. × pseudopolyphyllus	Fabaceae	1	Neonative	2003	Neo	Hc	S
Lupinus polyphyllus	Fabaceae	195	North America	c. 1900	H	Hc	S
Luzula luzuloides	Juncaceae	66	Europe	1871	G,Gs,H	Hc	S
Lycium barbarum	Solanaceae	872	China	1848	H	Ph	S
Lycium chinense	Solanaceae	198	China	1902	H	Ph	S
Lysichiton americanus	Araceae	285	North America	1947	H	Ge	S/V
Lysichiton camtschatcensis	Araceae	5	East Asia	1948	H	Ge	S/V
Lysimachia ciliata	Primulaceae	22	North America	1849	H	Hc	S/V
Lysimachia punctata	Primulaceae	1,310	Southeast Europe	1853	H	Hc	S/V
Lysimachia terrestris	Primulaceae	7	North America	1883	H	Hc	V
Lythrum junceum	Lythraceae	10	Mediterranean	1875	Bs,G,H,W	Hc	–
Macleaya cordata × microcarpa = M. × kewensis	Papaveraceae	18	Horticulture	1957	H	Hc	–
Mahonia aquifolium	Berberidaceae	999	North America	1874	A,H	Ph	S/V
Mahonia aquifolium × repens = M. × decumbens	Berberidaceae	10	Horticulture	1973	H	Ph	S/V
*Maianthemum bifolium	Asparagaceae	15	Europe	1597	?native	Ge	S/V
Maianthemum kamtschaticum	Asparagaceae	3	North America	1983	H	Ge	S/V
Malcolmia maritima	Brassicaceae	82	Balkans	1863	E,H	Th	S
Malope trifida	Malvaceae	2	Mediterranean	1928	Bs,W	Th	–
Malus atrosanguinea × niedzwetzkyana = M. × purpurea	Rosaceae	44	Horticulture	1982	H	Ph	–

Scientific name	Family	No. of hectads	Introduced from	First record	Vector	Life form	S/V
Malus baccata	Rosaceae	26	Siberia	1981	H	Ph	–
Malus hupehensis	Rosaceae	10	Asia	c. 1975	H	Ph	–
Malva alcea	Malvaceae	30	Europe	1905	H	Hc	S
Malva neglecta × sylvestris = M. × decipiens	Malvaceae	1	Neonative	2007	Neo	Hc	–
Malva nicaeensis	Malvaceae	8	Southern Europe	1859	As,G,Tn,W	Th	–
Malva olbia × thuringiaca = M. × clementii	Malvaceae	105	Horticulture	1862	H	Ph	–
Malva parviflora	Malvaceae	25	Southern Europe	1859	Bs,E,G,W	Th	–
*Malva pseudolavatera	Malvaceae	27	Europe	1859	?	Th	S
Malva pusilla	Malvaceae	30	Europe	1743	Bs,G,W	Th	–
Malva setigera	Malvaceae	55	Europe	1792	?	Th	S
Malva trimestris	Malvaceae	34	Mediterranean	1831	Bs,G,H,W	Th	–
Malva verticillata	Malvaceae	8	East Asia	1799	C,W	Th	–
Matricaria discoidea subsp. discoidea	Asteraceae	3,530	Widespread	1869	G	Th	S
Matricaria discoidea subsp. occidentalis	Asteraceae		Widespread	1894	G	Th	S
Matthiola incana	Brassicaceae	103	Europe	1806	H	Ch	S
Matthiola longipetala subsp. bicornis	Brassicaceae	40	Balkans	1904	H	Th	–
*Matthiola sinuata	Brassicaceae	18	Europe	1633	?native	Hc	S
Mauranthemum paludosum	Asteraceae	11	Southwest Europe	1930	H	Th	–
Medicago laciniata	Fabaceae	10	North Africa	1897	C,G,Tn,W	Th	–
Medicago littoralis	Fabaceae	1	Mediterranean	2001	B,S,W	Th	S
*Medicago minima	Fabaceae	54	Europe	1660	?native	Th	S
*Medicago polymorpha	Fabaceae	178	Europe	1662	?native	Th	S
Medicago praecox	Fabaceae	2	Mediterranean	1909	W	Th	S
Medicago sativa nothosubsp. varia	Fabaceae	94	Neonative	1686	Neo	Hc	S
Medicago sativa subsp. sativa	Fabaceae	913	Europe	1688	A,Bs,Gs,	Hc	S
Medicago truncatula	Fabaceae	5	Mediterranean	1863	G,Tn,W	Th	S
Melampyrum arvense	Orobanchaceae	9	Europe	1716	As,G	Th	S
Melilotus albus	Fabaceae	706	Europe	1822	A,Bs,W	Hc	S
Melilotus indicus	Fabaceae	178	Southern Europe	1852	Bs,F,G,S,W	Th	S
Melilotus officinalis	Fabaceae	1,001	Europe	1835	A,Bs,	Hc	S
Melilotus sulcatus	Fabaceae	6	Mediterranean	1859	Bs,W	Th	–
Melissa officinalis	Lamiaceae	789	Southern Europe	1763	F,H	Hc	S
Mentha aquatica × arvensis × spicata = M. × smithiana	Lamiaceae	156	Horticulture	1838	F	Hc	S/V
Mentha aquatica × spicata = M. × piperita	Lamiaceae	728	Neonative	1720	Neo	Hc	V
Mentha arvensis × spicata = M. × gracilis	Lamiaceae	303	Europe	1792	F	Hc	S/V
Mentha longifolia × spicata = M. × villosonervata	Lamiaceae	105	Horticulture	1934	F	Hc	V
Mentha longifolia × suaveolens = M. × rotundifolia	Lamiaceae	67	Europe	1900	F	Hc	S/V

Scientific name	Family	No. of hectads	Introduced from	First record	Vector	Life form	S/V
Mentha requienii	Lamiaceae	56	Corsica	1872	H	Hc	S/V
Mentha spicata × suaveolens = M. × villosa	Lamiaceae	1004	Horticulture	c. 1866	F	Hc	V
Mimulus cupreus × guttatus = M. × burnetii	Phrymaceae	66	Horticulture	1931	H	Hc	V
Mimulus cupreus × guttatus × luteus	Phrymaceae	18	Neonative	1974	Neo	Hc	V
Mimulus cupreus × luteus = M. × maculosus	Phrymaceae	18	Horticulture	1963	H	Hc	S/V
Mimulus guttatus	Phrymaceae	1,048	North America	1824	H	Hc	S/V
Mimulus guttatus × luteus = M. × robertsii	Phrymaceae	593	Neonative	1848	H	Hc	V
Mimulus luteus	Phrymaceae	58	Chile	1881	H	Hc	S/V
Mimulus moschatus	Phrymaceae	216	North America	1866	H	Hc	S/V
Mimulus peregrinus	Phrymaceae	2	Neonative	2011	Neo	Hc	S/V
Mirabilis jalapa	Nyctaginaceae	15	Central & South America	1961	H	Ph	–
Miscanthus sacchariflorus × sinensis = M. × giganteus	Poaceae	4	Agriculture	2003	Bm	Hc	–
Miscanthus sinensis	Poaceae	4	East Asia	1979	Bm,H	Hc	–
Misopates calycinum	Veronicaceae	5	Mediterranean	1904	Bs	Th	–
Monsonia brevirostrata	Geraniaceae	1	South Africa	1914	W	Th	–
Montia parvifolia	Montiaceae	1	North America	1913	H	Hc	S/V
Morus nigra	Moraceae	60	Asia	1961	F,H	Ph	S
Muehlenbeckia complexa	Polygonaceae	34	New Zealand	1909	H	Ph	S
Muscari armeniacum	Asparagaceae	870	Southeast Europe	1892	H	Ge	S/V
Muscari botryoides	Asparagaceae	14	Southern Europe	1962	H	Ge	S/V
Muscari comosum	Asparagaceae	35	Europe	1888	H	Ge	S/V
*Muscari neglectum	Asparagaceae	196	Europe	1776	?native	Ge	S/V
Myagrum perfoliatum	Brassicaceae	2	Mediterranean	1823	Bs,G	Th	–
*Mycelis muralis	Asteraceae	1,314	Europe	1633	?native	Hc	S
*Myosurus minimus	Ranunculaceae	146	Europe	1597	?native	Th	S
Myrica pensylvanica	Myricaceae	3	North America	1934	H	Ph	V
Myriophyllum aquaticum	Haloragaceae	406	South America	1960	H	Hy	S/V
Myrrhis odorata	Apiaceae	1,054	Europe	1715	F,H	Hc	S
Narcissus bicolor	Amaryllidaceae	11	Southwest Europe	1996	H	Ge	S/V
Narcissus bulbocodium	Amaryllidaceae	12	Southwest Europe	1905	H	Ge	S/V
Narcissus cyclamineus	Amaryllidaceae	18	Southwest Europe	1974	H	Ge	S/V
Narcissus cyclamineus × moschatus = N. × dichromus	Amaryllidaceae	2	Horticulture	1993	H	Ge	S/V
Narcissus cyclamineus × pseudonarcissus = N. × monochromus	Amaryllidaceae	8	Horticulture	1992	H	Ge	S/V
Narcissus cyclamineus × tazetta = N. × cyclazetta	Amaryllidaceae	10	Horticulture	1987	H	Ge	S/V
Narcissus hispanicus	Amaryllidaceae	250	Southwest Europe	1813	H	Ge	S/V
Narcissus jonquilla	Amaryllidaceae	2	Southwest Europe	1997	H	Ge	S/V
Narcissus jonquilla × pseudonarcissus = N. × odorus	Amaryllidaceae	5	Horticulture	1904	H	Ge	S/V
Narcissus macrolobus	Amaryllidaceae	2	Southwest Europe	1990	H	Ge	S/V

Scientific name	Family	No. of hectads	Introduced from	First record	Vector	Life form	S/V
Narcissus minor	Amaryllidaceae	16	Southwest Europe	1885	H	Ge	S/V
Narcissus moschatus	Amaryllidaceae	1	Southwest Europe	1995	H	Ge	S/V
Narcissus moschatus × poeticus = N. × boutignyanus	Amaryllidaceae	14	Horticulture	1993	H	Ge	S/V
Narcissus nobilis	Amaryllidaceae	11	Southwest Europe	1994	H	Ge	S/V
Narcissus obvallaris	Amaryllidaceae	59	Horticulture	1830	H	Ge	S/V
Narcissus papyraceus	Amaryllidaceae	6	Mediterranean	1971	H	Ge	S/V
Narcissus poeticus	Amaryllidaceae	433	Southern Europe	1779	H	Ge	S/V
Narcissus poeticus × pseudonarcissus = N. × incomparabilis	Amaryllidaceae	241	Horticulture	1711	H	Ge	S/V
Narcissus poeticus × tazetta = N. × medioluteus	Amaryllidaceae	95	Horticulture	1737	H	Ge	S/V
Narcissus pseudonarcissus × triandrus = N. × taitii	Amaryllidaceae	1	Horticulture	2005	H	Ge	S/V
Narcissus radiiflorus	Amaryllidaceae	7	Southern Europe	1975	H	Ge	S/V
Narcissus tazetta	Amaryllidaceae	37	Mediterranean	1876	H	Ge	S/V
Narcissus triandrus	Amaryllidaceae	10	Southwest Europe	1905	H	Ge	S/V
Nassella neesiana	Poaceae	1	South America	1916	W	Hc	–
Nassella tenuissima	Poaceae	3	Argentina	1958	W	Hc	–
Nectaroscordum siculum subsp. bulgaricum	Amaryllidaceae	6	Southeast Europe	1974	H	Ge	S/V
Nectaroscordum siculum subsp. siculum	Amaryllidaceae	2 (48)†	Mediterranean	1906	H	Ge	S/V
Nemesia strumosa	Scrophulariaceae	17	South Africa	1928	H	Th	–
Nemophila menziesii	Boraginaceae	1	North America	1928	H	Th	–
Nepeta nepetella × racemosa = N. × faassenii	Lamiaceae	141	Horticulture	1902	H	Hc	–
Nepeta racemosa	Lamiaceae	9	Caucasus	1961	H	Hc	–
Nertera granadensis	Rubiaceae	1	Southern hemisphere	1957	H	Hc	V
Neslia paniculata	Brassicaceae	11	Europe	1795	Bs,G,W	Th	–
Nicandra physalodes	Solanaceae	284	Peru	1860	Bs,G,H,O,W	Th	–
Nicotiana alata	Solanaceae	73	South America	1912	H	Th	–
Nicotiana alata × forgetiana = N. × sanderae	Solanaceae	99	Horticulture	1965	H	Th	–
Nicotiana forgetiana	Solanaceae	15	South America	1991	H	Th	–
Nicotiana rustica	Solanaceae	4	North America	1869	H	Th	–
Nicotiana tabacum	Solanaceae	11	Central & South America	1912	M,W	Th	–
Nigella damascena	Ranunculaceae	456	Southern Europe	1832	H	Th	S
Nonea lutea	Boraginaceae	18	Russia	1977	H	Th	–
Nothofagus alpina	Nothofagaceae	107	South America	1956	H	Ph	S
Nothofagus alpina × obliqua = N. × dodecaphleps	Nothofagaceae	2	Neonative	1978	Neo	Ph	–
Nothofagus obliqua	Nothofagaceae	120	South America	1956	H	Ph	S
Nothoscordum borbonicum	Amaryllidaceae	15	South America	1930	H	Ge	S/V
Nuphar advena	Nymphaeaceae	13	North America	1963	H	Hy	V

Scientific name	Family	No. of hectads	Introduced from	First record	Vector	Life form	S/V
Nymphaea × marliacea	Nymphaeaceae	49	Horticulture	1987	H	Hy	V
*Nymphoides peltata	Menyanthaceae	581	Europe	1570	?native	Hy	S/V
Ochagavia carnea	Bromeliaceae	2	Chile	1973	H	Ch	–
Odontites jaubertianus subsp. chrysanthus	Orobanchaceae	2	France	1965	?	Th	S
Odontites jaubertianus subsp. jaubertianus	Orobanchaceae	1	France	2006	?	Th	S
Oemleria cerasiformis	Rosaceae	7	North America	1977	H	Ph	S/V
Oenothera biennis	Onagraceae	606	Uncertain	c. 1650	H	Hc	S
Oenothera biennis × glazoviana = O. × fallax	Onagraceae	220	Neonative	1875	Neo	Hc	S
Oenothera glazioviana	Onagraceae	1,034	?Europe	1866	H	Hc	S
Oenothera parviflora	Onagraceae	8	North America	1905	H	Hc	–
Oenothera stricta	Onagraceae	85	Chile	1847	H	Hc	S
Olearia avicenniifolia	Asteraceae	5	New Zealand	1997	H	Ph	
Olearia avicenniifolia × moschata = O. × haastii	Asteraceae	33	New Zealand	1978	H	Ph	
Olearia macrodonta	Asteraceae	92	New Zealand	1947	H	Ph	S
Olearia paniculata	Asteraceae	10	New Zealand	1969	H	Ph	–
Olearia traversii	Asteraceae	4	Chatham Islands	1966	H	Ph	–
Omphalodes verna	Boraginaceae	53	Southeast Europe	1840	H	Hc	S/V
*Onobrychis viciifolia	Fabaceae	432	Europe	1597	?native	Hc	S
Ononis mitissima	Fabaceae	3	Mediterranean	1865	Bs	Th	–
Onopordum nervosum	Asteraceae	1	Southwest Europe	1963	H	Hc	S
Ornithogalum arabicum	Asparagaceae	1	Mediterranean	1990	H	Ge	S/V
Ornithogalum nutans	Asparagaceae	100	Europe	1805	H	Ge	S/V
*Ornithogalum umbellatum subsp. campestre	Asparagaceae	992	Europe	1548	?native	Ge	S/V
Ornithopus compressus	Fabaceae	4	Southern Europe	1893	S,Tn	Th	S
Ornithopus sativus	Fabaceae	2	Southwest Europe	1909	S	Th	S
Orobanche crenata	Orobanchaceae	2	Southern Europe	1845	As	Ge	S
Orontium aquaticum	Araceae	1	North America	1990	H	Hy	V
Oryzopsis miliacea	Poaceae	6	Southern Europe	1927	Bs,H,W	Hc	S
Oscularia deltoides	Aizoaceae	4	South Africa	1953	H	Ch	S/V
Osteospermum jucundum	Asteraceae	32	South Africa	1989	H	Ch	V
Oxalis articulata	Oxalidaceae	724	South America	1912	H	Ge	V
Oxalis corniculata	Oxalidaceae	1,019	Widespread	1585	H,I	Hc	S/V
Oxalis debilis	Oxalidaceae	185	South America	1900	H,I	Ge	V
Oxalis decaphylla	Oxalidaceae	4	Mexico	1967	I	Ge	V
Oxalis dillenii	Oxalidaceae	19	North America	1951	I	Hc	S
Oxalis exilis	Oxalidaceae	535	Antipodes	1913	H,I	Hc	S/V
Oxalis incarnata	Oxalidaceae	205	South Africa	1912	H,I	Th	S
Oxalis latifolia	Oxalidaceae	93	Central & South America	1921	H,I	Ge	V
Oxalis megalorrhiza	Oxalidaceae	6	Chile	1940	H	Hc	V
Oxalis pes-caprae	Oxalidaceae	20	South Africa	1901	H,I	Ge	V
Oxalis rosea	Oxalidaceae	15	Chile	1931	I	Th	S

Scientific name	Family	No. of hectads	Introduced from	First record	Vector	Life form	S/V
Oxalis stricta	Oxalidaceae	309	North America	1823	I	Hc	S/V
Oxalis tetraphylla	Oxalidaceae	12	Mexico	1917	H,I	Ge	V
Oxalis valdiviensis	Oxalidaceae	5	Chile	1882	H	Hc	S
Pachyphragma macrophyllum	Brassicaceae	5	Caucasus	1964	H	Hc	S/V
Pachysandra terminalis	Buxaceae	6	Japan	1968	H	Ch	–
Paeonia lactiflora	Paeoniaceae	2	East Asia	1978	H	Hc	–
Paeonia mascula	Paeoniaceae	21	Southern Europe	1805	H	Ge	V
Paeonia officinalis	Paeoniaceae	143	Southern Europe	1650	H	Hc	–
Pancratium maritimum	Amaryllidaceae	2	Mediterranean	1993	H	Ge	–
Panicum capillare	Poaceae	66	North America	1867	As,Bs,O,W	Th	–
Panicum dichotomiflorum	Poaceae	3	North America	1945	Bs,O,W	Th	–
Panicum miliaceum	Poaceae	453	Asia	1885	A,Bs,G,O	Th	–
Panicum schinzii	Poaceae	2	Africa	1918	Bs,G,W	Th	–
Papaver atlanticum	Papaveraceae	255	Morocco	1927	H	Hc	S
Papaver pseudoorientale	Papaveraceae	266	Southwest Asia	1884	H	Hc	S/V
Papaver somniferum subsp. setigerum	Papaveraceae	9	Southern Europe	1850	Bs,F,H	Th	S
*Parietaria judaica	Urticaceae	1,515	Europe	1548	?native	Hc	S
Parietaria officinalis	Urticaceae	3	Europe	1966	H,M	Hc	S
Parthenocissus inserta	Vitaceae	108	North America	1948	H	Ph	V
Parthenocissus quinquefolia	Vitaceae	315	North America	1927	H	Ph	V
Parthenocissus tricuspidata	Vitaceae	38	East Asia	1928	H	Ph	V
Paspalum dilatatum	Poaceae	1	Southern Europe	1924	Bs,W	Hc	–
Paspalum distichum	Poaceae	4	Widespread	1961	Bs,E,W	Hc	V
Passiflora caerulea	Passifloraceae	37	South America	1981	H	Ph	–
Pastinaca sativa subsp. sativa	Apiaceae	68	Agriculture	1905	F	Hc	–
Pastinaca sativa subsp. urens	Apiaceae	3	Europe	1992	?	Hc	S
Paulownia tomentosa	Paulowniaceae	14	China	1990	H	Ph	S
Pelargonium tomentosum	Geraniaceae	3	South Africa	1938	H	Ch	V
Peltaria alliacea	Brassicaceae	1	Europe	2006	H	Hc	S
Pentaglottis sempervirens	Boraginaceae	1,808	Southwest Europe	1724	H	Hc	S
Pericallis hybrida	Asteraceae	4	Horticulture	1971	H	Hc	–
Persicaria alpina	Polygonaceae	5	Europe	1909	H	Hc	–
Persicaria alpina × weyrichii = P. × fennica	Polygonaceae	1	Horticulture	1981	H	Hc	V
Persicaria amplexicaulis	Polygonaceae	234	Himalayas	1908	H	Hc	S
Persicaria campanulata	Polygonaceae	215	Himalayas	1933	H	Hc	S/V
Persicaria capitata	Polygonaceae	115	Himalayas	1968	H	Hc	–
Persicaria mollis	Polygonaceae	3	Himalayas	1956	H	Hc	S/V
Persicaria nepalensis	Polygonaceae	10	Himalayas	1963	Bs	Th	S
Persicaria pensylvanica	Polygonaceae	11	North America	1870	O	Th	–
Persicaria sagittata	Polygonaceae	2	North America	1889	G	Th	–
Persicaria wallichii	Polygonaceae	435	Himalayas	1917	H	Hc	S/V
Persicaria weyrichii	Polygonaceae	6	East Asia	1978	H	Hc	S/V
Petasites albus	Asteraceae	264	Europe	1843	H	Hc	V
Petasites fragrans	Asteraceae	1,756	North Africa	1835	H	Ge	V

Scientific name	Family	No. of hectads	Introduced from	First record	Vector	Life form	S/V
Petasites japonicus	Asteraceae	150	Japan	1924	H	Ge	V
Petrorhagia dubia	Caryophyllaceae	1	Southern Europe	1902	?,W	Th	S
*Petrorhagia prolifera	Caryophyllaceae	4	Europe	1650	?native	Th	S
Petrorhagia saxifraga	Caryophyllaceae	9	Europe	1893	H	Ch	S/V
Petunia axillaris × integrifolia = P. × hybrida	Solanaceae	163	Horticulture	1948	H	Th	–
Phacelia tanacetifolia	Boraginaceae	445	North America	1885	As,Gs,G,H	Th	S
Phalaris angusta	Poaceae	1	North & South America	1913	A,W	Th	–
Phalaris aquatica	Poaceae	111	Southern Europe	1906	A,E,G,W	Hc	S
Phalaris canariensis	Poaceae	851	North Africa	1632	A,Bs,G,W	Th	S
Phalaris minor	Poaceae	53	Southern Europe	1638	Bs,G,E,W	Th	S
Phalaris paradoxa	Poaceae	119	Southern Europe	1832	Bs,G,E,W	Th	S
Phaseolus coccineus	Fabaceae	11	Central & South America	1910	F	Hc	–
Phaseolus vulgaris	Fabaceae	2	South America	1904	Bs,F	Th	–
Philadelphus coronarius	Hydrangeaceae	355	Europe	1919	H	Ph	–
Philadelphus coronarius × microphyllus = P. 'Lemoinei'	Hydrangeaceae	152	Horticulture	1987	H	Ph	–
Philadelphus coronarius × microphyllus × pubescens = P. 'Virginalis'	Hydrangeaceae		Horticulture		H	Ph	–
Phlomis fruticosa	Lamiaceae	40	Mediterranean	1896	H	Ph	S
Phlomis russeliana	Lamiaceae	7	Turkey	1949	H	Hc	–
Phlox paniculata	Polemoniaceae	21	North America	1915	H	Hc	S
Phoenix dactylifera	Arecaceae	2	North Africa	1908	F	Ph	–
Phormium cookianum	Xanthorrhoeaceae	4	New Zealand	1920	H	Ch	S
Phormium tenax	Xanthorrhoeaceae	113	New Zealand	1898	H	Ch	S
Phuopsis stylosa	Rubiaceae	43	Caucasus	1905	H	Hc	–
Phygelius capensis	Scrophulariaceae	30	South Africa	1960	H	Ph	S
Physalis alkekengi	Solanaceae	79	Europe	1650	H	Hc	S/V
Physalis angulata	Solanaceae	3	Central & South America	1928	O,W	Th	–
Physalis ixocarpa	Solanaceae	11	North & South America	1915	Bs,W	Th	–
Physalis peruviana	Solanaceae	71	South America	1927	Bs,F,G,O,W	Hc	–
Physalis philadelphica	Solanaceae	7	North & South America	1971	F,G	Th	–
Physalis pubescens	Solanaceae	2	North & South America	1915	?	Hc	–
Physocarpus opulifolius	Rosaceae	32	North America	1909	H	Ph	–
Phyteuma scheuchzeri	Campanulaceae	3	Europe	1951	H	Hc	S
Phytolacca acinosa	Phytolaccaceae	56	Asia	1925	A,H	Hc	–
Phytolacca polyandra	Phytolaccaceae	6	China	1983	H	Hc	S
Picris hieracioides subsp. *grandiflora*	Asteraceae	2	Europe	1843	?	Hc	S
Picris hieracioides subsp. *villarsii*	Asteraceae	3	Europe	1935	?	Hc	S
Pilosella aurantiaca subsp. *aurantiaca*	Asteraceae	115	Europe	1793	H	Hc	S/V

Scientific name	Family	No. of hectads	Introduced from	First record	Vector	Life form	S/V
Pilosella aurantiaca subsp. carpathicola	Asteraceae	1,532	Europe	1853	H	Hc	S/V
Pilosella aurantiaca × officinarum = P. × stoloniflora	Asteraceae	9	Neonative	1956	Neo	Hc	S/V
Pilosella caespitosa	Asteraceae	14	Europe	1868	H	Hc	S/V
Pilosella caespitosa × lactucella = P. × floribunda	Asteraceae	2	Europe	1867	H	Hc	S/V
Pilosella flagellaris subsp. flagellaris	Asteraceae	50	Europe	1869	H	Hc	S/V
Pilosella praealta subsp. praealta	Asteraceae	15 (11)†	Europe	1899	H	Hc	S/V
Pilosella praealta subsp. thaumasia	Asteraceae	4	Europe	1918	H	Hc	S/V
Pimpinella peregrina	Apiaceae	1	Europe	2010	Gs	Hc	S
Pistia stratiotes	Araceae	19	Tropics	1983	H	Hy	–
Pittosporum crassifolium	Pittosporaceae	9	New Zealand	1939	H	Ph	S
Pittosporum tenuifolium	Pittosporaceae	35	New Zealand	1958	H	Ph	S
Plagiobothrys scouleri	Boraginaceae	12	North America	1974	Gs	Th	S
Plantago afra	Plantaginaceae	6	Southern Europe	1759	Bs,G	Th	–
Plantago arenaria	Plantaginaceae	11	Southern Europe	1820	As,Bs,W	Th	S
Platanus occidentalis × orientalis = P. × hispanica	Platanaceae	444	Horticulture	1939	H	Ph	–
Platanus orientalis	Platanaceae	22	Southwest Asia	1987	H	Ph	–
Plecostachys serpyllifolia	Asteraceae	1	South Africa	1995	H	Ch	–
Pleioblastus chino	Poaceae	2	Japan	1996	H	Ph	V
Pleioblastus pygmaeus	Poaceae	8	Japan	1993	H	Ph	V
Pleioblastus simonii	Poaceae	7	Japan	1965	H	Ph	V
Poa chaixii	Poaceae	113	Europe	1852	H	Hc	S
Poa flabellata	Poaceae	2	South America	1900	H	Hc	–
Poa palustris	Poaceae	60	Europe	1879	A,G,W	Hc	S
*Polycarpon tetraphyllum	Caryophyllaceae	40	Europe	1774	?native	Th	S
Polygonum arenarium	Polygonaceae	48	Mediterranean	1860	Bs,G	Th	–
Polygonum cognatum	Polygonaceae	1	Asia	1908	G,T	Ch	S
Polygonum patulum	Polygonaceae	4	Europe	1902	As,Bs,G,W	Th	–
Polypogon interruptus	Poaceae	1	North & South America	1955	G,W	Th	–
Polypogon maritimus	Poaceae	2	Southern Europe	1908	G,W	Th	–
Polypogon viridis	Poaceae	268	Southern Europe	1860	G,W	Hc	S/V
Pontederia cordata	Pontederiaceae	64	North America	1949	H	Hy	S
Populus alba	Salicaceae	1,518	Europe	1597	H	Ph	V
Populus alba × tremula = P. × canescens	Salicaceae	1,147	Europe	c. 1700	H	Ph	V
Populus balsamifera	Salicaceae	67	North America	1911	H	Ph	–
Populus balsamifera × deltoides = P. × jackii	Salicaceae	409	Horticulture	1876	H	Ph	V
Populus balsamifera × deltoides × nigra	Salicaceae	5	Neonative	1984	Neo	Ph	S
Populus balsamifera × trichocarpa = P. × hastata	Salicaceae	105	Horticulture	1977	H	Ph	–

Scientific name	Family	No. of hectads	Introduced from	First record	Vector	Life form	S/V
Populus deltoides × nigra = P. × canadensis	Salicaceae	1,572	Horticulture	1799	H	Ph	S
Populus deltoides × trichocarpa = P. × generosa	Salicaceae	14	Horticulture	1977	H	Ph	–
Populus laurifolia × nigra = P. × berolinensis	Salicaceae	10	Horticulture	1994	H	Ph	–
Populus trichocarpa	Salicaceae	378	North America	1935	H	Ph	V
Potentilla inclinata	Rosaceae	4	Europe	1836	H	Hc	–
Potentilla indica	Rosaceae	152	Asia	1879	H	Hc	S/V
Potentilla intermedia	Rosaceae	22	Russia	1859	G	Hc	–
Potentilla montana	Rosaceae	1	Southwest Europe	2006	H	Hc	S/V
Potentilla norvegica	Rosaceae	44	Europe	1856	Bs,G,H,S	Hc	S
Potentilla recta	Rosaceae	236	Southern Europe	1858	Gs,H	Hc	S
Potentilla rivalis	Rosaceae	1	North America	1976	?	Th	S
Poterium sanguisorba subsp. balearicum	Rosaceae	401	Southern Europe	1849	A	Hc	S
Pratia angulata	Campanulaceae	16	New Zealand	1930s	H	Hc	S/V
Primula auricula	Primulaceae	4	Europe	1880	H	Hc	–
Primula florindae	Primulaceae	18	Tibet	1957	H	Hc	S
Primula japonica	Primulaceae	16	Japan	1966	H	Hc	S
Primula juliae	Primulaceae	2	Caucasus	1995	H	Hc	S
Primula juliae × vulgaris = P. × pruhonicensis	Primulaceae	22	Horticulture	1987	H	Hc	S
Primula prolifera	Primulaceae	2	Asia	2007	H	Hc	S
Primula pulverulenta	Primulaceae	6	China	1966	H	Hc	S
Primula sikkimensis	Primulaceae	1	Asia	1921	H	Hc	–
Prunella grandiflora	Lamiaceae	2	Europe	1971	H	Hc	S
Prunella laciniata	Lamiaceae	14	Europe	1878	?	Hc	S
Prunella laciniata × vulgaris = P. × intermedia	Lamiaceae	20	Neonative	1913	Neo	Hc	S
Prunus cerasifera	Rosaceae	1,069	Southwest Asia	1918	H	Ph	S/V
Prunus cerasifera × spinosa = P. × simmleri	Rosaceae	1	Neonative	2006	Neo	Ph	S/V
Prunus domestica × spinosa = P. × fruticans	Rosaceae	238	Neonative	1909	Neo	Ph	S/V
Prunus dulcis	Rosaceae	29	Southwest Asia	1953	H	Ph	–
Prunus incisa	Rosaceae	4	Japan	1969	H	Ph	S
Prunus laurocerasus	Rosaceae	1,826	Southern Europe	1886	H	Ph	S/V
Prunus lusitanica	Rosaceae	682	Southwest Europe	1927	H	Ph	S
Prunus mahaleb	Rosaceae	10	Southern Europe	1936	H	Ph	S
Prunus pensylvanica	Rosaceae	2	North America	1986	H	Ph	S
Prunus persica	Rosaceae	21	China	1953	H	Ph	–
Prunus serotina	Rosaceae	103	North America	1853	H	Ph	S
Prunus serrulata	Rosaceae	80	East Asia	1968	H	Ph	–
Pseudofumaria alba	Papaveraceae	53	Europe	1957	H	Hc	S
Pseudofumaria lutea	Papaveraceae	1,333	Europe	1796	H	Hc	S
Pseudosasa japonica	Poaceae	242	Japan	1955	H	Ph	V

Scientific name	Family	No. of hectads	Introduced from	First record	Vector	Life form	S/V
Pseudoturritis turrita	Brassicaceae	7	Europe	1722	G,H	Hc	S
Pterocarya fraxinifolia	Juglandaceae	44	Caucasus	1980	H	Ph	V
Pterocarya fraxinifolia × stenoptera = P. × rehderiana	Juglandaceae	2	Horticulture	2005	H	Ph	–
Pulmonaria 'Mawson's Blue'	Boraginaceae	20	Horticulture	1951	H	Hc	V
Pulmonaria officinalis	Boraginaceae	674	Europe	1793	H	Hc	S/V
Pulmonaria rubra	Boraginaceae	33	Southern Europe	1969	H	Hc	S/V
Pyracantha coccinea	Rosaceae	271	Southern Europe	1901	H	Ph	S/V
Pyracantha coccinea × rogersiana	Rosaceae	9	Horticulture	1992	H	Ph	–
Pyracantha rogersiana	Rosaceae	43	China	1976	H	Ph	–
*Pyrus cordata	Rosaceae	5	Europe	1870	?native	Ph	S
Quercus canariensis	Fagaceae	5	Mediterranean	1970	H	Ph	–
Quercus cerris	Fagaceae	1,293	Southern Europe	1844	H	Ph	S
Quercus cerris × suber = Q. × crenata	Fagaceae	82	Horticulture	1964	H	Ph	S
Quercus coccinea	Fagaceae	26	North America	1939	H	Ph	–
Quercus ilex	Fagaceae	892	Southern Europe	1862	H	Ph	S
Quercus rubra	Fagaceae	650	North America	1942	H	Ph	S
Ramonda myconi	Gesneriaceae	3	Southwest Europe	1921	H	Hc	–
Ranunculus aconitifolius	Ranunculaceae	12	Europe	1885	H	Hc	S
Ranunculus marginatus	Ranunculaceae	4	Mediterranean	1950	Bs,G	Th	S
Ranunculus muricatus	Ranunculaceae	7	Southern Europe	1777	Bs,G,W	Th	S
*Ranunculus sardous	Ranunculaceae	403	Europe	1663	?native	Th	S
Raphanus raphanistrum subsp. landra	Brassicaceae	1	Mediterranean	1859	G	Hc	–
Raphanus raphanistrum subsp. maritimus × subsp. raphanistrum	Brassicaceae	1	Neonative	2010	Neo	Th	S
Rapistrum perenne	Brassicaceae	2	Europe	1859	Bs,G	Hc	S
Rapistrum rugosum subsp. linnaeanum	Brassicaceae	263	Southern Europe	1861	Bs,G,W	Hc	S
Reineckea carnea	Asparagaceae	1	East Asia	1984	H	Hc	–
Reseda alba	Resedaceae	64	Southern Europe	1826	Bs,G,H,W	Th	S
*Reseda lutea	Resedaceae	1,170	Europe	1597	?native	Hc	S
Reseda odorata	Resedaceae	8	Mediterranean	1793	H	Th	–
Reseda phyteuma	Resedaceae	3	Southern Europe	1907	G,W	Th	S
Rhagadiolus stellatus	Asteraceae	1	Southern Europe	1895	Bs,G,W	Th	–
Rhamnus alaternus	Rhamnaceae	6	Mediterranean	1908	H	Ph	S
Rheum palmatum	Polygonaceae	8	Northeast Asia	1964	H	Hc	–
Rheum × rhabarbarum	Polygonaceae	494	Horticulture	1960	F	Hc	V
*Rhinanthus angustifolius	Orobanchaceae	17	Europe	1724	?native	Th	S
Rhododendron groenlandicum	Ericaceae	14	North America	1860	H	Ph	S/V
Rhododendron luteum	Ericaceae	153	West Asia	1939	H	Ph	S
Rhododendron catawbiense × ponticum = R. × superponticum	Ericaceae	2,419	Neonative	1849	Neo	Ph	S/V
Rhododendron ponticum subsp. baeticum	Ericaceae		Southwest Europe	1845	H	Ph	S/V
Rhus copallina	Anacardiaceae	1	North America	2006	H	Ph	V
Rhus coriaria	Anacardiaceae	1	Southern Europe	2006	H	Ph	V

Scientific name	Family	No. of hectads	Introduced from	First record	Vector	Life form	S/V
Rhus typhina	Anacardiaceae	504	North America	1966	H	Ph	V
Ribes divaricatum	Grossulariaceae	2	North America	1998	F	Ph	–
Ribes nigrum	Grossulariaceae	1,811	Europe	1660	F	Ph	S
Ribes odoratum	Grossulariaceae	33	North America	1975	H	Ph	–
*Ribes rubrum	Grossulariaceae	1,989	Europe	1568	?native	Ph	S
Ribes sanguineum	Grossulariaceae	1,402	North America	1867	H	Ph	S/V
Ribes uva-crispa	Grossulariaceae	2,308	Europe	1763	F	Ph	S
Ricinus communis	Euphorbiaceae	6	Tropics	1865	H,O	Ph	–
Robinia pseudoacacia	Fabaceae	636	North America	1888	H	Ph	S
Rodgersia podophylla	Saxifragaceae	11	East Asia	1966	H	Ge	V
Romulea rosea	Iridaceae	1	South Africa	1969	H	Ge	–
Rorippa amphibia × austriaca = R. × hungarica	Brassicaceae	1	Neonative	1987	Neo	Hc	–
Rorippa austriaca	Brassicaceae	25	Europe	1905	G	Hc	S/V
Rorippa austriaca × sylvestris = R. × armoracioides	Brassicaceae	19	Neonative	1971	Neo	Hc	S/V
Rosa arvensis × rugosa = R. × paulii	Rosaceae	2	Horticulture	1990	H	Ph	–
Rosa canina × rugosa = R. × praegeri	Rosaceae	4	Neonative	1927	Neo	Ph	S/V
Rosa ferruginea	Rosaceae	105	Europe	1973	H	Ph	S
Rosa gallica	Rosaceae	10	Europe	1871	H	Ph	S
Rosa 'Hollandica'	Rosaceae	148	Horticulture	1955	H	Ph	S/V
Rosa luciae	Rosaceae	18	East Asia	1905	H	Ph	S/V
Rosa multiflora	Rosaceae	218	East Asia	1930	H	Ph	S
Rosa rugosa	Rosaceae	1,140	East Asia	1917	H	Ph	S/V
Rosa rugosa × spinosissima	Rosaceae	1	Neonative	1997	Neo	Ph	S/V
Rosa setigera	Rosaceae	3	North America	1902	H	Ph	S
Rosa virginiana	Rosaceae	47	North America	1887	H	Ph	S/V
Rosa × alba	Rosaceae	11	Horticulture	1905	H	Ph	S
Rosmarinus officinalis	Lamiaceae	129	Mediterranean	1969	F,H	Ph	S
Rostraria cristata	Poaceae	6	Mediterranean	1902	Bs,E,W	Th	–
Rubia tinctorum	Rubiaceae	4	Asia	1795	H	Ge	V
Rubus allegheniensis	Rosaceae	1	North America	1899	F	Ph	S/V
Rubus armeniacus	Rosaceae	628	Europe	1895	F	Ph	S/V
Rubus canadensis	Rosaceae	6	North America	1961	F	Ph	S/V
Rubus cockburnianus	Rosaceae	94	China	1972	H	Ph	S/V
Rubus cockburnianus × idaeus = R. × knappianus	Rosaceae	1	Neonative	2009	Neo	Ph	S/V
Rubus elegantispinosus	Rosaceae	81	Europe	1971	F	Ph	S/V
Rubus idaeus × phoenicolasius = R. × paxii	Rosaceae	1	Horticulture	1931	F	Ph	V
Rubus laciniatus	Rosaceae	290	Uncertain	1908	F	Ph	S/V
Rubus loganobaccus	Rosaceae	64	Agriculture	1938	F	Ph	S/V
Rubus odoratus	Rosaceae	13	North America	1914	H	Ph	S/V
Rubus odoratus × parviflorus = R. × fraseri	Rosaceae	3	Horticulture	1977	H	Ph	V
Rubus parviflorus	Rosaceae	22	North America	1913	H	Ph	S/V

Scientific name	Family	No. of hectads	Introduced from	First record	Vector	Life form	S/V
Rubus phoenicolasius	Rosaceae	25	East Asia	1907	F	Ph	S/V
Rubus spectabilis	Rosaceae	342	North America	1851	H	Ph	S/V
Rubus tricolor	Rosaceae	163	China	1976	H	Ch	S/V
Rudbeckia hirta	Asteraceae	38	North America	1917	H	Hc	–
Rudbeckia laciniata	Asteraceae	24	North America	1898	H	Hc	S/V
Rumex acetosa subsp. ambiguus	Polygonaceae	7	Uncertain	1988	F	Hc	S/V
Rumex confertus	Polygonaceae	4	Europe	1918	?	Hc	S
Rumex confertus × crispus = R. × skofitzii	Polygonaceae	0	Neonative	1954	Neo	Hc	–
Rumex confertus × obtusifolius = R. × borbasii	Polygonaceae	0	Neonative	1954	Neo	Hc	–
Rumex conglomeratus × frutescens = R. × wrightii	Polygonaceae	1	Neonative	1952	Neo	Hc	V
Rumex crispus × cristatus = R. × dimidiatus	Polygonaceae	9	Neonative	1949	Neo	Hc	S
Rumex crispus × frutescens = R. × mirabilis	Polygonaceae	1	Neonative	2002	Neo	Hc	V
Rumex crispus × obtusifolius × patientia	Polygonaceae	1	Neonative	2006	Neo	Hc	–
Rumex crispus × patientia = R. × confusus	Polygonaceae	3	Neonative	1938	Neo	Hc	–
Rumex cristatus	Polygonaceae	94	Europe	1918	As,Gs,G,H	Hc	S
Rumex cristatus × obtusifolius = R. × lousleyi	Polygonaceae	13	Neonative	1949	Neo	Hc	–
Rumex cristatus × palustris = R. × akeroydii	Polygonaceae	4	Neonative	1991	Neo	Hc	–
Rumex cristatus × patientia = R. × xenogenus	Polygonaceae	1	Neonative	2000	Neo	Hc	–
Rumex dentatus	Polygonaceae	6	Europe	1877	G,W	Hc	–
Rumex frutescens	Polygonaceae	8	South America	1913	?,W	Hc	S/V
Rumex frutescens × obtusifolius = R. × cornubiensis	Polygonaceae	1	Neonative	1994	Neo	Hc	V
Rumex obovatus	Polygonaceae	2	South America	1912	Bs,G,W	Hc	–
Rumex obtusifolius × patientia = R. × erubescens	Polygonaceae	4	Neonative	1938	Neo	Hc	S
Rumex patientia	Polygonaceae	23	Europe	1871	As,Gs,H	Hc	S
Rumex salicifolius	Polygonaceae	2	North America	1900	Bs,G	Hc	S/V
Rumex scutatus	Polygonaceae	15	Europe	1800	F,H	Ch	S/V
Ruschia caroli	Aizoaceae	2	South Africa	1959	H	Ch	V
Ruscus hypoglossum	Asparagaceae	6	Europe	1934	H	Ch	V
Ruta graveolens	Rutaceae	42	Europe	1928	H	Ph	–
Sagittaria latifolia	Alismataceae	26	North America	1941	H	Hy	S
Sagittaria rigida	Alismataceae	6	North America	1908	H	Hy	S
Sagittaria subulata	Alismataceae	2	North America	1962	H	Hy	S
Salix acutifolia	Salicaceae	29	Eurasia	1915	H	Ph	–
Salix alba × babylonica = S. × sepulcralis	Salicaceae	115	Horticulture	1886	H	Ph	–
Salix alba × babylonica × euxina = S. × pendulina	Salicaceae	74	Horticulture	1955	H	Ph	–

Scientific name	Family	No. of hectads	Introduced from	First record	Vector	Life form	S/V
Salix alba × euxina × pentandra = S. × meyeriana	Salicaceae	58	Neonative	1836	Neo	Ph	–
Salix alba × euxina × triandra = S. × alopecuroides	Salicaceae	3	Horticulture	1930	H	Ph	–
Salix alba × pentandra = S. × ehrhartiana	Salicaceae	34	Horticulture	1949	H	Ph	–
Salix aurita × caprea × viminalis = S. × stipularis	Salicaceae	38	Neonative	1789	Neo	Ph	–
Salix aurita × viminalis = S. × fruticosa	Salicaceae	75	Neonative	1887	Bm,Neo	Ph	S/V
Salix caprea × cinerea × viminalis = S. × calodendron	Salicaceae	144	Neonative	1773	Bm,Neo	Ph	–
Salix caprea × viminalis = S. × smithiana	Salicaceae	1,008	Neonative	1841	Bm,Neo	Ph	S/V
Salix cinerea × purpurea × viminalis = S. × forbyana	Salicaceae	72	Neonative	1841	Neo	Ph	–
Salix cinerea × repens × viminalis = S. × angusensis	Salicaceae	6	Neonative	1947	Neo	Ph	V
Salix cinerea × viminalis = S. × holosericea	Salicaceae	494	Neonative	1820	Neo	Ph	S/V
Salix daphnoides	Salicaceae	158	Northern Europe	1905	H	Ph	–
Salix elaeagnos	Salicaceae	90	Southern Europe	1914	H	Ph	–
Salix eriocephala	Salicaceae	5	North America	1972	H	Ph	V
Salix euxina	Salicaceae	98	Turkey	1808	H	Ph	–
Salix myrsinifolia × viminalis = S. × seminigricans	Salicaceae	1	Neonative	1985	Neo	Ph	–
Salix purpurea × repens × viminalis	Salicaceae	1	Neonative	2000	Neo	Ph	–
Salix purpurea × viminalis = S. × rubra	Salicaceae	164	Neonative	1805	Neo	Ph	S
Salix repens × viminalis = S. × friesiana	Salicaceae	8	Neonative	1897	Neo	Ph	V
Salix udensis	Salicaceae	24	East Asia	1960	H	Ph	–
Salpichroa origanifolia	Solanaceae	12	South America	1927	H	Hc	S
Salsola kali subsp. tragus	Amaranthaceae	3	Europe	1875	As,Bs,G,W	Th	–
Salvia glutinosa	Lamiaceae	5	Southern Europe	1896	H	Hc	S
Salvia nemorosa × pratensis = S. × sylvestris	Lamiaceae	1	Horticulture	1875	H	Hc	S
Salvia officinalis	Lamiaceae	30	Southern Europe	1905	F,H	Ph	–
*Salvia pratensis	Lamiaceae	59	Europe	1699	?native	Hc	S
Salvia reflexa	Lamiaceae	23	North America	1928	Bs,G,Gs,W	Th	–
Salvia sclarea	Lamiaceae	15	Southern Europe	1918	H	Hc	–
Salvia verticillata	Lamiaceae	23	Southern Europe	1857	H	Hc	S
Salvia viridis	Lamiaceae	21	Southern Europe	1859	H	Th	–
Sambucus canadensis	Caprifoliaceae	30	North America	1943	H	Ph	S/V
Sambucus racemosa	Caprifoliaceae	349	North temperate zone	1905	H	Ph	S
Sanguisorba canadensis	Rosaceae	10	North America	1848	H	Hc	S
Santolina chamaecyparissus	Asteraceae	40	Mediterranean	1902	H	Ch	–
Saponaria ocymoides	Caryophyllaceae	34	Southern Europe	1913	H	Ch	S

Scientific name	Family	No. of hectads	Introduced from	First record	Vector	Life form	S/V
Sarracenia purpurea	Sarraceniaceae	21	Europe	1906	H	Hc	S
Sasa palmata	Poaceae	202	Japan	c. 1946	H	Ph	V
Sasa veitchii	Poaceae	19	Japan	1962	H	Ph	V
Sasaella ramosa	Poaceae	42	Japan	1983	H	Ph	V
Satureja montana	Lamiaceae	14	Southern Europe	1809	F,H	Ch	S
Saxifraga × arendsii	Saxifragaceae	23	Horticulture	1987	H	Hc	V
Saxifraga cuneifolia	Saxifragaceae	13	Southern Europe	1932	H	Hc	V
Saxifraga cymbalaria	Saxifragaceae	41	Mediterranean	1865	I	Th	S
Saxifraga hirsuta × umbrosa = S. × geum	Saxifragaceae	20	Southwest Europe	1854	H	Hc	V
Saxifraga paniculata	Saxifragaceae	2	Europe	1988	H	Hc	–
Saxifraga rotundifolia	Saxifragaceae	6	Europe	1843	H	Hc	S
Saxifraga spathularis × umbrosa = S. × urbium	Saxifragaceae	694	Horticulture	1837	H	Hc	V
Saxifraga stolonifera	Saxifragaceae	4	East Asia	1876	H	Hc	S/V
Saxifraga umbrosa	Saxifragaceae	27	Southwest Europe	1792	H	Hc	S/V
Scabiosa atropurpurea	Dipsacaceae	14	Southern Europe	1842	H	Hc	S
Schizostylis coccinea	Iridaceae	14	South Africa	1979	H	Ge	V
Schkuhria pinnata	Asteraceae	2	Central & South America	1900	Bs,W	Th	–
Scilla bifolia	Asparagaceae	55	Europe	1921	H	Ge	S/V
Scilla bithynica	Asparagaceae	26	Turkey	1982	H	Ge	S/V
Scilla forbesii	Asparagaceae	206	Turkey	1968	H	Ge	S/V
Scilla liliohyacinthus	Asparagaceae	10	Southwest Europe	1964	H	Ge	S/V
Scilla luciliae	Asparagaceae	44	Turkey	1966	H	Ge	S/V
Scilla messeniaca	Asparagaceae	4	Greece	1979	H	Ge	S/V
Scilla peruviana	Asparagaceae	10	Mediterranean	1930	H	Ge	S/V
Scilla sardensis	Asparagaceae	47	Turkey	1981	H	Ge	S/V
Scilla siberica	Asparagaceae	137	Russia	1968	H	Ge	S/V
Scolymus hispanicus	Asteraceae	1	Southern Europe	1876	Bs	Hc	–
Scorpiurus muricatus	Fabaceae	10	Southern Europe	1859	Bs,I,Tn,W	Th	–
Scorzonera hispanica	Asteraceae	7	Europe	1956	F	Hc	–
Scrophularia peregrina	Scrophulariaceae	4	Mediterranean	1914	?	Th	S
Scrophularia scopolii	Scrophulariaceae	2	Southeast Europe	1923	?	Hc	S
*Scrophularia scorodonia	Scrophulariaceae	73	Western Europe	1689	?native	Hc	S
Scrophularia vernalis	Scrophulariaceae	130	Europe	1633	H	Hc	S
Scutellaria altissima	Lamiaceae	11	Europe	1929	H	Hc	S
Securigera varia	Fabaceae	156	Europe	1836	H	Hc	S
Sedum anacampseros	Crassulaceae	1	Southern Europe	1977	H	Hc	–
Sedum dasyphyllum	Crassulaceae	130	Europe	1724	H	Ch	S/V
Sedum hispanicum	Crassulaceae	19	Southeast Europe	1927	H	Ch	S/V
Sedum kamtschaticum	Crassulaceae	26	East Asia	1981	H	Ge	S/V
Sedum kimnachii	Crassulaceae	29	Mexico	1976	H	Ch	S/V
Sedum lydium	Crassulaceae	7	Turkey	1909	H	Ch	S/V
Sedum nicaeense	Crassulaceae	6	Mediterranean	1915	H	Ch	S/V
Sedum praealtum	Crassulaceae	3	Mexico	1920	H	Ph	S/V

Scientific name	Family	No. of hectads	Introduced from	First record	Vector	Life form	S/V
Sedum rupestre	Crassulaceae	1,237	Europe	1666	H	Ch	S/V
Sedum sexangulare	Crassulaceae	42	Europe	1763	H	Ch	S/V
Sedum spathulifolium	Crassulaceae	17	North America	1957	H	Ch	S/V
Sedum spectabile	Crassulaceae	294	East Asia	1930	H	Hc	–
Sedum spectabile × telephium = S. 'Herbstfreude'	Crassulaceae	21	Horticulture	1988	H	Hc	–
Sedum spurium	Crassulaceae	623	Caucasus	1905	H	Ch	S/V
Sedum stellatum	Crassulaceae	1	Mediterranean	1892	H	Th	S
Sedum stoloniferum	Crassulaceae	34	Caucasus	1908	H	Ch	–
Semiarundinaria fastuosa	Poaceae	2	Japan	1967	H	Ph	V
Sempervivum arachnoideum	Crassulaceae	2	Europe	1991	H	Ch	–
Sempervivum tectorum	Crassulaceae	293	Europe	1629	H	Ch	–
Senecio cambrensis	Asteraceae	9	Neonative	1948	Neo	Th	S
Senecio cineraria	Asteraceae	312	Mediterranean	1891	H	Ch	S
Senecio cineraria × erucifolius = S. × thuretii	Asteraceae	2	Neonative	1978	Neo	Ch	V
Senecio cineraria × jacobaea = S. × albescens	Asteraceae	271	Neonative	1901	Neo	Ch	S
Senecio doria	Asteraceae	6	Europe	1837	H	Hc	S
Senecio doronicum	Asteraceae	1	Europe	1977	H	Hc	S
Senecio glastifolius	Asteraceae	2	South Africa	1971	H	Hc	S
Senecio grandiflorus	Asteraceae	2	South Africa	1971	H	Hc	S
Senecio inaequidens	Asteraceae	147	South Africa	1836	V,W	Hc	S
Senecio minimus	Asteraceae	1	New Zealand	1913	W	Th	S
Senecio ovatus	Asteraceae	7	Europe	1886	H	Hc	S
Senecio sarracenicus	Asteraceae	106	Europe	1632	H	Hc	S
Senecio smithii	Asteraceae	37	South America	1920	H	Hc	S
Senecio squalidus	Asteraceae	1,399	Neonative	1794	H,Neo	Hc	S
Senecio squalidus × viscosus = S. × londinensis	Asteraceae	72	Neonative	1944	Neo	Hc	V
Senecio squalidus × vulgaris = S. × baxteri	Asteraceae	36	Neonative	1886	Neo	Th	–
Senecio sylvaticus × viscosus = S. × viscidulus	Asteraceae	5	Neonative	1947	Neo	Hc	–
Senecio vernalis	Asteraceae	15	Eastern Europe	1905	Gs	Th	–
Senecio viscosus	Asteraceae	1,614	Europe	1749	?	Th	S
*Serapias parviflora	Orchidaceae	1	Mediterranean	1989	?native	Ge	–
Setaria adhaerens	Poaceae	3	Old World tropics	1976	Bs,C,E,O,W	Th	–
Setaria faberi	Poaceae	5	East Asia	1975	Bs,G,O	Th	–
Setaria italica	Poaceae	57	China	1863	A,Bs,G	Th	–
Setaria parviflora	Poaceae	8	North America	1913	Bs,W	Hc	–
Setaria pumila	Poaceae	398	Widespread	1805	As,Bs,G,O,W	Th	–
Setaria verticillata	Poaceae	92	Widespread	1666	Bs,C,E,O,W	Th	–
Setaria viridis	Poaceae	388	Widespread	1666	As,Bs,E,G,O,W	Th	S
Sicyos angulatus	Cucurbitaceae	1	North America	1958	Bs,O	Th	–

Scientific name	Family	No. of hectads	Introduced from	First record	Vector	Life form	S/V
Sida spinosa	Malvaceae	4	Tropics	1876	Bs,O,W	Ch	–
Sidalcea candida	Malvaceae	5	North America	1991	H	Hc	–
Sidalcea malviflora	Malvaceae	47	North America	1940	H	Hc	–
Sigesbeckia orientalis	Asteraceae	1	Asia	1928	W	Th	–
Sigesbeckia serrata	Asteraceae	8	Central & South America	1873	G	Th	S
Silene alpestris	Caryophyllaceae	1	Europe	1974	H	Hc	S
Silene armeria	Caryophyllaceae	42	Europe	1718	H	Th	–
Silene baccifera	Caryophyllaceae	4	Europe	1570	?	Hc	S
Silene catholica	Caryophyllaceae	1	Italy	1857	H	Hc	S
Silene chalcedonica	Caryophyllaceae	25	Russia	1902	H	Hc	S
Silene coeli-rosa	Caryophyllaceae	12	Southwest Europe	1937	H	Th	–
Silene coronaria	Caryophyllaceae	507	Southeast Europe	1879	H	Hc	S
Silene dichotoma	Caryophyllaceae	1	Europe	1872	As,Bs,W	Th	–
Silene dioica × latifolia = S. × hampeana	Caryophyllaceae	1,170	Neonative	1853	Neo	Hc	S
Silene italica	Caryophyllaceae	6	Europe	1825	H	Hc	S
Silene multifida	Caryophyllaceae	1	Caucasus	1912	H	Hc	S
Silene muscipula	Caryophyllaceae	3	Europe	1905	Bs,G	Th	–
Silene pendula	Caryophyllaceae	2	Southern Europe	1883	H	Th	–
Silene schafta	Caryophyllaceae	2	Caucasus	1915	H	Hc	S
Silene vulgaris subsp. macrocarpa	Caryophyllaceae	4	Mediterranean	1913	?I	Hc	S
Sinacalia tangutica	Asteraceae	17	China	1903	H	Hc	S/V
Sinapis alba subsp. dissecta	Brassicaceae	4	Southern Europe	1859	Bs,W	Th	–
Sisymbrium altissimum	Brassicaceae	358	Europe	1858	Bs,Gs,W	Th	S
Sisymbrium erysimoides	Brassicaceae	3	Mediterranean	1916	W	Th	–
Sisymbrium irio	Brassicaceae	27	Europe	1650	G,W	Th	–
Sisymbrium loeselii	Brassicaceae	39	Europe	1883	Bs,G,W	Th	S
Sisymbrium orientale	Brassicaceae	697	Europe	1832	Bs,G,W	Th	S
Sisymbrium polyceratium	Brassicaceae	5	Southern Europe	c. 1785	B,G	Th	–
Sisymbrium strictissimum	Brassicaceae	7	Europe	1873	?	Hc	S
Sisymbrium volgense	Brassicaceae	4	Russia	1896	G	Hc	S/V
*Sisyrinchium bermudiana	Iridaceae	72	North America	1845	?native	Hc	S/V
Sisyrinchium californicum	Iridaceae	48	North America	1896	H	Hc	S/V
Sisyrinchium laxum	Iridaceae	3	South America	1970	H	Hc	–
Sisyrinchium montanum	Iridaceae	50	North America	1871	H	Hc	S/V
Sisyrinchium striatum	Iridaceae	115	South America	1922	H	Hc	–
Smyrnium perfoliatum	Apiaceae	23	Southern Europe	1932	H	Hc	S
Solanum carolinense	Solanaceae	1	North America	1977	O	Hc	–
Solanum chenopodioides	Solanaceae	14	South America	1935	?	Ch	S
Solanum diflorum	Solanaceae	6	South America	1997	Bs,H	Ch	–
Solanum laciniatum	Solanaceae	27	Australia	1920	H	Ch	S
Solanum lycopersicum	Solanaceae	724	Central & South America	1798	F	Th	–
Solanum nigrum subsp. schultesii	Solanaceae	38	Southern Europe	1946	?	Th	SW
Solanum nigrum × physalifolium = S. × procurrens	Solanaceae	3	Neonative	1985	Neo	Th	–

Scientific name	Family	No. of hectads	Introduced from	First record	Vector	Life form	S/V
Solanum physalifolium	Solanaceae	222	South America	1949	As,Bs,O,W	Th	S
Solanum rostratum	Solanaceae	70	North America	1886	Bs,G,O,W	Th	–
Solanum sarachoides	Solanaceae	47	South America	1897	?	Th	S
Solanum scabrum	Solanaceae	2	?Africa	1982	F	Th	–
Solanum sisymbriifolium	Solanaceae	7	South America	1922	As,Bs,O,W	Th	–
Solanum triflorum	Solanaceae	8	North America	1876	As,W	Th	S
Solanum tuberosum	Solanaceae	933	South America	1835	F	Ge	V
Solanum villosum	Solanaceae	11	Southern Europe	1877	Bs,O,W	Th	–
Soleirolia soleirolii	Urticaceae	1,018	Mediterranean	1917	H	Hc	S/V
Solidago canadensis subsp. altissima	Asteraceae	1,015	North America	1954	H	Hc	S/V
Solidago canadensis subsp. canadensis	Asteraceae		North America	1849	H	Hc	S/V
Solidago canadensis × virgaurea = S. × niederederi	Asteraceae	1	Neonative	1979	Neo	Hc	V
Solidago gigantea subsp. serotina	Asteraceae	726	North America	1916	H	Hc	S/V
Solidago graminifolia	Asteraceae	11	North America	1809	H	Hc	S/V
Solidago rugosa	Asteraceae	6	North America	1982	H	Hc	S/V
Soliva pterosperma	Asteraceae	1	South America	1960	S,V,W	Th	S
Sorbaria kirilowii	Rosaceae	17	China	1975	H	Ph	–
Sorbaria sorbifolia	Rosaceae	53	Asia	1971	H	Ph	S/V
Sorbaria tomentosa	Rosaceae	21	Himalayas	1970	H	Ph	S
Sorbus aucuparia × intermedia = S. × liljeforsii	Rosaceae	14	Neonative	1913	Neo	Ph	–
Sorbus aucuparia × scalaris = S. × proctoriana	Rosaceae	1	Neonative	2005	Neo	Ph	–
Sorbus croceocarpa	Rosaceae	97	Uncertain	1883	H	Ph	S
Sorbus decipiens	Rosaceae	4	Europe	1949	H	Ph	–
*Sorbus domestica	Rosaceae	23	Europe	1677	?native	Ph	S
Sorbus glabriuscula	Rosaceae	17	China	1981	H	Ph	S
Sorbus hybrida	Rosaceae	36	Scandinavia	1985	H	Ph	–
Sorbus intermedia	Rosaceae	914	Scandinavia	1908	H	Ph	S
Sorbus latifolia	Rosaceae	50	Europe	1866	H	Ph	S
Sorbus mougeotii	Rosaceae	8	Europe	2007	H	Ph	–
Sorghum bicolor	Poaceae	29	Africa	1890	Bs,F,W	Th	–
Sorghum halepense	Poaceae	58	North Africa	1924	Bs,O,W	Hc	–
Sparaxis grandiflora	Iridaceae	4	South Africa	1955	H	Ge	–
Spartina alterniflora	Poaceae	6	North America	1816	B	Hc	V
Spartina alterniflora × maritima = S. × townsendii	Poaceae	74	Neonative	1846	Neo	Hc	V
Spartina anglica	Poaceae	316	Neonative	1887	Neo	Hc	V
Spartina patens	Poaceae	1	North America	2005	H	Hc	V
Spartina pectinata	Poaceae	4	North America	1967	H	Hc	V
Spartium junceum	Fabaceae	133	Mediterranean	1905	H	Ph	S
Spergula morisonii	Caryophyllaceae	3	Europe	1943	I	Th	S
*Spergularia bocconei	Caryophyllaceae	18	Europe	1901	?native	Th	S
Spiraea alba	Rosaceae	42	North America	1876	H	Ph	S/V

Scientific name	Family	No. of hectads	Introduced from	First record	Vector	Life form	S/V
Spiraea alba × douglasii = S. × billardii	Rosaceae	190	Horticulture	1964	H	Ph	V
Spiraea alba × salicifolia = S. × rosalba	Rosaceae	135	Horticulture	1979	H	Ph	V
Spiraea canescens	Rosaceae	12	Himalayas	1919	H	Ph	–
Spiraea cantoniensis × trilobata = S. × vanhouttei	Rosaceae	47	Horticulture	1974	H	Ph	–
Spiraea chamaedryfolia subsp. ulmifolia	Rosaceae	7	Southeast Europe	1830	H	Ph	S
Spiraea douglasii subsp. douglasii	Rosaceae	73 (241)†	North America	1910	H	Ph	S/V
Spiraea douglasii subsp. menziesii	Rosaceae	5	North America	1957	H	Ph	S/V
Spiraea douglasii × salicifolia = S. × pseudosalicifolia	Rosaceae	480	Horticulture	1984	H	Ph	V
Spiraea japonica	Rosaceae	66	Japan	1909	H	Ph	–
Spiraea media	Rosaceae	3	Eastern Europe	1989	H	Ph	–
Spiraea multiflora × thunbergii = S. × arguta	Rosaceae	33	Horticulture	1967	H	Ph	–
Spiraea salicifolia	Rosaceae	214	Europe	1805	H	Ph	S/V
Spiraea tomentosa	Rosaceae	5	North America	1905	H	Ph	V
Stachys alpina	Lamiaceae	4	Europe	1897	?	Hc	S
Stachys annua	Lamiaceae	8	Europe	1830	G,O	Th	–
Stachys byzantina	Lamiaceae	283	Southwest Asia	1858	H	Hc	–
Stachys recta	Lamiaceae	4	Europe	1898	?	Hc	S
Staphylea pinnata	Staphyleaceae	33	Europe	1633	H	Ph	–
Sternbergia lutea	Amaryllidaceae	1	Mediterranean	1912	H	Ge	S/V
Stranvaesia davidiana	Rosaceae	63	China	1966	H	Ph	–
*Stratiotes aloides	Hydrocharitaceae	275	Europe	1626	?native	Hy	V
Sutera cordata	Scrophulariaceae	25	South Africa	1997	H	Ch	–
Symphoricarpos albus	Caprifoliaceae	2,723	North America	1863	A,H	Ph	V
Symphoricarpos albus × microphyllus × orbicularis = S. × doorenbosii	Caprifoliaceae	2	Horticulture	1992	H	Ph	V
Symphoricarpos microphyllus × orbicularis = S. × chenaultii	Caprifoliaceae	296	Horticulture	1974	H	Ph	V
Symphoricarpos orbiculatus	Caprifoliaceae	20	North America	1974	H	Ph	V
Symphytum asperum	Boraginaceae	46	Southwest Asia	1862	H	Hc	S
Symphytum asperum × caucasicum	Boraginaceae	1	?Southwest Asia	1994	H	Hc	–
Symphytum asperum × grandiflorum × officinale = S. × hidcotense	Boraginaceae	207	Horticulture	1979	H	Hc	S/V
Symphytum asperum × officinale = S. × uplandicum	Boraginaceae	2,242	Uncertain	1861	H	Hc	S
Symphytum asperum × officinale × tuberosum	Boraginaceae	16	Neonative	1994	Neo	Hc	V
Symphytum asperum × orientale = S. norvicense	Boraginaceae	4	?West Asia	1957	H	Hc	S
Symphytum bulbosum	Boraginaceae	24	Southern Europe	1893	H	Hc	S/V
Symphytum caucasicum	Boraginaceae	37	Caucasus	1879	H	Hc	S
Symphytum grandiflorum	Boraginaceae	381	Caucasus	1898	H	Hc	S/V

Scientific name	Family	No. of hectads	Introduced from	First record	Vector	Life form	S/V
Symphytum orientale	Boraginaceae	502	West Asia	1849	H	Hc	S
Symphytum tauricum	Boraginaceae	2	Southeast Europe	1854	H	Hc	–
*Symphytum tuberosum	Boraginaceae	609	Europe	1777	?native	Hc	S/V
Syringa vulgaris	Oleaceae	1,542	Southeast Europe	1879	H	Ph	V
Tagetes erecta	Asteraceae	12	Mexico	1969	H	Th	–
Tagetes minuta	Asteraceae	2	South America	1922	Bs,Co,G,W	Th	–
Tagetes patula	Asteraceae	56	Mexico	1933	H	Th	–
Tamarix africana	Tamaricaceae	6	Mediterranean	1963	H	Ph	–
Tamarix gallica	Tamaricaceae	245	Southwest Europe	1796	H	Ph	–
Tanacetum balsamita	Asteraceae	1	Caucasus	1801	F	Ph	–
Tanacetum macrophyllum	Asteraceae	15	Southeast Europe	1883	H	Hc	S/V
*Tanacetum vulgare	Asteraceae	1,910	Europe	1597	?native	Hc	S/V
Telekia speciosa	Asteraceae	52	Europe	1914	H	Hc	S
Tellima grandiflora	Saxifragaceae	498	North America	1866	H	Hc	S
Tetragonia tetragonioides	Aizoaceae	14	New Zealand	1911	F	Th	–
Tetragonolobus maritimus	Fabaceae	18	Europe	1875	?A,?Gs	Hc	S
Teucrium botrys	Lamiaceae	7	Europe	1844	?	Th	S
Thalictrum aquilegiifolium	Ranunculaceae	41	Europe	1905	H	Hc	–
Thalictrum delavayi	Ranunculaceae	1	China	1978	H	Hc	–
Thermopsis montana	Fabaceae	6	North America	1925	H	Hc	V
Thlaspi alliaceum	Brassicaceae	14	Europe	1923	?As	Th	S
Thymus vulgaris	Lamiaceae	60	Mediterranean	1928	F,H	Ch	S
Tilia × euchlora	Malvaceae	24	Caucasus	1982	H	Ph	–
Tilia tomentosa	Malvaceae	34	Southeast Europe	1938	H	Ph	–
Tolmiea menziesii	Saxifragaceae	317	North America	1916	H	Hc	V
Tordylium maximum	Apiaceae	3	Southern Europe	1670	?	Th	S
Trachelium caeruleum	Campanulaceae	9	Mediterranean	1892	H	Hc	–
Trachyspermum ammi	Apiaceae	5	Mediterranean	1921	F	Th	–
Trachystemon orientalis	Boraginaceae	157	Caucasus	1844	H	Hc	S/V
Tradescantia fluminensis	Commelinaceae	7	South America	1982	H	Ch	–
Tradescantia virginiana	Commelinaceae	26	North America	1965	H	Hc	–
Tragopogon porrifolius	Asteraceae	324	Mediterranean	1597	F	Hc	S
Tragopogon porrifolius × pratensis = T. × mirabilis	Asteraceae	21	Neonative	1909	Neo	Hc	S
Tragopogon pratensis subsp. pratensis	Asteraceae	85	Europe	1836	?	Hc	S
Tragus australianus	Poaceae	1	Australia	1953	W	Th	–
Tragus berteronianus	Poaceae	1	Africa	1953	W	Th	–
Trifolium alexandrinum	Fabaceae	7	Mediterranean	1897	Bs,Gs	Th	–
Trifolium angustifolium	Fabaceae	10	Southern Europe	1876	Bs,Tn,W	Th	–
Trifolium aureum	Fabaceae	23	Europe	1789	As,Bs,W	Th	S
Trifolium cernuum	Fabaceae	2	Southwest Europe	1917	S,W	Th	–
Trifolium echinatum	Fabaceae	2	Southeast Europe	1878	Bs,W	Th	–
Trifolium hirtum	Fabaceae	4	Southern Europe	1906	G,Tn,W	Th	–
Trifolium hybridum subsp. elegans	Fabaceae	13	Europe	1870	Bs,G,Tn,W	Hc	S

Scientific name	Family	No. of hectads	Introduced from	First record	Vector	Life form	S/V
Trifolium hybridum subsp. hybridum	Fabaceae	1,683	Europe	1762	A,Bs,G,Tn,W	Hc	S
Trifolium incarnatum subsp. incarnatum	Fabaceae	80	Southern Europe	1798	A,Tn	Th	S
Trifolium lappaceum	Fabaceae	2	Southern Europe	1862	Bs,G,Tn,W	Th	–
Trifolium pannonicum	Fabaceae	12	Southeast Europe	1876	A,B,Gs	Hc	–
Trifolium resupinatum	Fabaceae	36	Southern Europe	1830	Bs,E,G,Tn,W	Th	–
Trifolium stellatum	Fabaceae	3	Mediterranean	1804	?	Th	S
Trifolium tomentosum	Fabaceae	7	Mediterranean	1880	E,Tn,W	Th	–
Trigonella caerulea	Fabaceae	3	Mediterranean	1867	Bs,G,W	Th	–
Trigonella corniculata	Fabaceae	2	Mediterranean	1874	Bs	Th	–
Trigonella hamosa	Fabaceae	2	Mediterranean	1905	G,W	Th	–
Tripleurospermum inodorum × maritimum	Asteraceae	21	Neonative	1963	Neo	Hc	S
Trisetum flavescens subsp. purpurascens	Poaceae	8	Europe	1994	Gs	Hc	S
Tristagma uniflorum	Amaryllidaceae	139	South America	1921	H	Ge	V
Triteleia laxa	Amaryllidaceae	1	North America	1999	H	Ge	V
× Triticosecale rimpaui	Poaceae	42	Agriculture	1996	F	Th	–
Triticum durum	Poaceae	2	Agriculture	1853	F,G,H	Th	–
Triticum turgidum	Poaceae	32	Agriculture	1887	F,G	Th	–
Tropaeolum majus	Tropaeolaceae	439	Peru	1904	H	Th	–
Tropaeolum peregrinum	Tropaeolaceae	5	South America	1951	H	Hc	–
Tropaeolum speciosum	Tropaeolaceae	21	Chile	1950	H	Hc	V
Tulipa gesneriana	Liliaceae	335	Horticulture	1955	H	Ge	–
Tulipa saxatilis	Liliaceae	5	Crete	1976	H	Ge	S/V
Tulipa sylvestris	Liliaceae	76	Southern Europe	1790	H	Ge	S/V
Turgenia latifolia	Apiaceae	4	Southern Europe	1660	Bs,G	Th	–
Ugni molinae	Myrtaceae	1	South America	1995	H	Ph	S
Ulmus laevis	Ulmaceae	9	Europe	1943	H	Ph	V
Umbilicus oppositifolius	Crassulaceae	10	Caucasus	1933	H	Hc	S
Urochloa panicoides	Poaceae	3	Africa	1955	Bs,W	Th	–
Urtica membranacea	Urticaceae	5	Mediterranean	2006	?	Th	–
Vaccaria hispanica	Caryophyllaceae	22	Europe	1832	H	Th	–
Vaccinium corymbosum	Ericaceae	7	North America	1980	F	Hc	S
Vaccinium macrocarpon	Ericaceae	6	North America	1859	F	Ch	S
Valeriana phu	Valerianaceae	2	?Turkey	1907	H	Hc	S
Valeriana pyrenaica	Valerianaceae	135	Southwest Europe	1782	H	Hc	S
*Valerianella eriocarpa	Valerianaceae	21	Southern Europe	1837	?	Th	–
Vallisneria spiralis	Hydrocharitaceae	2	Widespread	1868	H	Hy	V
Verbascum blattaria	Scrophulariaceae	188	Europe	1629	H	Hc	–
Verbascum bombyciferum	Scrophulariaceae	10	Turkey	1948	H	Hc	–
Verbascum bombyciferum × phlomoides	Scrophulariaceae	1	Neonative	1976	Neo	Hc	–

Scientific name	Family	No. of hectads	Introduced from	First record	Vector	Life form	S/V
Verbascum chaixii subsp. austriacum	Scrophularicaeae	1	Europe	1907	G,H	Hc	S
Verbascum chaixii subsp. chaixii	Scrophularicaeae		Europe	1904	G,H	Hc	S
Verbascum densiflorum	Scrophularicaeae	105	Europe	1921	H	Hc	–
Verbascum nigrum × pulverulentum = V. × mixtum	Scrophularicaeae	8	Neonative	1800	Neo	Hc	–
Verbascum nigrum × pyramidatum	Scrophularicaeae	2	Neonative	1970	Neo	Hc	–
Verbascum nigrum × speciosum = V. × angulosum	Scrophularicaeae	1	Neonative	2001	Neo	Hc	–
Verbascum phlomoides	Scrophularicaeae	212	Europe	1836	H	Hc	–
Verbascum phlomoides × pulverulentum = V. × murbeckii	Scrophularicaeae	1	Neonative	2000	Neo	Hc	–
Verbascum phlomoides × thapsus = V. × kerneri	Scrophularicaeae	3	Neonative	1950	Neo	Hc	–
Verbascum phoeniceum	Scrophularicaeae	22	Southeast Europe	1880	Bs,H	Hc	–
*Verbascum pulverulentum	Scrophularicaeae	77	Europe	1670	?native	Hc	S
Verbascum pulverulentum × thapsus = V. godronii	Scrophularicaeae	7	Neonative	1933	Neo	Hc	–
Verbascum pulverulentum × virgatum	Scrophularicaeae	1	Neonative	2001	Neo	Hc	–
Verbascum pyramidatum	Scrophularicaeae	12	Caucasus	1922	H	Hc	–
Verbascum pyramidatum × thapsus	Scrophularicaeae	1	Neonative	1976	Neo	Hc	–
Verbascum speciosum	Scrophularicaeae	33	Southeast Europe	1909	H	Hc	S
Verbascum speciosum × thapsus = V. × duernsteinense	Scrophularicaeae	3	Neonative	1976	Neo	Hc	–
Verbascum thapsus × virgatum = V. × lemaitrei	Scrophularicaeae	4	Neonative	1892	Neo	Hc	–
Verbascum virgatum	Scrophularicaeae	225	Europe	1787	H	Hc	S
Verbena bonariensis	Verbenaceae	163	South America	1949	H	Hc	S
Verbena × hybrida	Verbenaceae	9	Horticulture	1961	H	Th	–
Verbena rigida	Verbenaceae	10	South America	1937	H	Hc	–
Veronica acinifolia	Veronicaceae	6	Southern Europe	1906	I	Th	S
Veronica austriaca subsp. teucrium	Veronicaceae	21	Europe	1903	H	Hc	S
Veronica barkeri	Veronicaceae	10	Chatham Islands	1995	H	Ph	–
Veronica brachysiphon	Veronicaceae	22	New Zealand	1996	H	Ph	–
Veronica crista-galli	Veronicaceae	49	Caucasus	1888	?	Th	S
Veronica dieffenbachii	Veronicaceae	14	Chatham Islands	1991	H	Ph	–
Veronica elliptica × salicifolia = H. × lewisii	Veronicaceae	27	New Zealand	1939	H	Ph	S
Veronica elliptica × speciosa = H. × franciscana	Veronicaceae	270	Horticulture	1904	H	Ph	S
Veronica filiformis	Veronicaceae	2,359	Caucasus	1838	H	Hc	V
Veronica longifolia	Veronicaceae	87	Europe	1910	H	Hc	S
Veronica longifolia × spicata	Veronicaceae	11	Horticulture	1991	H	Hc	S
Veronica paniculata	Veronicaceae	2	Europe	1962	H	Hc	–
Veronica peregrina	Veronicaceae	159	North & South America	1836	I	Th	–
Veronica persica	Veronicaceae	2,707	Southwest Asia	1825	As,Bs	TH	S

Scientific name	Family	No. of hectads	Introduced from	First record	Vector	Life form	S/V
*Veronica polita	Veronicaceae	1,011	Europe	1747	?native	Th	S
Veronica praecox	Veronicaceae	5	Europe	1933	?	Th	S
Veronica repens	Veronicaceae	2	Corsica	1874	?	Hc	–
Veronica salicifolia	Veronicaceae	122	New Zealand	1913	H	Ph	Ph
Viburnum lantana × rhytidophyllum = V. × rhytidophylloides	Caprifoliaceae	5	Horticulture	1977	H	Ph	–
Viburnum rhytidophyllum	Caprifoliaceae	66	China	1968	H	Ph	–
Viburnum sargentiae	Caprifoliaceae	2	Northeast Asia	2003	H	Ph	–
Viburnum tinus	Caprifoliaceae	286	Southern Europe	1916	H	Ph	S/V
Viburnum trilobum	Caprifoliaceae	2	North America	2007	H	Ph	–
Viburnum veitchii	Caprifoliaceae	1	China	2003	H	Ph	–
Vicia benghalensis	Fabaceae	6	Mediterranean	1868	As,G,W	Th	–
*Vicia bithynica	Fabaceae	64	Europe	1778	?native	Hc	S
Vicia hybrida	Fabaceae	1	Southern Europe	1670	G,Tn	Th	–
Vicia monantha	Fabaceae	1	Mediterranean	1908	G	Hc	–
Vicia narbonensis	Fabaceae	5	Southern Europe	1890	Bs,G,Tn	Th	–
Vicia pannonica	Fabaceae	8	Europe	1866	Gs	Th	S
Vicia tenuifolia	Fabaceae	35	Europe	1859	G	Hc	S
Vicia villosa	Fabaceae	97	Europe	1857	Bs,G,H, Tn,W	Th	S
Vigna radiata	Fabaceae	2	Asia	1971	F	Th	–
Vinca difformis	Apocynaceae	23	Southwest Europe	1984	H	Hc	V
Vinca major	Apocynaceae	1,631	Mediterranean	1650	H	Hc	S/V
Viola altaica × lutea × tricolor = V. × wittrockiana	Violaceae	630	Horticulture	1927	H	Th	S
Viola arvensis × lutea	Violaceae	2	Neonative	1865	Neo	Hc	V
Viola arvensis × tricolor = V. × contempta	Violaceae	105	Neonative	1884	Neo	Th	S
Viola cornuta	Violaceae	49	Southwest Europe	1878	H	Hc	S/V
*Vulpia unilateralis	Poaceae	21	Europe	1903	?native	Th	S
Watsonia borbonica subsp. ardernei	Iridaceae	2	South Africa	1971	H	Ge	S/V
Weigela florida	Caprifoliaceae	64	China	1928	H	Ph	–
Xanthium ambrosioides	Asteraceae	2	Argentina	1910	W	Th	–
Xanthium spinosum	Asteraceae	11	South America	1846	Bs,G,W	Th	–
Xanthium strumarium	Asteraceae	22	Widespread	1562	G,O,W	Th	–
Xerochrysum bracteatum	Asteraceae	13	Australia	1928	H	Th	–
Yucca gloriosa var. recurvifolia	Asparagaceae	54	North America	1983	H	Ph	–
Yushania anceps	Poaceae	23	India	1961	H	Ph	V
Zantedeschia aethiopica	Araceae	56	South Africa	1952	H	Hc	V
Zea mays	Poaceae	131	Central America	1876	A,F	Th	–
Zizania latifolia	Poaceae	4	East Asia	1916	H	Hy	V

TOTAL 1,809 plus 55 uncertain alien/native (marked with *)

† Subspecies often not separated by recorders; the figure in parentheses is the hectad total for the undivided species, excluding the records made for any of the individual subspecies.

List of Angiosperm (Flowering Plant) Archaeophytes Recorded in the British Isles

Appendix 2 is intended to provide a definitive list of the angiosperm archaeophytes of the British Isles. This preamble explains the criteria and conventions used in compiling the list.

CONVENTIONS AND ABBREVIATIONS

- Status (denizen, colonist, cultivated) is colour-coded in the appendix (defined in Chapter 1):

Denizen	Colonist	Cultivated

There are 87 denizens, 65 colonists and 45 cultivated taxa designated.
- Six taxa that are invasive at least in some places, as defined in Chapter 1, are given in boldface.
- Number of hectads (of which there are 3,859 in total) recorded from January 1987 to May 2012 is taken from the BSBI Distribution Database (http://bsbidb.org.uk).
- Usage: C (culinary); M (medicinal); O (ornamental); W (weedy).
- Life forms are given according to the Raunkiaer (1934) system, as defined in Table 4, with abbreviations as follows: Ch = chamaephyte; Ge = geophyte; Hc = hemicryptophyte; Hy = hydrophyte; Ph = phanerophyte; Th = therophyte.

- S/V status indicates whether the taxon reproduces in this country from seed
  (S) and/or from vegetative propagation (V). A dash in this column indicates
  that the taxon does not reproduce with us, and is therefore a casual or
  survivor.

Scientific name	English name	No. of hectads	Life form	S/V	Usage
Adonis annua	Pheasant's-eye	40	Th	–	W
Aegopodium podagraria	Ground-elder	3,187	Hc	S/V	C, M
Aethusa cynapium subsp. agrestis	Fool's Parsley	61	Th	S	W
Agrostemma githago	Corncockle	396	Th	–	W
Agrostis gigantea	Black Bent	1,536	Hc	S/V	W
Alcea rosea	Hollyhock	466	Hc	S	?M, O
Allium ampeloprasum	Wild Leek	110	Ge	S/V	C
Allium cepa	Onion	65	Ge	S/V	C
Allium porrum	Leek	31	Ge	–	C
Allium sativum	Garlic	12	Ge	V	C, M
Alopecurus myosuroides	Black-grass	1,012	Th	S	W
Anagallis arvensis subsp. foemina	Blue Pimpernel	71	Th	S	M, W
Anchusa arvensis	Bugloss	1,216	Th	S	W
Anethum graveolens	Dill	30	Th	–	C, M
Anisantha sterilis	Barren Brome	2,023	Th	S	W
Anthemis arvensis	Corn Chamomile	305	Th	S	W
Anthemis cotula	Stinking Chamomile	642	Th	S	W
Anthriscus cerefolium	Garden Chervil	18	Th	S	C, M
Apera spica-venti	Loose Silky-bent	175	Th	S	W
Arctium lappa	Greater Burdock	912	Hc	S	C, M
Armoracia rusticana	Horse-radish	1,502	Hc	V	C, M
Arnoseris minima	Lamb's Succory	0	Th	–	W
Artemisia absinthium	Wormwood	640	Ch	S	M
Artemisia vulgaris	Mugwort	2,246	Hc	S	C, M
Asparagus officinalis	Garden Asparagus	544	Ge	S/V	C, M
Atriplex hortensis	Garden Orache	114	Th	–	C, M
Avena fatua	Wild-oat	1,641	Th	S	W
Avena sativa	Oat	1,130	Th	S	C
Avena strigosa	Bristle Oat	53	Th	S	C
Ballota nigra	Black Horehound	1,299	Hc	S	M, W
Borago officinalis	Borage	880	Th	S	C, M
Brassica rapa subsp. campestris	Wild Turnip	263	Hc	S	C, W
Brassica rapa subsp. rapa	Turnip	50	Th	S	C
Briza minor	Lesser Quaking-grass	69	Th	S	W
Bromus secalinus	Rye Brome	313	Th	S	C, W
Bupleurum rotundifolium	Thorow-wax	24	Th	–	W
Calendula officinalis	Pot Marigold	837	Th	S	M, O
Camelina sativa	Gold-of-pleasure	84	Th	–	W, fibre, oil

Scientific name	English name	No. of hectads	Life form	S/V	Usage
Campanula rapunculus	Rampion Bellflower	19	Hc	S	C, O
Cannabis sativa	Hemp	228	Th	–	M, fibre, oil
Capsella bursa-pastoris	Shepherd's-purse	3,368	Th	S	C, M, W
Carum carvi	Caraway	60	Hc	S	C, M
Castanea sativa	Sweet Chestnut	1,752	Ph	S	C, M, wood
Caucalis platycarpos	Small Bur-parsley	0	Th	–	W
Centaurea calcitrapa	Red Star-thistle	16	Hc	S	?W
Centaurea cyanus	Cornflower	668	Th	S	M, W
Chaenorhinum minus	Small Toadflax	1,220	Th	S	W
Chelidonium majus	Greater Celandine	1,649	Hc	S	M
Chenopodium bonus-henricus	Good-King-Henry	774	Hc	S	C
Chenopodium ficifolium	Fig-leaved Goosefoot	824	Th	S	W
Chenopodium glaucum	Oak-leaved Goosefoot	70	Th	S	W
Chenopodium hybridum	Maple-leaved Goosefoot	172	Th	S	W
Chenopodium murale	Nettle-leaved Goosefoot	159	Th	S	W
Chenopodium polyspermum	Many-seeded Goosefoot	957	Th	S	W
Chenopodium urbicum	Upright Goosefoot	20	Th	S	W
Chenopodium vulvaria	Stinking Goosefoot	22	Th	S	W
Cicer arietinum	Chick Pea	9	Th	–	C
Cichorium intybus	Chicory	891	Hc	S	C, M
Conium maculatum	Hemlock	2,066	Hc	S	M, W
Coriandrum sativum	Coriander	101	Th	–	C, M
Crepis foetida	Stinking Hawk's-beard	3	Th	S	W
Cuminum cyminum	Cumin	2	Th	–	C
Descurainia sophia	Flixweed	370	Th	S	W
Diplotaxis tenuifolia	Perennial Wall-rocket	447	Ch	S	C, W
Echium plantagineum	Purple Viper's-bugloss	84	Hc	S	C, M
Erodium moschatum	Musk Stork's-bill	284	Th	S	O, ?W, scent
Eruca vesicaria subsp. sativa	Garden Rocket	86	Th	S	C
Eryngium campestre	Field Eryngo	14	Hc	S	M
Erysimum cheiranthoides	Treacle-mustard	697	Th	S	W
Erysimum cheiri	Wallflower	903	Ch	S	M, O, scent
Euphorbia exigua	Dwarf Spurge	731	Th	S	W
Euphorbia helioscopia	Sun Spurge	2,420	Th	S	W
Euphorbia lathyris	Caper Spurge	997	Ch	S	M, O
Euphorbia peplus	Petty Spurge	2,237	Th	S	W
Euphorbia platyphyllos	Broad-leaved Spurge	194	Th	S	W
Fallopia convolvulus	Black-bindweed	2,154	Th	S	W
Filago gallica	Small Cudweed	8	Th	S	W
Filago pyramidata	Broad-leaved Cudweed	19	Th	S	W
Foeniculum vulgare	Fennel	1,031	Hc	S	C, M, O
Fumaria densiflora	Dense-flowered Fumitory	204	Th	S	W
Fumaria officinalis subsp. officinalis	Common Fumitory	647 (2,020)*	Th	S	M, W
Fumaria officinalis subsp. wirtgenii	Common Fumitory	299	Th	S	M, W

Scientific name	English name	No. of hectads	Life form	S/V	Usage
Fumaria parviflora	Fine-leaved Fumitory	70	Th	S	W
Fumaria vaillantii	Few-flowered Fumitory	69	Th	S	W
Galeopsis angustifolia	Red Hemp-nettle	125	Th	S	W
Galeopsis segetum	Downy Hemp-nettle	0	Th	–	M, W
Galeopsis speciosa	Large-flowered Hemp-nettle	516	Th	S	W
Galium tricornutum	Corn Cleavers	15	Th	S	W
Geranium dissectum	Cut-leaved Crane's-bill	2,829	Th	S	W
Glebionis segetum	Corn Marigold	1,344	Th	S	W
Helminthotheca echioides	Bristly Oxtongue	1,238	Th	S	W
Hordeum distichon	Two-rowed Barley	978	Th	S	C
Hordeum murinum subsp. murinum	Wall Barley	1,459	Th	S	W
Hordeum vulgare	Six-rowed Barley	311	Th	S	C
Hyoscyamus niger	Henbane	363	Hc	S	M
Imperatoria ostruthium	Masterwort	108	Hc	S	M
Inula helenium	Elecampane	392	Hc	S	M, O
Isatis tinctoria	Woad	49	Hc	S	Dye
Juglans regia	Walnut	883	Ph	S	C, M, wood
Kickxia elatine	Sharp-leaved Fluellen	796	Th	S	W
Kickxia spuria	Round-leaved Fluellen	516	Th	S	W
Lactuca sativa	Garden Lettuce	31	Th	–	C
Lactuca serriola	Prickly Lettuce	1,133	Hc	S	M, W
Lamium album	White Dead-nettle	1,911	Hc	S/V	C, M, W
Lamium amplexicaule	Henbit Dead-nettle	1,205	Th	S	W
Lamium confertum	Northern Dead-nettle	257	Th	S	W
Lamium hybridum	Cut-leaved Dead-nettle	1,272	Th	S	W
Lamium purpureum	Red Dead-nettle	2,874	Th	S	W
Legousia hybrida	Venus's Looking-glass	335	Th	S	W
Lens culinaris	Lentil	10	Th	–	C
Lepidium campestre	Field Pepperwort	553	Th	S	C, W
Lepidium coronopus	Swine-cress	1,327	Th	S	W
Lepidium ruderale	Narrow-leaved Pepperwort	353	Th	S	W
Lepidium sativum	Garden Cress	137	Th	–	C
Linaria repens	Pale Toadflax	554	Hc	S/V	W
Linum usitatissimum	Flax	959	Th	S	M, fibre, oil
Lithospermum arvense	Field Gromwell	240	Th	S	W
Lolium temulentum	Darnel	20	Th	S	W
Lythrum hyssopifolia	Grass-poly	24	Th	S	W
Malus pumila	Apple	2,060	Ph	S	C, M, wood
Malva neglecta	Dwarf Mallow	990	Th	S	C, M, W
Malva sylvestris	Common Mallow	1,972	Hc	S	C, M, O, W
Matricaria chamomilla	Scented Mayweed	1,552	Th	S	M, W
Melilotus altissimus	Tall Melilot	1,023	Hc	S	M, W, fodder
Mentha spicata	Spear Mint	1,414	Hc	S/V	C, M

Scientific name	English name	No. of hectads	Life form	S/V	Usage
Mercurialis annua	Annual Mercury	763	Th	S	M, W
Mespilus germanica	Medlar	80	Ph	S	C
Misopates orontium	Weasel's-snout	305	Th	S	W
Myosotis arvensis	Field Forget-me-not	3,070	Th	S	W
Nepeta cataria	Cat-mint	221	Hc	S	M
Onopordun acanthium	Cotton Thistle	612	Hc	S	C, M, O, W
Papaver argemone	Prickly Poppy	423	Th	S	W
Papaver dubium	Long-headed Poppy	1,944	Th	S	W
Papaver hybridum	Rough Poppy	198	Th	S	W
Papaver lecoqii	Yellow-juiced Poppy	496	Th	S	W
Papaver rhoeas	Common Poppy	1,906	Th	S	M, O, W
Papaver somniferum subsp. somniferum	Opium Poppy	1,810	Th	S	C, M, O, W, oil
Petroselinum crispum	Garden Parsley	245	Hc	S	C, M
Pisum sativum	Garden Pea	157	Th	–	C
Polygonum arenastrum	Equal-leaved Knotgrass	2,445	Th	S	M, W
Polygonum rurivagum	Cornfield Knotgrass	257	Th	S	W
Portulaca oleracea	Common Purslane	44	Th	S	C
Prunus cerasus	Dwarf Cherry	599	Ph	S	C
Prunus domestica subsp. domestica	Plum	451 (1,859)*	Ph	S	C
Prunus domestica subsp. insititia	Damson, Bullace	447	Ph	S	C
Prunus domestica subsp. italica	Greengage	55	Ph	S	C
Pyrus communis	Pear	687	Ph	S	C, wood
Pyrus pyraster	Wild Pear	132	Ph	S	C
Ranunculus arvensis	Corn Buttercup	216	Th	S	W
Raphanus raphanistrum subsp. raphanistrum	Wild Radish	1,238	Th	S	C, W
Raphanus sativus	Garden Radish	167	Th	–	C
Reseda luteola	Weld	1,992	Hc	S	Dye
Roemeria hybrida	Violet Horned-poppy	0	Th	–	W
Rumex alpinus	Monk's-rhubarb	119	Hc	S/V	M
Salix alba	White Willow	1,991	Ph	S	M, O, wood
Salix alba × S. euxina = S. × fragilis	Crack Willow	2,338	Ph	V	M, O, wood
Salix triandra	Almond Willow	671	Ph	S	O, withy
Salix triandra × S. viminalis = **S. × mollissima**	Sharp-stipuled Willow	120	Ph	–	Biomass, withy
Salix viminalis	Common Osier	2,685	Ph	S	Withy
Sambucus ebulus	Dwarf Elder	366	Hc	S/V	M, dye
Saponaria officinalis	Soapwort	1,219	Hc	S/V	M, O, soap
Scandix pecten-veneris	Shepherd's-needle	224	Th	S	W
Secale cereale	Rye	139	Th	–	C
Sedum album	White Stonecrop	2,025	Ch	S/V	?M, O
Silene gallica	Small-flowered Catchfly	134	Th	S	W

Scientific name	English name	No. of hectads	Life form	S/V	Usage
Silene latifolia subsp. alba	White Campion	1,901	Hc	S	W
Silene noctiflora	Night-flowering Catchfly	300	Th	S	W
Silybum marianum	Milk Thistle	326	Hc	S	M, O, W
Sinapis alba subsp. alba	White Mustard	799	Th	S	C, W
Sinapis arvensis	Charlock	2,627	Th	S	C, W
Sisymbrium officinale	Hedge Mustard	2,676	Th	S	C, M, W
Smyrnium olusatrum	Alexanders	1,110	Hc	S	C
Spinacia oleracea	Spinach	31	Th	–	C
Stachys arvensis	Field Woundwort	1,074	Th	S	W
Tanacetum parthenium	Feverfew	2,343	Ch	S	M, O
Thlaspi arvense	Field Penny-cress	1,548	Th	S	C, W
Torilis arvensis	Spreading Hedge-parsley	119	Th	S	W
Trigonella foenum-graecum	Fenugreek	10	Th	–	C, M
Tripleurospermum inodorum	Scentless Mayweed	2,601	Th	S	W
Triticum aestivum	Bread Wheat	1,244	Th	S	C
Triticum species (T. dicoccon, T. compactum, T. monococcum, T. spelta)	Primitive wheats	3	Th	–	C
Urtica urens	Small Nettle	1,742	Th	S	C, M, W
Valerianella carinata	Keeled-fruited Cornsalad	729	Th	S	?C, W
Valerianella dentata	Narrow-fruited Cornsalad	231	Th	S	W
Valerianella rimosa	Broad-fruited Cornsalad	25	Th	S	W
Verbena officinalis	Vervain	726	Hc	S	M, W
Veronica agrestis	Green Field-speedwell	1,383	Th	S	W
Veronica hederifolia subsp. lucorum	Ivy-leaved Speedwell	1,383	Th	S	W
Veronica hederifolia subsp. hederifolia		1,408 (1,924)*	Th	S	W
Veronica triphyllos	Fingered Speedwell	4	Th	S	W
Vicia faba	Broad Bean	577	Th	–	C, fodder
Vicia sativa subsp. sativa	Common Vetch	855†	Th	S	Fodder
Vicia sativa subsp. segetalis	Common Vetch	1,630	Th	S	Fodder, W
Vinca minor	Lesser Periwinkle	1,358	Ch	S/V	M, O
Viola arvensis	Field Pansy	2,116	Th	S	?M, W
Vitis vinifera	Grape-vine	144	Ph	S	C
Vulpia myuros	Rat's-tail Fescue	1,129	Th	S	W

* Subspecies often not separated by recorders; the figure in parentheses is the hectad total for the undivided species, excluding the records made for any of the individual subspecies

† The figure for Vicia sativa subsp. sativa is likely to represent gross over-recording in error for subsp. segetalis

References

Abbott, R. J., Ashton, P. A. & Forbes, D. G. (1992). Introgressive origin of the radiate groundsel *Senecio vulgaris* L. var. *hibernicus* Syme: Aat-3 evidence. *Heredity* **68**, 426–35.

Abbott, R. J., Brennan, A. C., James, J. K., Forbes, D. G., Hegarty, M. J. & Hiscock, S. J. (2009). Recent hybrid origin and invasion of the British Isles by a self-incompatible species, Oxford ragwort (*Senecio squalidus* L., Asteraceae). *Biological Invasions* **11**, 1145–58.

Aikio, S., Duncan, R. P. & Hulme, P. E. (2010). Lag-phases in alien plant invasions: separating the facts from the artefacts. *Oikos* **119**, 370 8.

Allen, D. E. (1951). The flora of the Liverpool bombed sites. *Merseyside Naturalists' Association. Bird Report and Natural History Notes from the Liverpool Area* **1951**, 25–29.

Armstrong, C., Osborne, B., Kelly, J. & Maguire, C. M. (2009). Giant Rhubarb (*Gunnera tinctoria*) Invasive Species Action Plan. Prepared for NIEA and NPWS as part of Invasive Species Ireland. Accessible online at http://invasivespeciesireland.com/wp-content/uploads/2011/01/Gunnera–tinctoria–ISAP.pdf

Bailey, C. (1884). On the structure, the occurrence in Lancashire, and the source of origin of *Naias graminea*, Del., var. *Delilei*, Magnus. *Journal of Botany* **22**, 305–33.

Bailey, J. P. (2001). *Fallopia* × *conollyana* the railway yard knotweed. *Watsonia* **23**, 539–41.

Bailey, J. P. & Conolly, A. P. (2000). Prize-winners to pariahs – a history of Japanese Knotweed s. l. (Polygonaceae) in the British Isles. *Watsonia* **23**, 93–110.

Bailey, J. P. & Stace, C. A. (1992). Chromosome number, morphology, pairing, and DNA values of species and hybrids in the genus Fallopia (Polygonaceae). *Plant Systematics and Evolution* **180**, 29–52.

Baker, A. M., Thompson, J. D. & Barrett, S. C. H. (2000). Evolution and maintenance of stigma-height dimorphism in *Narcissus*, 1. Floral variation and style-morph ratios. *Heredity* **84**, 502–13.

Baker, H. G. (1974). The evolution of weeds. *Annual Review of Ecology and Systematics* **5**, 1–24.

Baker, J. G. (1863). *North Yorkshire: Studies of its Botany, Geology, Climate and Physical Geography*. Longman, Green, Longman, Roberts, & Green, London.

Bakker, P. & Boeve, E. (1985). *Stinzenplanten*. Zutphen, 's-Graveland.

Ballantyne, G. H. (1971). Ballast aliens in south Fife, 1820–1919. *Transactions of the Botanical Society of Edinburgh* **41**, 125–37.

Bangerter, E. B. (1967). A survey of *Calystegia* in the London area. *London Naturalist* **46**, 15–23.

Bangerter, E. B. & Kent, D. H. (1957). *Veronica filiformis* Sm. in the British Isles. *Proceedings of the Botanical Society of the British Isles* **2**, 197–217.

Bangerter, E. B. & Kent, D. H. (1965).
Additional notes on *Veronica filiformis*.
Proceedings of the Botanical Society of the British Isles **6**, 113–18.

Barrett, S. C. H., Colautti, R. I. & Eckert, C.
G. (2008). Plant reproductive systems
and evolution during biological invasion.
Molecular Ecology **17**, 373–83.

Barrett, S. C. H. & Harder, L. D. (2005). The
evolution of polymorphic sexual systems
in daffodils (*Narcissus*). *New Phytologist*
165, 45–53.

Baumel, A., Ainouche, M. L., Misset, M.
T., Gourret, J.-P. & Bayer, R. J. (2003).
Genetic evidence for hybridization
between the native *Spartina maritima*
and the introduced *Spartina alterniflora*
(Poaceae) in South-West France: *Spartina*
× *neyrautii* re-examined. *Plant Systematics
and Evolution* **237**, 87–97.

Beattie, E. P. (1962). Esparto grass aliens in
Fife (v. c. 85). *Proceedings of the Botanical
Society of the British Isles* **4**, 404–06.

Brenan, J. P. M. (1947). A contribution to the
adventive flora of Southampton. *Botanical
Society and Exchange Club of the British Isles.
Report for 1945* **13**, 106–12.

Brown, G. C. (1930). The alien plants of Essex.
Essex Naturalist **22**, 31–47.

Brown, J. M., Brummitt, R. K., Spencer, M.
& Carine, M. A. (2009). Disentangling
the bindweeds: hybridization and
taxonomic diversity in British *Calystegia*
(Convolvulaceae). *Botanical Journal of the
Linnean Society* **160**, 388–401.

Bucharova, A. & van Kleunen, M. (2009).
Introduction history and species
characteristics partly explain
naturalization success of North American
woody species in Europe. *Journal of
Ecology* **97**, 230–8.

Burbidge, F. W. & Colgan, N. (1902). A new
Senecio hybrid. *Irish Naturalists' Journal* **11**,
311–17, and *J. Bot.* **40**, 401–6.

Burges, R. C. L. (1946). Adventive flora of
Burton-on-Trent. *Botanical Society and
Exchange Club of the British Isles. Report for
1943–44* **12**, 815–19.

Callaway, R. M. & Aschehoug, E. T. (2000).
Invasive plants versus their new and
old neighbors: a mechanism for exotic
invasion. *Science* **290**, 521–3.

Callaway, R. M., Cipollini, D., Barto, K.,
Thelen, G. C., Hallett, S. G., Prati, D.,
Stinson, K. & Klironomos, J. (2008). Novel
weapons: invasive plant suppresses fungal
mutualists in America but not in its
native Europe. *Ecology* **89**, 1043–55.

Callaway, R. M., Thelen, G. C., Rodriguez, A.
& Holben, W. E. (2004). Soil biota and
exotic plant invasion. *Nature* **427**, 731–3.

Carvalheiro, L. G., Buckley, Y. M., Ventim,
R., Fowler, S. V. & Memmott, J. (2008).
Apparent competition can compromise
the safety of highly specific biocontrol
agents. *Ecology Letters* **11**, 690–700.

Center, T. D., Van, T. K., Rayachhetry, M.,
Buckingham, G. R., Dray, F. A., Wineriter,
S. A., Purcell, M. F. & Pratt, P. D. (2000).
Field colonization of the melaleuca snout
beetle (*Oxyops vitiosa*) in south Florida.
Biological Control **19**, 112–23.

Chater, A. O. (2010). *Flora of Cardiganshire.*
Privately published.

Chesson, P. L. (1985). Coexistence of
competitors in spatially and temporally
varying environments – a look at the
combined effects of different sorts of
variability. *Theoretical Population Biology*
28, 263–87.

Child, L. E. & Wade, P. M. (2000). *The Japanese
Knotweed Manual: The Management and
Control of an Invasive Alien Weed* (Fallopia
japonica). Packard Publishing, Chichester.

Clapham, A. R., Tutin, T. G. & Warburg, E.
F. (1952, 1962, 1987). *Flora of the British
Isles.* First, second and third editions.
Cambridge University Press, Cambridge.

Clement, B. & Touffet, J. (1990). Plant
strategies and secondary succession on
Brittany heathlands after severe fire.
Journal of Vegetation Science **1**, 195–202.

Clement, E. J. (2010). Gonocarpus – a native of
Ireland with a bipolar disjunction. *BSBI
News* **113**, 26–7.

Clement, E. J. & Foster, M. C. (1994). *Alien
Plants of the British Isles.* Botanical Society
of the British Isles, London.

Clement, E., Smith, D. P. J. & Thirlwell, I. R.
(2005). *Illustrations of Alien Plants of the
British Isles.* Botanical Society of the British
Isles, London.

Clifton, J. & Wheeler, J. (2012). *Conifer Moths of
the British Isles.* Privately published.

Coart, E., Vekemans, X., Smulders, M. J. M., Wagner, I., van Huylenbroeck, J., van Bockstaele, E. & Roldán-Ruiz, I. (2003). Genetic variation in the endangered wild apple (*Malus sylvestris* (L.) Mill.) in Belgium as revealed by amplified fragment length polymorphism and microsatellite markers. *Molecular Ecology* **12**, 845–57.

Conolly, A. P. (1977). The distribution and history in the British Isles of some alien species of *Polygonum* and *Reynoutria*. *Watsonia* **11**, 291–311.

Cook, C. D. K. & Urmi-König, K. (1983). A revision of the genus *Stratiotes*. *Aquatic Botany* **16**, 213–49.

Coombe, D. E. (1956). Biological flora of the British Isles: *Impatiens parviflora* DC. *Journal of Ecology* **44**, 701–13.

Cooper, J. E. (1914). Casual plants in Middlesex. *Journal of Botany* **52**, 127–31.

Cope, T. A. & Gray, A. J. (2009). *Grasses of the British Isles*. Botanical Society of the British Isles, London.

Cramp, S. & Perrins, C. M. (Eds) (1994). *Handbook of the Birds of Europe, the Middle East and North Africa. Volume 8*, pp.686–717. Oxford University Press, Oxford.

Crawley, M. J. (2005). *The Flora of Berkshire*. Brambleby Books, Harpenden.

Crawley, M. J. & Brown, S. L. (1995). Seed limitation and the dynamics of feral oilseed rape on the M25 motorway. *Proceedings of the Royal Society of London Series B, Biological Sciences* **259**, 49–54.

Crawley, M. J. & Brown, S. L. (2004). Spatially structured population dynamics in feral oilseed rape. *Proceedings of the Royal Society of London Series B, Biological Sciences* **271**, 1909–16.

Crawley, M. J., Johnston, A. E., Silvertown, J., Dodd, M., de Mazancourt, C., Heard, M. S., Henman, D. F. & Edwards, G. R. (2005). Determinants of species richness in the Park Grass Experiment. *American Naturalist* **165**, 348–362.

Critchfield, W. B. & Little, E. L. (1966). Geographic distribution of the pines of the world. *US Department of Agriculture, Miscellaneous Publication* **991**.

Cross, J. R. (1981). The establishment of *Rhododendron ponticum* in the Killarney

Oakwoods, SW Ireland. *Journal of Ecology* **69**, 807–824.

Cryer, J. (1920). Adventive plants on waste ground, Bradford, York, 1919. *Botanical Society and Exchange Club of the British Isles. Report for 1919* **5**, 719.

Cullen, J. (2011). Naturalised rhododendrons widespread in Great Britain and Ireland. *Hanburyana* **5**, 11–29.

Cunningham, M. H. & Kenneth, A. G. (1979). *Flora of Kintyre*. EP Publishing Limited, Wakefield.

Curtis, R. (1931). Adventive flora of Burton-on-Trent. *Botanical Society and Exchange Club of the British Isles. Report for 1930* **9**, 465–9.

Daehler, C. C. (2009). Short lag times for invasive tropical plants: evidence from experimental plantings in Hawai'i. *PLoS ONE* **4(2)**, e4462. doi:10.1371/journal. pone.0004462.

Dandy, J. E. (1958). *List of British Vascular Plants*. British Museum (Natural History) & Botanical Society of the British Isles, London.

Darwin, C. (1877). *The Different Forms of Flowers on Plants of the Same Species*. John Murray, London.

Davidson, A. M., Jennions, M. & Nicotra, A. B. (2011). Do invasive species show higher phenotypic plasticity than native species and, if so, is it adaptive? A meta-analysis. *Ecology Letters* **14**, 419–31.

Davies, W. J. (1961). Self-sown conifers. *Proceedings of the Botanical Society of the British Isles* **4**, 198–9.

Davis, M. B. (1987). Invasions of forest communities during the Holocene: beech and hemlock in the Great Lakes Region. In: *Colonization, Succession and Stability*, pp.373–93 (Eds Gray, A. J., Crawley, M. J. & Edwards, P. J.). Blackwell Scientific Publications, Oxford.

Defra (2003). *Review of Non-native Species Policy*. Defra, London.

Defra (2008). *The Invasive Non-Native Species Framework Strategy for Great Britain*. Defra, London.

Delannay, X. (1978). La gynodioecie chez les angiospermes. *Les Naturalistes Belges* **59**, 223–37.

Dickie, I. A., Yeates, G. W., St John, M. G., Stevenson, B. A., Scott, J. T., Rillig, M.

C., Peltzer, D. A., et al. (2011). Ecosystem service and biodiversity trade-offs in two woody successions. *Journal of Applied Ecology* **48**, 926–34.

Di Tomaso, J. M. (1998). Impact, biology, and ecology of saltcedar (*Tamarix* spp.) in the southwestern United States. *Weed Technology* **12**, 326–36.

Dlugosch, K. M. & Parker, I. M. (2008). Founding events in species invasions: genetic variation, adaptive evolution, and the role of multiple introductions. *Molecular Ecology* **17**, 431–49.

Druce, G. C. (1893). *L. repens × vulgaris. Botanical Exchange Club of the British Isles. Report for 1892* **1**, 380–1.

Druce, G. C. (1910). Notes on the flora of Middlesex. *Journal of Botany* **48**, 269–78.

Druce, G. C. (1917). Aliens at St. Philip's Marsh, Bristol. *Botanical Society and Exchange Club of the British Isles. Report for 1916* **4**, 512–14.

Druce, G. C. (1918). *Alchemilla argentea* G. Don. *Botanical Society and Exchange Club of the British Isles. Report for 1917* **5**, 20–6.

Druce, G. C. (1927). *The Flora of Oxfordshire*. Second edition. Clarendon Press, Oxford.

Druce, G. C. (1928). *British Plant List*. Second edition. T. Buncle & Co., Arbroath.

Dulberger, R. (1970). Tristyly in *Lythrum junceum*. *New Phytologist* **69**, 751–9.

Dullinger, S., Essl, F., Rabitsch, W., Erb, K.-H., Gingrich, S., Haberl, H., Huelber, K., et al. (2013). Europe's other debt crisis caused by the long legacy of future extinctions. *Proceedings of the National Academy of Sciences of the United States of America* **110**, 7342–7.

Dunlop, G. A. (1908). An annotated list of the alien plants of the Warrington District. *Memoirs and Proceedings of the Manchester Literary and Philosophical Society* **52(15)**, 1–27.

Dunn, P. H. (1979). Distribution of leafy spurge (*Euphorbia esula*) and other weedy *Euphorbia* spp. in the United States. *Weed Science* **27**, 509–16.

Dunn, S. T. (1905). *Alien Flora of Britain*. West, Newman & Co., London.

Easy, G. M. (1976). The flora and fauna of rubbish tips and waste places in Cambridgeshire. *Nature in Cambridgeshire* **19**, 23–31.

Edgar, E. & Connor, H. E. (2010). *Flora of New Zealand, Volume 2, Grasses*. Second edition. Manaaki Whenua Press, Lincoln, New Zealand.

Edwards, C. (2006). *Managing and Controlling Invasive Rhododendron*. Forest Research, Roslin.

Ellenberg, H. (1953). Physiologisches und ökologisches Verhalten derselben Pflanzenarten. *Bericht der Deutschen Botanischen Gesellschaft* **65**, 351–61.

Ellstrand, N. C. & Schierenbeck, K. A. (2000). Hybridization as a stimulus for the evolution of invasiveness in plants? *Proceedings of the National Academy of Sciences of the United States of America* **97**, 7043–50.

Elton, C. S. (1958). *The Ecology of Invasions by Animals and Plants*. John Wiley, New York.

Emmet, A. M. (Ed) (1988). *A Field Guide to the Smaller British Lepidoptera*. Second edition. British Entomological and Natural History Society, London.

Endara, M.-J. & Coley, P. D. (2011). The resource availability hypothesis revisited: a meta-analysis. *Functional Ecology* **25**, 389–98.

English Nature (1996). *Accessible Natural Green Space Standards in Towns and Cities: A Review and Toolkit for their Implementation*. English Nature Research Reports, Peterborough.

Eschen, R. & Williams, F. (2010). The annual cost of invasive species to the British economy quantified. *Aliens: The Invasive Species Bulletin* **31**, 47–51.

Evans, P. A., Evans, I. M. & Rothero, G. P. (2002). *Flora of Assynt*. Privately published.

Evans, W. & Evans, W. E. (1903–04). Alien plants near Edinburgh. *Annals of Scottish Natural History* **1903**, 174–9; **1904**, 236–40.

Fennell, M., Gallagher, T. & Osborne, B. (2010). Patterns of genetic variation in invasive populations of *Gunnera tinctoria*: an analysis at three spatial scales. *Biological Invasions* **12**, 3973–87.

Ferris, C., King, R. A. & Gray, A. J. (1997). Molecular evidence for the maternal parentage in the hybrid origin of *Spartina anglica* C. E. Hubbard. *Molecular Ecology* **6**, 185–7.

Firn, J., Moore, J. L., MacDougall, A. S., Borer, E. T., Seabloom, E. W., Hille Ris Lambers, J., Harpole, W. S., *et al.* (2011). Abundance of introduced species at home predicts abundance away in herbaceous communities. *Ecology Letters* **14**, 274–81.

Fitter, A. H. & Fitter, R. S. R. (2002). Rapid changes in flowering time in British plants. *Science* **296**, 1689–91.

Forseth, I. N. & Innis, A. F. (2004). Kudzu (*Pueraria montana*): history, physiology, and ecology combine to make a major ecosystem threat. *Critical Reviews in Plant Sciences* **23**, 401–13.

Foxcroft, L. C., Rouget, M., Richardson, D. M. & MacFadyen, S. (2004). Reconstructing 50 years of *Opuntia stricta* invasion in the Kruger National Park, South Africa: environmental determinants and propagule pressure. *Diversity and Distributions* **10**, 427–37.

Fraser, J. (1904–06). Alien plants near Edinburgh. *Annals of Scottish Natural History* **1904**, 106–13; **1905**, 96–103; **1906**, 100–05.

Fraser, J. (1907–11). Alien plants. *Annals of Scottish Natural History* **1907**, 37–42; **1908**, 101–06; **1909**, 40–4; **1910**, 43–6; **1911**, 99–102.

Fraser, J. (1912a). Alien plants. *Scottish Botanical Review* **1**, 39–41.

Fraser, J. (1912b). Alien plants. *Transactions and Proceedings of the Botanical Society of Edinburgh* **26**, 28–9.

Fraser, J. (1914a). Notes on some Scottish plants. *Transactions and Proceedings of the Botanical Society of Edinburgh* **26**, 234–5.

Fraser, J. (1914b). Some Galloway plants. *Dumfriesshire and Galloway Natural History and Antiquarian Society. Transactions and Journal of Proceedings 1913–14* **3(2)**, 29–34.

Fridley, J. D. (2012). Extended leaf phenology and the autumn niche in deciduous forest invasions. *Nature* **485**, 359–62.

Fridley, J. D., Stachowicz, J. J., Naeem, S., Sax, D. F., Seabloom, E. W., Smith, M. D., Stohlgren, T. J., Tilman, D. & Von Holle, B. (2007). The invasion paradox: reconciling pattern and process in species invasions. *Ecology* **88**, 3–17.

Fry, C. (2012). *The Plant Hunters: The Adventures of the World's Greatest Botanical Explorers.* André Deutsch, London.

Fyfe, V. C. (1950). The genetics of tristyly in *Oxalis valdiviensis. Heredity* **4**, 365–71.

Ganders, F. R. (1979). The biology of heterostyly. *New Zealand Journal of Botany* **17**, 607–35.

Gasso, N., Basnou, C. & Vila, M. (2010). Predicting plant invaders in the Mediterranean through a weed risk assessment system. *Biological Invasions* **12**, 463–76.

Gentle, C. B. & Duggin, J. A. (1997). Allelopathy as a competitive strategy in persistent thickets of *Lantana camara* L. in three Australian forest communities. *Plant Ecology* **132**, 85–95.

Gerarde, J. (1597). *The Herball or Generall Historie of Plantes.* John Norton, London.

Gibbons, D. W., Reid, J. B. & Chapman, R. A. (1993). *The New Atlas of Breeding Birds in Britain and Ireland: 1988–1991.* T. & A. D. Poyser, London.

Gilbert, M., Gregoire, J. C., Freise, J. F. & Heitland, W. (2004). Long-distance dispersal and human population density allow the prediction of invasive patterns in the horse chestnut leafminer *Cameraria ohridella. Journal of Animal Ecology* **73**, 459–68.

Gilbert, O. L. (1993). Regenerating Balsam Poplar (*Populus candicans* Ait.) × Black Poplar (*P. nigra* L.) (Salicaceae) at a site in Leeds. *Watsonia* **19**, 188–90.

Gilbert, O. L. (1995). Biological flora of the British Isles: *Symphoricarpos albus* (L.) S. F. Blake. *Journal of Ecology* **83**, 159–66.

Glasson, W. A. (1890). On the occurrence of foreign plants in West Cornwall. *Transactions of the Penzance Natural History and Antiquarian Society* **3**, 62–9, 145–7.

Godwin, H. (1956, 1975). *The History of the British Flora.* First and second editions. Cambridge University Press, Cambridge.

Goodall, D. W. (1952). Quantitative aspects of plant distribution. *Biological Reviews* **27**, 194–242.

Gray, A. J., Marshall, D. F. & Raybould, A. F. (1991). A century of evolution in *Spartina anglica. Advances in Ecological Research* **21**, 1–62.

Green, P. T., O'Dowd, D. J., Abbott, K. L., Jeffery, M., Retallick, K. & MacNally, R. (2011). Invasional meltdown: Invader-

invader mutualism facilitates a secondary invasion. *Ecology* **92**, 1758–68.

Green, T. (2005). Comment: is there a case for the Celtic Maple or the Scots Plane? *British Wildlife* **16(3)**, 184–8.

Grenfell, A. L. (1983). More on tan bark aliens. *BSBI News* **35**, 10.

Grenfell, A. L. (1985). More on tan bark aliens. *BSBI News* **39**, 6–7.

Grierson, R. (1931). Clyde casuals, 1916–1928. *Glasgow Naturalist* **9**, 5–51.

Grigson, G. (1955). *The Englishman's Flora*. Phoenix House, London.

Grime, J. P. (1979). *Plant Strategies and Vegetation Processes*. John Wiley, Chichester.

Grime, J. P., Hodgson, J. G. & Hunt, R. (1988). *Comparative Plant Ecology*. Unwin Hyman, London.

Grotkopp, E., Erskine-Ogden, J. & Rejmánek, M. (2010). Assessing potential invasiveness of woody horticultural plant species using seedling growth rate traits. *Journal of Applied Ecology* **47**, 1320–8.

Grotkopp, E. & Rejmánek, M. (2007). High seedling relative growth rate and specific leaf area are traits of invasive species: phylogenetically independent contrasts of woody angiospernis. *American Journal of Botany* **94**, 526–32.

Grubb, P. J. (1977). The maintenance of species-richness in plant communities: the importance of the regeneration niche. *Biological Reviews* **52**, 107–45.

Hall, P. M. (1941). Wool-aliens at Portchester. *Papers and Proceedings of the Hampshire Field Club and Archaeological Society* **15**, 76.

Hanson, C. G. (2000). Update on birdseed aliens (1985–1998). *Watsonia* **23**, 213–20.

Hanson, C. G. & Mason, J. L. (1985). Bird seed aliens in Britain. *Watsonia* **15**, 237–52.

Harris, D. R. (1996). *The Origins and Spread of Agriculture and Pastoralism in Eurasia: Crops, Fields, Flocks and Herds*. UCL Press, London.

Harris, S. A. (2002). Introduction of Oxford Ragwort, *Senecio squalidus* L. (Asteraceae), to the United Kingdom. *Watsonia* **24**, 31–43.

Harvey, J. H. (1996). Fritillary and martagon – wild or garden? *Garden History* **24**, 30–8.

Hayward, I. M. & Druce, G. C. (1919). *The Adventive Flora of Tweedside*. T. Buncle & Co., Arbroath.

Hernandez-Lopez, A., Rougerie, R., Augustin, S., Lees, D. C., Tomov, R., Kenis, M., Cota, *et al.* (2012). Host tracking or cryptic adaptation? Phylogeography of *Pediobius saulius* (Hymenoptera, Eulophidae), a parasitoid of the highly invasive horse-chestnut leafminer. *Evolutionary Applications* **5**, 256–69.

Higgins, S. I., Richardson, D. M., Cowling, R. M. & Trinder-Smith, T. H. (1999). Predicting the landscape-scale distribution of alien plants and their threat to plant diversity. *Conservation Biology* **13**, 303–13.

Hill, M. O., Roy, D. B., Mountford, J. O. & Bunce, R. G. H. (2000). Extending Ellenberg's indicator values to a new area: an algorithmic approach. *Journal of Applied Ecology* **37**, 3–15.

Hill, M. O. *et al.* (2005). Audit of non-native species in England. *English Nature Research Reports* 662. English Nature, Peterborough.

Hogg, J. (1867). On the ballast flora of the coasts of Durham and Northumberland. *Annals and Magazine of Natural History, Series 3*, **19**, 38–43.

Hollingsworth, M. L. & Bailey, J. P. (2000). Evidence for massive clonal growth in the invasive weed *Fallopia japonica* (Japanese Knotweed). *Botanical Journal of the Linnean Society* **133**, 463–72.

Hollingsworth, M. L., Hollingsworth, P. M., Jenkins, G. I., Bailey, J. P. & Ferris, C. (1998). The use of molecular markers to study patterns of genotypic diversity in some invasive alien *Fallopia* spp. (Polygonaceae). *Molecular Ecology* **7**, 1681–91.

Holm, L. G., Plucknett, D. L., Pancho, J. V. & Herberger, J. P. (1977). *The World's Worst Weeds: Distribution and Biology*. University of Hawai'i Press, Honolulu.

Holmes, T. P., Aukema, J. E., Von Holle, B., Liebhold, A. & Sills, E. (2009). Economic impacts of invasive species in forests past, present, and future. In: *Year in Ecology and Conservation Biology 2009*, pp.18–38 (Eds Ostfeld, R. S. & Schlesinger, W. H.). Wiley-Blackwell, Chichester.

Howes, F. N. (1974) *A Dictionary of Useful and Everyday Plants and their Common Names*. Cambridge University Press, Cambridge.

Howkins, C. (2003). *Sweet Chestnut – History, Landscape, People.* Chris Howkins, Surrey.

Hudson, P. J. (1992). *Grouse in Space and Time: The Population Biology of a Managed Gamebird.* Game Conservancy Limited, Fordingbridge.

Huenneke, L. F. & Vitousek, P. M. (1990). Seedling and clonal recruitment of the invasive tree *Psidium cattleianum* – implications for management of native Hawaiian forests. *Biological Conservation* 53, 199–211.

Hulme, P. E. (2012). Weed risk assessment: a way forward or a waste of time? *Journal of Applied Ecology* 49, 10–19.

Huntley, B., Allen, J. R. M., Collingham, Y. C., Hickler, T., Lister, A. M., Singarayer, J., Stuart, A. J., Sykes, M. T. & Valdes, P. J. (2013). Millennial climatic fluctuations are key to the structure of last glacial ecosystems. *PLoS ONE* 8(4), e61963. doi:10.1371/journal.pone.0061963

Huntley, B. & Birks, H. J. B. (1983). *An Atlas of Past and Present Pollen Maps for Europe: 0–13000 Years Ago.* Cambridge University Press, Cambridge.

Hutchinson, G. E. (1957). Concluding remarks. *Cold Spring Harbor Symposium in Quantitative Biology* 22, 415–57.

Hutchinson, G. E. (1965). *The Ecological Theater and the Evolutionary Play.* Yale University Press, New Haven, CT.

Huttanus, T. D., Mack, R. N. & Novak, S. J. (2011). Propagule pressure and introduction pathways of *Bromus tectorum* (Cheatgrass; Poaceae) in the Central United States. *International Journal of Plant Sciences* 172, 783–94.

Hyde, H. A. & Wade, A. E. (1934). *Welsh Flowering Plants.* National Museum of Wales, Cardiff.

Ingram, D. & Robertson, N. (1999). *Plant Disease. A Natural History.* Collins, London.

Isbell, F., Calcagno, V., Hector, A., Connolly, J., Harpole, W. S., Reich, P. B., Scherer-Lorenzen, M., et al. (2011). High plant diversity is needed to maintain ecosystem services. *Nature* 477, 199–202.

Ison, J. & Braithwaite, W. (2009). The status of some alien trees and shrubs in Britain. Unpublished BSBI report, accessible online at www.bsbi.org.uk/Trees.pdf

James, C. M., Wurzell, B. S. & Stace, C. A. (2000). A new hybrid between a European and a Chinese species of *Artemisia* (Asteraceae). *Watsonia* 23, 139–47.

Jarvis, A. (2011). The problem of invasive alien plant species. *BSBI News* 117, 44–5.

Jiang, L., Tan, J. & Pu, Z. (2010). An experimental test of Darwin's naturalization hypothesis. *American Naturalist* 175, 415–23.

Jobling, J. (1990). Poplars for wood production and amenity. *Forestry Commission Bulletin* 92.

John, D. M., Whitton, B. A. & Brook, A. J. (2002). *The Freshwater Algal Flora of the British Isles.* Cambridge University Press, Cambridge.

Kadereit, J. W. & Leins, P. (1988). A wind tunnel experiment on seed dispersal in *Papaver* L. sects. *Argemonidium* Spach and *Rhoeadium* Spach (Papaveraceae). *Flora* 181, 189–203.

Kay, Q. O. N. (1978). The role of preferential and assortive pollination in the maintenance of flower colour polymorphisms. In: *The Pollination of Flowers by Insects* (Ed. Richards, A. J.). Linnean Society of London, London.

Kay, Q. O. N. & Stevens, D. P. (1986). The frequency, distribution and reproductive biology of dioecious species in the native flora of Britain and Ireland. *Botanical Journal of the Linnean Society* 92, 39–64.

Keane, R. M. & Crawley, M. J. (2002). Exotic plant invasions and the enemy release hypothesis. *Trends in Ecology & Evolution* 17, 164–70.

Keller, R. P., Kocev, D. & Dzeroski, S. (2011). Trait-based risk assessment for invasive species: high performance across diverse taxonomic groups, geographic ranges and machine learning/statistical tools. *Diversity and Distributions* 17, 451–61.

Kent, D. H. (1956). *Senecio squalidus* L. in the British Isles, 1. Early records (to 1877). *Proceedings of the Botanical Society of British Isles* 2, 115–18.

Kent, D. H. (1975). *The Historical Flora of Middlesex.* Ray Society, London.

Kim, M., Cui, M.-L., Cubas, P., Gillies, A., Lee, K., Chapman, M. A., Abbott, R. J. & Coen, E. (2008). Regulatory genes control a key morphological and ecological trait transferred between species. *Science* **322**, 1116–19.

Kirschner, M. & Gerhart, J. (1998). Evolvability. *Proceedings of the National Axcademy of Sciences USA* **95**, 8420–7.

Lehmann, E. (1944). *Veronica filiformis* Sm., eine selbsterile Pflanze. *Jahrbücher für wissenschaftliche Botanik, Berlin* **91**, 395–403.

Lippincott, C. L. (2000). Effects of *Imperata cylindrica* (L.) Beauv. (Cogongrass) invasion on fire regime in Florida sandhill (USA). *Natural Areas Journal* **20**, 140–9.

Lousley, J. E. (1944). The pioneer flora of bombed sites in central London. *Botanical Society and Exchange Club of the British Isles. Report for 1941–42* **12**, 528–31.

Lousley, J. E. (1946). The flora of bombed sites in the city of London in 1944. *Botanical Society and Exchange Club of the British Isles. Report for 1943–44* **12**, 875–83.

Lousley, J. E. (1949b). Brockham Hill bomb crater. *London Naturalist* **28**, 29.

Lousley, J. E. (1953). The recent influx of aliens into the British flora. In: *The Changing Flora of Britain*, pp.140–59 (Ed. Lousley, J. E.). Botanical Society of the British Isles, London.

Lousley, J. E. (1961). A census list of wool aliens found in Britain, 1946–1960. *Proceedings of the Botanical Society of the British Isles* **4**, 221–47.

Lowe, A. J. & Abbott, R. J. (2003). A new British species of *Senecio* (Asteraceae), another hybrid derivative of *S. vulgaris* L. and *S. squalidus* L. *Watsonia* **24**, 375–88.

Mabey, R. (2010). *Weeds – How Vagabond Plants Gatecrashed Civilisation and Changed the Way We Think About Nature*. Profile Books, London.

McClintock, D. (1977). J. E. Lousley and alien plants in the British Isles. *Watsonia* **11**, 287–90.

McConnachie, A. J., Hill, M. P. & Byrne, M. J. (2004). Field assessment of a frond-feeding weevil, a successful biological control agent of red waterfern, *Azolla filiculoides*, in southern Africa. *Biological Control* **29**, 326–31.

MacDougall, A. S., Wilson, S. D. & Bakker, J. D. (2008). Climatic variability alters the outcome of long-term community assembly. *Journal of Ecology* **96**, 346–54.

McLean, B. (1997). *A Pioneering Plantsman: A. K. Bulley and the Great Plant Hunters*. Stationery Office, London.

Macpherson, P., Dickson, J. H., Ellis, R. G., Kent, D. H. & Stace, C. A. (1996). Plant status nomenclature. *BSBI News* **72**, 13–16.

Malecki, R. A., Blossey, B., Hight, S. D., Schroeder, D., Kok, L. T. & Coulson, J. R. (1993). Biological control of Purple Loosestrife. *Bioscience* **43**, 680–6.

Malik, A. (2007). Environmental challenge *vis à vis* opportunity: The case of water hyacinth. *Environment International* **33**, 122–38.

Marchant, C. J. (1963). Corrected chromosome numbers for *Spartina × townsendii* and its parent species. *Nature* **199**, 929.

Marsden-Jones, E.M. & Turrill, W.B. (1954). *British Knapweeds. A Study in Synthetic Taxonomy*. Ray Society, London.

Maskell, L. C., Firbank, L. G., Thompson, K., Bullock, J. M. & Smart, S. M. (2006). Interactions between non-native plant species and the floristic composition of common habitats. *Journal of Ecology* **94**, 1052–60.

Matthews, J. R. (1955). *Origin and Distribution of the British Flora*. Hutchinson University Library, London.

Medvecká, J., Kliment, J., Májeková, J., Halada, L., Zaliberová, M., Gojdičová, E., Feráková, V. & Jarolímek, I. (2012). Inventory of the alien flora of Slovakia. *Preslia* **84**, 257–309.

Melville, R. & Smith, R. L. (1928). Adventive flora of the Metropolitan area (1). *Botanical Society and Exchange Club of the British Isles. Report for 1927* **8**, 444–54.

Meyer, J. Y. & Florence, J. (1996). Tahiti's native flora endangered by the invasion of *Miconia calvescens* DC (Melastomataceae). *Journal of Biogeography* **23**, 775–81.

Millennium Ecosystem Assessment (2005). *Ecosystems and Human Well-being: Biodiversity Synthesis*. World Resources Institute, Washington, DC.

Milne, R. & Abbott, R. (2000). Origin and evolution of invasive naturalized material of *Rhododendron ponticum* L. in the British Isles. *Molecular Ecology* **9**, 541–56.

Minteer, B. A. & Collins, J. P. (2010). Move it or lose it? The ecological ethics of relocating species under climate change. *Ecological Applications* **20**, 1801–04.

Mitchell, A. (2001). *Collins Field Guide – Trees of Britain and Northern Europe*. Third edition. Collins, London.

Mitchell, C. E., Blumenthal, D., Jarosik, V., Puckett, E. E. & Pysek, P. (2010). Controls on pathogen species richness in plants' introduced and native ranges: roles of residence time, range size and host traits. *Ecology Letters* **13**, 1525–35.

Mitchell, C. E. & Power, A. G. (2003). Release of invasive plants from fungal and viral pathogens. *Nature* **421**, 625–7.

Moles, A. T., Falster, D. S., Leishman, M. R. & Westoby, M. (2004). Small-seeded species produce more seeds per square metre of canopy per year, but not per individual per lifetime. *Journal of Ecology* **92**, 384–96.

Müller, C., Kuppelwieser, H., von Arx, R. & Beyeler, R. (2001). *Vegetation Control on Railway Tracks and Grounds*. Swiss Agency for the Environment, Forests and Landscape (SAEFL), Bern.

Murphy, J. P. (1981). Senecio × albescens Burbidge & Colgan at Killiney, Co. Dublin: a seventy-eight years old population. *Watsonia* **13**, 303–11.

Musgrave, T., Gardner, C. & Musgrave, W. (1998). *The Plant Hunters: Two Hundred Years of Adventure and Discovery Around the World*. Ward Lock, London.

Naeem, S., Duffy, J. E. & Zavaleta, E. (2012). The functions of biological diversity in an age of extinction. *Science* **336**, 1401–06.

Ness, J. H., Rollinson, E. J. & Whitney, K. D. (2011). Phylogenetic distance can predict susceptibility to attack by natural enemies. *Oikos* **120**, 1327–34.

Nilsson, N. H. (1954). Über Hochkompleze Bastardverbindungen in der Gattung *Salix*. *Hereditas (Lund)* **40**, 517–22.

Nygren, A. (1967). Apomixis in the angiosperms. *Handbuch der Pflanzenphysiologie* **18**, 551–96.

Olliver, K. (2011). *Treating Japanese Knotweed on the Olympic Park*. Olympic Delivery Authority, London.

Orians, G. H. (1986). Site characteristics favoring invasions. In: *Ecology of Biological Invasions of North America and Hawai'i* (Eds Mooney, H. A. & Drake, J. A.). Springer Verlag, New York.

Ornduff, R. (1972). The beakdown of trimorphic incompatibility in *Oxalis* section *Corniculatae*. *Evolution* **26**, 52–65.

Page, C. N. (1982, 1997). *The Ferns of Britain and Ireland*. First and second editions. Cambridge University Press, Cambridge.

Palmer, J. R. (1977). Oil-milling adventive plants in north-west Kent 1973–6. *Transactions of the Kent Field Club* **6**, 85–90.

Panetta, F. D. & Cacho, A. J. (2012). Beyond fecundity control: which weeds are most containable? *Journal of Applied Ecology* **49**, 311–21.

Pappert, R. A., Hamrick, J. L. & Donovan, L. A. (2000). Genetic variation in *Pueraria lobata* (Fabaceae), an introduced, clonal, invasive plant of the southeastern United States. *American Journal of Botany* **87**, 1240–5.

Payne, R. M. (1978). The flora of walls in south-eastern Essex. *Watsonia* **12**, 41–6.

Pearman, D. A. (2007). 'Far from any house' – assessing the status of doubtfully native species in the flora of the British Isles. *Watsonia* **26**, 271–90.

Pearman, D. A. (2013). Late-discovered petaloid monocotyledons: separating the native and alien flora. *New Journal of Botany* **3**, 24–32.

Pearman, D. A., Preston, C. D., Rothero, G. P. & Walker, K. J. (2008). *The Flora of Rum: an Atlantic Island Reserve*. Privately published.

Pellant, M. (1996). *Cheatgrass: The Invader that Won the West*. Interior Columbia Basin Ecosystem Management Project, Idaho.

Perring, F. H. & Walters, S. M. (Eds) (1962). *Atlas of the British Flora*. Thomas Nelson and Sons, London.

Phalan, B., Onial, M., Balmford, A. & Green, R. E. (2011). Reconciling food production and biodiversity conservation: land sharing and land sparing compared. *Science* **333**, 1289–91.

Pilgrim, E. & Hutchinson, N. (2004). *Bluebells for Britain*. Plant Life, Salisbury.

Porley, R. D. & Matcham, H. W. (2003). The status of *Orthodontium gracile* in Britain and Ireland. *Journal of Bryology* **25**, 64–6.

Praeger, R. L. (1934). *The Botanist in Ireland*. Hodges, Figgis & Co., Dublin.

Presland, J. (2011). Indian Balsam – triffid or treat? *BSBI News* **116**, 60–8.

Preston, C. D. (1988). The spread of *Epilobium ciliatum* Raf. in the British Isles. *Watsonia* **17**, 279–88.

Preston, C. D. (2004). John Raven's report on his visit to the Hebrides. *Watsonia* **25**, 17–44.

Preston, C. D., Pearman, D. A. & Dines, T. D. (Eds) (2002). *New Atlas of the British and Irish Flora*. Oxford University Press, Oxford.

Preston, C. D., Pearman, D. A. & Hall, A. R. (2004). Archaeophytes in Britain. *Botanical Journal of the Linnean Society* **145**, 257–94.

Prevey, J. S., Germino, M. J. & Huntly, N. J. (2010). Loss of foundation species increases population growth of exotic forbs in sagebrush steppe. *Ecological Applications* **20**, 1890–1902.

Pugsley, D. J. (1941). A study of the colonisation and subsequent flora of coal dumps at Cwmbach, Aberdare. *Proceedings of the Swansea Scientific and Field Naturalists' Society* **2**, 159–77.

Pyšek, P., Danihelka, J., Sádlo, J., Chrtek, J., Chytrý, M., Jarošík, V., Kaplan, Z., et al. (2012). Catalogue of alien plants of the Czech Republic (2nd edition): checklist update, taxonomic diversity and invasion patterns. *Preslia* **84**, 155–255.

Pyšek, P., Sádlo, J. & Mandák, B. (2004). Catalogue of alien plants of the Czech Republic. *Preslia, Praha* **74**, 97–186.

Rabinowitz, D. (1981). Seven forms of rarity. In: *The Biological Aspects of Rare Plant Conservation* (Ed Synge, H.). pp.205–17. John Wiley, Chichester.

Raunkaer, C. C. (1934). *The Life Forms of Plants and Statistical Plant Geography*. Clarenden Press, Oxford.

Raven, J. & Walters, M. (1956). *Mountain Flowers*. Collins, London.

Rayner, J. F. (1924–25). The alien or adventive flora of Hampshire and the Isle of Wight. *Proceedings of the Isle of Wight Natural History Society* **1**, 166–75, 229–74.

Redfern, M. (2011). *Plant Galls*. Collins, London.

Rejmánek, M. & Pitcairn, M. J. (2003). When is eradication of exotic pest plants a realistic goal? In: *Turning the Tide: the*

Eradication of Invasive Species. Proceedings of the International Conference on Eradication of Island Invasives (Eds Veitch, C. R. & Clout, M. N.). IUCN, New York.

Reynolds, S. C. P. (2002). *A Catalogue of Alien Plants in Ireland*. National Botanic Gardens, Glasnevin.

Rich, T. C. G., Houston, L., Robertson, A. & Proctor, M. C. F. (2010). *Whitebeams, Rowans and Service Trees of Britain and Ireland*. Botanical Society of the British Isles, London.

Richards, C. L., Bossdorf, O., Muth, N. Z., Gurevitch, J. & Pigliucci, M. (2006). Jack of all trades, master of some? On the role of phenotypic plasticity in plant invasions. *Ecology Letters* **9**, 981–93.

Richardson, D. M., Allsopp, N., D'Antonio, C. M., Milton, S. J. & Rejmánek, M. (2000). Plant invasions – the role of mutualisms. *Biological Reviews* **75**, 65–93.

Ridley, H. N. (1930). *The Dispersal of Plants Throughout the World*. L. Reeve & Co., Ashford.

Rieseberg, L. H., Archer, M. A. & Wayne, R. K. (1999). Transgressive segregation, adaptation and speciation. *Heredity* **83**, 363–72.

Robinson, J. F. (1902). *The Flora of the East Riding of Yorkshire*. Transactions of the Hull Scientific and Field Naturalists' Club for the year 1902 2. A. Brown & Sons, Hull.

Rodwell, J. S. (2006). *National Vegetation Classification: Users' Handbook*. Joint Nature Conservation Committee, Peterborough.

Ronse, A. & Braithwaite, M. E. (2012). Seed 'for growing under trees': the source of wood lawn neophytes in the parkland of Scottish mansion houses. *New Journal of Botany* **2**, 149–54.

Room, P. M., Harley, K. L. S., Forno, I. W. & Sands, D. P. A. (1981). Successful biological-control of the floating weed *Salvinia*. *Nature* **294**, 78–80.

Rosser, E. M. (1955). A new British species of *Senecio*. *Watsonia* **3**, 228–32.

Rousseau, J. J. (1755). *On the Origin of the Inequality of Mankind*. Marc-Michel Rey, Holland.

Rumsey, F. J. & Spencer, M. (2012). Is *Equisetum ramosissimum* (Equisetaceae:

Equisetophyta) native to the British Isles? *Fern Gazette* **19**, 37–46.

Ryves, T. B. (1974). An interim list of the wool-alien grasses from Blackmoor, North Hants., 1969–1972. *Watsonia* **10**, 35–48.

Ryves, T. B. (1977). Notes on wool-alien species of *Lepidium* in the British Isles. *Watsonia* **11**, 367–72.

Ryves, T. B. (1988). Supplementary list of wool-alien grasses recorded from Blackmoor, North Hants., 1959–1976. *Watsonia* **17**, 73–79.

Ryves, T. B., Clement, E. J. & Foster, M. C. (1996). *Alien Grasses of the British Isles.* BSBI, London.

Salisbury, E. J. (1943). The flora of bombed areas. *Nature* **151**, 462–6.

Salisbury, E. J. (1961). *Weeds & Aliens.* Collins, London.

Sandwith, C. I. (1933). The adventive flora of the port of Bristol. *Botanical Society and Exchange Club of the British Isles. Report for 1932* **10**, 314–63.

Savidge, J. P., Heywood, V. H. & Gordon, V. (1963). *Travis's Flora of South Lancashire.* Liverpool Botanical Society, Liverpool.

Scherber, C., Crawley, M. J. & Porembski, S. (2003). The effects of herbivory and competition on the invasive alien plant *Senecio inaequidens* (Asteraceae). *Diversity and Distributions* **9**, 415–26.

Schrader, L., Kim, J. W., Ence, D., Zimin, A., Klein, A., Wyschetzki, K., Weichselgartner, T., et al. (2014) Transposable element islands facilitate adaptation to novel environments in an invasive species. *Nature Communications* **5**, 5495. doi:10.1038/ncomms6495.

Scott, W. & Palmer, R. (1987). *The Flowering Plants and Ferns of the Shetland Islands.* Shetland Times, Lerwick.

Scottish Executive (2004). *Scotland's Biodiversity: It's in Your Hands.* Scottish Executive, Edinburgh.

Scully, R. W. (1916). *Flora of County Kerry.* Hodges, Figgis & Co., Dublin.

Sell, P. D. (2007). Introduced 'look-alikes' and other difficult introduced plants in our Cambridgeshire flora. *BSBI News* **105**, 24–30.

Sell, P. D. & Murrell, G. (1996–2014). *Flora of Great Britain and Ireland. Volumes 2–5.* Cambridge University Press, Cambridge.

Shaw, R. H., Bryner, S. & Tanner, R. (2009). The life history and host range of the Japanese knotweed psyllid, *Aphalara itadori* Shinji: potentially the first classical biological weed control agent for the European Union. *Biological Control* **49**, 105–13.

Shaw, R. H., Tanner, R., Djeddour, D. & Cortat, G. (2011). Classical biological control of Japanese knotweed – lessons for Europe. *Biological Control* **51**, 552–8.

Shenstone, J. C. (1912). The flora of London building sites. *Journal of Botany* **50**, 117–24.

Sheppard, A. W., Shaw, R. H. & Sforza, R. (2006). Top 20 environmental weeds for classical biological control in Europe: a review of opportunities, regulations and other barriers to adoption. *Weed Research* **46**, 93–117.

Shigesada, N. & Kawasaki, K. (1997). *Biological Invasions: Theory and Practice.* Oxford University Press, Oxford.

Shimwell, D. W. (2006). A shoddy tale: perspectives on the wool alien flora of West Yorkshire in the twenty-first century. *Watsonia* **26**, 127–37.

Shimwell, D. W. (2009). Studies in the floristic diversity of Durham walls, 1958–2008. *Watsonia* **27**, 323–38.

Silvey, A. (2012). *The Plant Hunters: True Stories of Their Daring Adventures to the Far Corners of the Earth.* Farrar Straus Giroux, London.

Simberloff, D. (2011). How common are invasion-induced ecosystem impacts? *Biological Invasions* **13**, 1255–68.

Simpson, F. W. (1989). A hybrid oak. *Suffolk Natural History* **25**, 95.

Skellam, J. G. (1951). Random dispersal in theoretical populations. *Biometrika* **38**, 196–218.

Smith, C., Dawson, D., Archer, J., Davies, M., Frith, M., Hughes, E. & Massini, P. (2011). *London: Garden City? From Green to Grey; Observed Changes in Garden Vegetation Structure in London, 1998–2008.* London Wildlife Trust, London.

Smith, R. L. & Wade, A. E. (1939). Notes on the adventive flora of the Cardiff district. *Botanical Society and Exchange Club of the British Isles. Report for 1938* **12**, 72–83.

Stace, C. A. (1961). Some studies in *Calystegia*: compatibility and hybridization in *C. sepium* and *C. silvatica. Watsonia* **5**, 88–105.

Stace, C. A. (1965). Some studies in *Calystegia*, II. Observations on the floral biology of the British inland taxa. *Proceedings of the Botanical Society of the British Isles* **6**, 21–31.

Stace, C. A. (1991, 1997, 2010). *New Flora of the British Isles.* First, second and third editions. Cambridge University Press, Cambridge.

Stace, C. A., Preston, C. D. & Pearman, D. A. (2015). *Hybrid Flora of the British Isles.* Botanical Society of Britain and Ireland, London.

Stearns, S. C. (2000). Life history evolution: successes, limitations, and prospects. *Naturwissenschaften* **87**, 476–86.

Stewart, A., Pearman, D. A. & Preston, C. D. (1994). *Scarce Plants in Britain.* Joint Nature Conservation Committee, Peterborough.

Storrie, J. (1876). Notes on the ballast plants of Cardiff and neighbourhood. *Report and Transactions of the Cardiff Naturalists' Society* **8**, 141–5.

Storrie, J. (1886). Foreign plants found on the ballast heaps, near Cardiff and Penarth Docks. In: *The Flora of Cardiff, a Descriptive List of the Indigenous Plants Found in the District of the Cardiff Naturalists' Society.* Cardiff Naturalists' Society, Cardiff.

Stringer, C. (2006). *Homo Britannicus.* Penguin Books, London.

Sun, Y., Collins, A. R., Schaffner, U. & Muller-Scharer, H. (2013). Dissecting impact of plant invaders: do invaders behave differently in the new range? *Ecology* **94**, 2124–30.

Swale, E. M. F. (1962). Notes on some algae from the Reddish Canal. *British Phycological Journal* **2**, 174–5.

Thuiller, W., Gallien, L., Boulangeat, I., de Bello, F., Muenkemueller, T., Roquet, C. & Lavergne, S. (2010). Resolving Darwin's naturalization conundrum: a quest for evidence. *Diversity and Distributions* **16**, 461–75.

Thurston, E. (1929). The alien and British plants of Par and Charlestown harbours, Falmouth Docks and Eastern Green, Penzance. *Journal of the Royal Institution of Cornwall* **23**, 137–205.

Tilman, D. (1994). Competition and biodiversity in spatially structured habitats. *Ecology* **75**, 2–16.

Titman, D. (1976). Ecological competition between algae: experimental confirmation of resource-based competition theory. *Science* **192**, 463–5.

Tomlinson, P. & Hall, A. (1996). A review of the archaeological evidence for food plants from the British Isles: an example of the use of the Archaeobotanical Computer Database (ABCD). *Internet Archaeology* **1**. doi:10.11141/ia.1.5

Trimen, H. (1866). Exotic plants about London in 1865. *Journal of Botany* **4**, 147–51.

Tubbs, C. R. (1986). *The New Forest.* Collins, London.

Tutin, T.G., Heywood, V.H., Burges, N.A., Moore, D.M., Valentine, D.H., Walters, S.M. & Webb, D.A. (Eds) (1964–80). *Flora Europaea.* Cambridge University Press, Cambridge.

Vallejo-Marín, M. (2012). *Mimulus peregrinus* (Phrymaceae): a new British allopolyploid species. *PhytoKeys* **14**, 1–14.

van der Putten, W. H., Bardgett, R. D., Bever, J. D., Bezemer, T. M., Casper, B. B., Fukami, T., Kardol, et al. (2013). Plant-soil feedbacks: the past, the present and future challenges. *Journal of Ecology* **101**, 265–76.

Veen, M. v. d., Livarda, A. & Hill, A. (2008). New plant foods in Roman Britain – dispersal and social access. *Environmental Archaeology* **13**, 11–36.

Veltman, C. J., Nee, S. & Crawley, M. J. (1996). Correlates of introduction success in exotic New Zealand birds. *American Naturalist* **147**, 542–57.

Vera, F. W. M. (2000). *Grazing Ecology and Forest History.* CABI Publishing, Wallingford.

Vitousek, P. M. & Walker, L. R. (1989). Biological invasion by *Myrica faya* in Hawai'i – plant demography, nitrogen-fixation, ecosystem effects. *Ecological Monographs* **59**, 247–65.

Vivanco, J. M., Bais, H. P., Stermitz, F. R., Thelen, G. C. & Callaway, R. M. (2004). Biogeographical variation in community response to root allelochemistry: novel weapons and exotic invasion. *Ecology Letters* **7**, 285–92.

Wade, A. E. (1958). The history of *Symphytum asperum* Lepech. and *S.* × *uplandicum* Nyman in Britain. *Watsonia* **4**, 117–18.

Wade, A. E. & Smith, R. L. (1926). The adventive flora of the port of Cardiff. *Botanical Society and Exchange Club of the British Isles. Report for 1925* **7**, 999–1027.

Wade, A. E. & Smith, R. L. (1927). Additions to the adventive flora of the port of Cardiff. *Botanical Society and Exchange Club of the British Isles. Report for 1926* **8**, 181–3.

Wade, P. M., de Waal, L. C., Child, L. E., Dodd, F. S. & Darby, E. J. (1994). Control of invasive riparian and aquatic weeds. National Rivers Authority, R&D Project Record 294/7/W.

Walker, K. J. (2014). *Sarracenia purpurea* subsp. *purpurea* (Sarraceniceae) naturalised in Britain and Ireland: distribution, ecology, impacts and control. *New Journal of Botany* **4**, 33–41.

Waloff, N. & Richards, O. W. (1977). The effect of insect fauna on growth mortality and natality of broom, *Sarothamnus scoparius*. *Journal of Applied Ecology* **14**, 787–98.

Watson, H. C. (1847–59). *Cybele Britannica: or British Plants and their Geographical Relations*. Longman & Co., London.

Watson, H. C. (1873–74). *Topographical Botany*. Privately published.

Watson, M. C., Watt, M. S., Withers, T. M., Kimberley, M. O. & Rolando, C. A. (2011). Potential for *Cleopus japonicus* to control the weed *Buddleja davidii* in plantation forests in New Zealand. *Forest Ecology and Management* **261**, 78–83.

Watts, N. S. & Watts, G. D. (1979). Norwich bird-seed mixtures and the casual plants of Harford tip. *Transactions of the Norfolk and Norwich Naturalists' Society* **24**, 300–09.

Webb, D. A. (1985). What are the criteria for presuming native status? *Watsonia* **15**, 231–6.

Webb, D. A. & Halliday, G. (1973). The distribution, habitat and status of *Hypericum canadense* L. in Ireland. *Watsonia* **9**, 333–44.

Weiss, F. E. & Murray, H. (1909). On the occurrence and distribution of some alien aquatic plants in the Reddish Canal. *Memoirs and Proceedings of the Manchester Literary and Philosophical Society* **53**, 1–8.

Wheldon, J. A. (1912–14). Some alien plants of the Mersey Province. *Lancashire and Cheshire Naturalist* **5** (1912), 167–9, 211–16, 255–7, 337–41; **5** (1913), 362–64, 409–11, 425–7; **6** (1913), 32–5, 75–6, 83–6, 134–8, 174–7, 211–12, 248–50, 293–6, 332–5; **6** (1914), 371–5, 407–12, 463–4; **7** (1914), 33–6, 83–4, 119–22, 128–30.

White, J. W. (1912). *The Flora of Bristol*. John Wright & Sons, Bristol.

Wigginton, M. J. (Ed) (1999). *British Red Data Books, 1. Vascular Plants*. Third edition. Joint Nature Conservation Committee, Peterborough.

Wilkinson, M. J., Elliott, L. J., Allainguillaume, J., Shaw, M. W., Norris, C., Welters, R., Alexander, M., Sweet, J. & Mason, D. C. (2003). Hybridization between *Brassica napus* and *B. rapa* on a national scale in the United Kingdom. *Science* **302**, 457–9.

Williamson, M. (1999). Invasions. *Ecography* **22**, 5–12.

Williamson, M. (2002). Alien plants in the British Isles. In: *Biological Invasions: Economic and Environmental Costs of Alien Plant, Animal and Microbe Species* (Ed. Pimentel, D.). CRC Press, Boca Raton, FL.

Willis, A. J. (1960). *Juncus subulatus* Forsk. in the British Isles. *Watsonia* **4**, 211–17.

Wilmore, G. T. D. (2000). *Alien Plants of Yorkshire*. Yorkshire Naturalists' Union, York.

Wilson, A. K. (1938). The adventive flora of the East Riding of Yorkshire. *Occasional Papers of the Hull Scientific and Field Naturalists' Club* **1**.

Woodell, S. R. J. & Rosseter, J. (1959). The flora of Durham walls. *Proceedings of the Botanical Society of the British Isles* **2**, 257–73.

Young, W. (1936). A list of the flowering plants and ferns recorded from Fife and Kinross (v. c. 85). *Transactions of the Botanical Society of Edinburgh* **32**, 1–173.

Indexes

SPECIES INDEX

Page numbers in **bold** include illustrations.

GENERAL INDEX

Page numbers in **bold** include illustrations.